# Eucalyptus

## Medicinal and aromatic plants – industrial profiles

Individual volumes in this series provide both industry and academia with in-depth coverage of one major medicinal or aromatic plant of industrial importance.

Edited by Dr Roland Hardman

Volume 1
Valerian
*Edited by Peter J. Houghton*

Volume 2
Perilla
*Edited by He-ci Yu, Kenichi Kosuna and Megumi Haga*

Volume 3
Poppy
*Edited by Jenö Bernáth*

Volume 4
Cannabis
*Edited by David T. Brown*

Volume 5
Neem
*Edited by H.S. Puri*

Volume 6
Ergot
*Edited by Vladimír Křen and Ladislav Cvak*

Volume 7
Caraway
*Edited by Éva Németh*

Volume 8
Saffron
*Edited by Moshe Negbi*

Volume 9
Tea Tree
*Edited by Ian Southwell and Robert Lowe*

Volume 10
Basil
*Edited by Raimo Hiltunen and Yvonne Holm*

Volume 11
Fenugreek
*Edited by Georgios Petropoulos*

Volume 12
Gingko biloba
*Edited by Teris A. Van Beek*

Volume 13
Black Pepper
*Edited by P.N. Ravindran*

Volume 14
Sage
*Edited by Spiridon E. Kintzios*

Volume 15
Ginseng
*Edited by W.E. Court*

Volume 16
Mistletoe
*Edited by Arndt Büssing*

Volume 17
Tea
*Edited by Yong-su Zhen*

Volume 18
Artemisia
*Edited by Colin W. Wright*

Volume 19
Stevia
*Edited by A. Douglas Kinghorn*

Volume 20
Vetiveria
*Edited by Massimo Maffei*

Volume 21
Narcissus and Daffodil
*Edited by Gordon R. Hanks*

Volume 22
Eucalyptus
*Edited by John J.W. Coppen*

# Eucalyptus
The Genus *Eucalyptus*

Edited by
John J.W. Coppen

CRC Press is an imprint of the
Taylor & Francis Group, an **informa** business

CRC Press
Taylor & Francis Group
6000 Broken Sound Parkway NW, Suite 300
Boca Raton, FL 33487-2742

First issued in paperback 2019

© 2002 by Taylor & Francis Group, LLC
CRC Press is an imprint of Taylor & Francis Group, an Informa business

No claim to original U.S. Government works

ISBN-13: 978-0-415-27879-9 (hbk)
ISBN-13: 978-0-367-39618-3 (pbk)

This book contains information obtained from authentic and highly regarded sources. Reasonable efforts have been made to publish reliable data and information, but the author and publisher cannot assume responsibility for the validity of all materials or the consequences of their use. The authors and publishers have attempted to trace the copyright holders of all material reproduced in this publication and apologize to copyright holders if permission to publish in this form has not been obtained. If any copyright material has not been acknowledged please write and let us know so we may rectify in any future reprint.

Except as permitted under U.S. Copyright Law, no part of this book may be reprinted, reproduced, transmitted, or utilized in any form by any electronic, mechanical, or other means, now known or hereafter invented, including photocopying, microfilming, and recording, or in any information storage or retrieval system, without written permission from the publishers.

For permission to photocopy or use material electronically from this work, please access www.copyright.com (http://www.copyright.com/) or contact the Copyright Clearance Center, Inc. (CCC), 222 Rosewood Drive, Danvers, MA 01923, 978-750-8400. CCC is a not-for-profit organization that provides licenses and registration for a variety of users. For organizations that have been granted a photocopy license by the CCC, a separate system of payment has been arranged.

**Trademark Notice:** Product or corporate names may be trademarks or registered trademarks, and are used only for identification and explanation without intent to infringe.

**Visit the Taylor & Francis Web site at**
**http://www.taylorandfrancis.com**

**and the CRC Press Web site at**
**http://www.crcpress.com**

# Contents

*List of contributors* vii
*Preface to the series* ix
*Preface* xi

## PART 1
## General aspects  1

1 Botany of the eucalypts  3
  IAN BROOKER

2 Eucalyptus, water and the environment  36
  IAN R. CALDER

3 Eucalypts in cultivation: an overview  52
  JOHN W. TURNBULL AND TREVOR H. BOOTH

4 Genetic improvement of eucalypts: with special reference to oil-bearing species  75
  JOHN C. DORAN

5 Eucalyptus chemistry  102
  JOSEPH J. BROPHY AND IAN A SOUTHWELL

6 Distillation of eucalyptus leaf oils: theory and practice  161
  E.F.K DENNY

## PART 2
## Cultivation and production of eucalypts around the world: with special reference to the leaf oils  181

7 Cultivation and production of eucalypts in Australia: with special reference to the leaf oils  183
  GEOFFREY R DAVIS

8 Cultivation and production of eucalypts in the People's Republic of China: with special reference to the leaf oils  202
  SHAOXIONG CHEN

9 Cultivation and production of eucalypts in Africa:
with special reference to the leaf oils 216
PAUL A. JACOVELLI

10 Cultivation and production of eucalypts in South America:
with special reference to the leaf oils 239
LAÉRCIO COUTO

11 Cultivation and production of eucalypts in India:
with special reference to the leaf oils 251
S.S. HANDA, R.K. THAPPA AND S.G. AGARWAL

PART 3
Biological and end-use aspects 267

12 Chemistry and bioactivity of the non-volatile constituents
of eucalyptus 269
TAKAO KONOSHIMA AND MIDORI TAKASAKI

13 Antimicrobial activity of eucalyptus oils 291
STANLEY G. DEANS

14 Eucalyptus in insect and plant pest control: use as a mosquito
repellent and protectant of stored food products; allelopathy 304
PETER GOLOB, HIROYUKI NISHIMURA AND ATSUSHI SATOH

15 Chemical ecology of herbivory in eucalyptus: interactions
between insect and mammalian herbivores and plant essential oils 324
IVAN R. LAWLER AND WILLIAM J. FOLEY

16 Eucalyptus oil products: formulations and legislation 345
JUDI BEERLING, STEVE MEAKINS AND LES SMALL

17 Production, trade and markets for eucalyptus oils 365
JOHN J.W. COPPEN

18 Research trends and future prospects 384
ERICH V. LASSAK

Appendices 405

1 Sources of eucalyptus seed 407
2 Estimates of eucalypt plantations worldwide 408
3 Advice to a prospective new producer of eucalyptus oil
or other leaf extractive 412
4 Composition of some commercially distilled eucalyptus oils 414
5 Quality criteria and specifications of eucalyptus oils 416
6 Packaging and labelling requirements for the handling
and transportation of eucalyptus oils 421
7 Useful addresses 424
   *Subject index* 427
   *Eucalyptus species index* 440

# Contributors

S.G. Agarwal, Regional Research Laboratory, Jammu-Tawi 180 001, India.

Judi Beerling, Quest International, Kennington Road, Ashford, Kent TN24 0LT, UK.

Trevor H. Booth, CSIRO Forestry and Forest Products, PO Box E4008, Kingston ACT 2604, Australia.

Ian Brooker, Centre for Plant Biodiversity Research, CSIRO Plant Industry, GPO Box 1600, Canberra ACT 2601, Australia.

Joseph J. Brophy, Department of Chemistry, University of New South Wales, Kensington, NSW 2033, Australia.

Ian R. Calder, Centre for Land Use and Water Resources Research, Porter Building, University of Newcastle, Newcastle-upon-Tyne NE1 7RU, UK.

Shaoxiong Chen, China Eucalypt Research Centre, Renmin Dadao Zhong 30, Zhanjiang, Guangdong 524022, P.R. China.

John J.W. Coppen, 12 Devon Close, Rainham, Kent ME8 7LG, UK.

Laércio Couto, Departamento de Engenharia Florestal, Universidade Federal de Viçosa, 36571 Viçosa, MG, Brazil.

Geoffrey R. Davis, G.R. Davis Pty Ltd, PO Box 123, 29-31 Princes Street, Riverstone NSW 2765, Australia.

Stanley G. Deans, Aromatic & Medicinal Plant Group, Food Systems Division, SAC Auchincruive, South Ayrshire KA6 5HW, UK.

E.F.K. Denny, Denny, McKenzie Associates, PO Box 42, Lilydale, Tasmania 7268, Australia.

John C. Doran, CSIRO Forestry and Forest Products, PO Box E4008, Kingston ACT 2604, Australia.

William J. Foley, Division of Botany and Zoology, Australian National University, Canberra 0200, Australia.

Peter Golob, Food Security Department, Natural Resources Institute, University of Greenwich, Chatham Maritime, Kent ME4 4TB, UK.

S.S. Handa, Regional Research Laboratory, Jammu-Tawi 180 001, India.

Paul A. Jacovelli, 6 Clive Road, Preston, Lancs PR1 0AT, UK.

Takao Konoshima, Kyoto Pharmaceutical University, Misasagi, Yamashina-ku, Kyoto 607, Japan.

Erich V. Lassak, Phytochemical Services, 254 Quarter Sessions Road, Westleigh, NSW 2120, Australia.

Ivan R. Lawler, School of Tropical Environment Studies and Geography, James Cook University of Queensland, Douglas Q 4811, Australia.

Steve Meakins, Quest International, Kennington Road, Ashford, Kent TN24 0LT, UK.

Hiroyuki Nishimura, Department of Bioscience and Technology, School of Engineering, Hokkaido Tokai University, Sapporo 005-8601, Japan.

Atsushi Satoh, Department of Bioscience and Technology, School of Engineering, Hokkaido Tokai University, Sapporo 005-8601, Japan.

Les Small, Quest International, Kennington Road, Ashford, Kent TN24 0LT, UK.

Ian A. Southwell, NSW Agriculture, Wollongbar Agricultural Institute, Wollongbar, NSW 2477, Australia.

Midori Takasaki, Kyoto Pharmaceutical University, Misasagi, Yamashina-ku, Kyoto 607, Japan.

R.K. Thappa, Regional Research Laboratory, Jammu-Tawi 180 001, India.

John W. Turnbull, CSIRO Forestry and Forest Products, PO Box E4008, Kingston ACT 2604, Australia.

# Preface to the series

There is increasing interest in industry, academia and the health sciences in medicinal and aromatic plants. In passing from plant production to the eventual product used by the public, many sciences are involved. This series brings together information which is currently scattered through an ever increasing number of journals. Each volume gives an in-depth look at one plant genus, about which an area specialist has assembled information ranging from the production of the plant to market trends and quality control.

Many industries are involved such as forestry, agriculture, chemical, food, flavour, beverage, pharmaceutical, cosmetic and fragrance. The plant raw materials are roots, rhizomes, bulbs, leaves, stems, barks, wood, flowers, fruits and seeds. These yield gums, resins, essential (volatile) oils, fixed oils, waxes, juices, extracts and spices for medicinal and aromatic purposes. All these commodities are traded worldwide. A dealer's market report for an item may say 'Drought in the country of origin has forced up prices'.

Natural products do not mean safe products and account of this has to be taken by the above industries; which are subject to regulation. For example, a number of plants which are approved for use in medicine must not be used in cosmetic products.

The assessment of safe to use starts with the harvested plant material which has to comply with an official monograph. This may require absence of, or prescribed limits of, radioactive material, heavy metals, aflatoxin, pesticide residue, as well as the required level of active principle. This analytical control is costly and tends to exclude small batches of plant material. Large scale contracted mechanised cultivation with designated seed or plantlets is now preferable.

Today, plant selection is not only for the yield of active principle, but for the plant's ability to overcome disease, climatic stress and the hazards caused by mankind. Such methods as *in vitro* fertilization, meristem cultures and somatic embryogenesis are used. The transfer of sections of DNA is giving rise to controversy in the case of some end-uses of the plant material.

Some suppliers of plant raw material are now able to certify that they are supplying organically-farmed medicinal plants, herbs and spices. The European Union directive (CVO/EU No. 2092/91) details the specifications for the *obligatory* quality controls to be carried out at all stages of production and processing of organic products.

Fascinating plant folklore and ethnopharmacology leads to medicinal potential. Examples are the muscle relaxants based on the arrow poison, curare, from species of *Chondrodendron*, and the anti-malarials derived from species of *Cinchona* and *Artemisia*. The methods of detection of pharmacological activity have become increasingly reliable and specific, frequently involving enzymes in bioassays and avoiding the use of laboratory animals. By using bioassay linked fractionation of crude plant juices or extracts, compounds can be specifically targeted which, for example, inhibit blood platelet aggregation, or have anti-tumour, or anti-viral, or any other required activity. With the assistance of robotic devices, all the members of a genus may be readily screened. However, the plant material must be *fully* authenticated by a specialist.

The medicinal traditions of ancient civilisations such as those of China and India have a large armamentaria of plants in their pharmacopoeias which are used throughout South-East Asia. A similar situation exists in Africa and South America. Thus, a very high percentage of the World's population relies on medicinal and aromatic plants for their medicine. Western medicine is also responding. Already in Germany all medical practitioners have to pass an examination in phytotherapy before being allowed to practise. It is noticeable that throughout Europe and the USA, medical, pharmacy and health related schools are increasingly offering training in phytotherapy.

Multinational pharmaceutical companies have become less enamoured of the single compound magic bullet cure. The high costs of such ventures and the endless competition from 'me too' compounds from rival companies often discourage the attempt. Independent phytomedicine companies have been very strong in Germany. However, by the end of 1995, eleven (almost all) had been acquired by the multinational pharmaceutical firms, acknowledging the lay public's growing demand for phytomedicines in the Western World.

The business of dietary supplements in the Western World has expanded from the health store to the pharmacy. Alternative medicine includes plant-based products. Appropriate measures to ensure the quality, safety and efficacy of these either already exist or are being answered by greater legislative control by such bodies as the Food and Drug Administration of the USA and the recently created European Agency for the Evaluation of Medicinal Products, based in London.

In the USA, the Dietary Supplement and Health Education Act of 1994 recognised the class of phytotherapeutic agents derived from medicinal and aromatic plants. Furthermore, under public pressure, the US Congress set up an Office of Alternative Medicine and this office in 1994 assisted the filing of several Investigational New Drug (IND) applications, required for clinical trials of some Chinese herbal preparations. The significance of these applications was that each Chinese preparation involved several plants and yet was handled as a *single* IND. A demonstration of the contribution to efficacy, of *each* ingredient of *each* plant, was not required. This was a major step forward towards more sensible regulations in regard to phytomedicines.

My thanks are due to the staffs of Harwood Academic Publishers and Taylor & Francis who have made this series possible and especially to the volume editors and their chapter contributors for the authoritative information.

Roland Hardman

# Preface

The eucalypt, or gum tree, is such an established feature of the Australian landscape that it has left its mark in that most well-loved of Australian institutions, Waltzing Matilda, penned by 'Banjo' Patterson towards the end of the nineteenth century:

> Once a jolly swagman camped by a billabong,
> Under the shade of a coolibah tree[1] ...

For those for whom the association between coolibah tree and eucalyptus goes unrecognised the latter word may, instead, conjure up pictures of koala bears munching contentedly on the leaves of gum trees. And for yet others, eucalyptus may have medicinal connotations – whether through the use of eucalyptus-flavoured throat lozenges or chest rubs or through eucalyptus oil and the increasingly popular practice of aromatherapy. In truth, all these associations are genuine but the single, simple word 'eucalyptus' does not convey the complexity and diversity of all that it embraces, whether measured in terms of the uses to which it is put or the number of species to which it refers. Nor does it convey to the man or woman in the street the international nature of eucalyptus. Australia may be its natural home but its progeny have spread far and wide and the industries associated with it now span the globe. Whether cultivated as narrow tracts alongside roads, railways and canals in China and India, or as vast blocks of monoculture in Brazil and elsewhere, no continent, outside of Antarctica, has failed to be smitten by the lure of eucalyptus.

The genus *Eucalyptus*, which is native to Australia and some islands to the north of it, consists of over 800 species of trees.[2] This number continues to grow as new taxa are described. The trees grow under a wide range of climatic and edaphic conditions in their natural habitat and the very large and varied gene pool which can be drawn upon for planting purposes is one reason for the successful introduction of *Eucalyptus* into so many other countries in the world. Another reason is the fast-growing nature of eucalypts which makes them ideally suited to obtaining an economic return within a relatively short period of time. With the advantages have come some perceived disadvantages, however, and a fair measure of controversy. Not least the effects of eucalypts on the soil in terms of water and nutrient abstraction. Calder (Chapter 2) demonstrates that proper scientific research is now distinguishing between myth and reality and dispelling some of the misconceptions surrounding eucalyptus and the environment.

---

1 *Eucalyptus coolabah* or *E. microtheca*.
2 *Corymbia*, previously a sub-genus of *Eucalyptus*, has been elevated to the rank of a separate genus by Hill and Johnson. However, since at least one well-known species (*Corymbia citriodora*) is the source of a commercially important oil, which will continue to be traded under the name *Eucalyptus citriodora*, the original classification is retained in this book to facilitate the discussion.

The multipurpose nature of *Eucalyptus* and its cultivation for such end-uses as timber, pulp and fuelwood is generally well described and documented. Information on the medicinal and aromatic properties of eucalyptus, however, although well known, generally resides in the primary research literature or is scattered, much of it superficially, in a number of different media forms. This is particularly true of the commercial and practical aspects of production, both of the trees themselves and of the pharmaceutical and perfumery products which reach the marketplace (either the primary extractives, such as eucalyptus oil, or end-use products). It is these aspects, embracing everything from the land preparation and fertiliser prescriptions used in Africa and China for the cultivation of oil-bearing species to the chemistry and composition of eucalyptus oils, from an examination of world trade and prices for the oils and the increasingly burdensome demands being placed on producers and exporters by legislation and quality specifications to formulations used to make bath foams and hair conditioners, that are described and discussed in this book. Everything, in short, which should constitute an industrial profile. Some things, such as distillation, are taken for granted by their practitioners but, as Denny shows (Chapter 6), commercial practice is not always best practice and there is ample scope for improvement. Koalas (and possums) are not left out and what we learn about their eating habits and preferences (Chapter 15) may enable us to determine the role of essential oils and related compounds in conferring resistance to herbivores – with considerable economic consequences.

The use of eucalyptus as a commercial source of volatile oil forms the basis for much of the content of the book, but not all of it. The relative ease with which plant essential oils are distilled and analysed made them primary targets for early investigation and exploitation. It remains the case today that they are by far the largest group of eucalyptus extractives to be exploited for medicinal and fragrance or flavour purposes. However, research in recent years has uncovered a large number of pharmacologically active non-volatile compounds unique to eucalyptus, many of which – such as the euglobals and macrocarpals – form groups of structurally similar compounds. Their potential for use in the treatment of AIDS and cancer, amongst other things, ensures that they will be the subject of much more research in the years to come. Results to date, and an indication of the species of *Eucalyptus* which contain the most promising groups of compounds, are described in some detail.

The native Australian aborigines had long used eucalyptus. Not only its wood and bark for the fabrication of canoes, spears and boomerangs, and domestic items such as bowls and dishes, but its leaves, roots and other parts for medicinal purposes. The crushed leaves would inevitably have disclosed the presence of fragrant oils but it was not until 1788, the year of white settlement in Australia, that the first recorded distillation of eucalyptus oil was made. In 1852 a still was set up for its commercial production. Joseph Bosisto, who had emigrated from England four years earlier, established his still on Dandenong Creek, about 40 km southeast of Melbourne, with the encouragement of Ferdinand von Mueller, then Government Botanist in Victoria. The rest, as they say, is history.

Today, of the hundreds of species of *Eucalyptus* that have been shown to contain volatile oil in their leaves, only about a dozen are utilised, of which six account for the greater part of world oil production: *E. globulus*, *E. exserta*, *E. polybractea*, *E. smithii*, *E. citriodora* and *E. dives*. Australia is no longer the main producer of eucalyptus oil – this distinction belongs to China, by a long way – but it has maintained a highly efficient industry in the face of competition from more than a dozen countries. Medicinal-type eucalyptus oil – or its main constituent, 1,8-cineole (eucalyptol) – is an ingredient of many hundreds of pharmaceutical products and used for the treatment of ailments ranging from colds, coughs and congestion to sports injuries and muscle and joint pain, from insect bites to skin disorders. It is also used in an array of personal care products such as shampoos, bath foams, soaps and body lotions, not to mention cleaning agents. All this in addition to use of the neat oil, again in many and varied ways, around the home. Complementary to the medicinal oils are the perfumery eucalyptus oils, exemplified by the familiar lemon-scented oil of *E. citriodora*, used directly for fragrance purposes or as a source of citronellal.

The existence of chemical variants within the same species of *Eucalyptus* creates potential pitfalls for the analyst and producer alike but it also offers prospective benefits arising from plant selection and breeding. Continued screening of *Eucalyptus* species, together with research into aspects such as optimum species and provenance selection for locations outside their natural distribution, vegetative propagation, improved farm management practices and new applications, indicates that eucalyptus oil will continue to be an essential oil of major importance in world trade and usage and to the economy of many countries and communities in both the developed and developing world. Together with new avenues of research opened up by the discovery of pharmacologically active non-volatile constituents, and the prospect of treating or preventing a variety of important human diseases and ailments, eucalyptus may not yet have exhausted its capacity for surprising us at its multifarious nature.

Finally, I should like to thank all the contributors to this volume for their efforts and expertise, Dr Roland Hardman, Series Editor, for his support and encouragement throughout, and, not least, my family, particularly Eve, for their patience and understanding.

John J.W. Coppen

# Part 1
# General aspects

# 1 Botany of the eucalypts

*Ian Brooker*

## Introduction

There are probably few tree genera that have had as much written about them as the eucalypts. Indigenous to the Australasian region, they are now among the most widely cultivated of all plants, particularly in tropical and subtropical parts of the world. The multiple uses to which eucalypts are put, from construction timber to fuel and pulp, and from oils to amenity planting, make this plant genus one of the most valuable and widely used in the world.

To the average Australian they are a natural feature of the environment, where vast forests have been exploited for commercial purposes, often in the expectation that natural regeneration will sustain an industry more or less permanently. In the last thirty years, however, an awareness of the need to preserve the forests has become as much a political as a silvicultural necessity.

The perspective in many other countries, where the eucalypt is known as an alien plant, is different. But after generations of planting the eucalypt may be seen as almost part of the landscape, providing fuel, timber and ornamentation in places where the original forests have been razed beyond the possibility of natural regeneration. In a few countries the eucalypt is the dominant tree in plantations where vast numbers of even-aged trees of more or less identical habit and size occupy great areas. Their uniformity is often due to the surprisingly small choice of species, many subsequently cloned or produced by manipulated crossing to package the most desired characters for site adaptation and specific end use.

When it is considered that relatively few species of eucalypt form the basis of plantation industries, it may be questioned to what extent this large and greatly variable genus has been tested. It occurs in a vast assortment of forms and in sites which range from rainforest to alpine to desert. There is no doubt that among the 800 species of eucalypt which have now been described, species as yet untried will prove to be successful for a multitude of reasons, whether for their fuel, timber, fibre, oils or other chemicals, or merely for shelter and amenity.

## The discovery of the eucalypts

The popular conception of the discovery of *Eucalyptus* relates to the voyages of Captain James Cook in the Endeavour in the 1770s. Following the naming and exploration of New Zealand on his first voyage, the party arrived on the eastern coast of Australia. From their first principal landfall which Cook named Botany Bay, they sailed northwards, making plant collections under the guidance of Sir Joseph Banks. There are several extant eucalypt specimens from this first voyage, one of which from Botany Bay was described, but not recognised as a new genus, and was placed in the established genus *Metrosideros* by the botanist Joseph Gaertner in 1788.

On Cook's third voyage to the south seas, the botanist of the party, David Nelson, collected a specimen on Bruny Island to the south of the Tasmanian landmass. Taken back to England it was studied by the French botanist, Charles Louis L'Héritier de Brutelle, working at the British Museum, Natural History, where the specimen remains. He published it in 1788 as the single species of a new genus which he named *Eucalyptus*, from the Greek, *eu*, well, and *calyptos*, covered, alluding to the cap (operculum) that covered the stamens in the bud before anthesis. In the fifty years following Cook's voyages many more eucalypt species were discovered. These were found by settlers radiating from the new settlements on Port Jackson (Sydney), Hobart, Melbourne and Adelaide, and by naturalists on exploration trips around the continent, of which the best known are those of Labillardière in 1792, Robert Brown in 1801–1803, and that of the surveyor Allan Cunningham (1817–1822).

By 1867, when the first substantial classification of the genus *Eucalyptus* was published by George Bentham, 135 species were known. Since then hundreds more have been discovered, with periods of activity coinciding with the efforts of the principal eucalypt botanists operating at the various times. Notable eras of *Eucalyptus* publication have been the latter half of the nineteenth century (Ferdinand von Mueller), 1903–1933 (Joseph Henry Maiden) and 1934 (William Faris Blakely). An indication of the accumulation of named eucalypts is given in Figure 1.1, where periods of intensive activity can be seen. Considering that eucalypts are now recognised to number about 800, it appears that, on average, for every species published a synonym has also been published.

While eucalypts were largely brought to public awareness from the early nineteenth century onwards, there can be no doubt that eucalypts were known to Europeans well before that time. In 1699, the English navigator William Dampier landed on the west coast of Australia and made plant collections which are extant and held at Cambridge, England. There are no eucalypts among them (A.S. George pers. comm.) but it is unlikely that Dampier or his party could not have come across them, even though they landed in a very arid part of the continent. The earliest collected eucalypt is believed to have come from Ceram, an island of eastern Indonesia. It was probably *Eucalyptus deglupta* Blume, one of the few *Eucalyptus* species not indigenous to the Australian landmass but occurring in New Guinea, Indonesia and the southern Philippines. The

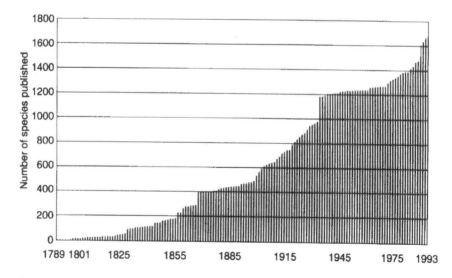

*Figure 1.1* Numbers of *Eucalyptus* species published by year.

specimen was not described and published at the time and its history remains a curiosity. The species was, however, formally described in 1849 by Carl Blume from a sterile specimen earlier given the manuscript name 'Populus deglubata'.

There is a possibility that a eucalypt species was reasonably well known before the Ceram specimen. *E. alba* Reinw. and *E. urophylla* Blake, the latter only described in 1977, are indigenous to Timor and some Indonesian islands and it is likely that these species were encountered and used by the colonising Portuguese, as well as the local people. Seed may also have been taken to Brazil, another colony of the Portuguese, from the early 1600s.

## The origins of *Eucalyptus*

Australia is the most arid of continents. Nevertheless, there are significant regions of very high rainfall along the eastern coast and in southwestern Tasmania. These areas support the remaining rainforests which, floristically, are in great contrast to the far more widely distributed 'typical' Australian landscapes dominated by *Eucalyptus* and *Acacia*. It was once assumed that the sclerophyll vegetation of most of the land surface was the autochthonous flora and that the rainforest was a later invader, probably from the north, across land bridges which were manifest at times of low sea level. The more recent theories of plate tectonics, supported by fossil evidence, show that the present landmass was once part of a single super-continent, Gondwana (Barlow 1981) comprising what is now known as New Guinea, Antarctica, India, Arabia, Africa, Madagascar and South America.

As Gondwana broke up in the Tertiary period, the Australian landmass drifted northwards from Antarctica. It is believed that in the early part of this drift the continent experienced high rainfall and was characterised by a relatively uniform rainforest vegetation. The landmass gradually became drier and the ancient soils lost their fertility. The new land environments became totally unsuited to rainforest and it contracted, largely to the eastern seaboard. However, by the evolution of new adaptable forms some plant families were able to occupy the areas of less certain rainfall and soils of diminishing fertility. Notable examples are *Myrtaceae*, which spawned the eucalypts, and *Mimosaceae*, the distinctive phyllodinous acacias. Both plant groups required morphological and physiological modifications to enable them to thrive. This may be seen in *Acacia* in the dominance among the hundreds of dryland species of the phyllode, a modification of the soft divided leaves of the probable rainforest precursors.

In the case of the eucalypts, the most conspicuous change for the survival of plants in the harsher environments of the open forests and shrublands can be seen in the leaves and inflorescences. In the closed forests there would have been far more competition for sunlight and the leaves are necessarily dorsiventral, held more or less horizontally, with the photosynthetic tissues facing upwards and stomata on the underside. For the most successful reproduction of primarily entomophilous plants, it is likely that the inflorescences should be presented as conspicuously as possible. This requires that the inflorescence be large and terminal on the crown (e.g. *E. calophylla* R. Br. ex Lindley, Figure 1.2). We cannot know what the precursor or precursors of *Eucalyptus* were, but from an analysis of the present rainforest and of the sclerophyllous vegetation of the majority of the Australian environment, *Protoeucalyptus* may have had the characters given in Table 1.1.

The contrast between the shadowy environment beneath the crowns of the rainforest and the intense light of the open forests and shrublands is immense in terms of available sunlight. An essential adaptation was the development of the isobilateral leaf which, in addition, tended to be held vertically, often presenting an edge to the sun with the stomata on both sides, a style of leaf that minimises the effects of direct sunlight and protects the stomata as much as possible.

*Figure 1.2* Terminal inflorescences (*E. calophylla*).

*Table 1.1* Suggested characters in *Protoeucalyptus*

| Character | Protoeucalyptus |
|---|---|
| Habit | Tree |
| Bark | Rough |
| Cotyledons | Reniform |
| Juvenile leaf phase | Short |
| Adult leaves | Dorsiventral |
| Leaf oil glands | Few and small |
| Inflorescences | Terminal |
| Sepals | Present |
| Petals | Present |
| Anthers | Versatile |
| Fruit | Slightly fleshy |
| Seeds | Hemitropous |

It is interesting to observe in leaf characters, however, the present day relationships between the current species of the more humid forests and their related taxa in the more arid lands. In the emergence of *Eucalyptus* and its development of a hierarchy of infra-generic forms, many taxonomic series have present-day species in the humid forests, with unmistakably related species that have been modified to adapt to the drier environments. This is notable in the bloodwoods, which have radiated over most of the continent. All the bloodwood species currently in the humid east and far southwest have dorsiventral leaves. None of the numerous species of the arid lands in this group has dorsiventral leaves. The situation of the northern tropics, with their short, very wet season and long dry one, is somewhat different, with predominantly dorsiventral species and some isobilateral ones. Examples of humid-climate, dorsiventrally-leaved species and dry-climate, isobilaterally-leaved species of the same taxonomic groups are shown in Table 1.2.

The situation with the inflorescence is somewhat different. While there has been an apparent retreat of the inflorescence from the outside of the crown to the leaf axils in most taxonomic groups (e.g. *E. pyrocarpa* L. Johnson & Blaxell, Figure 1.3), the bloodwood species in particular have retained the terminal inflorescence in all arid zone species. All bloodwood species, whether

*Table 1.2* Examples of related pairs of *Eucalyptus* species showing advance from the primitive dorsiventral leaf to the isobilateral leaf

| Taxonomic group | Dorsiventral leaves (humid regions) | Isobilateral leaves (dry regions) |
| --- | --- | --- |
| Red bloodwoods | E. polycarpa | E. terminalis |
| Yellow bloodwoods | E. leptoloma | E. eximia |
| Swamp gums | E. brookeriana | E. ovata |
| Apple boxes | E. angophoroides | E. bridgesiana |
| Boxes | E. rummeryi | E. microtheca |
| Ironbarks | E. paniculata | E. beyeri |
| White mahoganies | E. acmenoides | E. apothalassica |
| Stringybarks | E. muelleriana | E. macrorhyncha |

*Figure 1.3* Axillary inflorescences (*E. pyrocarpa*).

of the humid forests or of the inland arid regions, have the primitive terminal inflorescence (Figure 1.2). However, several hundred species of the other taxonomic groups which have radiated across the continent, and which occur in myriad environments, have the derived axillary inflorescence.

The bark of *Protoeucalyptus* was probably rough. This has been retained without exception in the eastern bloodwood species occupying humid regions. Desert species provide a contrast and the rough bark may vary from nil to rough over part or most of the trunk. The adaptive nature of rough bark is open to much speculation. It may appear to be protective but is scarcely a necessity as rough and smooth-barked species grow in association in many varying environments. The form of the bark, whether rough or smooth, remains a powerful aid in identifying taxonomic groups within the genus, although experience is required to distinguish the different types.

## Classification in the genus *Eucalyptus*

The genus *Eucalyptus*, which began in formal botanical terms as a single species, *E. obliqua* L'Hérit., in 1788, has now been recognised to comprise about 800 species, in numbers second only in Australia to *Acacia*. By 1800 only a few species had been described. These were, as expected, all from around the new colony at Port Jackson (Sydney). Most were named by Joseph Smith working at the British Museum. They include many of the most valuable timber species which, then as now, must have been the most important in a pioneer settlement for construction and fuel, namely, *E. saligna* Smith, *E. resinifera* Smith, *E. pilularis* Smith, *E. capitellata* Smith and *E. paniculata* Smith.

In the first half of the nineteenth century, land and sea-based expeditions resulted in the discovery of many more new eucalypts. Groups of related species soon became evident, e.g. the stringybarks and the ironbarks. Consequently, a comprehensive classification of the 135 known species into these recognisable categories, as formal series and subseries, was published in 1867 by George Bentham of the Royal Botanic Gardens, Kew, England. He did not visit Australia but was greatly assisted by the endeavours of the great scientist Ferdinand von Mueller, working from Melbourne, who sent Bentham plant material on which to base his studies.

Bentham devised a system that divided the eucalypts into five taxonomic series based on stamen characters. For each group he further discussed other characters such as bark, habit, etc., although these latter appear not to be essentially diagnostic. His largest series was divided into subseries based on the structure of the inflorescence and, to a lesser extent, the leaves. Overall, Bentham's system of classification was a brave attempt to order the eucalypts but his groupings have little relevance today.

The sixty years following Bentham saw the major researches in *Eucalyptus* of Joseph Henry Maiden and William Faris Blakely working at the Royal Botanic Gardens, Sydney. Their collaboration culminated in the monumental 'A Critical Revision of the genus *Eucalyptus*' by Maiden (1903–1933) which was a descriptive and historical treatment of all eucalypt species known to the time, together with detailed analyses of certain organs, e.g. the anthers, cotyledons and seeds. Maiden produced no comprehensive classification but this was remedied by his co-worker, Blakely, who published in 1934 his 'Key to the Eucalypts' – an intricately constructed system treating 606 species and varieties, bearing some resemblance to Bentham's classification but improved by the reference to many more characters, such as ontogenic development of the leaves, the inflorescences, buds, fruits and habit.

Blakely's work marks the high point of the discipline of purely comparative morphology, although his stated aim was to place species 'in the most natural position to each other and kindred genera'. With no instruction as to his methodology, it is not possible to determine if his estimations of similarities were unquestionable and intentional expressions of homology to indicate natural affinities. There has been no formal, comprehensive classification since Blakely's Key.

Few new eucalypt species were published immediately following Blakely and in the twenty-five years after World War II, and a period of stability in numbers appeared to have set in. It was about this time that a new, but not formalised, classification was published by Lindsay Pryor of the Australian National University, Canberra, and Lawrie Johnson of the Royal Botanic Gardens, Sydney (Pryor and Johnson 1971). Pryor was a strong advocate of the importance of breeding barriers in assessing natural affinities. The authors stated that 'there are several completely reproductively isolated groups within *Eucalyptus* and these conform to the subgenera of our system'. Pryor and Johnson's system is explicitly phylogenetic and divided the genus into seven subgenera (Table 1.3). Their taxonomic sections, series, subseries and superspecies were not qualified by diagnostic data but the classification became the benchmark for a host of subsequent eucalypt studies.

Pryor and Johnson's classification was deliberately extracodical (informal and outside the Code of Botanical Nomenclature). The system was hierarchical (sections, series, etc.) and contrasts with

Table 1.3 The classification of Pryor and Johnson (1971) with examples of well-known species in the taxa

| Subgenera (Pryor and Johnson 1971) | Well-known species and species groups |
| --- | --- |
| Blakella | Ghost gums (e.g. *E. tessellaris*) |
| Corymbia | Bloodwoods (e.g. *E. citriodora*) |
| Eudesmia | Includes *E. miniata* and *E. baileyana* |
| Gaubaea | Comprises *E. curtisii* and *E. tenuipes* |
| Idiogenes | *E. cloeziana* only |
| Monocalyptus | White mahoganies (e.g. *E. acmenoides*), stringybarks (e.g. *E. globoidea*), blackbutts (e.g. *E. pilularis*), ashes (e.g. *E. obliqua*), peppermints (e.g. *E. dives*, *E. radiata*) |
| Symphyomyrtus | Red mahoganies (e.g. *E. robusta*), red gums (e.g. *E. camaldulensis*), mallees (e.g. *E. polybractea*), gums (e.g. *E. globulus*), boxes (e.g. *E. polybractea*), ironbarks (e.g. *E. staigeriana*) |

Table 1.4 Proposed new classification of the genus *Eucalyptus* (Brooker unpubl.)

Subgenus *Angophora*
Subgenus *Corymbia* sensu Pryor and Johnson 1971
Subgenus *Blakella* sensu Pryor and Johnson 1971
Subgenus (*E. curtisii*)
Subgenus (*E. guilfoylei*)
Subgenus *Eudesmia* sensu Pryor and Johnson 1971
Subgenus *Symphyomyrtus* sensu Pryor and Johnson 1971
Subgenus (*E. raveretiana*, *E. brachyandra*, *E. howittiana*, *E. deglupta*)
Subgenus (*E. microcorys*)
Subgenus (*E. tenuipes*)
Subgenus *Idiogenes* sensu Pryor and Johnson 1971 (*E. cloeziana*)
Subgenus (*E. rubiginosa*)
Subgenus *Eucalyptus* (= *Monocalyptus* in Pryor and Johnson 1971)

the formal classification used by Chippendale (1988), where only one infra-generic rank (series) was used. This latter work, however, made possible the formal reference of all species published up to 1988 to taxonomic series, which is generally required in systematic papers.

The most recent major contribution to classification in the genus *Eucalyptus* was made by Hill and Johnson (1995) who formally published a new genus, *Corymbia*, comprising two of the subgenera of the Pryor and Johnson classification (1971), namely, *Corymbia* and *Blakella*. The rationale for this step was based on cladistic analyses of all the traditional eucalypt components at the general 'subgenus' level and is corroborated by molecular studies of Udovicic *et al.* (1995).

Some ambiguity, however, remains over the relationship of the closely related genus *Angophora* Cav. A recent study on the relationships within *Eucalyptus* concluded that *Angophora*, *Corymbia* and *Blakella* form a monophyletic group (Sale *et al.* 1996). Perhaps all these recent studies should be considered merely as hypotheses that will contribute ultimately to an optimum system with further researches. My current preference is for a single genus consisting of thirteen subgenera (Table 1.4), namely, the five major subgenera of Pryor and Johnson (1971), the genus *Angophora*, a subgenus comprising the four tropical species with the small fruit, and six single-species subgenera – a system embracing demonstrable morphological distinctions.

While comparative morphology is the basis for estimating natural affinities and resulting classification, the value of characters used varies greatly. Features such as bark in eucalypts have been traditionally used in keys and descriptions, but bark as a character is only of medium reliability as its constant exposure to the elements results in attrition and change of colour. Internal characters protected from outside influences are of higher reliability. In this respect, essential oils have been considered as possible aids to testing natural affinities. Little success has been achieved and it may be concluded that the developmental pathways for morphology and for essential oils are not closely associated within the eucalypt plant.

In contrast, essential oils have been useful in distinguishing related species. This is easily demonstrated in the distinctive lemon-scented *E. citriodora* Hook. when compared with the closely related *E. maculata* Hook. The latter species does not contain citronellal and so lacks the lemon fragrance. Another instance is the distinction between *E. ovata* Labill., which has a very low oil content, and the related *E. brookeriana* A. Gray, which is rich in oil, particularly 1,8-cineole (Brooker and Lassak 1981).

Complicating the situation is the well known phenomenon in eucalypts of the variation in essential oil composition between individual trees and races of trees. In the 1920s, Penfold and Morrison reported the existence of several races in the broad-leaved peppermint which they named *E. dives* Schau. 'Type', *E. dives* 'var. A', 'var. B' and 'var. C', according to their broad chemical groupings (e.g. Penfold and Morrison 1929; see also Penfold and Willis 1961). The latter variety was of greatest commercial importance since it was particularly high in 1,8-cineole. Today, these different types are regarded simply as examples of what may be a wide range of possible chemical variants or chemotypes and not given formal names, although the old nomenclature sometimes still persists.

## The eucalypt plant

### Habit

The plantation eucalypt is notably a fast-growing tree of good form (e.g. *E. maculata*, Figure 1.4), suitable for modern harvesting practices and intended ultimately for use as pole timber or a source of fibre. The species used for these purposes are remarkably few in number. They mostly originate in the wetter forests of the eastern seaboard of the Australian continent, e.g. *E. grandis* W. Hill ex Maiden, *E. saligna*, *E. globulus* Labill., *E. tereticornis* Smith, *E. nitens* (Deane & Maiden) Maiden, *E. dunnii* Maiden and *E. smithii* R. Baker. These species are tall trees from the natural forested areas and altogether number about fifty, and include some from the far southwest of Western Australia, namely, *E. marginata* Donn ex Smith and *E. diversicolor* F. Muell.

In drier regions, and often on less fertile soils, smaller trees of woodland habit and habitat dominate (e.g. *E. kumarlensis* Brooker, Figure 1.5). There are about 250 species of predominantly woodland form and they occupy the hills, slopes, plateaus and mountains, and the heavier soils of the plains. Notable among these is the widespread river red gum, *E. camaldulensis* Dehnh., which is a large spreading tree of freshwater streams across most of the continent. Its adaptability to many differing habitats has made it one of the most widely planted trees in other countries, where it is chosen for its fast growth, hard timber and excellent fuel.

Most of the interior of Australia is arid, and *Acacia* is the dominant genus, but eucalypt species are usually to be found in these regions along the seasonal streams and in the rocky hills. In the south, particularly, there are extensive sandplains of low fertility. These support the very

*Figure 1.4* Forest tree with long bole and small terminal crown (*E. maculata*).

*Figure 1.5* Woodland tree with short bole and large spreading crown (*E. kumarlensis*).

numerous mallee species (about 180 in all) although a few occur in eastern uplands, e.g. *E. approximans* Maiden and *E. gregsoniana* L. Johnson & Blaxell. The mallee is the dominant form over wide areas of tall shrubland. As a mature plant it is easily recognised by the multi-stemmed habit (e.g. *E. livida* Brooker & Hopper, Figure 1.6) and the presence of a lignotuber which is either buried just below the soil surface or is exposed through weathering of the soil.

*Figure 1.6* Mallee with multiple stems (*E. livida*).

*Figure 1.7* Seedling of *E. dwyeri* showing small lignotubers which have formed in the axils of fallen cotyledons.

## *The lignotuber*

Most *Eucalyptus* species develop lignotubers. They begin as swellings in the axils of the cotyledons (e.g. *E. dwyeri* Maiden & Blakely, Figure 1.7). As the seedling gets older the swellings form in the axils of several of the lower leaf pairs. They are usually conspicuous on the young stem as they necessarily form decussately. The swellings coalesce and often engulf the upper part of the root. The whole bulbous mass becomes lignified and is a store of dormant shoots which develop and produce new stem structures when the upper part of the plant is lost through cutting, fire, grazing, etc. In very old eucalypts the lignotuber may be massive.

The lignotuber is a vital organ in the regeneration of trees or mallees and its evolution may be a particular response to ancient and constant fire regimes. Not all eucalypts form lignotubers and the regenerative capacity of the plant is an important factor in the choice of species for plantations in which repeated harvests for oil-bearing foliage are the requirement. Species best suited to a plantation and harvesting programme are those which are lignotuberous and guarantee coppice regeneration. Because of the large number of *Eucalyptus* species and the task required to detect lignotubers in all taxa and provenances, there is no comprehensive list of species which form lignotubers.

Most eucalypts form lignotubers, yet many of the best known commercial species are non-lignotuberous, e.g. *E. regnans* F. Muell., *E. delegatensis* Smith, *E. pilularis*, *E. grandis* and *E. nitens* (Jacobs 1955). These are species of wet forests of southeastern Australia which regenerate prolifically from seed, especially after site disturbance. In these species the base of the stem or trunk becomes swollen and is a store of dormant buds which respond much like the true lignotuber and result in coppice regrowth from a felled tree base. *E. diversicolor*, also non-lignotuberous, occupies high rainfall country of southwestern Australia, but the same regenerative process occurs in species of the much drier country of Western Australia, e.g. *E. astringens* (Maiden) Maiden. This species is one of several that are both non-lignotuberous and of distinctive 'mallett' habit, in which the trunk branches low and produces a characteristic crown of steep branches (Brooker and Kleinig 1990).

*E. camaldulensis*, which has the widest natural distribution of all eucalypt species, is variable in lignotuber formation. Populations in the northern half of the Australian continent are reported to be lignotuberous while southern populations are non-lignotuberous (Pryor and Byrne 1969), although recent seedling studies (CSIRO unpubl.) indicate that stress on the young plant is as important a stimulus for lignotuber formation as genetic predisposition.

## Bark

The bark character in *Eucalyptus* is one of the most difficult to assess and describe. It has long been recognised that many species shed their outer bark each year and that others retain the dead bark. The eucalypt tree puts on an annual increment of bark tissue. Regular decortication results in smooth bark, while retention of dead bark results in rough bark, but the colour, form and thickness of the bark types varies conspicuously with the taxonomic grouping and with the age of the tree, its health, the season, etc. Recognition of the various bark types can only be achieved by regular field experience but, when learnt, it provides the botanist, naturalist and forester with a powerful tool in the recognition of species and taxonomic categories.

It is also important to bear in mind that most of the bark descriptions in texts have been made from trees growing in their natural environments, and that different conditions, particularly in other countries where eucalypts are planted, may result in different responses of colour and texture.

### Smooth bark

The outer dead bark of some species and species groups sheds in slabs at a particular time of the year. This results in the sudden exposure of the underbark which is a different colour and soon changes with weathering until the trunk is an even grey or pale pink (e.g. *E. citriodora*, Figure 1.8). Such a trunk contrasts strongly with a rough-barked species (e.g. *E. macrorhyncha* F. Muell., Figure 1.9).

In the red gums (e.g. *E. camaldulensis*, Figure 1.10), and the grey gums, the outer bark sheds in irregular patches at different times during the year. The freshly-exposed underbark is usually

*Figure 1.8* Bark completely shed leaving a smooth bole (*E. citriodora*).

*Figure 1.9* Bark retained forming a rough bole (*E. macrorhyncha*).

*Figure 1.10* Mottled smooth bark (*E. camaldulensis*).

brightly coloured yellow or coppery, which fades to light grey and then weathers to dark grey. The trunks can be spectacular immediately after decortication, and even over a whole year a mottled effect of three colours can be seen. Other species shed their bark in ribbons, e.g. *E. viminalis* Labill. and *E. globulus*, and some of these species hold the partly shed ribbons hanging in the crown, where they accumulate season after season and give rise to the name 'ribbon gums' (Figure 1.11).

*Rough bark*

When the dead bark accumulates year by year it dries by the loss of soft tissue and the fibres remain. The resulting rough bark has many forms. The outer layers usually weather to grey overlying the red-brown unexposed inner rough bark, which will be exposed later by the attrition of the outer fibres. The fibres may form long strings that can be pulled off, as in the stringybarks (e.g. *E. macrorhyncha*, Figure 1.9). The rough bark may break into flakes or 'squares' so characteristic of the bloodwoods (e.g. *E. calophylla*, Figure 1.12). In a large group of species the rough bark becomes infused with kino, a natural exudate, resulting in extremely hard bark that is usually deeply furrowed, as in the ironbarks (e.g. *E. cullenii* Cambage, Figure 1.13). A few species have hardened rough bark but, unlike the ironbarks, lack the characteristic deep furrowing over the whole trunk. These are the compact bark species, e.g. *E. fraxinoides* Deane & Maiden.

It is not possible to summarise adequately the rough bark types in *Eucalyptus* in so short a space, as it is too difficult to treat each form for which the conventional terms do not apply, e.g. in the many mallee species with some rough bark, and the red gums, swamp gums and mountain gums which are mostly smooth-barked but which have varying amounts of rough bark which has accumulated at the base of the trunk or stems.

*Figure 1.11* Ribbons of bark (*E. globulus*).

*Figure 1.12* Tessellated bark (*E. calophylla*).

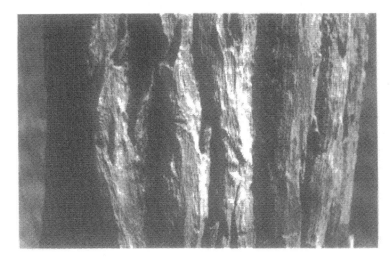

*Figure 1.13* Ironbark (*E. cullenii*).

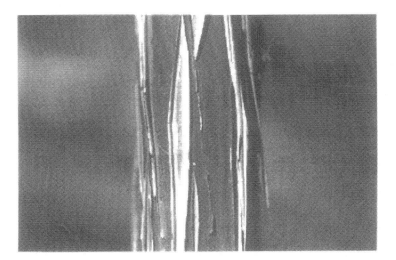

*Figure 1.14* Minniritchi bark (*E. caesia*).

A small group of species has probably the most beautiful rough bark of all. These are the minniritchi group in which the thin, outer, dark red-brown bark only partly sheds longitudinally (e.g. *E. caesia* Benth., Figure 1.14). The edges curl and expose fresh green bark, providing partly rough, partly smooth bark on the green and rich red-coloured stems.

## Seeds and germination

The seed morphology of eucalypts is extremely variable. Shape, size, colour and surface ornamentation are strongly inherited traits and indicative of taxonomic groups (Boland *et al*. 1991). Perhaps the most distinctive seed is that of most of the bloodwoods in which the body of the seed is basal and subtends a terminal wing. The seed of most species is flattened-ellipsoidal with a ventral hilum although in some large groups the seed has a terminal hilum. This occurs in all 'subgenus *Eucalyptus*', e.g. *E. obliqua*, and parts of 'subgenus *Symphyomyrtus*', namely, series *Annulares*, e.g. *E. pellita*, series *Fimbriatae-lepidotae*, e.g. *E. propinqua*, and series *Exsertae*, e.g. *E. tereticornis*.

There is no endosperm and the bulk of the seed is the embryo, dominated by the cotyledons, which are flat and appressed in ghost gums (*Blakella*, Pryor and Johnson 1971) but in all other species are variously folded across each other. The germinant consists of a radical, root collar, hypocotyl, cotyledons, epicotyl and terminal growth meristem. Germination is epigeal.

## Seedlings and leaf ontogeny

The primitive cotyledon shape is reniform and this form of cotyledon occurs widely in the genus (Maiden 1933). The extreme modification is the bisected cotyledon formed by emargination, resulting in a Y-shaped structure. A large number of species have cotyledons shaped between these extremes and are usually described as bilobed, although the distinction between the bilobed and the reniform is often blurred.

The normal pattern throughout the life of a eucalypt is to form leaves decussately and this arrangement is seen most characteristically in the seedling. The leaves are formed in pairs on opposite sides of the four-sided stem. Successive pairs, as assessed by the axes from the leaf tips of a pair through the stem, are at right angles to each other. Exceptions are the trigonous stems and resultant simple spiral leaf pattern in *E. lehmannii* (Schauer) Benth. and related species (Carr and Carr 1980) and the spirally-leaved seedlings, based on a five-sided stem, of several mallee species, e.g. *E. oleosa* F. Muell. ex Miq. (Brooker 1968).

While opposite leaf pairs continue to be formed at the growth tips of the adult plant, the leaf arrangement is modified in all species which develop adult leaves, apart from the exceptions above. The leaf-bearing axis elongates unequally on opposite sides such that the leaf pairs become separated and the arrangement appears to be alternate (Jacobs 1955). The leaves are usually petiolate and pendulous and therefore able to become positioned for optimum use of light. This false alternation may be a simpler way of producing the same effect as a true spiral.

The eucalypt plant is notably heterophyllous and the seedling leaves usually contrast with the adult leaves in terms of shape, petiolation, disposition in relation to the axis, colour, etc. Blake (1953) suggested a somewhat arbitrary set of categories to describe the progression of the leaf forms through the life cycle of the plant, namely, seedling, juvenile, intermediate and adult leaves. Some species seldom progress beyond the juvenile phase and are reproductively mature without advancing to the adult phase. Hence the flower buds, flowers and fruit form in the axils of juvenile leaves which exclusively make up the crown of a mature tree (e.g. *E. cinerea* F. Muell. ex Benth.). Most species in this category may form intermediate and adult leaves. A few species have never been seen in the adult phase, e.g. *E. crenulata* Blakely ex Beuzev. and *E. kruseana* F. Muell., although both of these rare and isolated species hybridise with co-occurring species, *E. ovata* DC. and *E. loxophleba* Benth., respectively, and produce offspring with leaves advanced from the juvenile state.

## The adult leaves

### Leaf form and coloredness

The mature crown of most eucalypts consists of adult leaves. These are usually petiolate, pendulous and lanceolate. Those of most species are concolorous (isobilateral). The truly pendulous leaf is frequently asymmetrical, being oblique at the base on the underside of the vertically hanging blade. Discolorous (dorsiventral) leaves occur in about fifty species and tend to be held more horizontally. The discolorous leaf is seen most prominently in the bloodwood species of humid regions, e.g. *E. polycarpa* F. Muell., and all of the eastern blue gums, e.g. *E. grandis*, red mahoganies, e.g. *E. pellita* F. Muell., and the grey gums, e.g. *E. propinqua* Deane & Maiden. An anomalous species of obscure affinity, but widely planted, is *E. cladocalyx* F. Muell. It occurs in relatively dry regions of South Australia and is strongly discolorous in the leaf. The fact that no species with dorsiventral leaves has formed the basis of an essential oil industry indicates that the increase in oil content and the changes in oil composition have proceeded in the evolutionary advance of the genus.

The distinction between the two leaf types is a strong diagnostic feature, but a few species may be difficult to categorise in this respect, e.g. most stringybarks are slightly discolorous. *E. muelleriana* A. Howitt., which grows in mesic areas, is the most distinctive of them.

*Leaf venation*

Leaf venation can be a strong character for the identification of eucalypts. Basically, the midrib subtends the side veins, between which varying densities of further reticulation occur, i.e. in the form of tertiary and quaternary veining. The primitive pattern appears to be the strongly pinnate form in which the side veins leave the midrib at a wide angle. This is seen in all bloodwoods (e.g. *E. calophylla*, Figure 1.15). The most modified form of venation is that seen in snow gums, e.g. *E. pauciflora* Sieber ex Spreng. and one or two peppermint species, e.g. *E. willisii* Ladiges,

*Figure 1.15* Bloodwood leaf (*E. calophylla*) showing wide-angled side veins and oil glands, mostly one per areole.

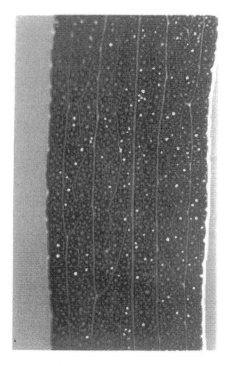

*Figure 1.16* Leaf of peppermint (*E. willisii*) showing parallel side veins.

Humphries & Brooker (Figure 1.16). A great variety of patterns occurs between these extremes. In a very few species, particularly narrow-leaved species, the side veins cannot be seen at all (e.g. *E. approximans*).

## Oil glands

*Eucalyptus* leaves are well known for their aromatic oils. Most texts refer to the invariable presence of oil glands in the leaves although some species are reported as having obscure glands. In fact many species have no visible glands at all. This situation occurs most commonly in the dryland and desert bloodwood species, e.g. *E. terminalis* F. Muell., and ghost gums, e.g. *E. tessellaris* F. Muell. Bloodwood species in more humid regions have oil glands, which are usually small and appear discretely situated in the middle of the very small areoles. Only one bloodwood of high rainfall areas, *E. ficifolia* F. Muell. of far southwestern Western Australia, appears to lack glands. It can occur sympatrically with *E. calophylla*, which is always glandular, making the presence or absence of glands a useful key character for these species, which are often held to be difficult to distinguish when not in flower. Outside the bloodwood and ghost gum groups, very few species in wetter areas consistently lack visible leaf glands, although *E. fasciculosa* F. Muell. (Figure 1.17) is an exception. The great majority of eucalypt species have conspicuous oil glands in the leaves.

The formation and development of oil glands in the eucalypt leaf has been discussed by Boland *et al.* (1991) on the basis of the studies of Carr and Carr (1970) on the formation of glands in the cotyledon or hypocotyl of the embryo. The process of gland formation begins from an epidermal cell which, on division, forms casing and epithelial cells for the gland. The cavity that becomes the oil gland itself derives from the development of an intercellular space that

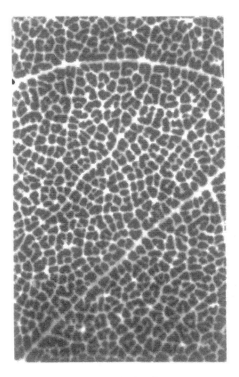

*Figure 1.17* Part of a leaf of *E. fasciculosa* showing dense reticulation without visible oil glands.

*Figure 1.18* Section of a leaf of *E. kochii* showing oil glands close under both epidermises.

increases in volume and matures as an ovoid or globular cavity (e.g. *E. kochii* Maiden & Blakely, Figure 1.18). The epidermal origin of the glands is reflected in the mature leaf, and in a leaf cross-section the glands are seen to occupy a zone just under the current epidermal layer of one or both surfaces.

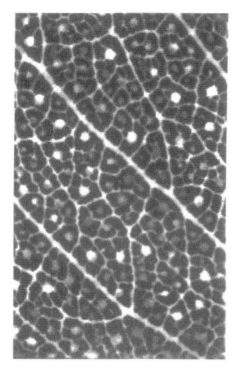

*Figure 1.19* Part of a leaf of *E. kochii* showing dense reticulation and intersectional oil glands.

In the mature adult leaf of most species the oil glands are very conspicuous. Their size, shape and colour, and their apparent spatial association with veinlets, result in a multitude of patterns which are useful in the identification of species. The glands may appear to be at the intersections of veinlets (e.g. *E. kochii*, Figure 1.19), situated singly in the areoles (e.g. *E. calophylla*, Figure 1.15) or few to many within the areoles (e.g. *E. microcorys* F. Muell., Figure 1.20), or a combination of both. In a few species they are so numerous and dense that they obscure the leaf venation completely (e.g. *E. mimica* Brooker & Hopper unpubl., Figure 1.21). Within a species, while the leaf venation pattern remains unaltered and is the more fundamental diagnostic feature, gland presence and density can vary. For the purposes of field estimation, oil gland density estimated by visual inspection is no certain indicator of total oil content.

## The eucalyptus oil industry in Australia

The English began a settlement at Port Jackson in 1788. Among the settlers' first objectives would have been the need to provide shelter and fuel. For these purposes trees of the native forests were felled in great numbers for their timber. It would not have taken long for the aromatic nature of the leaves to be noticed. The oil was loosely referred to as peppermint because of the supposed resemblance to that of the English peppermint, but the two oils are now known to be quite different chemically.

In 1789 a sample of oil was sent to England. It is believed to have been steam-distilled from the Sydney Peppermint, *Eucalyptus piperita* Smith, and there was obviously a presumed association of the effective medicinal properties of the *E. piperita* oil with those of the peppermint herb

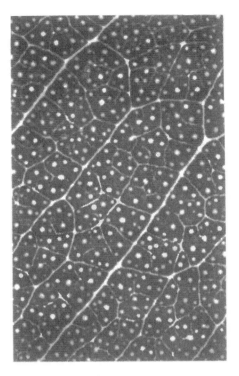

*Figure 1.20* Part of a leaf of *E. microcorys* showing moderate to sparse reticulation and numerous island oil glands.

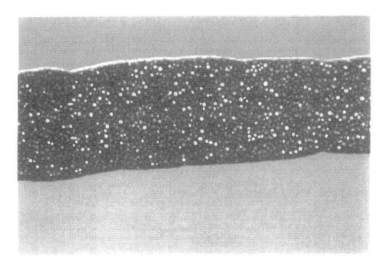

*Figure 1.21* Leaf of *E. mimica* showing very numerous oil glands obscuring side veins.

of England. A naval surgeon of the First Fleet, Denis Considen, considered that eucalyptus oil was 'much more efficacious in removing all cholicky complaints' (Boland *et al.* 1991).

As European occupation of the new colony extended further north, west and south, numerous new eucalypts were found, many of which were much richer in oils than the Sydney Peppermint.

Taxonomically the name 'peppermint' became traditionally used for a different group of species, also in 'subgenus *Eucalyptus*', and it was on these that the first eucalyptus oil industry was founded.

Around 1852 or 1854, a distillation plant was established near Dandenong, east of Melbourne, using the narrow-leaved peppermint, *E. radiata* Sieber ex Spreng. (Boland *et al.* 1991). This species is one of a small group of eucalypts noted for their oils and which have been exploited for commercial purposes. They still form the basis of small industries in tableland and mountain country of southeastern Australia. For optimum exploitation of the various species it was recognised that provenance was important in the choice of leaf material for distillation. This is now known to be due to the existence of different chemotypes within a species. There are notable phellandrene and cineole-rich forms of *E. radiata*, and piperitone and cineole-rich forms of *E. dives*. Despite the early successful use of these peppermint species, the industry is now largely based on entirely unrelated species.

Settlement progressed beyond the mountains and tablelands of southeastern Australia in the early nineteenth century and onto the inland plains, now mostly cleared and used for agriculture. Pockets of these new lands were unsuitable for cropping because of their infertility or stony soils but large areas of mallee species were available for exploitation. One in particular, blue mallee, *E. polybractea* R. Baker, was found to be rich in cineole and eucalyptus oil industries were founded on this species over 100 years ago in New South Wales and Victoria.

The mallee plant is peculiarly suited for 'agricultural' practice. It is small, becoming mature after a few years growth at about 1–2 m height, although old uncut or unburnt specimens can exceed 10 m. Its size and precocious maturity makes it ideally suited to mechanical harvesting. The propensity for regeneration from the lignotuber allows the virtual perpetual regrowth of the same biotype and it is harvested at approximately eighteen-month intervals. The ideal species will have a dense crown for high biomass, high cineole yield, strong regeneration capacity and fast growth. Selection and cloning will maximise these advantages for plantation industries.

A new and interesting development in oil mallee farming is being tested in southern Western Australia. There, vast areas of wheatland have become unproductive through salination, following extensive clearing of the natural vegetation. A few eucalypt species are being tested for growth on affected sites as a means of lowering the water table and, hence, reducing salinity. As a bonus, it is hoped to harvest the plants at suitable intervals and extract the essential oils. The species in question, *E. kochii*, *E. horistes* L. Johnson & K. Hill, *E. polybractea* and *E. loxophleba* Benth. subsp. *lissophloia* L. Johnson & K. Hill, have been chosen for their high oil content and expected adaptation to saline soils (Bartle 1997).

Mallee species seldom grow naturally in pure stands. They usually occur in association with one or more other eucalypt species of similar habit, but quite unrelated. Where natural stands were harvested, as in the early years of the industry, the value of these was enhanced by the culling of the less useful species. The most successful oil-producing centres are now going over to plantations, particularly of *E. polybractea*, established either from seed or from vegetatively reproduced high-yielding individuals.

Few species have been tested for the heritability of eucalyptus oil composition and yield. A trial of a high-yielding species from Western Australia, *E. kochii*, had a family heritability index of 0.83 based on $2\frac{1}{2}$-year-old progeny (Barton *et al.* 1991). Another trial on the commercially valuable and widely planted *E. camaldulensis* from Petford in northern Queensland had a family heritability of 0.62 (Doran and Matheson 1994), indicating that for these two species of widely differing affinities, growth habits and environments, heritability for essential oil factors is high.

In contrast to the Australian experience, eucalyptus oil industries in other countries have often begun as byproducts of eucalypt plantations established for other purposes, e.g. timber or pulp production, where large volumes of oil-bearing leaf material are available which would otherwise go to waste. This is the case in Spain and Portugal. Mallees have received little attention as commercial trees outside Australia because of their unpromising size and unsuitability for pole timber.

## *Eucalyptus* species used in oil production

In the following digests, arranged alphabetically, *Eucalyptus* species that are currently exploited for oil production are described. The first group comprises those species of modest or major current commercial significance, while the second group are those of lesser economic importance or those which have significant potential but which are, as yet, not fully exploited.

A number of other species have been used for oil production in the past but are not described in detail here. Neither are those few which are still very occasionally distilled. None are likely to assume greater importance in the future. Such species include the following:

1. *E. cneorifolia* DC., a mallee of Kangaroo Island off South Australia, which is still used for local cottage industries in the production of 1,8-cineole.
2. *E. elata* Dehnh., a peppermint common along streams in high rainfall country from west of Sydney, southwards in the subcoastal ranges to eastern Victoria, which contains high but variable amounts of piperitone.
3. *E. leucoxylon* F. Muell., a taxon with several subspecies which occurs in southern South Australia and western and central parts of Victoria and usually contains 1,8-cineole.
4. *E. macarthurii* Deane & Maiden, a tree of the Southern Tablelands of New South Wales which was once harvested for geranyl acetate.
5. *E. nicholii* Maiden & Blakely, a favourite ornamental tree of the Northern Tablelands of New South Wales harvested for 1,8-cineole.
6. *E. salmonophloia* F. Muell., a handsome tree of relatively dry country of southern Western Australia also harvested for 1,8-cineole.
7. *E. sideroxylon* Cunn. ex Woolls, an ironbark widespread on the drier slopes and plains associated with the Great Dividing Range in southeast Queensland and New South Wales which contains a cineole-rich oil. A distinctive form of *E. sideroxylon* from coastal New South Wales and eastern Victoria, once regarded as subsp. *tricarpa*, is now recognised as a separate species, *E. tricarpa* (L. Johnson) L. Johnson & Hill.
8. *E. viridis* R. Baker, a mallee of the wheatlands of western New South Wales and northern Victoria which grows with or near *E. polybractea*, and which contains a similar oil and is often harvested indiscriminately with it (Weiss 1997).

Remarkably few species are used worldwide as a commercial source of eucalyptus oil. Boland *et al.* (1991) state that of the 'more than 600 species of *Eucalyptus* probably fewer than twenty have ever been exploited commercially'. Coppen and Hone (1992) state that 'about a dozen species are utilised in different parts of the world, of which six account for the greater part of world production of eucalyptus oils'. Inventory surveys in the last fifteen years, however, have revealed several untried species that could be exploited. It is likely that a compromise may be sought between industries relying on high biomass availability in relatively low-yield tree species and those based on species with higher biomass per plant, high yield of desired oils and ease of mechanical harvesting. Spontaneous regeneration from lignotubers will be an important property of a desirable species.

## Species of commercial significance in essential oil production

### E. citriodora Hook. – Lemon-scented Gum

This species is one of the best known eucalypts, widely planted in Australia and elsewhere. It is unusual among eucalypts of subtropical origin in that it thrives in southern Australia among winter rainfall regimes. The trees are notably tall and erect with completely smooth bark of uniform colour, which contrasts with the closely related *E. maculata*, with its spotted, smooth bark and southern distribution. The juvenile leaves are typical of the bloodwood group in being large, alternate, peltate and hairy. The adult leaves of lemon-scented gum are narrowly lanceolate and on crushing reveal their most conspicuous character, the lemon scent.

*E. citriodora* belongs to a large taxonomic group, the bloodwoods (subgenus *Corymbia*), which are notable for the comparatively large, thick-walled, urceolate fruits. Within this group it is closely related to only two other species, *E. maculata* and *E. henryi* Blake, and together they are distinguished by the axillary, elongated, compound inflorescences, usually confined to the axils of the upper leaves, and by the flattened-ellipsoidal, red-black seeds with a cracked seedcoat and ventral hilum. This smooth-barked group is closest to the yellow bloodwoods, e.g. *E. peltata* Benth., which differ by the completely rough bark and the early loss of the outer operculum of the flower bud.

The leaves of bloodwood species are distinguished by the great number of secondary (side) veins and their wide angle to the midrib, which approaches 85 degrees. The side veins terminate at the intramarginal vein beyond which, as is typical in narrow-leaved bloodwoods, there is a single line of areoles parallel to the leaf margin. In *E. citriodora* the next degree of veining between the side veins occurs in about 2–4 rows, making a very clear, densely reticulate pattern. The oil glands occur spatially isolated within the areoles, usually single.

The main constituent of the oil of *E. citriodora* is citronellal, which gives the characteristic lemon scent. Although the oil has recently found commercial application as the active ingredient of some mosquito repellents (J. Coppen pers. comm.), it is used mainly for perfumery purposes. It is produced mainly in China, India and Brazil. Only one other *Eucalyptus* species has a lemon-scented oil, the citral-rich *E. staigeriana* F. Muell. ex Bailey (Lemon-scented Ironbark). This is unrelated to the Lemon-scented Gum and is easily distinguished from it by the completely coarse rough bark.

### E. dives Schau. – Broad-leaved Peppermint

This species is a small to medium-sized tree of relatively dry sites for peppermints, often occurring on sunny, northern slopes of hills with shallow, often stony soils. Its natural distribution is on the tablelands and lower mountains of southeastern Australia. The bark is typical peppermint, i.e. finely fibrous, not coarse and stringy, and is indistinguishable from that of *E. radiata*. The juvenile leaves are the most conspicuous element of the species. *E. dives* regrows readily on roadsides from lignotubers and the young plants produce many pairs of opposite, ovate, glaucous, juvenile leaves. The adult leaves are the broadest for the peppermint group and are lanceolate to broadly lanceolate, glossy green.

*E. dives* belongs to the monocalypts ('subgenus *Eucalyptus*'), a large subgenus distinguished by the axillary inflorescences, buds with a single operculum and seeds that are cuboid, pyramidal or elongate with a terminal hilum. The peppermints are characterised in the monocalypts by the many pairs of opposite leaves of the seedling. The buds and fruit are probably the largest for the peppermints of mainland Australia. The non-glaucous *E. nitida* J.D. Hook. of Tasmania has larger buds and fruit.

The secondary venation of the leaves of *E. dives* is conspicuous and terminates at a distinct intramarginal vein which is well removed from the edge. A minor, less distinct, intramarginal vein runs between the principal intramarginal vein and the edge. The reticulation is sparsely to only moderately dense through lack of tertiary and further veining, the areoles in the broader leaf being correspondingly larger than those in *E. radiata*. The oil glands are numerous per areole and do not appear to be associated with veinlets when the leaf is viewed fresh with transmitted light. They are approximately circular in outline and mostly green with some white in colour.

Several chemotypes occur in *E. dives* and there are 1,8-cineole, phellandrene and piperitone variants. In South Africa, around 150–180 t of the piperitone-rich oil was produced annually in the late 1980s (Coppen and Hone 1992) and although it is no longer used for the production of menthol, as it once was, it continues to be distilled for flavour and fragrance applications (J. Coppen pers. comm.).

### E. exserta F. Muell. – Queensland Peppermint

This is a small to medium-sized tree or mallee belonging to the red gum group. It is notable for the linear juvenile leaves and hard rough bark. The fruit of the entire red gum group are recognised by the hemispherical base and very prominent raised disc and exserted valves. The seeds of *E. exserta* are typical of most of the red gums, e.g. *E. tereticornis*, in being elongated, black, toothed around the edges and with a terminal hilum. They contrast strongly with those of the widely planted red gum, *E. camaldulensis*, which are yellow and double-coated.

*E. exserta* is widespread in eastern Queensland, particularly on low stony rises. The red gums divide into four taxonomic groups distinguished by seed, juvenile leaf, fruit and bark characters. To the north of its natural distribution the related *E. brassiana* Blake occurs as far as New Guinea, while to the south another related species, *E. morrisii* R. Baker, is endemic to low rocky hills of western New South Wales.

The leaves of *E. exserta* are slightly glossy, green to greyish green, and have the typical red gum pattern of moderately dense reticulation. Tertiary venation is fine with fairly large areoles and numerous oil glands per areole. Most glands appear round and green when the leaf is viewed fresh with transmitted light.

*E. exserta* has been planted in China for many years for its timber and leaf oils.

### E. globulus Labill. (four subspecies) – Southern Blue Gums

This is one of the most widely cultivated species around the world. Typically, it is a tall, erect forest tree with mostly smooth bark. The characteristic prolonged juvenile leaf phase is one of its most notable features, making it one of the easiest eucalypts to identify. In this phase the leaf pairs remain opposite and sessile to the sapling stage and are conspicuously broad and glaucous. The adult leaves are strikingly different in being long, falcate, glossy green. The juvenile leaves are often seen in clear contrast to the adult leaves when they appear as new growth on old branches within the crown.

*E. globulus* sens. lat. occurs disjunctly over a wide area of relatively high rainfall parts of southeastern Australia. It has created considerable confusion taxonomically, having been variably split into four or five species. It belongs to the large section *Maidenaria*, which may be diagnosed by the concolorous leaves, axillary inflorescences, juvenile leaves sessile and opposite for many pairs, and seeds prominently lacunose with a ventral hilum. Currently, *E. globulus* is regarded as four subspecies, one of which, subsp. *globulus* from Tasmania and southern, coastal Victoria, is the principal form grown in other countries.

Tasmanian Blue Gum (subsp. *globulus*) is one of the few eucalypt species with a single bud to the inflorescence. The buds are large, sessile, warty and glaucous. The other subspecies are subsp. *bicostata* (Maiden, Blakely and J. Simm.) Kirkpatr., which has three buds to the inflorescence and is distributed in high country from northern New South Wales to southern Victoria, subsp. *pseudoglobulus* (Naudin ex Maiden) Kirkpatr., which has three buds to the inflorescence and occurs in east coastal Victoria and possibly the northern end of Flinders Island in Bass Strait, and subsp. *maidenii* (F. Muell.) Kirkpatr., which has seven buds and occurs in coastal hills of eastern Victoria and far southeastern New South Wales.

The leaves have a moderately dense, fairly distinct reticulation formed by the tertiary and fine quaternary veinlets between the side veins. The oil glands occur singly or in twos and threes within the areoles and can be seen with transmitted light through a fresh leaf. The glands are roughly circular in outline and are mostly green with some white. Some glands touch veinlets or appear to terminate ultimate veinlets within an areole.

*E. globulus* is the principal source of eucalyptus oil in the world. It is utilised for such purposes in China, Spain, Portugal, India, Argentina, Brazil and Chile. The major constituent of the oil is 1,8-cineole.

E. polybractea *R. Baker (syn.* E. fruticetorum *(F. Muell. ex Miq.) R. Baker)* –
Blue-leaved Mallee

This is the principal species used in Australia for essential oil production. It is a mallee, naturally adapted to relatively low rainfall and infertile soils on plains and low hills inland from the Great Dividing Range in Victoria and New South Wales. As a mallee it matures as a tall, multi-stemmed shrub. It has fibrous rough bark over the lower part of the stems and the leaves are bluish or greyish green. The juvenile leaves are opposite for a few pairs, then alternate, shortly petiolate, lanceolate to linear, green or greyish green in colour. Adult leaves are notably bluish. The inflorescences are terminal panicles.

*E. polybractea* belongs to the box group of eucalypts, most of which are characteristically rough-barked and have terminal inflorescences. The boxes divide into tree forms and mallees. Among the latter, *E. polybractea* is closest to Green Mallee, *E. viridis*, which occurs over a similar distributional range. Green Mallee is distinguished by the green, glossy leaves, which are linear and held erect.

The leaf venation of Blue-leaved Mallee is common to many box species and quite different from monocalypts and *Maidenaria* species. The side veins are very acute, particularly towards the base of the leaf. Intramarginal veins are present while tertiary veining is dense, somewhat obscure and erose, resulting in elongated, oblique areoles which accommodate many discrete oil glands (Figure 1.22). There is no clear reticulation outside the intramarginal veins. The glands are large and very irregular in outline.

Australia is the only country which utilises *E. polybractea* for oil production, relying as it does mainly on 'cleaned' areas of natural stands, supplemented by modest areas of plantations. The centres of production are at West Wyalong in western New South Wales and near Inglewood in northern Victoria.

E. radiata *Sieber ex Spreng. (syn.* E. australiana *Baker & Smith,* E. phellandra *Baker & Smith)* –
Narrow-leaved Peppermint

This species is typically a small to medium-sized tree of relatively dry forests and woodlands. The bark is rough over the whole trunk, fibrous and interlaced, though not as coarse as in the

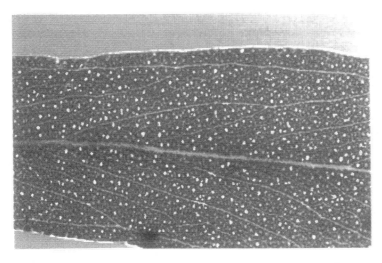

*Figure 1.22* Leaf of *E. polybractea* showing dense, obscure reticulation and numerous, very irregular oil glands, both features typical of the mallee boxes.

stringybarks. The juvenile leaves remain opposite for many pairs, are sessile, lanceolate and green. In morphology the young plants resemble *E. viminalis*, which may grow naturally in the vicinity, but the two species can be distinguished by the stronger fragrance, emanating from the more abundant oil glands, and by the reduced amount of leaf reticulation in *E. radiata*. The mature crown is usually dense and attractive with small, narrow leaves which are green to dark green and glossy.

Narrow-leaved Peppermint is a monocalypt ('subgenus *Eucalyptus*') and belongs to one of the taxonomically most difficult groups in this subgenus. Most peppermints are relatively narrow-leaved and the common name mainly has relevance to the contrasting, broader-leaved, *E. dives*. Baker and Smith (1915) published a new species, *E. australiana*, which they considered was a mainland form of the otherwise Tasmanian endemic, *E. amygdalina* Labill. They were apparently unaware that the long established name, *E. radiata*, accounted for their new species in conventional morphological terms. Blakely (1934) stated that *E. australiana* 'yields an excellent pharmaceutical oil far superior to that of *E. radiata*, due mainly to the presence of a large amount of cineole'. This nomenclatural anomaly illustrates the problem of trying to establish taxa from differences in their chemistry alone. While the 'chemical' species may be disregarded from the purely taxonomic point of view, it emphasises that provenance is an important aspect in assessing the value of a species for oil production. *E. phellandra*, another mainland taxon, is similarly regarded now as a high-phellandrene form of *E. radiata*.

One other accepted taxon has been conventionally associated with *E. radiata*, namely, *E. robertsonii* Blakely. Over the relatively wide natural distribution of *E. radiata*, populations extend from the typical tableland regions to the mountains of far southeastern New South Wales. Here, the species takes on a taller habit, the leaves are somewhat bluish and the buds are slightly glaucous. This form was reduced to a subspecies of *E. radiata* (Johnson and Blaxell 1973) but later reinstated as a species in its own right (Johnson and Hill 1990). In the latter paper Johnson and Hill split typical *E. radiata* and erected the northern form as *E. radiata* subsp. *sejuncta*. The new taxon differs from subsp. *radiata* by the broader juvenile leaves and appears to be richer in α-pinene than 1,8-cineole (Brophy unpubl.).

The inflorescences of the peppermints invariably have more, often many more, than seven buds. These are pedicellate, clavate, and minutely warty. The fruits usually have a thick rim and a conspicuous reddish brown disc. A peppermint species with which *E. radiata* may be confused is *E. elata*, the River Peppermint, which has only the basal part of the trunk rough and has very numerous buds such that the flowers are a characteristic ball of white. A peppermint species with notably narrower leaves, and a favoured ornamental, is *E. pulchella* Desf., which is a Tasmanian endemic and mostly smooth-barked.

The leaf reticulation in *E. radiata* is sparse with the side veins at a prominently acute angle (<30° to the midrib). There is usually an inner strong and outer weak intramarginal vein. This overall pattern is typical of the peppermints, although *E. willisii* from southern, coastal Victoria can have side veins parallel to the midrib, as seen in snow gums, e.g. *E. pauciflora*. The oil glands occur many to an areole without obvious association with the veinlets. The glands are roughly circular in outline and a mixture of green and white in colour when viewed fresh with transmitted light.

*E. radiata* was once exploited in Australia for oil production, and still is to a small extent. It is also grown for this purpose in a small way in South Africa and has shown promise in other parts of Africa such as Tanzania (J. Coppen pers. comm.). There are several recognised chemotypes. Boland *et al.* (1991) cite two, one being rich in 1,8-cineole and the other in phellandrene/piperitone. Recent commercial interest has focused on the cineole-rich form.

E. smithii *R. Baker – Gully Gum*

This is a medium-sized to tall, erect forest tree of southeastern Australia, where it occurs on the coastal plains and in valleys of the eastern fall of the Southern Tablelands of New South Wales. Small mallee forms are found on very rocky crags of mountain country of far eastern Victoria. In habit and bark the trees are easily confused with the unrelated *E. badjensis* Beuzev. & Welch and *E. elata* in that all three are tall trees with basal, compacted, brown-black rough bark. Juvenile leaves are also similar, being sessile, opposite, lanceolate and green for many pairs. However, inflorescences clearly differentiate the three: *E. smithii* is 7-flowered, *E. badjensis* is 3-flowered and *E. elata* has many more than 7 flowers.

*E. smithii* is classified in the section *Maidenaria* which divides into two large groups, one 3-flowered and the other 7-flowered. It is difficult to associate it taxonomically with any other species to establish natural affinity, although it has superficial similarities to the two mentioned above.

The leaves of Gully Gum are usually long and narrow. Side veins are clear with lesser side veins running more or less between. The intra-veinal zone is broken by relatively weak tertiary veinlets and some even less distinct quaternary veinlets. Oil glands are fairly numerous, one to few per areole, without giving the immediate impression that the leaves are highly glandular. The glands are green and yellowish.

*E. smithii* is widely grown in South Africa for timber purposes and in the eastern Transvaal it is harvested for cineole-rich leaf oil (Coppen and Hone 1992).

E. staigeriana *F. Muell. ex Bailey – Lemon-scented Ironbark*

This is a small to medium-sized tree, usually of poor form in its natural habitat which is granite or sandstone hills of Cape York Peninsula in far northern Queensland. The bark is typical ironbark, i.e. hard, black and furrowed. The crown is conspicuously bluish due to the blue-green to glaucous, usually broad, adult leaves. The inflorescences are terminal panicles. Buds and fruit are small and inconspicuous.

The numerous ironbark species belong to three taxonomic groups. *E. staigeriana* is in the largest group which is distinguished by the early loss of the outer operculum and stamens which are all fertile. Both of these characters contrast with the buds of the better known ironbark species, *E. sideroxylon*.

The adult leaves are relatively small with very dense reticulation formed by the tertiary and quaternary veining. The oil glands are of two types, either small, green and single in the areoles (some areoles lacking glands) or fewer, larger and whitish, occurring at the intersections of quaternary veinlets.

*E. staigeriana* is one of two eucalypt species that are notable for their lemon-scented oils (see *E. citriodora* above) and is cultivated for the production of citral-rich oil, chiefly in Brazil. Weiss (1997) states that it is also grown for such purposes in Guatemala, the Seychelles and the former Zaire (now the Democratic Republic of the Congo).

## *Species of lesser importance or with potential for essential oil production*

### E. camaldulensis Dehnh. – River Red Gum

This is the most widespread species in Australia and one that is cultivated throughout the world in tropical and subtropical regions. It is notably a species of freshwater stream banks, whether of the major rivers or the inland rivers which only flow after heavy rain. It frequently has a heavy butt and widely spreading crown. Basically a smooth-barked species, it usually has an accumulation of unshed or imperfectly shed rough slab bark on the bottom 1 m of the trunk. As with all the red gums, the juvenile leaves are petiolate and only the first 4 or 5 pairs remain opposite. In River Red Gum the juvenile leaves are lanceolate, which contrast with the ovate leaves of most of the related species. The adult leaves are mid-green, slightly bluish green or yellowish green and vary from dull to slightly glossy.

*E. camaldulensis* belongs to the large series *Exsertae* and, as might be expected in a species of such widespread natural distribution, there is considerable variation in morphological characters. This has been recognised in the taxonomic treatment which has resulted in many named varieties. The principal split, which is recognised today more for convenience than any fixed morphological distinction, is into two varieties, var. *camaldulensis*, which is the form of the Murray-Darling river systems of southeastern Australia, most easily distinguished by the strongly beaked operculum, and var. *obtusa* Blakely, which is the name loosely applied to all other provenances, the name referring to the obtuse operculum. The seeds are distinctive in being yellow to yellow-brown, cuboid and having a double seedcoat and terminal hilum.

The leaf reticulation in River Red Gum is moderately dense in southern Australian occurrences while the northern River Red Gums have a denser, more closed pattern. The quaternary veining is relatively weak and ultimate fine veinlets may appear to end within an areole, such that the overall pattern is not finite. Oil glands are usually numerous and conspicuous and may be several per areole in leaves of southern provenances, becoming fewer per areole in northern specimens. The glands are approximately circular in outline and are usually a mixture of green, yellow and white when viewed fresh with transmitted light (Figure 1.23). Very rarely do the leaves appear glandless.

*Figure 1.23* Leaf of *E. camaldulensis* showing dense reticulation and island oil glands, both features typical of the red gums.

It is likely that there are many chemotypes. One well-known provenance, Petford in northern Queensland, for example, has two forms, one rich in 1,8-cineole and the other in sesquiterpenes. Because of its widespread cultivation it may well become a future source of eucalyptus oil.

E. cinerea F. Muell. ex Benth. – Argyle Apple

This is one of the most widely planted ornamental eucalypts in southern Australia. It is considered highly attractive for its mature crown of glaucous, ovate juvenile leaves, and is also much favoured in dried flower arrangements. It is only one of several species, however, that are reproductively mature in the juvenile leaf phase. It is one of the mostly easily recognised eucalypt species by the combination of the grey crown and thick, reddish brown, completely rough bark. It is further distinguished by its 3-flowered inflorescences and funnel-shaped fruits.

Argyle Apple belongs to the large section *Maidenaria*, many species of which are 3-flowered. From a distance it could only be confused among species of southern Australia with *E. risdonii* J.D. Hook., one of the completely unrelated monocalypts ('subgenus *Eucalyptus*') which is easily distinguished on closer inspection by the connate leaf pairs. Species of northern Australia may have similar crowns to *E. cinerea*, e.g. *E. melanophloia* F. Muell., which is an ironbark with an entirely different hard, furrowed, rough bark, and *E. pruinosa* Schauer, which has much larger leaves and terminal inflorescences.

The side veins of the leaves of *E. cinerea* meet the intramarginal vein at indentations, the intramarginal vein itself being well removed from the leaf edge. With the side veins the tertiary veining makes distinctive polygons which are themselves divided within by fine quaternary veining to form the ultimate areoles. Within the areoles the oil glands are one to few, small, whitish and mostly discrete, with some contiguous to the veinlets.

The leaves of *E. cinerea* contain a cineole-rich oil and it was once used for oil production in Australia. It has recently been cultivated for such purposes in Zimbabwe (Boland *et al.* 1991) although production has only amounted to a few tonnes annually (Coppen and Hone 1992).

### E. kochii Maiden & Blakely subsp. kochii *(Maiden & Blakely) Brooker – Oil Mallee*

This is a small to medium-sized mallee found only in the northern wheatbelt of Western Australia where it occurs largely as roadside remnants of once dense mallee scrub. The bark is rough. The dull, bluish grey, lanceolate juvenile leaves contrast with the glossy green, linear to narrowly lanceolate adult leaves which make up a dense crown of shining foliage.

*E. kochii* belongs to a large taxonomic series that occurs widely in southern Australia in relatively dry country, although most of the species are concentrated in the west of the continent. *E. kochii* consists of two subspecies in the Western Australian wheatbelt and adjacent arid country: subsp. *kochii*, which is the typical, western form, rich in 1,8-cineole, and subsp. *plenissima*, the eastern form, which extends into desert areas and has a much lower oil content despite the subspecific name.

The tertiary veining of *E. kochii* forms a dense, finite reticulum. The oil glands are very numerous and large, are irregular in shape and appear at the intersections of the veinlets (Figure 1.19).

Oil Mallee is currently being tested as a producer of 1,8-cineole by the CSIRO and the New South Wales Department of Agriculture at Condobolin in mid-western New South Wales. Early results show that *E. kochii* maintains a high production of 1,8-cineole per unit weight of leaf, but it is likely that the species will only produce sufficient biomass on soils which are lighter than those where the trial is being conducted (Milthorpe *et al.* 1998).

### E. olida *Johnson & Hill*

This is a small to medium-sized tree, recently discovered and rare, known only from two natural populations on the Northern Tablelands of New South Wales. It is rough-barked, has prominently petiolate, broadly ovate, pendulous, dull blue-green juvenile leaves. The adult leaves are dull to slightly glossy, green to grey-green. There are usually many clavate flower buds per inflorescence.

*E. olida* belongs to the blue ash group of species (in 'subgenus *Eucalyptus*'), so named from the colour of the juvenile leaves, which contrasts with the bright green juvenile leaves of the green ash species, e.g. *E. obliqua*. It shares the tight, grey, fibrous bark of *E. andrewsii* Maiden, the blue ash species common in the same part of New South Wales.

The leaves are sparsely to moderately reticulate with very acute side veins typical of the ash group. The areoles are often indistinct due to the indeterminate tertiary veining. Oil glands are numerous per areole.

The discovery of *E. olida* aroused considerable interest when it was found to contain a high proportion of *E*-methyl cinnamate in the leaves, and the species is now exploited commercially as a source of the chemical in Australia.

## Acknowledgements

I am grateful to Garry Brown for production of the Figures and Kirsten Cowley for preparing the data for Figure 1.1.

## References

Baker, R.T. and Smith, H.G. (1915) *Eucalyptus australiana*, sp. nov. and its essential oil. *J. Proc. Roy. Soc. N.S.W.*, 49, 514–525.

Barlow, B.A. (1981) *The Australian Flora: Its Origin and Evolution.* In A.S. George (ed.), *Flora of Australia*, Vol. 1, Australian Government Publishing Service, Canberra.

Bartle, J.R. (1997) Information Brochure, Farm Forestry Unit, Western Australian Department of Conservation and Land Management, Perth.

Barton, A.F.M., Cotterill, P.P. and Brooker, M.I.H. (1991) Heritability of cineole yield in *Eucalyptus kochii*. *Silvae Genetica*, 40, 37–38.

Blake, S.T. (1953) Botanical contributions of the Northern Territory Regional Survey. I. Studies on northern Australian species of *Eucalyptus*. *Aust. J. Bot.*, 1, 185–352.

Blakely, W.F. (1934) *A Key to the Eucalypts*, Government Printer, Sydney.

Boland, D.J., Brophy, J.J. and House, A.P.N. (eds) (1991) *Eucalyptus Leaf Oils. Use, Chemistry, Distillation and Marketing*, Inkata Press, Melbourne.

Brooker, M.I.H. (1968) Phyllotaxis in *Eucalyptus socialis* F. Muell. and *E. oleosa* F. Muell. *Aust. J. Bot.*, 16, 455–468.

Brooker, M.I.H. and Kleinig, D.A. (1990) *Field Guide to Eucalypts*, Vol. 2, *South-western and Southern Australia*, Inkata Press, Melbourne.

Brooker, M.I.H. and Lassak, E. (1981) The volatile leaf oils of *Eucalyptus ovata* Labill. and *E. brookerana* A.M. Gray (Myrtaceae). *Aust. J. Bot.*, 29, 605–615.

Carr, D.J. and Carr, S.G.M. (1970) Oil glands and ducts in *Eucalyptus* L'Hérit. 2. Development and structure of oil glands in the embryo. *Aust. J. Bot.*, 18, 191–212.

Carr, D.J. and Carr, S.G.M. (1980) The *Lehmannianae*: a natural group of Western Australian eucalypts. *Aust. J. Bot.*, 28, 523–550.

Chippendale, G.M. (1988) *Myrtaceae – Eucalyptus, Angophora*. In A.S. George (ed.), *Flora of Australia*, Vol. 19, Australian Government Publishing Service, Canberra.

Coppen, J.J.W. and Hone, G.A. (1992) *Eucalyptus Oils: A Review of Production and Markets*, NRI Bulletin 56, Natural Resources Institute, Chatham, U.K.

Doran, J.D. and Matheson, A.C. (1994) Genetic parameters and expected gains from selection for monoterpene yields in Petford *Eucalyptus camaldulensis*. *New Forests*, 8, 155–167.

Hill, K.D. and Johnson, L.A.S. (1995) Systematic studies in the eucalypts. 7. A revision of the bloodwoods, genus *Corymbia* (Myrtaceae). *Telopea*, 6, 185–504.

Jacobs, M.R. (1955) *Growth Habits of the Eucalypts*, Commonwealth Government Printer, Canberra.

Johnson, L.A.S. and Blaxell, D.F. (1973) New taxa and combinations in *Eucalyptus* – II. *Contributions from the N.S.W. National Herbarium*, 4, 379–383.

Johnson, L.A.S. and Hill, K.D. (1990) New taxa and combinations in *Eucalyptus* and *Angophora* (Myrtaceae). *Telopea*, 4, 37–108.

Maiden, J.H. (1903–1933) *A Critical Revision of the Genus Eucalyptus*, Government Printer, Sydney.

Milthorpe, P.L., Brooker, M.I.H., Slee, A. and Nicol, H.I. (1998) Optimum planting densities for the production of eucalyptus oil from blue mallee (*Eucalyptus polybractea*) and oil mallee (*E. kochii*). *Industrial Crops Prods.*, 8, 219–227.

Penfold, A.R. and Morrison, F.R. (1929) The occurrence of a number of varieties of *Eucalyptus dives* as determined by chemical analyses of the essential oils. Part 3. *J. Proc. Roy. Soc. N.S.W.*, 63, 79–84.

Penfold, A.R. and Willis, J.L. (1961) *The Eucalypts*, World Crop Series, Leonard Hill, London, and Interscience, New York.

Pryor, L.D. and Byrne, O.R. (1969) Variation and taxonomy in *Eucalyptus camaldulensis*. *Silvae Genetica*, 18, 64–71.

Pryor, L.D. and Johnson, L.A.S. (1971) *A Classification of the Eucalypts*, Australian National University Press, Canberra.

Sale, M.M., Potts, B.M., West, A.K. and Reid, J.B. (1996) Relationships within *Eucalyptus* (Myrtaceae) using PCR-amplification and southern hybridisation of chloroplast DNA. *Aust. Syst. Bot.*, 9, 273–282.

Udovicic, F., McFadden, G.I. and Ladiges, P.Y. (1995) Phylogeny of *Eucalyptus* and *Angophora* based on 5S rDNA spacer sequence data. *Molecular Phylogenetics and Evolution*, 4, 247–256.

Weiss, E.A. (1997) *Essential Oil Crops*, CAB International, New York.

# 2 Eucalyptus, water and the environment

*Ian R. Calder*

## Environmental concerns about eucalyptus

Traditionally, foresters have always believed their forests to be a benign influence on the environment. Sure in this belief, forestry programmes have been advocated, even where there are little or no commercial returns from the forests, solely for their 'environmental' benefits.

Yet not everybody shares the forester's sanguine views, especially when forestry plantations contain fast growing trees such as eucalypts. In many parts of the world concerns have been expressed over the environmental and social effects of large scale planting of eucalypts for industrial and social forestry applications. One of the key issues in this controversy concerns the hydrological effects of the afforestation. Essentially these fears are, as expressed in India by Vandana Shiva *et al.* (1982) and Vandana Shiva and Bandyopadhyay (1983, 1985), that eucalypts are voracious consumers of water and are likely to deplete water resources. When one considers that there are many well documented reports that eucalypts have been used to drain marshes – near Rome in the eighteenth or nineteenth century (Ghosh *et al.* 1978), and in Uganda (see Nshubemuki and Somi 1979) and in Israel (Saltiel 1965) in the twentieth century – and that eucalypt plantations are currently being used as 'water pumps' to deliberately lower water tables in parts of Australia that are experiencing salinity problems (Greenwood 1992), it is easy to understand how these fears have arisen.

In the absence of hard evidence to the contrary, speculation by the press and by some local environmental groups raised the controversy to such a pitch that in parts of Karnataka state in southern India farmers ripped out eucalyptus seedlings from government nurseries and plantations. Yet other farmers in Karnataka saw eucalypts as beneficial – they viewed them as a valuable source of income and were keen to plant them on their fields. And eucalyptus trees can have other environmental and economic benefits. As providers of a fast-growing source of timber, firewood and pulp they can help to reduce the pressure on the few remaining indigenous forests as wood sources and thus aid conservation efforts. Through saving foreign exchange on the importation of pulp they have obvious economic benefits. On a very much smaller scale, benefits, welcome nonetheless, accrue to the pharmaceutical and fragrance industries in India (and elsewhere) from the production of eucalyptus oil. However, an article in a popular weekly science magazine in 1988, 'The tree that caused a riot' (Joyce 1988), illustrates clearly the uncertainty and the worries that existed at that time.

To some extent the publication of recent research findings on the environmental impact of eucalyptus plantations in India has defused the conflict situation between foresters and environmentalists. Yet although some of the worries of the environmental groups may have been without substance it is clear now that the forester's conceptions of the benefits of these plantations were also often flawed.

In South Africa, eucalyptus plantations have also been the centre of a controversy which has revolved around the question of water use. Here, eucalypts are generally planted on the wetter hills and the concern is that less water is then available for supply to industries and mines, farmers downstream who want water for irrigation, and, often further downstream, the game parks and nature reserves which require minimum flows in the rivers to sustain wildlife and the local ecology. South Africa has carried out some of the most detailed and definitive studies of water use from forests and commercial eucalypt plantations of any country in the world. The results have not suffered obfuscation through water interests and forestry interests residing in different government ministries since, in South Africa, both reside within the Department of Water Affairs and Forestry. This has led to an innovative approach to land and water management and proposals that land uses with a high water requirement such as forests and sugar cane plantations should be subject to an 'interception levy' in compensation for the water that would otherwise be available under different land uses.

In both India and South Africa, and also in other countries with a high proportion of eucalypt plantations, the pressing question is not so much 'What are the environmental impacts?' but 'How can these plantations and other land uses be managed in an integrated and sustainable manner to the benefit of economics, socio-economics, the ecology and the water resources of the basin and basin inhabitants?'.

This chapter reviews some of the myths that are still being propagated in relation to the impacts of forests on the environment and the research that has been carried out in India and South Africa on the impacts of eucalypt plantations. It also outlines the new approach that is being developed worldwide to manage land and water resources in an integrated way. Although little of the research has been directed specifically at eucalypts being grown for medicinal or perfumery oils (or any other extractive), all of what is discussed here has relevance to such situations. In many cases the species used for oil production are similar, or identical, to those used for timber and pulp and the research findings would be expected to be equally applicable. Where 'waste' leaf from eucalypts felled for pulp is distilled then this will certainly be true. Where oil production is achieved through harvesting of trees which are still growing, however, this will result in a decrease in leaf area of the living trees, and it would be expected that growth rates, transpiration rates and interception losses would all be decreased. Water use from these plantations would then be expected to be significantly less than those from plantations grown for timber or pulp, which would have more extensive canopies. Although the author is not aware of any studies which have quantified this effect they would, in principle, be easy to carry out through the application of modern measurement technologies and modelling methods.

## Forest hydrology myths

There still remains much folklore, and many myths, about the role of forests and their relationship with the environment and hydrology which hinders rational land use decision-making. Foresters have sometimes been suspected of deliberately propagating some of these forest hydrology myths. Pereira (1989) states:

> The worldwide evidence that high hills and mountains usually have more rainfall and more natural forests than do the adjacent lowlands has, historically, led to confusion of cause and effect. Although the physical explanations have been known for more than 50 years, the idea that forests cause or attract rainfall has persisted. The myth was created more than a century ago by foresters in defence of their trees ... The myth was written into the textbooks and became an article of faith for early generations of foresters.

The overwhelming hydrological evidence supports Pereira's view that forests are not generators of rainfall yet this 'myth', like many others in forest hydrology, may contain a modicum of truth that prevents it from being totally laid to rest. The oft propagated views that 'forests increase rainfall' and that 'forests increase runoff, regulate flows, reduce erosion, reduce floods, sterilise water supplies and improve water quality', when scrutinised (Calder 1996, 1998, 1999), are mostly seen to be either exaggerated or untenable. For some we still require research to understand the full picture.

In relation to eucalyptus plantations, the principal concerns have been about their effects on total runoff and low flows and on erosion. Some of the research that has been carried out to investigate these relationships, which is now providing the scientific evidence which can replace the former folklore and beliefs which have for so long bedevilled rational land use decisions, is reviewed below. The review focuses mainly on studies in Australia, India and South Africa which have used both detailed process-based measurements carried out at the plot scale and measurements at the scale of whole catchments.

## Eucalytpus, evaporation and runoff

### Evaporative processes

It is convenient to differentiate the total evaporative loss from vegetation into its two principal components, interception and transpiration. This is because the processes involved are essentially different. Interception, the evaporation of water from the wet outer surfaces of leaves during and after rainfall, is simply a physical process, whereas transpiration, which involves the uptake of water by plant roots and transfer through the leaves, involves physiological processes, and is thus subject to much more complex control mechanisms.

### Rainfall interception by eucalypts

Evaporative losses of intercepted water occur both during the rainfall event itself and afterwards, from water stored on the leaves, branches and trunks of trees, when they are then constrained by the water storage capacity of the vegetation.

The average evaporation rates from wet trees tend to be much higher, say 2–5 times, than those from wet, shorter vegetation (Calder 1979, 1990) because they present a relatively rough surface to the wind and are more efficient in generating the forced eddy convection which, under the majority of meteorological conditions, is the dominant mechanism responsible for the vertical transport of water vapour from the leaves into the atmosphere. Little information is available on differences between species with respect to evaporation rates under wet conditions and it is usually assumed that evaporation rates from wet vegetation of the same height are similar, irrespective of species.

The leaf interception storage capacity, however, a function of the leaf surface area and its water-holding capacity, does vary widely between tree species. The highest storage capacities have been reported for tropical rainforest trees (Herwitz 1985), although it is believed that these high storages are largely caused by storage on the woody parts (trunks and branches) of these tall trees. *Eucalyptus* spp. are likely to fall into the lower end of the range of tree storage capacities.

In regions with relatively high intensity, short duration storms (i.e. tropical monsoon or convective storms, as opposed to relatively low intensity, long duration storms typically associated with frontal rainstorms) it is expected that the proportion of intercepted water lost during

Table 2.1 Comparative interception ratios from *Eucalyptus* and other tree species[a]

| Location | Species | Average annual rainfall (mm) | Interception (% of rainfall) | Reference |
|---|---|---|---|---|
| India (Dehra Dun) | 'E. hybrid' | 1670 | 12 | George 1978 |
|  | *Pinus roxburghii* | 1670 | 27 | Dabral and Subba Rao 1968 |
| India (Nilgiris) | *E. globulus* | 1150 | 22 | Samraj *et al*. 1982 |
|  | *Acacia mearnsii* | 1150 | 25 | Samraj *et al*. 1982 |
|  | Shola forest | 1150 | 34 | Samraj *et al*. 1982 |
| Israel | *E. camaldulensis* | 700 | 15 | Karschon and Heth 1967 |
| Australia (Lidsdale, NSW) | *E. regnans* | 810 | 11 | Smith 1974 |
|  | *Pinus radiata* | 810 | 19 | Smith 1974 |
| Australia (Maroonda, NSW) | *E. obliqua* | 1200 | 15 | Feller 1981 |
|  | *Pinus radiata* | 1200 | 25 | Feller 1981 |
|  | *E. regnans* | 1660 | 19 | Feller 1981 |

[a] Brookes and Turner (1964), Duncan *et al*. (1978), Prebble and Stirk (1980), Dunin *et al*. (1988) and Westman (1978) have reported interception ratios from eucalypts in Australia in the range 11–23 per cent but they did not give values from other tree species to allow comparisons to be made under the same climatic regimes.

rainfall is small compared with that lost from storage after the cessation of rainfall. Interception losses will therefore be primarily determined by canopy capacities and it is reasonable to expect that they will, under these circumstances, be lower from *Eucalyptus* spp. than from other tree species. This conclusion is supported by the results of many interception studies carried out in India and Australia (Table 2.1).

Rain drop size is important since it affects the wetting response of, and interception losses from, vegetation (Calder *et al*. 1986). Up to ten times as much rain may be required to achieve the same degree of canopy wetting for tropical convective storms, with large drop sizes, than would be necessary for the range of smaller drop sizes usually encountered for frontal rain (in, say, the UK). Vegetation type is also important in determining the wetting response since large-leafed vegetation would be expected to generate larger drop sizes than smaller leafed, and these large drops would be relatively inefficient in wetting up lower layers in the canopy. Results from studies using rainfall simulators also show that the final degree of canopy saturation varies with drop size, being greater for drops of smaller size. *Eucalyptus* species, with average sized leaves, are also about average in terms of their wetting up response and canopy storage, and also in terms of their effects on splash-induced erosion (see later).

*Transpiration from eucalypts*

Transpiration losses from *Eucalyptus* spp., as for most other vegetation types, are determined principally by

1. Climatic demand which is related to the prevailing radiation, atmospheric humidity deficit, temperature and wind speed.
2. Physiological response mechanisms which control stomatal apertures – the small pores in the leaf surface – in response to environmental conditions (both past and present). Of particular importance are the mechanisms which close stomata in response to increasing soil water stress and increasing atmospheric humidity deficits. These mechanisms have been shown to

operate on a number of plant and tree species, both on leaf scale (e.g. Lange et al. 1971) and on tree and forest scale (e.g. Jarvis 1976, Stewart and de Bruin 1983).
3   Canopy structure, particularly leaf area index.
4   Availability of soil water to the roots.

A knowledge of the stomatal response mechanisms, together with a knowledge of rooting patterns and soil water availability, is therefore crucial to the understanding of water use and water use efficiency in any particular environment.

## Research findings from Portugal and Western Australia

### Physiological controls

Pereira et al. (1986, 1987) carried out studies of the water relations and water use efficiency of *Eucalyptus globulus* trees growing at a site in Portugal. This species is widely grown in Portugal for pulp and 'waste' leaf from the felled trees is sometimes distilled for oil. They found that both photosynthetic carbon assimilation and leaf water use efficiency were highest in the spring, reduced in winter and lowest by mid-summer when severe drought conditions prevailed. The maximum rates of net photosynthesis were higher than most values reported for temperate zone woody plants (Korner and Cochrane 1985), slightly higher than those reported for *E. pauciflora* in its natural habitat in southern Australia (Slatyer and Morrow 1977), but broadly similar to those for *E. microcarpa* in New South Wales (Attiwill and Clayton-Greene 1984). Pereira et al. (1986) also found that *E. globulus* was effective in controlling water loss when the evaporative demand of the air was greatest. Maximum diurnal transpiration occurred at the time of maximum vapour pressure deficit in winter, but before maximum vapour pressure deficits were reached in summer.

### Physiological and water use studies

The water use and physiological studies of *Eucalyptus* spp. carried out in Western Australia are among the most detailed and comprehensive process studies carried out on any group of tree species. This is also a region where large scale planting of oil-bearing eucalypts is currently being undertaken with a view to controlling salinity, while at the same time producing an economic return through oil production (Bartle 1994). The results of these studies are therefore very relevant to such operations although they are not always easy to interpret. What is certain is that not only do different *Eucalyptus* spp. show very different stomatal response mechanisms, but rooting patterns may also vary widely between species and between sites. Furthermore, at sites where roots are able to extract water directly from shallow water tables, very high transpiration rates may result.

Colquhoun et al. (1984) carried out detailed studies on the daily and seasonal variation of stomatal resistance and xylem pressure potential in *E. marginata* and *E. calophylla*, the two major tree species of natural jarrah forest, and four other *Eucalyptus* species which were being used for reforestation. They found that the physiological responses exhibited by these different species, which all experienced the same meteorological conditions and were growing in close proximity, could be classified into three types:

1   *E. marginata* and *E. calophylla* showed little stomatal control of water loss and leaf resistances remained low throughout the study period. There was no evidence that stomatal resistance was correlated with atmospheric vapour pressure deficit.

2   *E. maculata*, *E. resinifera* and *E. saligna* all showed some stomatal control. For a typical summer day, stomatal resistances were low in the morning but increased through the day, broadly in line with increasing vapour pressure deficit (and atmospheric demand). However, no causal relationship was established and the stomatal regulation may have been caused by other factors.

3   *E. wandoo* exhibited stomatal control (as in 2 above) but also developed xylem pressure potentials far lower than all other species studied. The authors inferred that this species, which originated from the lower rainfall and shallower soil regions of the Darling range in Western Australia, may have been able to impose higher root suctions and extract more water from the soil than the other species studied.

Carbon et al. (1981), in a study of water stress and transpiration rates from natural jarrah forest growing in five different catchments in the Darling range, found little difference between *E. marginata* and *E. calophylla* in their transpiration responses (subsequently confirmed by Colquhoun et al. 1984) and little difference between sites, except in late summer when transpiration rates were generally higher from sites with permanent water tables. In another study, Carbon et al. (1982) found that on the Swan coastal plain, Western Australia, *Pinus pinaster* plantations consumed the most water, water use by the perennial pasture and native jarrah forest was similar but less than that of the pines, and annual pasture used the least.

Greenwood and Beresford (1979) investigated the transpiration rates from individual juvenile trees of different *Eucalyptus* spp. planted at three different sites in different rainfall regimes in the Hotham valley, southeast of Perth, Western Australia. They found considerable variation in rates between species at different sites and the species with the highest rate was different at each site: *E. globulus* was highest at Bannister, the 850 mm (annual rainfall) site, *E. cladocalyx* at Dryandra, the 500 mm site, and *E. wandoo* at Popanyinning, the 420 mm site. Interestingly, they also inferred that, because at Popanyinning the average transpiration per unit leaf area over all species had increased three-fold between early and late summer, the roots of the two-year-old trees had reached the water table which was at a depth of between 3 m and 5 m.

On the other hand, Greenwood et al. (1982) reported results of evaporation studies on a regenerating forest of *E. wandoo* on land which had formerly been cleared for agriculture (at the Bingham river site near Collie in southwest Australia) and found that the uptake of water by the trees was limited mainly to the unsaturated zone and that the trees did not extract a substantial quantity of water from the phreatic aquifer. The water table was located at a depth of approximately 9 m at this site and subsequent excavations showed that no large roots penetrated deeper than 1.2 m. They concluded that neither the regrowth rooting system nor that of the original forest could have penetrated far enough to draw water from the aquifer.

Greenwood et al. (1985) have reported further studies on what were then seven-year-old trees, at Bannister, the high rainfall site in the Hotham valley. The studies were made on two plantations, one upslope of, and the other immediately above, a saline seep. Annual evaporation from pasture in that area was 390 mm regardless of the position on slope. Annual evaporation (interception plus transpiration) from trees at the upslope site was 2300 mm for *E. maculata* and 2700 mm for both *E. globulus* and *E. cladocalyx*; at midslope the values were 1600 mm for *E. wandoo*, 1800 mm for *E. leucoxylon* and 2200 mm for *E. globulus*. They concluded that to support these high rates, which exceeded the annual rainfall of 680 mm by about a factor of four, the trees must be extracting water directly from groundwater. Soil coring studies which revealed roots at 6 m, 1 m below the water table, supported this hypothesis. (N.B. evaporation rates of this magnitude exceed the rate that can be supported by the input of solar radiation alone and imply considerable advection of heat from the air mass moving over the forest; this is indeed possible and has been verified for wet forests, see Calder 1985.)

## Research findings from India

### Water use studies

To answer the questions raised about the environmental impacts of fast growing tree plantations, and to devise ways in which some of the adverse effects could be minimised, studies of the hydrology of eucalyptus plantations, indigenous forest and an annual agricultural crop have been carried out in Karnataka (see Calder 1994 for a summary of the findings). Measurements were made at four main sites of the meteorology, the plant physiology, soil water status, rainfall interception and the direct water uptake of individual trees using tracing measurements. Measurements were also made of the growth rates of the trees. A deuterium tracing method, developed during the project, proved to be a very effective and powerful method for determining transpiration rates of whole trees and revealed a surprisingly 'tight' and simple relationship between the individual tree transpiration rates for young eucalypts and the cross-sectional areas of their trunks. This in turn led to a simple water use and growth (WAG) model (Calder 1992) that is applicable in India and other water-limited parts of the semi-arid tropics. The WAG model provides a framework for the estimation of the water use from eucalyptus plantations in relation to age, spacing and growth rate.

The model also provides a framework for investigating the mechanisms that control the growth rates of the plantations in relation to water use, what is termed the water use efficiency. Through assignment of values, based on marginal costs, of the water consumed and the value per unit volume of the timber produced it is possible to assess the economic returns of the plantation set against the real costs of the water consumed. These water use calculations need also to build in interception losses and any other losses from the understorey or bare soil beneath the tree canopy. One feature of eucalyptus plantations is that although water consumption is high, transpirational water use efficiency is also high, so that for a given amount of water transpired the volume of timber produced is probably as high as that from any tree species. Bearing in mind that interception losses and understorey losses are essentially a 'fixed overhead' loss per year, not usefully employed, the high growth rates of eucalyptus species mean that these overheads are a small proportion of the total water consumed. On a plot basis the water use efficiency is probably as high, if not higher, than that of any other tree species.

The recognition that water is a valuable resource in its own right, and that forests generally evaporate more water than other crops, provides a powerful incentive for trying to improve the water use efficiency of plantation forests. Both growth rates and water use efficiency of eucalyptus plantations in the dry zone in India are low by world standards. To some extent climatic factors, which are not amenable to manipulation, are responsible for the low water use efficiencies. Nevertheless, it is believed that there is still great potential for improvements through, for example, species selection, removing nutrient and water stress, and improved silvicultural practices such as optimal spacing and weeding.

In experiments carried out within a Controlled Environment Facility at Hosakote, near Bangalore, *Eucalyptus camaldulensis*, *E. grandis* and an indigenous species, *Dalbergia sissoo*, were grown under different conditions of water and nutrient stress. Results from the first two years of the experiment showed up to five-fold increases in growth rates on plots which received both water and fertiliser treatments compared with the control plots. Although water applications alone lead to large increases in growth rate there was no evidence to show that purely transpirational water use efficiency was improved (Calder *et al.* 1993).

*Water resource implications*

The hydrological studies carried out in southern India on plantations of exotic tree species, indigenous forest, and an agricultural crop showed a varied and complex pattern of hydrological impacts. The results can be summarised as follows:

1. At two of the dry zone sites, the water use of young *Eucalyptus* plantation on medium depth soil (3 m) was no greater than that of the indigenous dry deciduous forest.
2. At these sites, the annual water use of eucalyptus and indigenous forest was equal to the annual rainfall (within the experimental measurement uncertainty of about 10 per cent).
3. At all sites, the annual water use of forest was higher than that of annual agricultural crops (about two times higher than finger millet).
4. At the dry zone site which had deep soil (>8 m), the water use, over the three (dry) years of measurement, was greater than the rainfall. A later experiment showed that *Eucalyptus camaldulensis* roots can penetrate the soil at a rate exceeding 2.5 m per year and are able to extract and evaporate an extra 400–450 mm of water in addition to the annual input of rainfall (Calder *et al.* 1997). Observations in South Africa (Dye 1996) also confirm deep rooting behaviour in *E. grandis*.
5. Unlike the Australian studies referred to earlier (Greenwood *et al.* 1985) there was no evidence of root abstraction from the water table at any of the sites.
6. Although the water use of eucalyptus plantations is much higher than that of other tree species, the water use efficiency, expressed on a plot basis, is also much higher. For the same amount of water consumed, on a plot basis, a greater return in terms of useful biomass will be achieved from the eucalyptus plantations.

*Sustainable management systems*

From the research experience outlined earlier, it is suggested that the potentially adverse aspects of plantation forest practices can be curtailed through adoption of the following:

1. Rotation – Where soil water 'mining' occurs (i.e. where the roots penetrate successively deeper layers in the soil from the day of planting) one strategy which might prove advantageous, particularly on deep soils, would be to rotate *Eucalyptus* plantation with agricultural crops. A five-year period under an agricultural crop should allow the soil water reserves, depleted by say ten years of forestry, to be replenished. From studies in other arid zones of the world, there is evidence that deep-rooted trees bring nutrients from deep soil layers to the surface. If this is true of eucalyptus species in India, then there would be dual benefits from rotation: the trees would replace nutrients the agricultural crops remove, whilst the agricultural crops would replace water that the trees have removed.
2. Patchwork forestry – The forest water use results indicate that recharge to the groundwater under large areas of either plantation or indigenous forest in the dry zone in India is likely to be small and will not, on average, be more than 10 per cent of the rainfall. However, if plantation forests were grown as a 'patchwork', interspersed with annual agricultural crops, many of the adverse effects on the water table would be alleviated as up to half the annual rainfall should be available for recharge under the agricultural crops.

3   Irrigation – In theory it would be possible to optimise a 'patchwork' design with irrigated areas of forestry. It should be possible to grow the same volume of timber, using irrigation, on one-fifth to one-tenth of the usual land area, leaving the rest for rainfed agriculture. There may also be economic advantages of this type of scheme. If eucalypts grown for pulp were located close to the pulp mills transport costs would be minimised and this should substantially reduce production costs.

## Catchment studies in South Africa

In South Africa, commercial forests consist almost entirely of exotic species, principally pines and eucalypts, and form a large and important industry with plantations occupying around 1.2 m ha (Bands *et al.* 1987). However, the forests require at least 800 mm of rain to grow at economic rates but only 20 per cent of the country receives this amount and most of this is in mountainous areas. These mountainous areas are therefore under pressure both for afforestation and for water gathering. For such a water-short country as South Africa, with an average annual rainfall over the whole country of only 400 mm, the provision of water from the uplands, maximal in quantity and quality, is of vital concern. Nowhere are the conflicts between forestry and water interests more extreme than in South Africa.

Understanding of the interrelationships between land use, hydrology and the environment is essential for the sustainable and multi-use management of South Africa's mountainous areas and a hydrological research programme centred at Jonkershoek is providing that understanding. Catchment studies have provided unequivocal evidence for the reductions in streamflow that will occur as a result of afforestation with commercial eucalyptus and pine species (Bands *et al.* 1987).

The studies also destroy the forester's myth of forests 'attracting rain'. Quoting from Bands *et al.* (1987):

> Forests are associated with high rainfall, cool slopes or moist areas. There is some evidence that, on a continental scale, forests may form part of a hydrological feedback loop with evaporation contributing to further rainfall. On the Southern African subcontinent, the moisture content of air masses is dominated by marine sources, and afforestation will have negligible influence on rainfall and macroclimates. The distribution of forests is a consequence of climate and soil conditions – not the reverse.

In addition to the South African catchment studies of water use from *Eucalyptus* complementary studies have been carried out in Australia (e.g. Williamson and Bettenay 1979, Lima 1984) and India (Mathur *et al.* 1976). Although these may be sufficient to allow comparisons to be drawn concerning relative water use of *Eucalyptus* spp. in the regions where the studies were carried out, extrapolation of the results to other areas is difficult because the mechanisms responsible for enhanced losses can seldom be determined from catchment studies alone.

However, the general pattern shown by these studies is similar to that found in the majority of catchment studies which have compared runoff from forested and unforested catchments worldwide, namely, that the runoff is usually less from the forested catchment (Hibbert 1967, Bosch and Hewlett 1982).

## Water use summary

A new understanding has been gained in recent years of the evaporation from forests in both the dry and wet regions of the world based on process studies. These studies, which have used a wide variety of techniques, indicate decreased runoff from areas under forests as compared with areas under shorter crops. They also indicate that in wet conditions interception losses will be higher from forests than shorter crops, primarily because of increased atmospheric transport of water vapour from their aerodynamically rough surfaces. In dry (drought) conditions transpiration from forests is likely to be greater because of the generally increased rooting depth of trees as compared with shorter crops and their consequent greater access to soil water.

The new understanding indicates that in both very wet and very dry climates evaporation from forests is likely to be higher than that from shorter crops and that consequently runoff will be decreased from forested areas. Occasional exceptions to this rule exist but can generally be explained by unusual circumstances (e.g. Greenwood 1992).

## Eucalyptus and seasonal flow

Although it is possible, with only a few exceptions, to draw general conclusions with respect to the impacts of forests, including those of eucalypts, on annual flow, the same cannot be claimed for the impacts of forests on the seasonal flow regime. Different, site-specific (often competing) processes may be operating and the direction, let alone the magnitude of the impact, may be difficult to predict for a particular site.

### Theoretical considerations and observations

From theoretical considerations it would be expected that

1 Increased transpiration and increased dry period transpiration will increase soil moisture deficits and reduce dry season flows.
2 Increased infiltration under (natural) forest will lead to higher soil water recharge and increased dry season flows.
3 For cloud forests, increased cloud water deposition may augment dry season flows.

Observations from South Africa indicate that increased dry period transpiration is reducing dry season flows, in line with expectations. Van Lill *et al.* (1980), reporting studies at Mokobulaan in the Transvaal, showed that afforestation of grassland with *Eucalyptus grandis* reduced annual flows by 300–380 mm, with most of the reduction occurring during the wet summer season. More recently, Scott and Smith (1997), analysing results from five of the South African catchment studies, concluded that percentage reductions in low (dry season) flow as a result of afforestation were actually greater than the reduction in annual flow. Scott and Lesch (1997) also report that on the Mokobulaan research catchments under *E. grandis* the streamflow completely dried up by the ninth year after planting; the eucalypts were clearfelled at age sixteen years but perennial streamflow did not return for another five years. They attribute this large lag time to very deep soil moisture deficits generated by the eucalypts which require many years of rainfall before field capacity conditions can be established and recharge of the groundwater aquifer and perennial flows can take place.

Other reports from South Africa relate to the impacts on flow of self-seeded trees (invaders) originating from plantations of exotic species. It has been standard forestry practice in South Africa to avoid the deliberate planting of trees in the riparian zones of afforested catchments, to reduce the risk of soil erosion close to the stream channel and to avoid any increase in water use by riparian vegetation. Invaders are no respecters of forestry practice and often spread rapidly into these riparian areas. It has been demonstrated that removal of infestations of self established riparian trees can have huge effects on streamflow. Dye and Poulter (1995) have shown that removal of a strip of self-sown *Pinus patula* and *Acacia mearnsii* along a 500 m riparian zone at Kalmoesfontein increased streamflow by 120 per cent. Even more dramatic are reports (DWAF 1996) of a stream in Mpumalanga that, before the clearing of a 500 m strip of riparian *Eucalyptus grandis*, disappeared within 50 m of entering the stand. About three weeks after clearing the eucalypts the stream was visible for 200 m in the stand and after one month it was running through the stand. It was postulated that it had taken about a month for the stream and rainfall to recharge the water table, restoring a perennial stream from a dry streambed.

The hydrological dangers from invading trees are not just local in their impact. It has been calculated (Le Maitre *et al.* 1996) that, unless curtailed, invaders will eventually reduce the water supply to Cape Town by 30 per cent. It has also been shown that the cost of water from the best dam option is several-fold more expensive than the cost of water yielded through clearing the invading aliens (DWAF 1996).

*Summary*

The observations from South Africa on the impacts of eucalypt plantations on seasonal flows demonstrate the potential severity of the impacts and the care that must be taken in weighing up the pros and cons of establishing such plantations.

The complexity of the competing processes affecting dry season flows indicates that detailed, site-specific models will be required to predict impacts. In general, the role of *Eucalyptus* and other tree species in determining the infiltration properties of soils, as they affect the hydrological functioning of catchments through surface runoff generation, recharge and high and low flows and catchment degradation, remains poorly understood. Modelling approaches which are able to take into account forest influences and soil physical properties will be required to predict these site-specific impacts.

## Eucalypts and erosion

The role of eucalypt plantations and other forest plantations in relation to soil erosion remains controversial. If foresters are under suspicion for propagating the myth that forests are the cause of high rainfall in upland areas then there may be equal suspicions raised regarding the oft cited universal claims of the benefits of forests in relation to reduced erosion. As with impacts on seasonal flows, the impacts on erosion are likely to be site specific, and again, many, and often competing processes, are likely to be operating.

Conventional theory and observations indicate these benefits of afforestation:

1 Reduced incidence of surface runoff and reduced erosion transport due to high infiltration rates in natural, mixed forests.
2 Enhanced slope stability, which tends to reduce erosion, due to reduced soil water pressure and the binding effect of tree roots.

As Figure 2.1 illustrates, however, there may also be adverse effects:

1 Pre-planting, maintenance and harvesting activities may contribute to such things as soil compaction, increased surface flow and gully formation.
2 Reduced slope stability, which tends to increase erosion, due to windthrow of trees and the weight of the tree crop.
3 Splash-induced erosion from drops falling from the leaves of forest canopies.

*Figure 2.1* Erosion under eucalypts planted on steep slopes in Tamil Nadu, India (photo: I. Calder).

The benefits gained by afforesting degraded and eroded catchments will be very dependent on the situation and the management methods employed and afforestation should not necessarily be seen as a quick panacea. The choice of tree species will also be important in any programme designed to reduce erosion and catchment degradation, especially as it is now known that splash-induced erosion is related to tree species.

Forests can also influence soil erosion by altering the drop size distribution of the incident rainfall. Contrary to popular belief, forest canopies do not necessarily 'protect' the soil from raindrop impacts. In recent years there has been new understanding of the importance and significance of this effect, particularly as it relates to the choice of tree species to minimise erosion. In some cases there is actually the potential for increased erosion from drops falling from forest canopies. Although the importance of species in determining drop size and erosive impacts has not always been well understood, results of recent studies involving the measurement of dropsize spectra have advanced our knowledge in this area. In particular they have shown that

1. Below-canopy drop size is independent of the raindrop size falling on the top of the canopy.
2. The below-canopy drop size spectrum is a 'characteristic' of the species and varies widely between species. This can result in large differences in the potential for erosion; kinetic energies of drops falling from *Tectona grandis* can be as much as nine times greater than those from *Pinus caribaea* and about four times greater than *Eucalyptus camaldulensis*.

The canopy of the tree is not the only canopy affecting drop size modification and erosion. The presence of an understorey canopy, close to the ground where drops cannot reach terminal velocities, is very effective in ameliorating splash-induced erosion. Where the understorey has been removed through fire or biological competition the potential for erosion is very much increased (Figure 2.1).

## Conclusions

As a result of comprehensive research programmes hard evidence now exists on the hydrological impacts of eucalyptus plantations. The information has been hard won through detailed studies in

India, South Africa, South America and Australia. The results do not show *Eucalyptus* species to be quite as harmful as they have often been portrayed by environmentalists but neither do they show them to be without hydrological disbenefits. What they do show is a complex pattern of interactions, some of which may be seen as beneficial and others as adverse. Although the researchers have done their work, and major advances in knowledge have been achieved, the use of this knowledge for land management purposes, and in the better and more productive management of eucalypt plantations, has been less spectacular. This is perhaps unsurprising as the issues are complex and the tools for holistic resource management such as Decision Support Systems — which can assist with assessing the economic, hydrological and ecological tradeoffs associated with land-use decision making (O'Connell 1995, O'Callaghan 1996) — have only recently been developed and the inclusion of socio-economic evaluation methods has yet to be achieved.

*Eucalyptus* species are beneficial in having high economic value, both for timber and pulp and for oil production. Also beneficial are their high plot water use efficiencies; they probably produce more biomass per unit of water evaporated on a plot basis than any other tree species. Furthermore, their high growth rates make them very attractive plantation species, and in some countries such as Australia their ability to reduce and hold down water tables is a very great advantage in salinity control. Offsetting these advantages is the high water consumption that needs to be considered, in an integrated way, with other economic, socio-economic and environmental factors. The challenge for future forest managers, whether the forest is geared towards production of oil, timber or pulp, is to design sustainable forestry systems for eucalypts which minimise some of the adverse hydrological impacts that have been identified, whilst maximising economic benefits and being compatible with the social and economic needs of the local people.

# References

Attiwill, P.M. and Clayton-Greene, K.A. (1984) Studies of gas exchange and development in a subhumid woodland. *J. Ecol.*, 72, 285–294.

Bands, D.P., Bosch, J.M., Lamb, A.J., Richardson, D.M., van Wilgen, B.W., van Wyk, D.B. and Versfeld, D.B. (1987) *Jonkershoek Forestry Research Centre Pamphlet 384*, Dept. Environment Affairs, Pretoria, South Africa.

Bartle, J. (1994) New horizons for forestry: tree crops for the wheatbelt. *Newsletter of the Institute of Foresters of Australia Inc.*, 35(2), 4–7.

Bosch, J.M. and Hewlett, J.D. (1982) A review of catchment experiments to determine the effects of vegetation changes on water yield and evapotranspiration. *J. Hydrol.*, 55, 3–23.

Brookes, J.D. and Turner, J.S. (1964) Hydrology and Australian forest catchments. In *Water Resources Use and Management, Proc. Australian Academy of Sciences Symp.*, Canberra, 1963, Melbourne University Press, Melbourne, pp. 390–398.

Calder, I.R. (1979) Do trees use more water than grass? *Water Services*, 83, 11–14.

Calder, I.R. (1985) What are the limits on forest evaporation? – a comment. *J. Hydrol.*, 82, 179–192.

Calder, I.R. (1990) *Evaporation in the Uplands*, John Wiley & Sons, Chichester, UK.

Calder, I.R. (1992) A model of transpiration and growth of *Eucalyptus* plantation in water-limited conditions. *J. Hydrol.*, 130, 1–15.

Calder, I.R. (1994) *Eucalyptus, Water & Sustainability. A Summary Report*, ODA Forestry Series No. 6, Institute of Hydrology, Wallingford, UK.

Calder, I.R. (1996) Water use by forests at the plot and catchment scale. *Commonwealth For. Rev.*, 75, 19–30.

Calder, I.R. (1998) *Review Outline of Water Resource and Land Use Issues*, SWIM Paper 3, International Irrigation Management Institute, Colombo, Sri Lanka.

Calder, I.R. (1999) *The Blue Revolution – Land Use and Integrated Water Resources Management*, Earthscan, London.

Calder, I.R., Hall, R.L. and Prasanna, K.T. (1993) Hydrological impact of eucalypt plantations in India. *J. Hydrol.*, 150, 635–648.

Calder, I.R., Rosier, P.T.W., Prasanna, K.T. and Parameswarappa, S. (1997) *Eucalyptus* water use greater than rainfall input – a possible explanation from southern India. *Hydrol. Earth Systems Sci.*, 1(2), 249–256.

Calder, I.R., Wright, I.R. and Murdiyarso, D. (1986) A study of evaporation from tropical rainforest – West Java. *J. Hydrol.*, 89, 13–33.

Carbon, B.A., Bartle, G.A. and Murray, A.M. (1981) Patterns of water stress and transpiration in jarrah (*Eucalyptus marginata* Donn ex Sm.) forests. *Aust. For. Res.*, 11, 191–200.

Carbon, B.A., Roberts, F.J., Farrington, P.F. and Beresford, J.D. (1982) Deep drainage and water use of forests and pastures grown on deep sands in a Mediterranean environment. *J. Hydrol.*, 55, 53–64.

Colquhoun, I.J., Ridge, R.W., Bell, D.T., Loneragan, W.A. and Kuo, J. (1984) Comparative studies in selected species of *Eucalyptus* used in rehabilitation of the northern jarrah forest, western Australia. I. Patterns of xylem pressure potential and diffusive resistance of leaves. *Aust. J. Bot.*, 32, 367–373.

Dabral, B.G. and Subba Rao, B.K. (1968) Interception studies in chir and teak plantations – new forest. *Indian Forester*, 94, 541–551.

Duncan, H.P., Langford, K.J. and O'Shaughnessy, P.J. (1978) A comparative study of canopy interception. In *Hydrology Symp. Papers*, Canberra, September 1978, Institution of Engineers, pp. 150–154.

Dunin, F.X., O'Loughlin, E.M. and Reyenga, W. (1988) Interception loss from eucalypt forest: lysimeter determination of hourly rates for long term evaluation. *Hydrol. Processes*, 2, 315–329.

DWAF (1996) *The Working for Water Programme*, Ministry of Water Affairs and Forestry, Cape Town, South Africa.

Dye, P.J. (1996) Climate, forest and streamflow relationships in South African afforested catchments. *Commonwealth For. Rev.*, 75, 31–38.

Dye, P.J. and Poulter, A.G. (1995) A field demonstration of the effect on streamflow of clearing invasive pine and wattle trees from a riparian zone. *S. Afr. For. J.*, (173), 27–30.

Feller, M.C. (1981) Water balances in *Eucalyptus regnans*, *E. obliqua* and *Pinus radiata* forests in Victoria. *Aust. For.*, 44(3), 153–161.

George, M. (1978) Interception, stemflow and throughfall in a *Eucalyptus* hybrid plantation. *Indian Forester*, 104, 719–726.

Ghosh, R.C., Kaul, O.N. and Subba Rao, B.K. (1978) Some aspects of water relations and nutrition in *Eucalyptus* plantations. *Indian Forester*, 104, 517–524.

Greenwood, E.A.N. (1992) Water use by eucalypts – measurement and implications for Australia and India. In I.R. Calder, R.L. Hall and P.G. Adlard (eds), *Growth and Water Use of Forest Plantations, Proc. Internat. Symp.*, Bangalore, February 1991, Wiley, Chichester, UK, pp. 270–289.

Greenwood, E.A.N. and Beresford, J.D. (1979) Evaporation from vegetation in landscapes developing secondary salinity using the ventilated-chamber technique. I. Comparative transpiration from juvenile *Eucalyptus* above saline groundwater seeps. *J. Hydrol.*, 42, 369–382.

Greenwood, E.A.N., Beresford, J.D., Bartle, J.R. and Barron, R.J.W. (1982) Evaporation from vegetation in landscapes developing secondary salinity using the ventilated-chamber technique. IV. Evaporation from a regenerating forest of *Eucalyptus wandoo* on land formerly cleared for agriculture. *J. Hydrol.*, 58, 357–366.

Greenwood, E.A.N., Klein, L., Beresford, J.D. and Watson, G.D. (1985) Differences in annual evaporation between grazed pasture and *Eucalyptus* species in plantations on a saline farm catchment. *J. Hydrol.*, 78, 261–278.

Herwitz, S.R. (1985) Interception storage capacities of tropical rainforest canopy trees. *J. Hydrol.*, 77, 237–252.

Hibbert, A.R. (1967) Forest treatment effects on water yield. In W.E. Sopper and H.W. Lull (eds), *Forest Hydrology, Proc. Internat. Symp.*, Pergamon, Oxford, UK.

Jarvis, P.G. (1976) The interpretation of the variations in leaf water potential and stomatal conductance found in canopies in the field. *Phil. Trans. Roy. Soc. London, Ser. B*, 273, 593–610.

Joyce, C. (1988) The tree that caused a riot. *New Scientist*, 18 February, 54–59.

Karschon, R. and Heth, D. (1967) The water balance of a plantation of *Eucalyptus camaldulensis* Dehn. In *Contributions on Eucalyptus in Israel*, Vol. III, National Univ. and Institute of Agriculture, Kiriat Hayim, Israel, pp. 7–34.

Korner, C. and Cochrane, P. (1985) Stomatal responses and water relations for *Eucalyptus pauciflora* in summer along an elevational gradient. *Oecologia*, 66, 443–455.

Lange, O.L., Losch, R., Schulze, E.D. and Kappen, L. (1971) Responses of stomata to changes in humidity. *Planta*, 100, 76–86.

Le Maitre, D.C., van Wilgen, B.W, Chapman, R.A. and McKelly, D.H. (1996) Invasive plants and water resources in the Western Cape Province, South Africa: modelling the consequences of a lack of management. *J. Appl. Ecol.*, 33, 161–172.

Lima, W.P. (1984) *The Hydrology of Eucalypt Forests in Australia – A Review*, Instituto de Pesquisas e Estudos Florestais, Univ. Sao Paulo, Brazil.

Mathur, H.N., Ram Babu, J.P. and Bakhshish Singh (1976) Effects of clearfelling and reforestation on runoff and peak rates in small watersheds. *Indian Forester*, 102, 219–226.

Nshubemuki, L. and Somi, F.G.R. (1979) *Water Use by the Eucalypts – Observations and Probable Exaggerations*, Tanzania Silviculture Technical Notes (New Series) 44.

O'Callaghan, J.R. (1996) *Land Use. The Interaction of Economics, Ecology and Hydrology*, Chapman & Hall.

O'Connell, P.E. (1995) Capabilities and limitations of regional hydrological models. In *Scenario Studies for the Rural Environment*, Kluwer Academic, Netherlands, pp. 143–156.

Pereira, H.C. (1989) *Policy and Practice in the Management of Tropical Watersheds*, Westview Press, Colorado, USA.

Pereira, J.S., Tenhunen, J.D. and Lange, O.L. (1987) Stomatal control of photosynthesis of *Eucalyptus globulus* Labill. trees under field conditions in Portugal. *J. Exp. Bot.*, 38, 1678–1688.

Pereira, J.S., Tenhunen, J.D., Lange, O.L., Beyschlag, W., Meyer, A. and David, M.M. (1986) Seasonal and diurnal patterns in leaf gas exchange of *Eucalyptus globulus* trees growing in Portugal. *Can. J. For. Res.*, 16, 177–184.

Prebble, R.E. and Stirk, G.B. (1980) Throughfall and stemflow on silverleaf ironbark (*Eucalyptus melanophloia*) trees. *Aust. J. Ecol.*, 5, 419–427.

Saltiel, M. (1965) The afforestation possibilities of Israel from a hydrological point of view. *La-Yaaran*, 15, 61–64.

Samraj, P., Haldorai, B. and Henry, C. (1982) Conservation forestry. In *25 Years Research on Soil and Water Conservation in Southern Hilly High Rainfall Regions*, Monograph No. 4, Central Soil and Water Conservation Research and Training Institute, Dehra Dun, India, pp. 153–199.

Scott, D.F. and Lesch, W. (1997) Streamflow responses to afforestation with *Eucalyptus grandis* and *Pinus patula* and to felling in the Mokobulaan experimental catchments, South Africa. *J. Hydrol.*, 199, 360–377.

Scott, D.F. and Smith, R.E. (1997) Preliminary empirical models to predict reduction in total and low flows resulting from afforestation. *Water SA*, 23, 135–140.

Slatyer, R.O. and Morrow, P.A. (1977) Altitudinal variation in the photosynthetic characteristics of snow gum, *Eucalyptus pauciflora* Sieb. ex Spreng. I. Seasonal changes under field conditions in the Snowy Mountains area of south-eastern Australia. *Aust. J. Bot.*, 25, 1–20.

Smith, M.K. (1974) Throughfall, stemflow and interception in pine and eucalypt forest. *Aust. For.*, 36, 190–197.

Stewart, J.B. and de Bruin, H.A.R. (1983) Preliminary study of dependence of surface conductance of Thetford forest on environmental conditions. In B.A. Hutchinson and B.B. Hicks (eds), *The Forest–Atmosphere Interaction, Proc. Forest Environmental Measurement Conf.*, Oak Ridge, Tennesse, October 1983, D. Reidel, Dordrecht, Netherlands, pp. 91–104.

van Lill, W.S., Kruger, F.J. and van Wyk, D.B. (1980) The effects of afforestation with *Eucalyptus grandis* Hill ex Maiden and *Pinus patula* Schlecht. et Cham. on streamflow from experimental catchments at Mokobulaan, Transvaaal. *J. Hydrol.*, 48, 107–118.

Vandana Shiva and Bandyopadhyay, J. (1983) *Eucalyptus* – a disastrous tree for India. *The Ecologist*, 13, 184–187.

Vandana Shiva and Bandyopadhyay, J. (1985) *Ecological Audit of Eucalyptus Cultivation*, The English Book Depot, Dehra Dun, India.

Vandana Shiva, Sharatchandra, J.C. and Bandyopadhyay, J. (1982) Social forestry – no solution within the market. *The Ecologist*, 12, 158–168.

Westman, W.E. (1978) Inputs and cycling of mineral nutrients in a coastal subtropical eucalypt forest. *J. Ecol.*, 66, 513–531.

Williamson, D.R. and Bettenay, E. (1979) Agricultural land use and its effect on catchment output of salt and water – evidence from Southern Australia. *Prog. Wat. Tech.*, 11, 463–480.

# 3 Eucalypts in cultivation: an overview

*John W. Turnbull and Trevor H. Booth*

## Introduction

Australia's natural forests and woodlands are dominated by eucalypts. They extend over 28 million hectares from temperate latitudes (42°S) in Tasmania to the tropics (11°S) in Queensland. Beyond Australia small areas of natural eucalypt forests occur in tropical Indonesia, Papua New Guinea and the Philippines. Within this vast geographic range there is great ecological variability. Eucalypts fringe the sea coast in many places yet extend to an altitude of 2960 m in the mountains of Timor. In Tasmania the varnished gum, *Eucalyptus vernicosa*, is a sprawling shrub while not far away in the Florentine Valley the giant mountain ash (*E. regnans*) is one of several species that can exceed 70 m in height. The genetic diversity of the genus *Eucalyptus* is very great with over 700 species having been described and many of the species have great intra-specific variation. In 1988 the official Flora of Australia recognised 513 species (Chippendale 1988) but intensive taxonomic studies since then have resulted in descriptions of almost 200 new species (e.g. Brooker and Hopper 1991, Hill and Johnson 1992, Pryor *et al.* 1995).

When Australia was first settled in the late eighteenth century, eucalypts were used for farm buildings, fencing and fuelwood. The medicinal value of the essential oils extracted from the leaves of some species was also recognised. Elsewhere, eucalypts were regarded as botanical curiosities and were soon cultivated in botanical gardens and private arboreta in Europe. Once in cultivation, the potential of some species to grow rapidly and produce a variety of timber and non-timber forest products was recognised. Their seeds were sought after throughout the world wherever the winter climate was not too severe and they quickly deserved the title 'the emigrant eucalypts' coined by Zacharin (1978). Some were planted for their ornamental value, others for land reclamation. Farmers planted them for windbreaks and to produce posts and poles. In countries such as Brazil and South Africa, eucalypts were planted along railway lines to provide fuel for wood-burning locomotives. The reasons for planting eucalypts have changed significantly over time, and the end uses to which the species have been put are diverse. Today, eucalypts provide sawn-timber, plywood, fibreboard, mine props, pulp for paper and rayon, poles, firewood, charcoal, essential oils, honey, and shade and shelter (Hillis and Brown 1978). Less conventional uses include the production of plant growth regulators, tannin extracts, industrial chemical additives, adhesives and fodder additives (Song 1992).

The genus has many favourable characteristics including high growth rates, wide adaptability to soils and climate, ease of management through coppicing, valuable wood properties and absence of 'weediness'. From the hundreds of species in the natural forests many have been introduced and tested for their adaptability and utility but only a relatively few have become widely cultivated. In China, for example, over 200 species have been introduced but fewer than ten are now cultivated on a significant scale. This is also the situation elsewhere and just four species, *E. camaldulensis*, *E. globulus*, *E. grandis* and *E. tereticornis*, dominate plantations globally. Eucalypts used in the commercial production of essential oils (mainly those rich in cineole, citronellal, phellandrene or piperitone) include *E. globulus*, *E. polybractea*, *E. radiata*, *E. smithii*, *E. elata*, *E. citriodora*, *E. dives* and *E. kochii* (Small 1981, Boland *et al*. 1991, Coppen and Hone 1992).

This chapter describes the extent of eucalypt planting, the use of eucalypts to improve farm income and services, species/site selection and models which have been developed to try and optimise this (particularly for oil-bearing eucalypts), and some aspects of their growth characteristics and management technologies. The cultivation of commercially important oil-yielding species in particular parts of the world is described in detail in other chapters.

## Large-scale planting of eucalypts

### Size and location of main plantings

Reliable global estimates for areas of planted eucalypts are difficult to obtain. Statistical information is incomplete and there is confusion between new areas of planting and coppice regeneration of existing plantations. In places such as China and India many eucalypts are planted by farmers in very small woodlots, or in lines around fields and along waterways and railways, so accurate area estimates are extremely difficult to make. Even in Brazil, where eucalypts are planted in large blocks for industrial wood production, the estimates range from 3.2 million ha (EMBRAPA 1996) to over 5 million ha (Barros and Novais 1996). Nevertheless, a global estimate of 13.4 million ha of plantations (Davidson 1995 and Appendix 2) gives a good indication of the extent and popularity of eucalypts in cultivation.

In the tropics there were over 10 million ha of eucalypt plantations at the end of 1990 (FAO 1993), principally in tropical America (4 million ha) and in Asia (5 million ha). The American statistic is dominated by Brazil, where there are 3.2 million ha (EMBRAPA 1996), and in tropical Asia India has over 4 million ha including trees in various configurations on farms (Davidson 1995). In addition, there are substantial plantations in countries with more temperate climates including Australia 287,000 ha (Wood *et al*. 1999), Chile 300,000 ha (Flynn and Shield 1999), China 600,000 ha (Wang *et al*. 1994), Morocco 200,000 ha (Marien 1991), Portugal 500,000 ha (Pereira *et al*. 1996), South Africa 600,000 ha (Flynn and Shield 1999) and Spain 400,000 ha (Soares 1994). Disaggregated country statistics, including plantation ownership and utilisation for wood products, have been assembled by Flynn and Shield (1999).

Increased eucalypt plantation areas are projected in several countries. Plantings will continue in Brazil but not at the very high rates of the recent past as there will be more effort to increase the productivity and quality of existing areas. Both China and India have active reforestation programmes and, although there has been variable success and acceptance of eucalypts in parts of India, the great demand for wood will undoubtedly ensure that planting continues.

Eucalypt planting has accelerated in recent years in several tropical countries. Large areas of *E. camaldulensis* have been established in Thailand and Vietnam. There are also 43,000 ha of high yielding clonal plantations in the Congo (Bouillet *et al*. 1999) and in Vietnam. However, extensive planting of eucalypts in the lowland humid tropics has been inhibited by the incidence of pathogens and insect pests which reduce productivity. Only a few eucalypt species, such as *E. urophylla*, *E. deglupta* and *E. pellita*, appear to be adapted to hot, humid conditions (Werren 1991). There are also active plantation programmes in the more temperate areas. Australian plantations increased from 125,000 to 287,000 ha between 1994 and 1998 (Cromer *et al*. 1995,

Wood et al. 1999), mainly using *E. globulus* and *E. nitens*. In Chile the plantation area of eucalypts, mainly *E. globulus* and, more recently, *E. nitens*, increased from 185,000 ha in 1990 to over 300,000 ha in 1998 (Anon. 1990, Flynn and Shield 1999), and in Uruguay about 300,000 ha have been planted since 1988. These trends suggest that eucalypt plantation establishment through the 1990s will have resulted in a total area of over 15 million ha in 2000.

It is inevitable that natural forests will continue to be cleared for agriculture and other purposes and that planted eucalypts will have a major role, often expanding, in providing wood products, improved standards of living, and environmental protection in many countries.

*Industrial plantations*

The industrial use of eucalypts is expanding rapidly as large areas of plantations are being established to provide medium- to low-density short-fibre pulp for paper. The plantation-grown wood is usually harvested after 5–10 years and provides a uniform material with high brightness, and good opacity and bulk, all of which make the pulp very suitable for the production of copying, printing, writing and tissue papers. Demand for these products is rising, and with it the demand for eucalypt pulp, such that most of the internationally traded bleached hardwood pulp is now coming from eucalypts.

Huge areas of eucalypt plantations for industrial wood have been established by private companies in countries such as Brazil, Chile, Uruguay, Portugal and South Africa (Attiwill and Adams 1996, Campinhos 1999, Turnbull 1999). These plantations are in large compact blocks close to the production plant to minimise transportation costs. They usually comprise a single species and use genetically selected stock, often clones, which is managed very intensively with clean weeding, fertilising and pest control. These management practices are all designed to maximise production of very uniform wood. Leaves are harvested from some industrial pulpwood plantations of *E. globulus* to produce cineole-rich oil as a by-product.

Most research into plantation management has been undertaken to enhance the productivity and economic viability of these large-scale plantations. Some of the technologies developed can be readily transferred to small-scale planting or can be adapted, but as farm planting often offers the opportunity of producing more diverse products using more flexible management regimes, it is likely that additional research will be required to develop a full range of technologies appropriate for eucalypt planting by smallholders.

## Growing eucalypts on farms

*Why plant eucalypts?*

It is not the purpose of this section to describe the role of trees on farms. Others such as Hall et al. (1972), Cremer (1990) and Arnold and Dewees (1995) have done this already. The aim is to stress that farmers already plant eucalypts and others will do so when markets and other conditions are favourable.

Farmers are increasingly diversifying their farming systems by tree planting to minimise risk and provide a range of products and services. Deep-rooted species such as *Grevillea robusta* are very compatible with agricultural crops but eucalypts are not always the most desirable trees to plant on farms as many species have shallow root systems which compete effectively with agricultural crops. Nevertheless, there are situations where eucalypts are grown mixed with crops and the severity of the competition is reduced by applying different management strategies. Although many eucalypts have sparse crowns and permit high levels of radiation to reach the ground, the canopy can be manipulated by pruning or coppicing. Removal of the crown biomass has the effect of reducing the root mass and this diminishes root competition and releases nutrients into the soil as the roots decompose. In the drier parts of Malawi, where some farmers grow *E. camaldulensis* amongst their crops, the branches are pruned off until only the stem remains when the trees begin to compete with the maize or other crops.

Population growth in the developing world has led to a greater demand for fuelwood, poles and other wood products. In many countries where fuelwood is the main source of energy for the rural population there is a substantial shortfall in supply. Natural forests have felt the pressure and are being heavily exploited, but the scale of wood shortages in many countries is such that these forests alone cannot meet the needs. By the year 2010, consumption of fuelwood and poles will have increased substantially in rural areas, especially in India, Kenya, Nigeria and other densely populated countries of Asia and Africa where much of the natural forest has disappeared or is severely degraded (FAO 1991). Increasingly, farmers and other land holders are becoming involved in providing both wood for profit and subsistence from new on-farm plantings.

Farmers plant trees for a variety of reasons including shade and shelter; wood products; non-wood products such as fruit, fodder, honey, essential oils; and savings and security. They are often more interested in trees yielding products that can be sold to provide a cash income. The main reason for some farmers choosing to plant *E. globulus* in Yunnan province, China, is the quick cash return they get from harvesting leaves for cineole oil production, and in India and elsewhere branches and litter for fuel may be valued secondary products, especially by women (e.g. Nesmith 1991). Tree planting for shade, shelter and the marking of farm boundaries is also important for some farmers. The use of trees as a living bank account, to be harvested when there is a need for cash, is widespread.

Choosing the species which best provides profitable and readily marketed products, or meets the immediate needs of the farmer, is a critical first step in the decision-making process. Fast growing species which are disease-free and easily managed are usually preferred because they give an early return on the capital invested and do not divert the farmer from other activities. Some eucalypts meet these criteria and are popular with farmers in many parts of the world.

## *On-farm planting worldwide*

Eucalypts, established to a very large extent by farmers, dominate the rural landscapes in Ethiopia, China, India, Peru, Rwanda and elsewhere. In Ethiopia, farmers plant *E. globulus* on small areas of land and subsequently the plots may be managed to yield a variety of products including leaves and small branches for fuelwood, poles or posts for house building and other farm uses, and large poles and timber for sale. Farmers who have insufficient land to have woodlots nevertheless often grow a few trees which they can sell to buy food when their crops are exhausted. Many people in Ethiopia are absolutely dependent on eucalypts as a source of fuel and house building material. Unlike farmers in China, India and Peru they have not exploited the potential of *E. globulus* to yield cineole-rich oil.

## *People's Republic of China*

In southern China, eucalypts are a prominent and important part of the rural landscape. They are favoured because they grow relatively fast and tolerate poor, degraded soils. There are many dispersed plantings on farms and it is estimated that some one billion trees are planted beside farmhouses and along roads and waterways (Wang *et al*. 1994). The most widely used species have been *E. exserta*, *E. citriodora*, *E. globulus* and a hybrid of unknown origin referred to as

'*E. leizhou* No. 1', but these are being replaced by more productive species such as *E. grandis*, *E. urophylla* and hybrids between these two species. The wood is used for posts and poles, pulpwood, furniture, farm tools and fuel. Essential oils and tannins are extracted from the leaves and the flowers support honey production. Oil extraction began in 1958 and since 1970 has developed rapidly; China now dominates the international eucalyptus oil market (Song 1992, Coppen and Hone 1992). The eucalypts play an important part in crop production by providing shelter from typhoons in coastal areas and make a significant contribution to the quality of life and income of farmers.

*India*

Farmers in India have been growing eucalypts on an increasing scale over the past fifteen years. As the Indian population approaches one billion, so there is an immense requirement for wood for both industrial and domestic use. The annual demand for industrial wood is three times the estimated production (Kumar 1991). Economic benefits have generally come from eucalypt plantations, although excessive planting by farmers in Haryana and the Punjab has now resulted in a local glut of fuelwood and small-sized poles, leading to falling prices. Farmers who planted early made large profits while those planting later did not earn similar rewards and some would have been better growing agricultural crops (Dewees and Saxena 1995, Saxena and Vishwa Ballabh 1995).

*E. tereticornis*, known locally as 'Eucalyptus hybrid' or 'Mysore gum', is the preferred species for planting in India. It is used extensively for village woodlots, on farms and along the boundaries of roads, canals and railways. It has also been widely planted on what are broadly termed 'wastelands' or degraded lands. Its popularity is based on its capacity for rapid growth, resistance to cattle browsing, great adaptability to varying environmental conditions, ready market at profitable prices and suitability of the wood for fuel, rural housing, etc. (Rajan 1987, Shah 1988). Farmers have established block plantings and windbreaks on the margins of the fields and sometimes grow eucalypts in combination with agricultural crops. The potential of essential oil yielding eucalypts, such as *E. citriodora*, grown in combination with crops has been highlighted by Shiva *et al*. (1988), and in nearby Nepal selected clones of *E. camaldulensis*, which have high levels of cineole in the leaves, have been suggested for planting on small farms (White 1988). Eucalypts are often planted in India to meet future contingencies and have functioned as savings banks of the rural poor.

In India, as in some other countries, there is continuing controversy over the use of eucalypts. There are questions about the diversion of cropland from food production to forestry, reduction in employment and the utilisation of common lands for eucalypt growing by the more powerful members of the community (Saxena and Vishwa Ballabh 1995). While from the farmer's or entrepreneur's viewpoint, eucalypt growing makes economic sense, the landless poor may endure increased hardship and reduced welfare when farmers grow eucalypts (Gregersen *et al*. 1989). Although the debate is frequently framed in terms of the beneficial and negative ecological impacts of eucalypts, the major problem in fact relates to land availability, tenure and management (Poore and Fries 1985). Hydrological studies show a complex series of interactions, some of which may be seen as beneficial and others as adverse (Calder 1994). According to an analysis by Raintree and Lantican (1993),

> the real crux of the political controversy surrounding eucalyptus planting in India was the opportunity cost of a social forestry programme which generated such dramatic benefits for the relatively better off segments of society while leaving the originally targeted beneficiaries of the Social Forestry programme without benefit.

Clearly, if eucalypts are to realise their full potential to provide benefits to a community there will need to be much more careful analysis of ecological, economic and social factors before extensive planting is undertaken.

*Australia*

In some countries, farmers are urged by their governments to plant trees as part of their farming systems for environmental protection reasons. In the southwest of Western Australia deforestation has resulted in rising watertables which bring saline ground water to the surface. Deep-rooted perennials, especially trees and shrubs, can use excess water and become part of the salinity management strategy (e.g. Marcar *et al*. 1995). In Western Australia it is estimated that 3 million ha of trees and shrubs are necessary to manage the state's salinity problem successfully. Commercial wood plantations of eucalypts and pines, shrubby forage crops and new tree crops such as oil-producing mallee eucalypts are being promoted to achieve this target (Agriculture Western Australia 1996). Combinations of fodder trees and eucalypts may be planted in an agroforestry system to provide animal forage, essential oils and salinity control (Eastham *et al*. 1993). The government of Western Australia is supporting financially the establishment of a eucalyptus oil industry based on mallee species such as *E. kochii* (subspp. *kochii* and *plenissima*), *E. horistes*, *E. polybractea* and *E. angustissima*. Although most eucalypts are currently grown for a relatively limited medicinal oil market, the Western Australians are aiming to sell their oils for use as a range of industrial solvents (Bartle 1994, Bartle *et al*. 1996).

*The future*

There is increasing evidence that in future more industrial wood will be produced in small-scale plantations by farmers (Cossalter 1996). Plantations have already been established by farmers primarily to supply wood to industrial enterprises. In Spain, there is an example of a marriage between industrial and social forestry with farmers cooperating with a private pulp and paper company, Celulosas de Asturias S.A., to grow eucalypts on their land to sell to the mill, while also making money from the production of honey and essential oils (CEASA 1994). Similar schemes operate in Brazil (e.g. de Freitas 1995), the Philippines (Morrison and Bass 1992), India (Chinniah Sekar *et al*. 1996) and elsewhere.

There are some basic requirements which must be fulfilled if farmers are to grow trees profitably. There must be a good species/management technology available, a potential market, roads or other infrastructure to enable cheap transport of the product, clear tenure rights on the land and trees, and a source of credit for those who need it. If these conditions are met, farmers need little persuasion to plant trees. While some prefer to plant native trees, the majority will grow the species which are most profitable or provide the services they require, irrespective of the origin of the tree. The extensive planting of eucalypts in India for poles and fuelwood when the prices were favourable is evidence of this, although many were subsequently disappointed when prices fell (Saxena and Vishwa Ballabh 1995). If secure and profitable markets can be developed for eucalyptus oils then entrepreneurial farmers will quickly take advantage of the commercial opportunity.

## Species selection and growth prediction

When considering high oil-yielding eucalypts for planting it is useful to know which areas are suitable for particular species and how well they are likely to grow on specific sites. This section outlines some of the methods available for answering these questions for important oil-producing eucalypts such as *E. citriodora*, *E. globulus* and *E. camaldulensis* (Petford provenance).

## Climatic requirements

Climate is an important influence on tree growth but climatic data are frequently not available for forest sites, which are often remote from meteorological stations. Fortunately, mean climatic conditions can now be estimated reliably for most locations around the world. Mathematical relationships have been developed which can estimate mean monthly climatic conditions for factors such as maximum temperature, minimum temperature, precipitation, solar radiation and evaporation (e.g. Hutchinson *et al.* 1984). These interpolation relationships allow the range of climatic conditions experienced by particular trees to be analysed. For example, a program called BIOCLIM developed by Nix, Busby and Hutchinson takes in geocoded information on species' natural distributions (i.e. sets of observations of latitude, longitude and elevation) and produces information on species' climatic requirements (see Busby 1991 for summary).

### Data from natural vs exotic locations

One of the first published applications of BIOCLIM was to analyse the natural distribution of *E. citriodora* (Booth 1985), an important oil-yielding eucalypt. The program estimated twelve climatic factors for each of eighty-four natural locations of *E. citriodora*. The ranges of these factors provided an estimate of the species' climatic environment in Australia. For example, mean annual temperature was in the range 18–24°C, the mean maximum temperature of the hottest month was in the range 28–36°C and the mean annual rainfall was in the range 250–1100 mm. Some of these estimates must be treated with caution. The lower end of the rainfall range, for example, was surprisingly low. Trees may be just surviving in such an environment or may be growing where additional moisture is available, such as by a river. Despite problems of interpretation like this the information can be used to suggest areas outside Australia which might be suitable for growing the species.

Studying natural distributions can provide a first impression of a species' climatic requirements, but many species can thrive under conditions which are somewhat different from those within their natural range. Additional information from trials also helps to interpret extreme values, such as the low rainfall observation mentioned above. An analysis of *E. citriodora* plantation sites in Africa illustrated some of these differences (Booth *et al.* 1988). The precise locations of thirty-two trials of *E. citriodora*, including seventeen successful trials, in ten countries were obtained and climatic conditions estimated. Five of the mean annual temperatures recorded at the successful plantations were slightly below the range recorded from the natural distribution, whilst the mean annual rainfall was in the range 380–1580 mm, with a mean of 882 mm.

Information from many successful trials around the world has been used to develop the descriptions of the climatic requirements of *E. citriodora*, *E. globulus* (particularly *E. globulus* subsp. *globulus*) and *E. camaldulensis* (northern provenances) and this is illustrated in Table 3.1.

### Climatic mapping programs

Whilst descriptions such as those given in Table 3.1 are useful for suggesting areas where the species might be grown, it is difficult to visualise where such areas occur within a particular country. The climatic interpolation relationships mentioned above have also helped to develop a family of PC-based programs known as 'climatic mapping programs' (see Booth 1996a for several

*Table 3.1* Climatic requirements of some oil-producing eucalypts (after Booth and Pryor 1991)

|  | E. globulus subsp. globulus | E. citriodora | E. camaldulensis (*Petford provenance*) |
| --- | --- | --- | --- |
| Mean annual rainfall (mm) | 550–1500 | 650–2500 | 250–2500 |
| Rainfall regime[a] | s, u/b, w | s, u/b, w | s |
| Dry season[b] (months) | 0–7 | 0–7 | 2–8 |
| Mean max temperature, hottest month (°C) | 19–30 | 28–39 | 28–40 |
| Mean min temperature, coldest month (°C) | 2–12 | 8–22 | 6–22 |
| Mean annual temperature (°C) | 9–18 | 17–28 | 18–28 |
| Absolute minimum temperature (°C) | −8 | −3 | na[c] |

a s = summer, u/b = uniform/bimodal, w = winter.
b Dry season length is number of consecutive months for which rainfall is <40 mm.
c Not applicable (frost should not be a problem in areas indicated by other factors).

*Figure 3.1* Regions of China (dark-shaded areas) which are climatically suitable for *Eucalyptus globulus* subsp. *globulus*.

examples). These programs can take in descriptions of climatic ranges similar to those given in Table 3.1 and generate maps showing climatically suitable and unsuitable areas. The programs include a moveable marker which can be positioned over any location. This allows the detailed climatic data for that particular location to be examined and compared with the current description of a species' requirements. Figure 3.1 shows example output from a program for China indicating areas likely to be climatically suitable for *E. globulus* subsp. *globulus*. The output is based on interpolated climatic data estimated for nearly 100,000 locations across China. Many other climatic mapping programs have been developed at CSIRO Forestry and Forest Products in collaboration with researchers in other countries. These include programs for individual countries such as Australia, Thailand, Philippines, Indonesia, Laos, Vietnam and Zimbabwe, as well as major regions such as Africa and Latin America and, at a coarser scale, for the whole world (Booth 1996a,b).

## Growth prediction

### Relationship between site and environmental factors

Being able to identify where particular trees will grow is useful, but many people need, in addition, an approximate indication of how well particular trees will grow at specific sites. In ideal situations trees of the same species will already be widely grown within the area of interest. In this case it may be possible to collect data on growth as well as environmental conditions and to develop predictive relationships to assess potential planting sites. For example, Inions (1991) has described the development of site index curves for *E. globulus* plantations in Western Australia, where site index is related to top height at five years of age. Height is used as an indication of performance as it is relatively little affected by differences in stand density and top height is the average of the tallest forty trees per hectare. Curves of top height development by age can be plotted so that site index can be estimated for stands of any age, and this enables the performance of stands of different ages to be directly compared.

Using data from fifty-one sites, Inions (1991) developed the following equation relating site index to environmental factors:

$$S = B_0(S_1 \cdot S_2)^B$$

where

$S_1 = B_0 - B_1(\text{SILT50} \cdot \text{BD50} \cdot \text{RAIN\_DQ}) + B_2(\text{SILT50} \cdot \text{BD50} \cdot \text{RAIN\_DQ} \cdot \text{TEMP\_HM})$
$\quad - B_3 \cdot \text{RAIN} + B_4 \cdot \text{RAIN\_LM} - B_5\{(\text{SILT50} \cdot \text{BD50}) + (\text{CLAY50} \cdot \text{BD50})\}$

$S_2 = \exp\{B_0 - B_1(\text{NIT10})^{-1} + B_2(\text{NIT10})^{-2} - B_3(\text{P10})^{-1} + B_4 \cdot \text{SILT10}\}$

and

$S$ is site index (m)
$B_0, B_1, B_2, B_3$ and $B_4$ are coefficients (given in Inions 1992)
SILT50 is silt content at 50 cm soil depth (%, wt)
BD50 is bulk density at 50 cm soil depth ($\text{g cm}^{-3}$)
RAIN_DQ is rainfall in the driest quarter (mm)
TEMP_HM is highest monthly temperature (°C)
RAIN is total annual rainfall (mm)
RAIN_LM is rainfall in the driest month (mm)
CLAY50 is clay content at 50 cm soil depth (%, wt)
NIT10 is nitrate content at 10 cm soil depth ($\mu\text{g g}^{-1}$)
P10 is phosphorus content at 10 cm soil depth ($\mu\text{g g}^{-1}$)
SILT10 is silt content at 10 cm soil depth (%, wt)

The values of the coefficients have been omitted here to simplify the equation. Their actual values are irrelevant to most readers as they are only useful for predicting growth within the area where they were collected, that is, the southwest of Western Australia. Although the relationship is complex it can be seen that temperature, rainfall, soil water-holding capacity and soil nutrition are important factors determining tree growth. Other authors have developed predictive relationships relating the growth of oil-yielding eucalypts to environmental conditions. For example, Janmahasatien *et al.* (1996) describe relationships for *E. camaldulensis* in Thailand. However, Inions (1991, 1992) also gathered independent data from a further eighteen sites to test the equation given above. These sites had observed site indices ranging from about 10 to 22 m. The residual statistics generated by the equation were 0.38 with a standard deviation of 1.71 ($r^2 = 0.73$).

Regression relationships, such as those prepared by Inions (1991, 1992), relate tree growth directly to environmental factors such as rainfall and soil nutrients. In recent years considerable effort has gone into developing process-based models which attempt to simulate important processes related to growth such as light interception, photosynthesis and respiration. Relatively simple models such as ProMod (Battaglia and Sands 1997, Sands *et al.* 2000) and 3-PG (Landsberg and Waring 1997, Landsberg *et al.* 2001) have been developed to predict the suitability of particular sites for particular species. For example, the ProMod model was initially developed using data from nine *E. globulus* plots in Tasmania and Western Australia and was validated using data from nineteen *E. globulus* plots in northern Tasmania. Regression of predicted against observed mean annual increment (MAI) for the nineteen validation plots produced an $r^2$ of 0.81, compared with an $r^2$ of 0.91 for the research plots.

## Leaf biomass and oil production

Most growth prediction work has used measures such as height and volume, which are directly or indirectly related to wood production. Factors such as site index and MAI should also be related to leaf production, at least in the early stages of growth before canopy closure, and hence should be expected to bear some relation to oil yield. However, predicting actual oil yield, which is a function of both leaf biomass and leaf oil concentration, is more difficult. Doran (1991) concluded that the effects of environmental factors on the oil concentration in leaves are 'likely to involve a highly complex network of variables, and very little is known about them'. In addition to environmental factors he noted that important influences on oil concentration include genetics, type and age of leaf, and extraction techniques. Doran (1992) further concluded that the extent and direction of variation in oil concentration of *E. camaldulensis* depended very much on the individual tree. Though it may currently be impossible to predict oil yield for potential sites it should at least be possible in some areas to use relationships similar to those described by Inions (1991, 1992), Janmahasatien *et al.* (1996) and Battaglia and Sands (1997) to identify sites where oil-yielding species such as *E. globulus* and *E. camaldulensis* can grow well. However, relationships such as these can only be developed in areas where the species are already widely grown in plantations or trials. There is need for an evaluation system which can use minimal data to provide tentative predictions in those areas (the vast majority) where high oil-yielding species are not already widely grown.

## Other predictive models

The Plantgro model (Hackett 1991) was developed by CSIRO to provide coarse predictions of the growth of lesser-known crops. The model is run using individual plant, climate and soil files, and the program makes it easy for users to develop their own files to meet their particular requirements. The program produces summary predictions of likely growth patterns as well as detailed evaluations of limitations due to light, temperature, moisture and twelve important soil factors. Plantgro uses simple 'notional relationships', which are two-dimensional relationships indicating which conditions are most suitable for growth and those which are less suitable. Liebig's (1855) Law of the Minimum is used to combine limitations. In simple terms this states that the most limiting factor determines plant performance, that is, favourable factors do not compensate for unfavourable factors. Overall conditions are evaluated on a scale of 0–9 where 0 indicates ideal conditions (i.e. no limitations) and nine indicates the greatest possible limitations.

Fryer (1996) described a test of Plantgro which related predictions of site potential for *E. camaldulensis* in Central America to actual growth measurements ($r^2 = 0.63$ for nine locations spread across six countries). He also included the Plantgro plant file which he used to describe the environmental requirements of *E. camaldulensis*. In the same book Davidson (1996) describes a general method for developing Plantgro plant files for forest trees. Information from tree trial databases such as the TROPIS database being developed by the Centre for International Forestry Research (CIFOR) (Vanclay 1996), and from tree growth databases such as TREDAT maintained by CSIRO (Brown *et al.* 1989), can assist in the development of files describing the requirements of particular trees. Pawitan (1996) describes how Plantgro was integrated with a geographical information system to provide a broadscale evaluation of site potential for many tree species, including *E. camaldulensis*, across 40 per cent of Indonesia. As part of this work, Plantgro's limitation ratings were related to growth curves to produce estimates of wood production for use in economic analyses.

In conclusion, it can be said that climatic mapping programs can assist in identifying regions likely to be suitable for growing oil-yielding eucalypts, whilst statistical methods or simple models can be used to help to suggest how well particular trees are likely to grow at specific sites. More work is needed to develop methods for predicting likely oil yields of particular species and provenances at potential sites, as well as to determine the often substantial variations in growth, survival and adaptability between provenances of particular species.

Whilst existing yield prediction methods are useful, unexpected problems such as pests and diseases may arise in particular areas. For example, work at CSIRO Forestry and Forest Products has developed methods to predict risks of diseases such as *Cylindrocladium quinqueseptatum*, a leaf blight which affects some provenances of *E. camaldulensis* in parts of the world such as Vietnam (Booth *et al.* 2000). However, trials should always be established in a particular region before large-scale plantations are established and the socio-economic impacts of plantation establishment on the local population should also be carefully assessed.

## Nutritional factors and the growth of eucalypts

When managers have chosen the species they want to provide the desired products, their management options to improve growth are largely restricted to those which improve the availability of water, nutrients and other resources such as light. Sub-optimal levels of any of these factors will reduce growth and productivity. Many eucalypts have been cultivated because they grow rapidly and this characteristic is related to the unusual bud system of the leafy shoot in this genus (Jacobs 1955, FAO 1981, Florence 1996). In the axil of each leaf is a naked bud which enables continuous growth above a critical temperature, providing water and nutrients are not limiting. This feature is associated with very high rates of growth and productivity in eucalypt plantations (Beadle and Inions 1990). Ultimately, it is the amount of radiation at the plantation site which sets limits to growth but weather and soil constraints will reduce potential growth. Low temperatures and low rainfall limit the planting of eucalypts in many areas but constraining nutritional factors can often be readily manipulated in eucalypt cultivation.

Eucalypts grow naturally in a diversity of ecosystems and the quality of sites varies enormously throughout their distribution. The evolution of species on high nutrient, high water availability has produced different morphological and physiological attributes among eucalypts. Species that have adapted to nutrient-deficient and drought-prone sites often show limited ability to respond to additional nutrients in the form of fertiliser and this can reduce their sustainability for high-yielding, short-rotation, plantations (Kriedemann and Cromer 1996). Species trials have demonstrated species with high potential for rapid growth and, when there are minimal constraints, species such as *E. grandis*, *E. globulus* and *E. urophylla* are among the fastest growing and the most responsive to added nutrients (e.g. Turvey 1995). Other species with more conservative strategies and lower potential growth rates may be planted if they yield particularly desirable products. The oil-producing eucalypts such as E. *radiata*, E. *dives* and E. *polybractea* are in this category.

Studies of nutrient cycling, and interventions using inorganic fertiliser, received little attention until the development of fast-growing plantations. Evans (1992) suggests the reasons for increased interest in tree nutrition, especially in the tropics, are:

1. The high nutrient demands of fast-growing plantations and the high potential for nutrient depletion when grown on short rotations on infertile soils in the moist tropics.
2. Fertilising is more economic for trees grown on short rotations.
3. Nutrient inputs assist establishment and uniformity of growth where site heterogeneity is high.
4. Correction of specific micronutrient deficiencies, such as boron, can significantly improve productivity.

All these reasons apply in eucalypt cultivation and much research is in progress to define factors that limit eucalypt productivity, particularly nutrient availability and uptake in relation to other factors (e.g. Cromer *et al*. 1995). When eucalypts are grown for leaf oil production the export of nutrients from the site will be high, and at short intervals, so the potential for nutrient depletion of the soil is particularly great.

Managers need reliable methods of predicting nutrient requirements for short-rotation eucalypt plantations over a range of soil types and climates. However, until recently, systematic gathering of soil information and its interpretation has not been given the attention it deserves for efficient planning of tree planting operations. Soil management, like land preparation and harvesting, needs to be site-specific and there are no general solutions which will produce maximum yields (Erasmus 1991).

Information is accumulating on eucalypt nutrition (e.g. Attiwill and Adams 1996) and some generalisations can be made. The majority of eucalypts will not grow well on highly calcareous sites or heavy textured soils with high pH (Pryor 1976). It is also the case that many eucalypts go through their life cycle in Australia on soils with very low nutrient status, especially phosphorus. Under these conditions they have evolved to be very efficient users of nutrients with effective internal re-distribution mechanisms (Adams 1996). Many fertiliser experiments have demonstrated that young eucalypts usually respond to a greater or lesser degree to improved soil nutritional status providing other factors are not limiting.

Nutrient deficiencies in eucalypts have many causes, including a low rate of nutrient cycling, interactions with other nutrients, inadequate mycorrhizal associations and excessive weed competition. Manifestation of visual symptoms takes the form of poor tree vigour, leaf chlorosis and other colour changes, and malformed organs. Deficiency symptoms of nitrogen, phosphorus and potassium, highly mobile nutrients, appear first in old leaves and gradually extend towards the shoot apex. In contrast, less mobile or immobile nutrients such as calcium, sulphur, boron, iron, copper and zinc are first expressed in young leaves and progress to mature leaves (Dell *et al*. 1995). Boron deficiency, which commonly occurs in eucalypts, results in tip death and multiple leaders; severely deficient trees may become prostrate. An illustrated guide to nutrient deficiency, toxicity and non-nutritional symptoms of E. *globulus*, E. *grandis*, E. *pellita* and E. *urophylla* is provided by Dell *et al*. (1995). Impaired or abnormal growth of eucalypt leaves and shoots can also be the result of nutrient toxicities or non-nutrient factors such as plant pathogens, herbicides, air pollutants, genetic factors or environmental stresses, so great care must be exercised when using visual symptoms to identify a nutritional problem. Additional tests may be needed to confirm the diagnosis. Plant analysis is often the preferred method of diagnosing nutrient deficiencies and the following approaches have been used (Mead 1984):

1   Comparison of nutrient concentrations in healthy and unhealthy plants.
2   Estimation of plant nutrient status from a knowledge of relationships between concentrations of nutrients in plant tissues and plant growth.
3   Examination of nutrient balances within plants through calculation of nutrient rations.
4   Biochemical techniques.

The application of fertiliser can correct a specific deficiency and aid establishment and/or stimulate growth. Fertilisers are expensive and so there needs to be careful diagnosis of the nutritional problem before wide-scale application. Several approaches can be tried, including on-site fertiliser trials, foliar analysis, soil analysis and pot trials under controlled conditions. Techniques for diagnosis have progressed to the point that nutrient deficiencies can usually be detected and growth on unfertilised sites can often be predicted. However, assessment of nutrient status and prediction of the magnitude of response to fertiliser for advanced management planning requires more research (Goncalves et al. 1997).

Fertiliser trials in the field are slow and relatively expensive and must take into consideration soil variability and the representativeness of the trial site. They must also include the nutrient(s) which are deficient and hence are rarely used as the primary diagnostic tool (Evans 1992). Foliar analysis is only useful in detecting deficiencies after the trees are planted, whereas soil analysis or pot trials using soil from the field site can be used to determine the nutritional status of land before planting. Most research indicates that maximum response to fertilisers occurs if the application is within the first three months of planting. The extent of the response may be confounded by other factors such as the degree of weed control and the site preparation technique (Turvey 1995).

There is concern that the establishment of densely planted, even-aged and genetically homogeneous clonal eucalypt plantations designed to maximise wood production will bring increased risk of disease epidemics. However, the greater economic investment and returns in such plantations allows more intensive control of diseases, including selection and breeding for disease resistance (Turnbull 2000).

## Vegetative propagation

The selection and cloning of plantation crops such as rubber and oil palm has resulted in very substantial increases in latex and oil yields (Tan 1987, Hardon et al. 1987). Cloning of species of *Populus*, *Cryptomeria* and a few other trees has been practised for centuries but the application of vegetative propagation techniques to eucalypts only began to receive serious attention in the 1950s and 1960s. It is mainly since the late 1970s that eucalypts have been mass propagated vegetatively for use in plantations. The use of clones selected for high yields of essential oils appears to be the way forward for improving productivity in this industry as substantial between-tree variability in oil traits has been demonstrated. Some *E. camaldulensis* trees have leaves with almost double the oil content compared to 'average' trees in the same population (Doran and Williams 1994).

### Advantages of vegetative propagation

What is the attraction of clonal multiplication if most eucalypts can be raised easily from seeds? The reason is that cloning is a technology which enables almost all of the genetic gain generated in selection and breeding to be captured and utilised in commercial plantations (Nikles 1993). Common techniques of cloning are cuttings, layering, budding and grafting. More recent approaches involve micropropagation through the application of cell and tissue culture technologies. Tissue culture offers the potential for very rapidly multiplying selected eucalypt genotypes (Hartney and Svensson 1992).

Vegetative propagation is widely used to manage eucalypt breeding populations, and cuttings, grafts and micropropagules have all found application. However, their use is on a relatively small scale compared to the mass propagation by cuttings which is being applied extensively in clonal forestry. The term clonal forestry refers to the use of a number of tested, selected and identified clones (Eldridge *et al*. 1993). The rise of, and enthusiasm for, clonal forestry with eucalypts has mirrored the development of short-rotation, fast-growing eucalypt plantations to provide raw material for the pulp and paper industries. Since the early 1970s, millions of hectares of these plantations have been established, initially from seeds and, after 1980, with increasing use of rooted cuttings.

Early research on rooting eucalypt cuttings in Australia and North Africa indicated that it is increasingly difficult to root cuttings as the parent plant ages (Pryor and Willing 1963, Franclet 1963) although there is considerable variation between species in their ability to root from cuttings. One of the easiest species to root is the tropical *E. deglupta* (Davidson 1974) but others, such as *E. globulus*, have remained a challenge to the present day (Sasse 1995). Major advances in developing rooted cuttings from mature trees for commercial plantations were made in the Congo by researchers from the Centre Technique Forestier Tropical (CTFT) between 1969 and 1973 (Martin and Quillet 1974). Substantial increases in growth and wood production were achieved, initially with naturally occurring hybrids in plantations, and subsequently with artificially manipulated crosses (Martin 1991).

A Brazilian private company, Aracruz Forestal, recognised the potential role of clonal forestry in increasing productivity in its eucalypt pulpwood plantations. The company's research indicated that gains in volume production and wood quality could be achieved using hybrid clones, for example, *E. grandis* × *E. urophylla*, and in 1979 Aracruz decided to gradually replace its plantations derived from seeds with vegetatively propagated clonal plantations (Campinhos 1993). By 1990, nearly 126 million rooted cuttings of selected hybrid clones had been planted. Since putting its cuttings programme on an operational scale Aracruz has established 85 per cent of its area with clonal plantations and the remaining 15 per cent with seedlings (Campinhos and Silva 1990). The clonal option combined with intensive cultural practices has resulted in very substantial increases in productivity and wood quality and significant savings in harvesting costs (Campinhos 1993, 1999). The amount of product (bleached wood pulp) grown on each hectare of forest has almost doubled, from 5.9 to 10.9 t/ha/yr. At the same time, clonal plantations have provided a saving of about 22 per cent in harvesting and logging costs through tree uniformity, branch-free stems and high survival rate.

The use of cuttings has become an integral part of operations for many companies in a number of countries since Aracruz Forestal in Brazil and CTFT/Unité d'Afforestation Industrielle du Congo in West Africa initiated their large-scale clonal eucalypt plantations. In Morocco, vegetative propagation techniques have played an important role in transferring the results of tree improvement into a plantation programme which uses 2–3 million selected clones of *E. camaldulensis* and *E. camaldulensis* × *E. grandis* hybrids each year (Marien 1991). In Colombia, an operational improvement in wood yield of 40 per cent has been achieved with clones of *E. grandis* compared to seedlings since a clonal programme was initiated in 1986. At this site, and elsewhere, the matching of clones to specific site conditions is an important strategy (Wright 1995). South Africa, too, has been at the forefront of the application of clonal techniques to industrial plantations. For example, Mondi Forests have 35 ha of clone banks which are harvested to produce 8–10 million rooted cuttings annually (Herman *et al.* 1991) and here there is also careful matching of clones and sites.

*Factors which influence success*

A number of factors determine the success of clonal forestry, for example, the genetic quality of the clones, site-genotype interactions, silvicultural practices and nursery operations (Adendorff and Schön 1991). It is important that while producing high quality planting stock the costs are kept to a minimum. The Aracruz experience is that rooted cuttings cost about 40 per cent more than seedlings but the gains in productivity are so large that this difference is insignificant (Campinhos 1993).

The most important propagation characteristic of a species or clone is its rooting ability, and the cuttings of some eucalypts, for example, *E. globulus* and *E. nitens*, are very difficult to root. Failure to achieve a high level of rooting ability can result in many promising genotypes being excluded from the clonal programme. For example, if the threshold of rooting ability is set at 70 per cent then about 90 per cent of potential genotypes in *E. globulus* are excluded (Wilson 1992). *E. globulus* is such an important species for plantations that a major research effort has been directed to developing a satisfactory technique to achieve a high rooting (Sasse 1995). Celulose Beira Industrial (CELBI) in Portugal has developed a propagation system to root stem cuttings of *E. globulus* (Wilson 1992). However, the cuttings of *E. globulus* develop different root systems to seedlings and clonal growth has been variable. Further development of current propagation systems is required before widespread use of *E. globulus* cuttings can be recommended (Sasse and Sands 1995) although micropropagation of *E. globulus* and *E. nitens* is now a viable option (Ruaud 1996).

It must be emphasised that the mass propagation of cuttings is only worthwhile when there are selected genotypes to propagate. The techniques used on a commercial scale include the setting up and management of clone banks of outstanding individuals and mass propagation of cuttings from the clone banks. The techniques for the establishment, rejuvenation of clones, preparation of cuttings, rooting and hardening off of rooted cuttings are described by Eldridge *et al.* (1993). These techniques are those used in propagating eucalypt cuttings for large industrial plantations but similar practices can be applied to achieve improvement in smaller scale operations. It is at this smaller scale that most eucalyptus oil-producing plantations are likely to operate.

*Vegetative propagation by small-scale producers*

The small-scale forestry sector faces the difficulty of having to obtain selected material to propagate. It is unlikely that small-scale operations can justify their own tree breeding programmes so they must make arrangements to use material developed by larger companies or government agencies, or by joining tree improvement cooperatives where the costs of tree breeding and selection can be shared. Small-scale operations will generally have limited staff resources and low technology facilities so the propagation system is likely to use cuttings rather than micropropagation by tissue culture. However, small-scale production of tissue cultured plantlets has produced propagation material of *E. grandis* × *E. urophylla* hybrids for plantations in southern China. In the tropics, only minimal improvements in facilities and work standards may be needed to convert a conventional seedling nursery into a cuttings nursery. Technical aspects of the mass multiplication of eucalypts on a small scale are discussed by Kijkar (1991) and White (1993). Those considering using a clonal strategy are recommended to undertake a simple marginal benefit/cost analysis of the type described by Walker and Haines (1998) before making a final decision.

Eucalyptus oil can be produced either as a by-product of a wood industry or as a crop in its own right. In Australia and South Africa, growers have specialised oil-producing plantations but in Spain and Portugal 'waste' leaf comes from *E. globulus* grown for pulpwood. If produced as a by-product it is less likely that increasing the quantity or quality of the oil will be a major part of the breeding programme.

Attempts to improve yield and productivity of oil-producing eucalypts such as *E. radiata* by selection and cloning have been frustrated by the difficulty in producing clonal material vegetatively, although micropropagation is more promising (Donald 1991). Tissue culturing techniques have been applied to *E. kochii* subsp. *kochii* in Western Australia and the possibilities appear extremely good for producing cineole-rich oil in the tropics from selected clones of *E. camaldulensis* for which the vegetative propagation technology is relatively simple (Doran and Williams 1994).

## Coppice management

Coppice is the oldest silvicultural system known, dating back to Neolithic times in Britain. It was also used by the Greeks and Romans and extensively practised in Europe in the Middle Ages (Matthews 1989). The coppice system of management starts by felling a tree which was originally planted. When felled near ground level, many species produce shoots from the stump. These shoots come from dormant buds on the side of the stump or from adventitious buds in the cambial layer at the edge of the cut surface (Evans 1992). Many of the millions of hectares of eucalypts established as plantations are managed as coppice. Plantations managed only for the production of essential oils are perhaps the most extreme form of coppice, with leaf-bearing shoots being harvested mechanically or manually from an early age.

Why use a coppice system to regenerate the crop? Firstly, it is a very simple and reliable system which is cheaper, and requires less expertise, than using seed or cuttings. Initial growth is faster in the earliest stages and it is ideal where many small- to medium-sized plots are required for pulpwood or fuel. The system is generally applied to short-rotation crops, so producing early economic returns, and this is beneficial to resource-poor small growers. In addition, the variety of habitat provided by different stages of coppice can be beneficial to wild plants and animals so there is a conservation value (Matthews 1989). On the negative side, coppice grown and harvested on short rotations removes substantial quantities of nutrients from the site. The young shoots can also be attractive for browsing by animals and may be prone to frost damage in some localities. It may also be considered aesthetically less desirable than taller forest because of its uniformity and lower canopy level.

Many eucalypts have excellent coppicing powers. The shoots arise from dormant buds in the inner bark or from buds in the lignotuber at the base of the stem. Most eucalypts produce lignotubers which consist of a mass of vegetative buds and associated vascular tissue with substantial food reserves. When the tops of seedling or young trees are destroyed by fires, drought or grazing, vigorous new shoots develop from the lignotuber (Pryor 1976). The absence of lignotubers in some species, for example, *E. deglupta* and *E. regnans*, partly explains their poor coppicing ability.

Management of eucalypt coppice has been described by FAO (1981) and Evans (1992). The ability to produce coppice is very variable and is determined mainly by the species and condition of the stump but the season of felling may also have an effect. The eucalypt stump is best cut low (10–12 cm) with a chainsaw or bow saw to provide a smooth and sloping surface. Shoots growing from the base of the stump are generally more stable than those higher up. Stump diameter is also important in determining the productivity of the stump. Larger stumps produce more vigorous coppice although thick bark and very large stumps may inhibit sprouting. In *E. camaldulensis*, stool productivity increased with increasing diameter except for the largest 5 per cent. One year after cutting, large stools (mean diameter 19.5 cm) supported coppice with four times the basal area of coppice from small stools (mean diameter 7.5 cm) (Kondas 1982 in Evans 1992).

A stump produces many shoots and whether or not these coppice shoots are thinned before harvesting depends on the product being grown. For pole production it is usual to reduce the number of shoots to 2–3 in the first year and 1–2 in the second year. Thinning increases the value of the remaining stems but reduces total biomass production, although this may not be significant, for example, Toral *et al.* (1988). Coppice reduction is an expensive operation and is only carried out if it is essential. It is not carried out in plantations where the total above-ground biomass is harvested to produce leaf oil, and is rarely undertaken for fuelwood production. Normally, the yield from the first coppice crop is higher than the seedling crop, but thereafter it declines with each cropping (Evans 1992). However, Barros and Novais (1996) report that in Brazil the first coppice crop may have a 50 per cent reduction in productivity on some sites due to soil nutrient depletion, especially of phosphorus. Coppicing can have a significant short-term effect on local hydrology due to high water use as the above-ground biomass develops rapidly. Stream flow was greatly reduced in the two years following coppicing of *E. grandis* and *E. camaldulensis* in northern India but the effect disappeared in the third year (Bruijnzeel 1997).

The number of rotations that can be grown without replanting depends on the rate of stump mortality and reduction in vigour of older trees. *E. globulus* in India has been coppiced successfully on a ten-year rotation for over 100 years (FAO 1981). As a general rule, three or four coppice rotations can be harvested before total replanting needs to take place, as annual stump mortality is often in the range 5–10 per cent. Pulpwood plantations in Brazil are managed on a 6–9 year rotation followed by one or two coppice crops in which no thinning is practised. As more intensive management is applied the stock is replaced each rotation with improved genetic material from breeding programmes. In Swaziland, although oil production has now ceased, some areas of *E. smithii* managed specifically for oil were still being harvested in the early 1990s, twenty years or more after the first cut. A rotation of approximately sixteen months was used (Coppen and Hone 1992). Some of the lessons from general coppicing techniques will be relevant for growers coppicing eucalypts for essential oil production but the regular and heavy harvesting of leaves in oil-yielding plantations may require the development and application of more specialised technologies to ensure sustained production.

# References

Adams, M.A. (1996) Distribution of eucalypts in the Australian landscapes: landforms, soils, fire and nutrition. In P.M. Attiwill and M.A. Adams (eds), *Nutrition of Eucalypts*, CSIRO, Australia, pp. 61–76.

Adendorff, M.W. and Schön, P.P. (1991) Root strike and root quality: the key to commercial success. In A.P.G. Schönau (ed.), *Intensive Forestry: The Role of Eucalypts, Proc. IUFRO Symp.*, Durban, September 1991, Vol. 1, Southern African Institute of Forestry, Pretoria, pp. 30–38.

Agriculture Western Australia (1996) *Western Australian Salinity Action Plan*, Government of Western Australia, Perth.

Anon. (1990) Around the world: Chile. *Commonwealth For. Rev.*, 69, 212.

Arnold, J.E.M. and Dewees, P.A. (eds) (1995) *Tree Management Strategies: Responses to Agricultural Intensification*, Oxford University Press, Oxford, UK.

Attiwill, P.M. and Adams, M.A. (eds) (1996) *Nutrition of Eucalypts*, CSIRO, Australia.

Barros, N.F. and Novais, R.F. (1996) Eucalypt nutritional and fertiliser regimes in Brazil. In P.M. Attiwill and M.A. Adams (eds), *Nutrition of Eucalypts*, CSIRO, Australia, pp. 335–355.

Bartle, J. (1994) New horizons for forestry: tree crops for the wheatbelt. *Newsletter of the Institute of Foresters of Australia Inc.*, 35(2), 4–7.
Bartle, J.R., Campbell, C. and White, G. (1996) Can trees reverse land degradation? In *Farm Forestry and Plantations: Investing in Future Wood Supply, Proc. Australian Forest Growers Conf.*, Mount Gambier, Australia, September 1996, pp. 68–75.
Battaglia, M. and Sands, P. (1997) Modelling site productivity of *Eucalyptus globulus* in response to climate and site factors. *Aust. J. Plant Physiol.*, 24, 831–850.
Beadle, C.H. and Inions, G. (1990) Limits to growth of eucalypts and their biology of production. In J. Dargavel and N. Semple (eds), *Prospects for Australian Plantations*, Australian National Univ., Canberra, pp. 199–215.
Boland, D.J., Brophy, J.J. and House, A.P.N. (eds) (1991) *Eucalyptus Leaf Oils: Use, Chemistry, Distillation and Marketing*, ACIAR/CSIRO, Inkata Press, Melbourne.
Booth, T.H. (1985) A new method to assist species selection. *Commonwealth For. Rev.*, 64, 241–249.
Booth, T.H. (1996a) The development of climatic mapping programs and climatic mapping in Australia. In T.H. Booth (ed.), *Matching Trees and Sites, Proc. Internat. Workshop*, Bangkok, March 1995, ACIAR Proceedings No. 63, pp. 38–42.
Booth, T.H. (ed.) (1996b), *Matching Trees and Sites, Proc. Internat. Workshop*, Bangkok, March 1995, ACIAR Proceedings No. 63.
Booth, T.H., Jovanovic, T., Old, K.M. and Dudzinski, M.J. (2000) Climatic mapping to identify high-risk areas for *Cylindrocladium quinqueseptatum* leaf blight on eucalypts in mainland South East Asia and around the world. *Environmental Pollution*, 108, 365–372.
Booth, T.H., Nix, H.A., Hutchinson, M.F. and Jovanovic, T. (1988) Niche analysis and tree species introduction. *For. Ecol. Manage.*, 23, 47–59.
Booth, T.H. and Pryor, L.D. (1991) Climatic requirements of some commercially important eucalypt species. *For. Ecol. Manage.*, 43, 47–60.
Bouillet, J.P., Nzila, J.D., Ranger, J., Laclau, J.P. and Nizinski, G. (1999) *Eucalyptus* plantations in the equatorial zone, on the coastal plains of the Congo. In E.K.S. Nambiar, C. Cossalter and A. Tiarks (eds), *Site Management and Productivity in Tropical Plantation Forests*, Center for International Forestry Research, Bogor, Indonesia, pp. 13–21.
Brooker, M.I.H. and Hopper, S. (1991) New series, subseries, species and subspecies of *Eucalyptus* (Myrtaceae) from Western Australia and South Australia. *Nuytsia*, 9, 1–68.
Brown, A.G., Wolf, L.J., Ryan, R.A. and Voller, P. (1989) TREDAT: a tree crop database. *Aust. For.*, 52, 23–29.
Bruijnzeel, L.A. (1997) Hydrology of forest plantations in the tropics. In E.K.S. Nambiar and A.G. Brown (eds), *Management of Soil, Nutrients and Water in Tropical Plantation Forests*, ACIAR Monograph No. 34, ACIAR/CSIRO/CIFOR, Canberra, pp. 125–167.
Busby, J.R. (1991) BIOCLIM – a bioclimatic analysis and prediction system. In C.R. Margules and M.P. Austin (eds), *Nature Conservation: Cost Effective Biological Surveys and Data Analysis*, CSIRO, Melbourne, pp. 64–68.
Calder, I.R. (1994) *Eucalyptus, Water and Sustainability*, ODA Forestry Series No. 6, Institute of Hydrology, Wallingford, UK.
Campinhos, E. (1993) A Brazilian example of a large scale forestry plantation in a tropical region: Aracruz. In J. Davidson (ed.), *Recent Advances in Mass Clonal Multiplication of Forest Trees for Plantation Programmes, Proc. Regional Symp.*, Field Document No. 4, RAS/91/004, Food and Agriculture Organization of the United Nations, Los Banos, Philippines, pp. 46–59.
Campinhos, E. (1999) Sustainable plantations of high yield eucalypt trees for production of fiber: the Aracruz case. *New Forests*, 17, 129–143.
Campinhos, E. and Silva, E.C. (1990) Development of the *Eucalyptus* tree of the future. Paper presented at ESPRA Spring Conf., Seville, Spain, 1990.
CEASA (1994) Intelligent fibre. *World Paper*, November/December.
Chinniah Sekar, Vinaya Rai, R.S., Devaraj, P. and Senthilnathan, S. (1996) Investment in pulpwood plantations in Tamil Nadu, India: a paradigm to resource poor farmers. In M.J. Dieters, A.C. Matheson, D.G. Nikles, C.E. Harwood and S.M. Walker (eds), *Tree Improvement for Sustainable Tropical Forestry*,

*Proc. QFRI-IUFRO Conf.*, Caloundra, Australia, October/November 1996, Queensland Forestry Research Institute, Gympie, pp. 435–436.

Chippendale, G.M. (1988) *Flora of Australia*, Vol. 19, *Myrtaceae – Eucalyptus, Angophora*, Australian Government Publishing Service, Canberra.

Coppen, J.J.W. and Hone, G.A. (1992) *Eucalyptus Oils: A Review of Production and Markets*, NRI Bulletin 56, Natural Resources Institute, Chatham, UK.

Cossalter, C. (1996) Addressing constraints to the development of plantation forestry in the tropics: a role for tree improvement. In M.J. Dieters, A.C. Matheson, D.G. Nikles, C.E. Harwood and S.M. Walker (eds), *Tree Improvement for Sustainable Tropical Forestry, Proc. QFRI-IUFRO Conf.*, Caloundra, Australia, October/November 1996, Queensland Forestry Research Institute, Gympie, pp. 282–287.

Cremer, K.W. (ed.) (1990) *Trees for Rural Australia*, CSIRO Division of Forestry and Forest Products, Inkata Press, Melbourne.

Cromer, R., Smethurst, P., Turnbull, C., Misra, R., LaSala, A., Herbert, A. and Dimsey, L. (1995) Early growth of eucalypts in Tasmania in relation to nutrition. In B.M. Potts, N.M.G. Borralho, J.B. Reid, R.N. Cromer, W.N. Tibbits and C.A. Raymond (eds), *Eucalypt Plantations: Improving Fibre Yield and Quality, Proc. CRCTHF-IUFRO Conf.*, Hobart, Australia, February 1995, Cooperative Research Centre for Temperate Hardwood Forestry, Hobart, pp. 331–335.

Davidson, J. (1974) Reproduction of *Eucalyptus deglupta* by cuttings. *New Zealand J. For. Sci.*, 4, 191–203.

Davidson, J. (1995) Ecological aspects of *Eucalyptus* plantations. In K. White, J. Ball and M. Kashio (eds), *Proc. Regional Expert Consult. on Eucalyptus*, Bangkok, October 1993, Vol. 1, Food and Agriculture Organization of the United Nations, Regional Office for Asia and the Pacific, Bangkok, pp. 35–72.

Davidson, J. (1996) Developing Plantgro plant files for forest trees. In T.H. Booth (ed.), *Matching Trees and Sites, Proc. Internat. Workshop*, Bangkok, March 1995, ACIAR Proceedings No. 63, pp. 93–96.

De Freitas, M. (1995) Planted forests in Brazil. In *Caring for the Forest: Research in a Changing World. Proc. IUFRO 20th World Congress*, Tampere, Finland, August 1995, Vol. 2, pp. 147–154.

Dell, B., Malajczuk, N. and Grove, T.S. (1995) *Nutrient Disorders of Eucalypts*, ACIAR Monograph 31, ACIAR, Canberra.

Dewees, P.A. and Saxena, N.C. (1995) Wood product markets as incentives for farmer tree growing. In J.E.M. Arnold and P.A. Dewees (eds), *Tree Management Strategies: Responses to Agricultural Intensification*, Oxford University Press, Oxford, UK, pp. 198–241.

Donald, D.G.M. (1991) *Eucalyptus* species as an oil source in South Africa. In A.P.G. Schönau (ed.), *Intensive Forestry: The Role of Eucalypts, Proc. IUFRO Symp.*, Durban, September 1991, Vol. 2, Southern African Institute of Forestry, Pretoria, pp. 985–989.

Doran, J.C. (1991) Commercial sources, uses, formation and biology. In D.J. Boland, J.J. Brophy and A.P.N. House (eds), *Eucalyptus Leaf Oils. Use, Chemistry, Distillation and Marketing*, ACIAR/CSIRO, Inkata Press, Melbourne, pp. 11–25.

Doran, J.C. (1992) Variation in and Breeding for Oil Yields in Leaves of *Eucalyptus camaldulensis*, PhD Thesis, Australian National Univ., Canberra.

Doran, J.C. and Williams, E.R. (1994) Fast-growing *Eucalyptus camaldulensis* clones for foliar-oil production in the tropics. *Commonwealth For. Rev.*, 73, 261–266.

Eastham, J., Scott, P.R., Steckis, R.A., Barton, A.F.M., Hunter, L.J. and Sudmeyer, R.J. (1993) Survival, growth and productivity of tree species under evaluation for agroforestry to control salinity in the Western Australian wheatbelt. *Agrofor. Syst.*, 21, 223–237.

Eldridge, K.G., Davidson, J., Harwood, C. and van Wyk, G. (1993) *Eucalypt Domestication and Breeding*, Clarendon Press, Oxford, UK.

EMBRAPA (1996) Brazil has strategic reserve of eucalyptus (in Portuguese). *Folha da Floresta*, (8), 6–8.

Erasmus, D. (1991) The gathering and utilization of soil information for commercial forestry planning. In A.P.G. Schönau (ed.), *Intensive Forestry: The Role of Eucalypts, Proc. IUFRO Symp.*, Durban, September 1991, Vol. 1, Southern African Institute of Forestry, Pretoria, pp. 405–417.

Evans, J. (1992) *Plantation Forestry in the Tropics*, 2nd edn., Clarendon Press, Oxford, UK.

FAO (1981) *Eucalypts for Planting*, FAO Forestry Series No. 11, Food and Agriculture Organization of the United Nations, Rome.

FAO (1991) *Wood and Wood Products: Forestry Statistics Today for Tomorrow. 1961–1989 to 2010*, Food and Agriculture Organization of the United Nations, Rome.

FAO (1993) *Forest Resources Assessment 1990: Tropical Countries*. FAO Forestry Technical Paper 112, Food and Agriculture Organization of the United Nations, Rome.

Florence, R.G. (1996) *Ecology and Silviculture of Eucalypt Forests*, CSIRO, Australia.

Flynn, B. and Shield, E. (1999) *Eucalyptus: Progress in Higher Value Utilization. A Global Review*, Robert Flynn and Associates, and Economic Forestry Associates, Tacoma, USA, and Annerley, Australia.

Franclet, A. (1963) Improvement in reafforestation of *Eucalyptus* by vegetative propagation (in French). In *Proc. FAO First World Consult. Forest Genetics and Tree Improvement*, FAO/FORGEN-63, Vol. 2, 5/5, pp. 1–8.

Fryer, J.H. (1996) Site sampling and performance prediction for *Eucalyptus camaldulensis* in Central America. In T.H. Booth (ed.), *Matching Trees and Sites. Proc. Internat. Workshop*, Bangkok, March 1995, ACIAR Proceedings No. 63, pp. 112–117.

Goncalves, J.L.M., Barros, N.F., Nambiar, E.K.S. and Novais, R.F. (1997) Soil and stand management for short-rotation plantations. In E.K.S. Nambiar and A.G. Brown (eds), *Management of Soil, Nutrients and Water in Tropical Plantation Forests*, ACIAR Monograph No. 34, ACIAR/CSIRO/CIFOR, Canberra, pp. 379–417.

Gregersen, H., Draper, S. and Elz, D. (1989) *People and Trees – The Role of Social Forestry in Sustainable Development*, The World Bank, Washington DC.

Hackett, C. (1991) *Plantgro: A Software Package for the Coarse Prediction of Plant Growth*, CSIRO, Melbourne.

Hall, N.H., Boden, R.W., Christian, C.S., Condon, R.W., Dale, F.A., Hart, A.J., Leigh, J.H., Marshall, J.K., McArthur, A.G., Russell, V. and Turnbull, J.W. (1972) *The Use of Trees and Shrubs in the Dry Country of Australia*, Australian Government Publishing Service, Canberra.

Hardon, J.J., Corley, R.V.H. and Lee, C.H. (1987) Breeding and selecting oil palm. In A.J. Abbott and R.K. Atkin (eds), *Improving Vegetatively Propagated Crops*, Academic Press, London, pp. 64–78.

Hartney, V.J. and Svensson, J.G.P. (1992) The role of micropropagation for Australian tree species. In F.W.G. Baker (ed.), *Rapid Propagation of Fast Growing Woody Species*, CAB International, Wallingford, UK, pp. 7–28.

Herman, B., Denison, N.P., Friedman, G., Le Roux, J.J. and McKellar, D. (1991) Biotechnology/vegetative propagation in Mondi. In A.P.G. Schönau (ed.), *Intensive Forestry: The Role of Eucalypts, Proc. IUFRO Symp.*, Durban, September 1991, Vol. 1, Southern African Institute of Forestry, Pretoria, pp. 94–103.

Hill, K.D. and Johnson L.A.S. (1992) Systematic studies in the eucalypts – 5 new taxa and combinations of *Eucalyptus* (Myrtaceae) in Western Australia. *Telopea*, 4, 561–634.

Hillis, W.E. and Brown, A.G. (eds) (1978) *Eucalypts for Wood Production*, CSIRO, Melbourne.

Hutchinson, M.F., Booth, T.H., McMahon, J.P. and Nix, H.A. (1984) Estimating monthly mean values of daily total solar radiation for Australia. *Solar Energy*, 32, 277–290.

Inions, G. (1991) Relationships between environmental attributes and the productivity of *Eucalyptus globulus* in southwest Western Australia. In P. Ryan (ed.), *Productivity in Perspective, Proc. Third Australian Forest Soils and Nutrition Conf.*, Melbourne, pp. 116–132.

Inions, G. (1992) Studies on the Growth and Yield of Plantation *Eucalyptus globulus* in Southwest Western Australia, PhD Thesis, Univ. Western Australia.

Jacobs, M.R. (1955) *Growth Habits of the Eucalypts*, Government Printer, Canberra.

Janmahasatien, S., Viriyabuncha, C. and Snowdon, P. (1996) Soil sampling and growth prediction in Thailand. In T.H. Booth (ed.), *Matching Trees and Sites. Proc. Internat. Workshop*, Bangkok, March 1995, ACIAR Proceedings No. 63, pp. 101–106.

Kijkar, S. (1991) *Producing Rooted Cuttings of* Eucalyptus camaldulensis: *Handbook*. Asean-Canada Forest Tree Seed Centre, Muak-Lek, Saraburi, Thailand.

Kondas, S. (1982) Mysore Gum coppice growth – vigour, productivity and regulation of cutting. In P.B.L. Srivastava *et al.* (eds), *Tropical Forests – Source of Energy Through Optimization and Growth*, pp. 317–325. (In Evans 1992.)

Kriedemann, P.E. and Cromer, R.N. (1996) The nutritional physiology of the eucalypts – nutrition and growth. In P.M. Attiwill and M.A. Adams (eds), *Nutrition of Eucalypts*, CSIRO, Australia, pp. 109–121.

Kumar, V. (1991) Eucalyptus in the forestry scene of India. In A.P.G. Schönau (ed.), *Intensive Forestry: The Role of Eucalypts, Proc. IUFRO Symp.*, Durban, September 1991, Vol. 2, Southern African Institute of Forestry, Pretoria, pp. 1105–1116.

Landsberg, J.J. and Waring, R.H. (1997) A generalised model of forest productivity using simplified concepts of radiation-use efficiency, carbon balance and partitioning. *For. Ecol. Manage.*, 95, 209–228.

Landsberg, J.J, Waring, R.H. and Coops, N.C. (2002) Performance of the forest productivity model 3-PG applied to a wide range of forest types. *For. Ecol. Manage.* (in press).

Liebig, J. (1855) *Die Grundsatze der Agriculturchemie mit Rücksicht die in England Angestellten Untersuchungen*, F. Vieweg & Son, Braunschweig.

Marcar, N.E., Crawford, D., Leppert, P., Jovanovic, T., Floyd, R. and Farrow, R. (1995) *Trees for Saltland – A Guide to Selecting Native Species for Australia*, CSIRO, Australia.

Marien, J.N. (1991) Clonal forestry in Morocco: propagation and maturation problems. In A.P.G. Schönau (ed.), *Intensive Forestry: The Role of Eucalypts, Proc. IUFRO Symp.*, Durban, September 1991, Vol. 1, Southern African Institute of Forestry, Pretoria, pp. 126–132.

Martin, B. (1991) Ecology and the economy – synergy in eucalypt plantations in the Congo. In A.P.G. Schönau (ed.), *Intensive Forestry: The Role of Eucalypts, Proc. IUFRO Symp.*, Durban, September 1991, Vol. 2, Southern African Institute of Forestry, Pretoria, pp. 1117–1127.

Martin, B. and Quillet, G. (1974) Propagation by cuttings of forest trees in the Congo: results of trials carried out at Pointe Noire from 1969 to 1973 (in French). *Bois Forêts Tropiques*, (154), 41–57.

Matthews, J.D. (1989) *Silvicultural Systems*, Oxford University Press, Oxford, UK.

Mead, D.J. (1984) Diagnosis of nutrient deficiencies in plantations. In G.D. Bowen and E.K.S. Nambiar (eds), *Nutrition of Plantation Forests*, Academic Press, New York, pp. 259–291.

Morrison, E. and Bass, S.M.J. (1992) What about the people? In C. Sargent and S. Bass (eds), *Plantation Politics*, Earthscan, London, pp. 92–120.

Nesmith, C. (1991) Gender, trees and fuel: social forestry in West Bengal, India. *Human Organization*, 50, 337–348.

Nikles, D.G. (1993) Conservation and use of genetic diversity in improvement programmes with industrial forest tree species. In J. Davidson (ed.), *Recent Advances in Mass Clonal Multiplication of Forest Trees for Plantation Programmes, Proc. Regional Symp.*, Food and Agriculture Organization of the United Nations, Los Banos, Philippines, pp. 83–109.

Pawitan, H. (1996) The use of Plantgro in forest plantation planning in Indonesia. In T.H. Booth (ed.), *Matching Trees and Sites, Proc. Internat. Workshop*, Bangkok, March 1995, ACIAR Proceedings No. 63, pp. 97–100.

Pereira, J.S., Tomé, M., Madeira, M., Oliveira, A.C., Tomé, J. and Almeida, M.H. (1996) Eucalypt plantations in Portugal. In P.M. Attiwill and M.A. Adams (eds), *Nutrition of Eucalypts*, CSIRO, Australia, pp. 371–387.

Poore, M.E.D. and Fries, C. (1985) *The Ecological Effects of Eucalyptus*, FAO Forestry Paper No. 59, Food and Agriculture Organization of the United Nations, Rome.

Pryor, L.D. (1976) *Biology of Eucalypts*, Studies in Biology, No. 61, Edward Arnold, London.

Pryor, L.D., Williams, E.R. and Gunn, B.V. (1995) A morphometric analysis of *Eucalyptus urophylla* and related taxa with description of two new species. *Aust. Syst. Bot.*, 8, 57–70.

Pryor, L.D. and Willing, R.R. (1963) The vegetative propagation of *Eucalyptus* – an account of progress. *Aust. For.*, 27, 52–62.

Raintree, J.B. and Lantican, C. (1993) Forestry economics research and MPTS development. In Songkram Thammincha, Ladawan Puangchit and H. Wood (eds), *Forestry Economics Research in Asia*, Faculty of Forestry, Kasetsart Univ., Thailand, and IDRC, Canada, pp.15–31.

Rajan, B.K.C. (1987) *Versatile Eucalyptus*, Diana Publications, Bangalore, India.

Ruaud, J.-N. (1996) Towards vegetative propagation of temperate eucalypts. *Onwood*, CSIRO Research Update No. 12, 3.

Sands, P.J., Battaglia, M. and Mummery, D. (2000) Application of process-based models to forest management: experience with PROMOD, a simple plantation productivity model. *Tree Physiology*, 20, 383–392.

Sasse, J. (1995) Problems with propagation of *Eucalyptus globulus* by stem cuttings. In B.M. Potts, N.M.G. Borralho, J.B. Reid, R.N. Cromer, W.N. Tibbits and C.A. Raymond (eds), *Eucalypt Plantations: Improving Fibre Yield and Quality, Proc. CRCTHF-IUFRO Conf.*, Hobart, Australia, February 1995, CRC for Temperate Hardwood Forestry, Hobart, pp. 319–320.

Sasse, J. and Sands, R. (1995) Root system development in cuttings of *Eucalyptus globulus*. In B.M. Potts, N.M.G. Borralho, J.B. Reid, R.N. Cromer, W.N. Tibbits and C.A. Raymond (eds), *Eucalypt Plantations: Improving Fibre Yield and Quality, Proc. CRCTHF-IUFRO Conf.*, Hobart, Australia, February 1995, CRC for Temperate Hardwood Forestry, Hobart, pp. 299–303.

Saxena, N.C. and Vishwa Ballabh (1995) Farm forestry and the context of farming systems in South Asia. In N.C. Saxena and Vishwa Ballabh (eds), *Farm Forestry in South Asia*, Sage, New Delhi, pp. 24–50.

Shah, S.A. (1988) Experiences with eucalypts in social forestry in India – lessons learnt. In *Proc. Internat. Forestry Conf. Australian Bicentenary*, Albury, 1988, Vol. 5, 15 pp.

Shiva, M.P., Prakash, O.M., Singh, S. and Jaffer, R. (1988) Role of eucalypts in agroforestry and essential oil production potential. *Indian Forester*, 114, 776–783.

Small, B.E.J. (1981) The Australian *Eucalyptus* oil industry – an overview. *Aust. For.*, 44, 170–177.

Soares, J. (1994) The role of *Eucalyptus globulus* biomass production system for the pulp and paper industry in the economy of Portugal and the EEC. In J.S. Pereira and H. Pereira (eds), *Eucalyptus for Biomass Production: The State of the Art*, CEC, ISA, Lisbon, pp. 283–286.

Song, Y. (1992) Utilization of eucalypts in China. *Appita*, 45, 382–383.

Tan, H. (1987) Strategies in rubber tree breeding. In A.J. Abbot and R.K. Atkin (eds), *Improving Vegetatively Propagated Crops*, Academic Press, London, pp. 64–78.

Toral, I., Rosende, B.R. and de Pablo, B.G. (1988) Evaluation of thinning in *Eucalyptus globulus* (Labill.) thicket in Region V (in Spanish). *Viencia Investigacion Forestal*, 2(5), 1–11.

Turnbull, J.W. (1999) Eucalypt plantations. *New Forests*, 17, 37–52.

Turnbull, J.W. (2000) Economic and social importance of eucalypts. In P.J. Keane, G.A. Kile, F.D. Podger and B.N. Brown (eds), *Diseases and Pathogens of Eucalypts*, CSIRO, Melbourne.

Turvey, N.D. (1995) *Afforestation on Imperata Grasslands in Indonesia: Results of Industrial Tree Plantation Research Trials at Teluk Sirib on Palau Laut, Kalimantan Selatan*, ACIAR Technical Report No. 33.

Vanclay, J.K. (1996) TROPIS, the Tree Growth and Permanent Plot Information System. In *Growth Studies in Tropical Moist Africa, IUFRO Conf.*, Kumasi, Ghana, November 1996, www.cgiar.org/cifor/research/tropis/tropisdetail.html (accessed April 1997).

Walker, S.M. and Haines, R.J. (1998) Evaluation of clonal strategies for tropical acacias. In J.W. Turnbull, H.R. Crompton and K. Pinyopusarerk (eds), *Recent Developments in Acacia Planting, Proc. Internat. Workshop*, Hanoi, Vietnam, October 1997, ACIAR Proceedings No. 82, Canberra, pp. 197–202.

Wang, H., Jiang, Z. and Yan, H. (1994) Australian trees grown in China. In A.G. Brown (ed.), *Australian Tree Species Research in China, Proc. Internat. Workshop*, Zhangzhou, Fujian Province, P.R. China, November 1992, ACIAR Proceedings No. 48, pp. 19–25.

Werren, M. (1991) *Eucalyptus* plantation development in Sumatra. In A.P.G. Schönau (ed.), *Intensive Forestry: The Role of Eucalypts, Proc. IUFRO Symp.*, Durban, September 1991, Vol. 2, Southern African Institute of Forestry, Pretoria, pp. 1160–1166.

White, K.J. (1988). *Eucalyptus* on small farms. In D. Withington, K.G. MacDicken, C.B. Sastry and N.R. Adams (eds), *Multipurpose Tree Species for Small-Farm Use, Proc. Internat. Workshop*, Pattaya, Thailand, November 1987, Winrock International Institute for Agricultural Development and IDRC of Canada, Bangkok, pp. 86–96.

White, K.J. (1993) Small scale vegetative multiplication of *Eucalyptus* and its use in clonal plantations. In J. Davidson (ed.), *Recent Advances in Mass Clonal Multiplication of Forest Trees for Plantation Programmes, Proc. Regional Symp.*, Field Document No. 4, RAS/91/004, Food and Agriculture Organization of the United Nations, Los Banos, Philippines, pp. 60–75.

Wilson, P.J. (1992) The development of new clones of *Eucalyptus globulus* and *E. globulus* hybrids by stem cuttings propagation. In *Mass Production Technology for Genetically Improved Fast Growing Forest Tree Species, Proc. IUFRO Symp.*, Bordeaux, September 1992, Vol. 1, pp. 379–386.

Wood, M., Howell, C. and Jones, M. (1999) Australia's national plantation inventory: an interim update. In *Australian Forest Products Statistics*, September Quarter, Australian Bureau of Agricultural and Resource Economics, Canberra, pp. 1–2.

Wright, J.A. (1995) Operational gains and constraints with clonal *Eucalyptus grandis* in Colombia. In B.M. Potts, N.M.G. Borralho, J.B. Reid, R.N. Cromer, W.N. Tibbits and C.A. Raymond (eds), *Eucalypt Plantations: Improving Fibre Yield and Quality, Proc. CRCTHF-IUFRO Conf.*, Hobart, Australia, February 1995, CRC for Temperate Hardwood Forestry, Hobart, pp. 308–310.

Zacharin, R.F. (1978) *Emigrant Eucalypts*, Melbourne University Press, Melbourne.

# 4 Genetic improvement of eucalypts
## With special reference to oil-bearing species

*John C. Doran*

## Introduction

Domestication of eucalypts for oil production in plantations is in its infancy. Tree breeders have only just started to make genetic changes to planting stock and in many instances seed for plantations still comes from natural stands. Eucalypts are largely outbreeding and genetically highly variable, which represents a huge opportunity for the tree breeder, whose main task is to exploit this variability through exploration, evaluation, selection and breeding. In this situation, large gains in heritable traits such as various growth and oil characteristics can be achieved simply and cheaply using relatively unsophisticated procedures. This is in marked contrast to many agricultural crops (e.g. barley, wheat and rice) that have been domesticated for thousands of years, are in many cases inbreeding species, and where changes such as polyploidy and mutation need to be artificially induced by breeders.

Tree breeding is worthwhile only if the subsequent economic returns are greater than its implementation costs. In the following discussion it will be assumed that developing better genotypes for oil production through selection and breeding is cost-effective. However, it is recognised that this argument may well be hard to sustain at times of low eucalyptus oil prices. Many of the principal oil-producing species, for example, *Eucalyptus globulus* (China), *E. smithii* (Southern Africa) and *E. citriodora* (China, Brazil, India), are grown primarily for their wood; eucalyptus oil is a valuable, but relatively minor, additional product. Improvement of characteristics affecting wood production such as growth rate, form and wood quality will always be a tree breeder's first priority when working with these species. A benefit/cost analysis of incorporating selection for oil traits in such a programme may well return a negative result. Economic analysis should be applied to all breeding work to ensure that the benefits outweigh the cost of the programme.

Namkoong *et al.* (1980) suggested that genetic improvement programmes can usefully be grouped into three classes: those requiring low, medium or high intensity effort. The choice between these alternatives depends upon the resources (i.e. physical, financial and human) available and the potential economic benefits of the tree plantations (Harwood 1996). Economics may well dictate that a relatively simple, low-input strategy is appropriate for the genetic improvement of oil traits in eucalypts, and this will be the premise for further discussion in this chapter. Nevertheless, successful tree breeding, whether simple or complicated, needs scientific and technical expertise and a commitment to the provision of necessary resources in the long term. These resources will certainly include access to chemical analytical equipment such as gas chromatography (GC) and the expertise of an organic chemist when breeding for oil traits. The expense of screening large numbers of plants for oil content and composition is a consideration when determining an appropriate breeding strategy and plan.

Tree improvement programmes aim to develop new plantings superior to their predecessors in one or several key economic traits. The *modus operandi* of most contemporary programmes is to start with a carefully chosen breeding strategy implemented through a dependent breeding plan. The breeding strategy provides a philosophy of the management of genetic improvement while the breeding plan prescribes the 'nuts and bolts' for implementing the selected strategy. Typically, the plan includes a set of objectives and a flow chart of what is to be done each month for several years ahead, and is subject to revision every 2–5 years (Eldridge et al. 1993).

This chapter has been prepared to serve as a guide to eucalyptus oil producers interested in improving the yield and quality of oil through selection and breeding. As will be seen, oil traits in eucalypts have been found to be under moderate to strong genetic control, with substantial levels of genetic variation, and this will facilitate the capture of useful gains from relatively simple, low-input strategies appropriate to the present industry position of a static market and low prices. The chapter highlights the principal features of breeding strategies for improving oil traits in eucalypts and some of their key determinants, drawing on examples from the main oil-producing species where appropriate. In doing so, extensive use is made of three practical texts that are highly recommended to any newcomer to eucalypt breeding: Eldridge et al. (1993), Williams and Matheson (1994) and Cotterill and Dean (1990). These in turn introduce the reader to other literature important to understanding tree breeding and forest genetics.

## Basic concepts

Most worthwhile breeding strategies recommend starting with a well-adapted population with a broad genetic base. This base population is then subjected to a particular method and intensity of selection. Selected trees are then mated to maximise long-term genetic gain and minimise the effects of inbreeding within the limits imposed by human and economic resources.

Selection and mating are key activities in breeding. They accumulate genes which influence yield and adaptation, steadily increasing over successive generations the frequency of superior genotypes. Every successful breeding strategy, therefore, requires efficient methods of selecting superior genotypes. These methods include the progeny tests in which the selection is carried out, appropriate measurement techniques and selection technology (e.g. selection indices). Mating can be done by open pollination or controlled pollination, carefully minimising the potential for inbreeding and allowing for genetic material from other sources to be incorporated. In pursuing its principal functions of efficient selection and mating, a strategy should aim to assess the variation within a species, generate information about it and ensure that genetic resources and variation for future selections are conserved (Barnes 1987, Matheson 1990).

The cyclic or recurrent nature of the selecting, testing and mating processes which are part of an overall breeding strategy is illustrated in Figure 4.1. Every effective breeding strategy involves maintaining a hierarchy of three major types of population which can continue to meet the demand for genetically improved planting stock for the fourth population, the production population (i.e. commercial plantations) (Griffin 1989a, Matheson 1990, Eldridge et al. 1993). These populations are the base, breeding and propagation populations. Awareness of the concept of maintaining distinct types of population within the cycle is essential in planning the operations of genetic improvement (Libby 1973), even if circumstances dictate that some populations are combined in the one planting.

## Breeding objectives

When oil production is the sole objective of a eucalypt plantation, the usual goal of producers is to maximise the yield of high quality oil if it is economically feasible to do so. The yield of oil

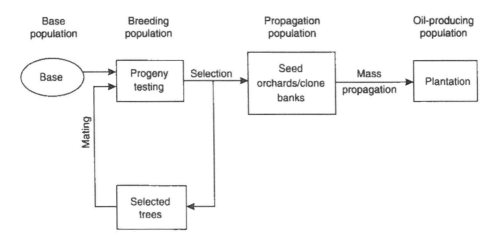

*Figure 4.1* Activity cycle for breeding and the hierarchy of the four populations relevant to breeding strategies and operations.

from a given area of land depends on the species planted, available leaf biomass and the concentration of oil in those leaves at that point in time. Oil composition usually determines quality and will be an important consideration if it is liable to variation in the species under cultivation and affects the marketability or price of the oil produced.

A typical breeding objective for an oil-producing species would therefore be to increase the yield of high quality oil per unit area. This can be done by improving the yield of leaves, and the concentration of oil in those leaves, and ensuring that the quality of the oil produced maximises saleability and price.

## Choice of base material

Correct choice of base material is critical to the success of breeding programmes. There are many examples in eucalypt breeding where breeding has begun with a poor choice of species or provenance and, in others, where the genetic base was much too limited (Eldridge *et al.* 1993). In the following discussion it is assumed that the choice of preferred species for oil production has already been made. The aim, then, is to point out important factors for consideration in acquiring suitable germplasm of a particular species with which to establish breeding populations. Examples pertaining to the main oil-producing species in use today (i.e. *E. citriodora, E. dives, E. exserta, E. globulus, E. polybractea, E. radiata, E. smithii* and *E. staigeriana*), as well as some with high potential (e.g. *E. camaldulensis* and *E. kochii*), are given wherever possible.

Knowledge of the genetic structure of natural populations is important in determining strategies for seed collections for breeding programmes. Studies of the amount and distribution of genetic diversity in eucalypt species, as assessed by allozyme variation, have shown that, on average, 18 per cent of the genetic diversity in the twenty-five species studied to date occurs between populations (Moran 1992). The species with most genetic differentiation between populations are those with regional (i.e. moderately extensive) distributions but with small disjunct populations. Regionally-distributed eucalypts had, on average, a higher proportion of variation between populations (25 per cent) than localised or widespread (14 per cent) species. Of the regionally-distributed species, those with disjunct distributions show greater population differentiation (39 per cent) than those with continuous distributions (11 per cent). Generally, however, most of the allozyme variation occurred within, rather than between, populations.

*Genetic resources*

After defining the objectives of the breeding programme, the next task in tree breeding is to establish a breeding population. Where selection aims to enhance oil production the natural inclination is to start by making selections in local plantations of the preferred oil-yielding species to establish a breeding population. In many instances this is a perfectly good strategy. However, there are some pitfalls in relying solely on existing plantations ('land races') as the primary source of selections for breeding populations. The genetic history of many eucalypt plantings is obscure or unknown. Many have a narrow genetic base and have been derived from a provenance that gives mediocre performance in the environment of the planting site. This situation has existed with many species in numerous countries and trials of broadly based seed collections from the natural provenances of a particular eucalypt have often shown better adaptability and faster growth rate than the local land race, even after genetic improvement of the land race (see examples in Eldridge *et al*. 1993).

Any new programme should start with a thorough review of provenance performance within the region of planting. It may well be found that, given the uncertain origins of the land race and the lack of systematic provenance testing in the past, it is preferable to start again, virtually, by introducing broadly based provenance collections from natural stands of the target species. If there is a clear indication that some provenances or regions-of-provenance are better than others, collections can be concentrated in those areas. These, and collections representing the best elements of the local land race, can then be used to establish large provenance-progeny trials. Such trials have multiple functions including ranking provenances and land races, serving as breeding populations for the first generation, providing resources for selection and breeding activities, and as commercial seed orchards. Several breeding strategies developed for wood-producing *Eucalyptus* species have advocated this approach, for example, *E. globulus* in China (Raymond 1988) and *E. camaldulensis* in Thailand (Raymond 1991).

*Choice of trees within species*

*Chemical forms*

An extreme type of variation within species, commonly found in *Eucalyptus* and in a wide variety of other plant families, is the occurrence of 'chemical forms'. Penfold and Willis (1953) described these as

> plants in naturally occurring populations which cannot be separated on morphological evidence, but which are readily distinguished by marked differences in chemical composition of their essential oils.

They do not differ qualitatively but show marked discontinuous, quantitative variation (Hellyer *et al*. 1969).

Penfold and Morrison (1927) first reported such forms in *E. dives* and called the variants 'physiological forms'. In order to distinguish four separate and distinct oil forms in the species they called them 'Type', 'Variety A', 'Variety B' and 'Variety C' in order of discovery. The use of

*Table 4.1* The range in abundance of four key compounds in the oils of six chemical forms in a population of *Eucalyptus radiata* (Johnstone 1984)

| 1,8-Cineole (%) | α-Phellendrene (%) | Piperitone (%) | Terpinen-4-ol (%) |
|---|---|---|---|
| 2–12 | 4–27 | 21–55 | 2–26 |
| 7–27 | 3–33 | 0.8–10.5 | 2–37 |
| 4–27 | 3–23 | 0–19 | 12–36 |
| 9–23 | 1–7 | 1–6 | 14–28 |
| 30–60 | 3–20 | 0–6 | 2–23 |
| 58–76 | 1–20 | 1–3.5 | 1–6 |

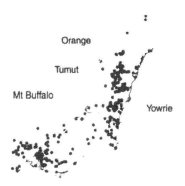

*Figure 4.2* The natural distribution (black dots) of *Eucalyptus radiata* in southeastern Australia as plotted from authenticated botanical specimens, showing the four regions Mt Buffalo, Orange, Tumut and Yowrie (circled) where Johnstone (1984) found most trees to give foliar essential oils of the high-cineole form.

this terminology has now been discontinued in favour of simply referring to the different types as chemical variants, chemical forms, chemovars or chemotypes, and highlighting the major oil component, for example '*E. dives* (cineole variant)' (e.g. see Lassak 1988). Chemical forms do not appear to be the result of site differences, seasonal variation (Simmons and Parsons 1987), leaf ageing effects or hybridisation.

Chemical forms, as highlighted in Chapter 5, abound in several of the principal oil-producing eucalypts: *E. camaldulensis* has five chemical forms, *E. citriodora* four forms, *E. dives* five forms and *E. radiata* six forms. The compositional types of *E. radiata* are shown in Table 4.1. Each form may occur in separate, distinct populations (chemical races), but they can also be present together on the one site and maintain their chemical integrity despite interbreeding. Differences in chemical form do not appear to be associated with differences in oil yield.

In a study of fifty *E. radiata* trees in a single population containing three distinct chemical forms, Whiffin and Bouchier (1992) concluded that the forms appear to be the result of the actions of the enzymes which control terpenoid biosynthesis modifying a basic monoterpene pool. In *E. camaldulensis*, a high-spathulenol form was reported to occur at a frequency of one in ten trees amongst high-1,8-cineole forms at Petford in northern Queensland (Doran and Brophy

1990). In a progeny trial of Petford families, the high-spathulenol forms could not be distinguished from the high-1,8-cineole forms until plants were 18–25 months old, suggesting that diversion of the monoterpene pool is linked to maturation processes and the activation of specific enzymes (Doran 1992).

The potential importance of chemical forms, either positively as a rich source of desirable chemicals or negatively as adversely affecting the quality of oil, has been noted by Hillis (1986). In choosing populations within a species as the genetic resources for a tree-breeding programme, it is preferable to concentrate on populations containing a single chemical form. In *E. radiata*, for example, Johnstone (1984) identified only four populations, from many sampled throughout the range of the species, where most trees were of the commercially valuable high-cineole form (Figure 4.2). These are the obvious populations to target in seed collections for base populations of this chemical form of the species. However, such a strategy may not be feasible when dealing with species or provenances where desirable (that is commercially valuable) and non-desirable chemical forms co-occur, as is the case with Petford *E. camaldulensis*. When dealing with populations with multiple forms, seed trees of the desirable, commercial oil type should be selected and segregating progeny removed from the breeding population as soon as they can be confidently identified.

*Provenances*

When breeding a species for oil production, the primary goals are to maximise the yield of leaves that contain the highest possible concentration of oil of a composition which gains a premium price in the market place. The yield of leaves is usually highly correlated with diameter growth.

There is now ample evidence that much faster growth and greater oil yields of eucalypts in plantations can be obtained through careful selection of provenance and genetic improvement (Eldridge *et al.* 1993). In a study of provenance variation in *E. globulus*, Doran and Saunders (1993) found highly significant differences between provenances, and amongst families within provenances, for three principal oil traits (concentration of oil in the leaf, concentration of 1,8-cineole in the leaf and cineole content of the oil) and for stem diameter at breast height in trees aged 1.75 years (juvenile foliage) and 2.75 years (mature foliage) in provenance/progeny trials in New South Wales and Victoria respectively. The results of the younger trial are summarised in Figure 4.3. Families from the Otways/Lorne and Jeeralang provenances consistently ranked among the best in leaf oil concentrations and were among the fastest growing sources. These would be the provenances favoured to provide candidates for breeding populations should breeding *E. globulus* for oil production in southern Australia be contemplated.

*Families*

Phenotypic (genotype + environment) variation between trees in natural populations in such traits as oil concentration and oil composition is very large. Frequency diagrams for oil concentration and 1,8-cineole content in regions-of-provenance of two important oil-yielding species, *E. polybractea* from West Wyalong and *E. radiata* from Nerrigundah-Yowrie, are given in Figure 4.4. As such traits are highly heritable, it is desirable to screen trees in the native stands for these traits and select as seed trees for breeding populations only those that rank amongst the best for both traits. Breeding programmes for oil production using mallee eucalypts, including *E. polybractea* (Western Australia), and *E. radiata* (Australia and South Africa) have employed this strategy (Donald 1980, 1991, Bartle 1994, Doran *et al.* 1998).

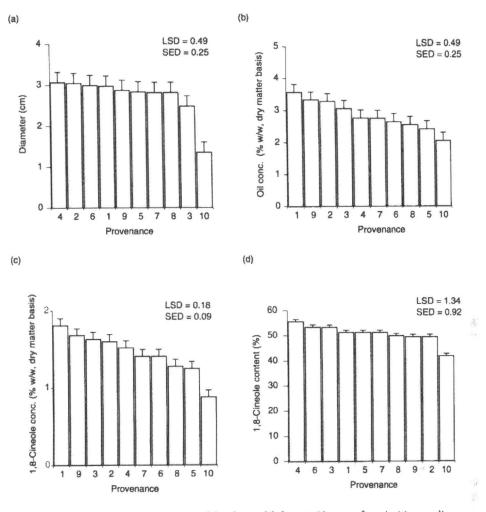

*Figure 4.3* Variation amongst provenances of *Eucalyptus globulus* at 1.75 years of age in (a) stem diameter, (b) oil concentration in leaves, (c) 1,8-cineole concentration in leaves, and (d) 1,8-cineole content of oil. The caps above the bars represent the standard error of the difference (SED) between provenances. Provenance identification: 1 = Otways, 2 = Jeeralang, 3 = West Coast, 4 = Cape Barren, 5 = King Island, 6 = Flinders Island, 7 = Moogara, 8 = Geeveston, 9 = Lorne, 10 = South Gippsland.

*Progeny*

Even when trees have been selected in the natural stands for such traits as oil concentration and cineole content, their open-pollinated seed will give progeny which are highly variable in these traits. This variation provides further opportunity for selection between the breeding population and propagation population phases of the breeding strategy. Figure 4.5 illustrates the variation in oil yield per tree in one family of a southern New South Wales provenance of *E. radiata* (Doran *et al.* 1998).

Typically, there is substantial variation in commercial oil traits at all levels of the selection hierarchy in most, if not all, of the main oil-producing species. A commonly repeated scenario is

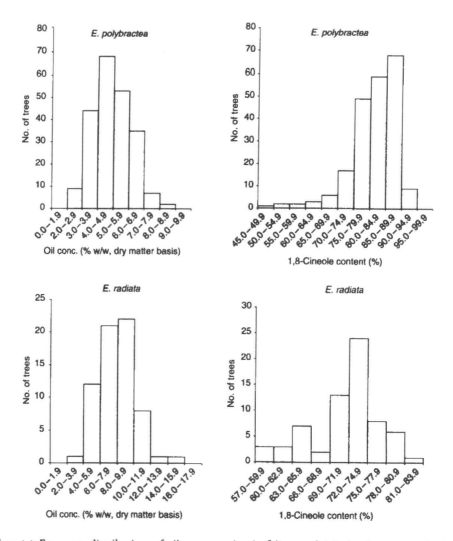

*Figure* 4.4 Frequency distributions of oil concentration in foliage and 1,8-cineole content of oil in *Eucalyptus polybractea* from West Wyalong, New South Wales, and *E. radiata* from the Nerrigundah-Yowrie region of southeastern New South Wales.

that reported for *E. globulus* (Doran and Saunders 1993) and illustrated in Figure 4.6. The high level of variation in oil concentration in all the provenances sampled can be seen. This variation provides the opportunity to select trees for this trait from almost the entire natural range of the species, despite the presence of significant provenance and family within-provenance variation.

## Biological characteristics influencing choice of breeding strategy

### Breeding system and inbreeding depression

Eucalypts have hermaphrodite, protandrous flowers and are pollinated by insects or birds (Griffin 1989a). They reproduce by a mixed mating system, with both outcrossing (where the

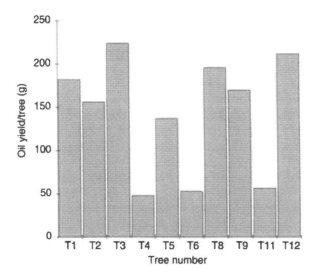

*Figure 4.5* Variation in oil yield per tree in a replicate of one family in a 23-month-old *Eucalyptus radiata* progeny trial.

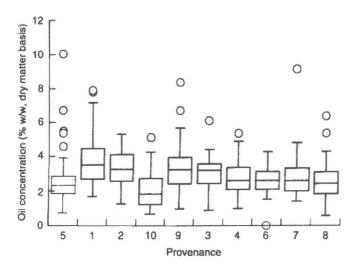

*Figure 4.6* Variation in oil concentration in juvenile leaves in a 1.75-year-old *Eucalyptus globulus* provenance-progeny trial in southeastern New South Wales. The box represents the interquartile range, the whiskers represent the outerquartile range and the circles indicate outliers. The horizontal line within each box is the provenance median value. Provenance names are given in Figure 4.3.

pollen from one tree fertilises the flowers of another tree) and selfing (pollination of an individual tree or clone with its own pollen) (Moran 1992, Moran and Bell 1983). The proportions of outcrosses and inbreeds reported in seed collected from natural populations range from 45 per cent outcrossing in one population of *E. pellita* (House and Bell 1996) to 97 per cent outcrossing in *E. camaldulensis* (P.A. Butcher pers. comm.).

Controlled pollination experiments on *E. regnans* (Eldridge and Griffin 1983) showed that selfed (inbred) seed produced seedlings which grew markedly slower than those from open-pollinated seed from the same parent trees. At age nine years, height was 12 per cent less, and stem diameter 19 per cent less, for trees raised from selfed seed compared with those raised from open-pollinated seed. This is equivalent to a reduction in wood volume of nearly 50 per cent in the selfed trees at this age, compared to open-pollinated seed (which would itself include about 25–30 per cent of selfed seeds and 70 per cent of outcrossed seeds). A few years later nearly all the selfed trees in the trial were suppressed and many were dead. This reduced performance of selfed seed has been shown to hold true for other eucalypts, such as *E. camaldulensis* and *E. grandis*, and for many other tree species (Sedgley and Griffin 1989).

Inbreeding results not only from self-fertilisation, but also from the mating of closely related individuals, such as trees descended from the same mother. Adjacent trees in natural forests will often be closely related. Levels of inbreeding can be reduced in well-designed seed orchards or seed production areas, in which many unrelated genotypes are represented and adjacent trees are not related. In a seed orchard of *E. regnans*, the outcrossing rate was estimated to be 91 per cent, compared with 74 per cent in a nearby natural population (Moran *et al*. 1989).

Cases of a progressive narrowing of the genetic base and associated deterioration in performance are frequent in the history of overseas introductions of eucalypts. For example, a limited introduction of *E. nitens* to South Africa in 1930 produced a land race which was so severely inbred after several generations that it was discarded in favour of new importations of seed from Australia (Purnell 1986).

In summary, selfing in eucalypts has detrimental effects on growth rate and seed set. Seed orchards should have a broad genetic base and adequate cross-pollination between trees to minimise self-fertilisation.

*Genetic control of oil traits*

Several authors have proposed single gene control of the production of particular monoterpenes (e.g. Squillace *et al*. 1980), usually with a dominant/recessive pair of alleles. However, these conclusions have been questioned because of invalidity of the assumption of independence of each variable (Birks and Kanowski 1988). Based on a review of inheritance studies of coniferous resin, Birks and Kanowski (1988) concluded that there was little, if any, unambiguous evidence of simple Mendelian inheritance of resin components. The oil traits of most interest for selection, such as oil concentration and yield of 1,8-cineole, are most likely governed by a complex of genes (Doran 1992).

In breeding programmes for oil production in eucalypts, the main traits for selection are quantitative traits for which the variation is continuous, such as growth rate and oil concentration. Thus the genetics of breeding eucalypts for oils is predominantly quantitative genetics.

Recent studies of the oils of various species of *Eucalyptus* using quantitative genetic methods have indicated moderate to strong genetic control of oil concentration and composition. Available estimates of genetic parameters for commercial traits in eucalypts are summarised below.

*Genetic parameters*

In manipulating quantitative characters the plant breeder deals with the nature, magnitude and inter-relationships of genetic and non-heritable variation (Allard 1960). Defining genetic variation is one of the first tasks of tree breeding. This variation is expressed by estimates of genetic parameters, each of which is 'a numerical quantity which specifies a population in respect to some characteristic' (Allard 1960).

Reliable, unbiased estimates of genetic parameters from well-designed field trials are essential to permit a realistic appraisal of rate and magnitude of improvement by selection and to compare possible alternative strategies (Eldridge *et al.* 1993).

*Additive and non-additive genetic variance*   Genetic variance ($V_G$) is made up of two parts: an additive part ($V_A$), due to average effects of genes, and a non-additive part ($V_{NA}$), due to such deviations from additivity as dominance (the masking effect of one allele by another allele at the same gene locus) and epistasis (interaction of genes at different loci) (Matheson 1990). Thus, $V_G = V_A + V_{NA}$. The gains from recurrent selection are due to the accumulation of genes which act independently (additive effects). Non-additive effects caused by interactions between genes are unavailable in future generations as these effects break down with random assortment and recombination by meiosis during sexual reproduction. Gains from non-additive effects can be exploited only by controlled pollination to produce outstanding specific crosses, or clonal propagation of outstanding individuals.

Non-additive genetic variance for a particular trait, using the concept of 'specific combining ability' (SCA), can be estimated from full-sib progeny tests where controlled pollinations have been carried out in a pattern best suited for such estimates, for example, diallels (all possible combinations of crosses) or partial diallels (a subset of all possible crosses) (Matheson 1990). Here, the expected value for trait $X$ in progeny of the cross of parents A and B is calculated as the population mean for all crosses + general combining ability (GCA) of parent A (i.e. the difference between the population mean and the mean for all crosses involving parent A) + GCA of parent B. The difference between the actual value for trait $X$ and the calculated expected value is the SCA, resulting from non-additive gene action. These deviations from the expected value may be positive or negative; the positive effect is usually of most interest to the tree breeder.

The level of SCA for growth traits in *Eucalyptus* has been discussed by Eldridge *et al.* (1993). There are very few estimates available of SCA for oil traits in this genus, mainly because of the lack of suitable trials. Preliminary estimates of GCA and SCA effects for 1,8-cineole yield in twelve-month-old *E. camaldulensis* progenies in a full-sib progeny test indicated that additive variation exceeded non-additive variation by a ratio of 6:1 (Doran 1992). The three control-pollinated families with the greatest 1,8-cineole yields ranked highest, apparently because at least one of the parents was a good general combiner and moderately high levels of SCA were also present. This result mirrors what has been determined for growth traits, where additive genetic variance is usually greater than the non-additive component, but useful amounts of the latter are present for utilisation in clonal programmes.

*Heritability*   Heritability ($h^2$) in the narrow sense of the term is the ratio of additive genetic variance ($V_A$) to the total phenotypic variance ($V_P$, where $V_P = V_A + V_{NA} + V_E$) (Falconer 1989). $V_E$ is the environmental variance. Put simply, it is a measure of how strongly a trait is influenced by genetics and how much by environment (Hanson 1963). Genetic gain from selection and breeding is potentially high when heritability is high. Heritability estimates are needed if selection is to be assisted by use of an 'index' and are very useful for estimating potential genetic gains from alternative breeding strategies. They are calculated by parent-offspring regression or from variance components estimated by statistical analysis of progeny test data. Ideally, progeny trials test many families and are designed to minimise $V_E$. Eldridge *et al.* (1993) list several factors that need qualification when heritability estimates are used, including acknowledgement that each estimate pertains only to the particular population and environment from which the data used for the calculation were derived.

Estimates for heritability ($h_i^2$ narrow sense, individual heritability) for the principal traits affecting commercial production of eucalyptus oil are given in Table 4.2.

The principal eucalyptus oil traits in the species indicated appear to be under moderate to strong genetic control, with mean individual heritability estimates of 0.11–0.54 for oil concentration in the leaves, 0.27–0.61 for cineole concentration in the leaves and 0.37–0.61 for cineole content of the oil. Most of these estimates are based on relatively small numbers of families and come from open-pollinated progeny trials, sometimes of a design that is not ideal for estimating heritabilities (e.g. Barton et al. 1991, Doran and Saunders 1993). Much remains to be done before reliable estimates of the heritability of oil traits are available for the major oil-producing eucalypts. However, confidence in the available estimates is boosted by their similarity to heritabilities reported for an oil-producing species from a closely related genus, *Melaleuca alternifolia*, the source of Australian tea tree oil. In *M. alternifolia*, heritabilities for oil concentration in the leaves were 0.67 and 0.51, and for cineole content of the oil 0.37, when estimated in

Table 4.2 Individual ($h_i^2$) and family ($h_f^2$) heritability estimates from open-pollinated progeny tests for commercial traits in some *Eucalyptus* species used in oil production[a]

| Trait | Species[b] | $h_i^2$ | $h_f^2$ | Age (yr) | No. of families | Reference |
|---|---|---|---|---|---|---|
| Cineole conc. in leaf[c] | CML | 0.53[d] | 0.62 | 3.75 | 19 | Doran and Matheson (1994) |
| | | 0.61[e] | | 3.75 | 19 | |
| | GLB | 0.34 | | 2.75 | 80 | Doran and Saunders (1993) |
| | | 0.27 | | 1.75 | 80 | |
| | KOC | | 0.83 | 3.8 | 50 | Barton et al. (1991) |
| | POB | 0.36 | | 2.4 | 11 | Grant (1997) |
| Cineole content of oil[f] | GLB | 0.48 | | 2.75 | 80 | Doran and Saunders (1993) |
| | | 0.37 | | 1.75 | 80 | |
| | POB | 0.61 | | 2.4 | 11 | Grant (1997) |
| Oil conc. in leaf[c] | CML | 0.54[d] | | 3.75 | 19 | Doran and Matheson (1994) |
| | | 0.42[e] | | 3.75 | 19 | |
| | GLB | 0.11 | | 2.75 | 80 | Doran and Saunders (1993) |
| | | 0.27 | | 1.75 | 80 | |
| | POB | 0.36 | | 2.4 | 11 | Grant (1997) |
| Stem volume per tree | GLB | 0.19 | | 6.0 | 45 | Volker et al. (1990) |
| Tree height | GLB | 0.12 | | 6.0 | 45 | Volker et al. (1990) |
| | | 0.29 | | 4.0 | 36 | Borralho et al. (1992) |
| Stem diameter[g] | GLB | 0.24 | | 6.0 | 45 | Volker et al. (1990) |
| Stem sectional area[g] | GLB | 0.13 | | 4.0 | 36 | Borralho et al. (1992) |
| Stem form | GLB | 0.22 | | 6.0 | 45 | Volker et al. (1990) |

a A coefficient of relatedness of 0.4 was used in calculating $h_i^2$ for oil traits.
b CML = *E. camaldulensis*, GLB = *E. globulus*, KOC = *E. kochii*, POB = *E. polybractea*.
c , w/w (dry matter basis).
d, e Mtao and Forest Hill, respectively, Zimbabwe.
f Relative abundance, .
g Measured at breast height.

two well-designed provenance-progeny trials in northern New South Wales (Butcher *et al.* 1996, Doran *et al.* 1997).

The oil yield to be expected from a plantation is the product of leaf biomass and oil concentration. Leaf biomass of an individual tree is typically highly correlated with stem diameter and height, as demonstrated by the significant correlations between the two indirect measures of leafiness, crown surface area and crown density, and stem diameter and height for *E. camaldulensis* shown in Table 4.3a. Heritability estimates for stem diameter and height for eucalypts generally fall in the low-to-moderate range, as exemplified by the figures for *E. globulus* given in Table 4.2. Nevertheless, useful gains can be made by selection and breeding in these traits and they should be included along with oil characteristics in the list of traits for selection in breeding programmes for oil production.

*Genetic and phenotypic correlation* Genetic correlations ($r_g$) are calculated as the correlation of the breeding values of two traits and express the extent to which these traits are influenced by the same genes (Falconer 1989). Genetic correlations are important for predicting the magnitude and direction of response in one trait to selection for another, as well as for a number of other useful purposes (Williams and Matheson 1994). Genetic correlations have some influence on the best way to structure breeding populations. For example, where there are strongly negative genetic correlations between two traits, it is very difficult to breed for them both in the one population and it may be necessary to divide the population. Strong positive genetic correlations, on the other hand, are useful in reducing the number of traits for selection, as selection for one of the correlated traits will be effective in improving the other trait. Phenotypic correlations ($r_p$), which include genetic and environmental components, are simply calculated from the corresponding measurements of two traits and are of much less importance to the tree breeder. Sometimes there is a large difference between genetic and phenotypic correlation coefficients,

*Table 4.3a* Estimates of genetic (above the diagonal) and phenotypic (below the diagonal) correlations between growth and oil traits for 3.75-year-old Petford *Eucalyptus camaldulensis* at Mtao, Zimbabwe (Doran and Matheson 1994)

|  | Tree height | Stem diameter[a] | Crown surface area | Crown density | Cineole conc. in leaf[b] | Oil conc. in leaf[b] |
|---|---|---|---|---|---|---|
| Tree height |  | 0.818 (0.139)[c] | 0.666 (0.202) | 0.481 (0.303) | 0.044 (0.295) | −0.481 (0.260) |
| Stem diameter[a] | 0.769 |  | 0.816 (0.171) | 0.477 (0.348) | −0.100 (0.358) | −0.456 (0.333) |
| Crown surface area | 0.618 | 0.751 |  | 0.467 (0.344) | −0.104 (0.341) | −0.466 (0.311) |
| Crown density | 0.421 | 0.482 | 0.445 |  | 0.233 (0.371) | −0.139 (0.385) |
| Cineole conc. in leaf[b] | 0.041 | −0.071 | −0.079 | 0.185 |  | 0.803 (0.109) |
| Oil conc. in leaf[b] | −0.427 | −0.350 | −0.365 | −0.096 | 0.806 |  |

a Measured at breast height.
b , w w (dry matter basis).
c Figures in parentheses indicate standard errors for genetic correlations.

even to the extent that one might show a positive association between traits while the other is negative (Eldridge et al. 1993).

There are few examples of genetic correlations for oil traits in the literature. Table 4.3a shows estimates of phenotypic and genetic correlations from a progeny test of E. *camaldulensis* (Doran and Matheson 1994). The estimated genetic and phenotypic correlations between growth traits and 1,8-cineole yield were small with large standard errors for the genetic correlation coefficients. Although these estimates lacked precision, it appears that these characters are weakly associated and selection for one is unlikely to affect the other, either adversely or favourably. However, oil concentration, as estimated by the concentration of monoterpenoids in leaves, and growth traits showed moderate and unfavourable genetic and phenotypic correlations. These genetic correlations had very large standard errors and were considered unreliable by the authors. Since maximising yields of 1,8-cineole was the more important breeding objective, no adjustment of the breeding strategy to cater for the adverse genetic correlation was advocated. 1,8-Cineole yield and oil concentration were, as expected, strongly associated, with an $r_g$ of 0.80 (se = 0.11) and an $r_p$ of 0.81. The positive genetic and phenotypic correlations between stem diameter and height and the crown traits of surface area and crown density were considered beneficial because total oil production depends not only on oil concentration in the leaves but also on total leaf biomass.

Table 4.3b shows estimates of phenotypic and genetic correlations from a progeny test of E. *polybractea* (Grant 1997). Here there was a strong positive genetic correlation between oil yield per tree and leaf biomass ($r_g$ of 0.85). The large negative genetic correlations between oil traits and the average area of individual leaves are in agreement with other data, indicating that the larger-leaved forms of the species are poorer oil producers than the narrow-leaved forms.

In a provenance/progeny trial of *Melaleuca alternifolia*, Butcher et al. (1996) reported a strong negative genetic correlation ($r_g$ of −0.42) between oil yield and plant dry weight. In breeding to improve oil production in M. *alternifolia*, and faced with an apparent adverse genetic correlation, these authors recommended combined index selection (Cotterill and Dean 1990) with a Kempthorne type restriction (Cotterill and Jackson 1981, Cotterill and Dean 1990) imposed on plant dry weight to prevent decline in this trait as a means of maximising genetic gains. In another trial of this species, basal stem diameter, a trait closely correlated with plant dry weight, showed no adverse genetic correlation with oil yield ($r_g$ of −0.14), reinforcing the qualification that genetic parameters are specific to the population of trees from which the data were collected and the conditions of the planting site (Doran et al. 1997).

*Table 4.3b* Estimates of genetic (above the diagonal) and phenotypic (below the diagonal) correlations between growth and oil traits for 2.4-year-old *Eucalyptus polybractea* at West Wyalong, New South Wales, Australia (Grant 1997)

|  | Oil conc. in leaf[a] | Cineole conc. in leaf[a] | Cineole content of oil[b] | Oil yield per tree | Area per leaf | Leaf biomass |
|---|---|---|---|---|---|---|
| Oil conc. in leaf[a] |  | 0.919 | 0.379 | 0.738 | −0.295 | −0.174 |
| Cineole conc. in leaf[a] | 0.917 |  | 0.684 | 0.512 | −0.361 | −0.016 |
| Cineole content of oil[b] | 0.317 | 0.648 |  |  | −0.338 |  |
| Oil yield per tree | 0.494 | 0.428 | 0.155 |  | 0.148 | 0.846 |
| Area per leaf | −0.342 | −0.455 | −0.449 | 0.007 |  | 0.395 |
| Leaf biomass | 0.027 | −0.009 | −0.065 | 0.836 | 0.215 |  |

a , w/w (dry matter basis).
b Relative abundance, .

*Juvenile–mature correlation (age–age correlation)*   Juvenile–mature correlation, also known as age–age correlation, is a form of trait–trait genetic correlation (Eldridge *et al*. 1993). When there is a large positive genetic correlation between a trait in young trees and the same trait in older ones, selection can be made at an early age and the rate of genetic gain is correspondingly improved.

Attempts to select eucalypts for oil characteristics in the nursery and at very young ages in plantations have been largely unsuccessful. Barton *et al*. (1991) found no qualitative variation between parents and progeny in *E. kochii*, but they did report significant quantitative variation with the age of the plant. Juvenile leaf from six-month-old seedlings grown in the glasshouse was lower in cineole and higher in terpene hydrocarbon yields than leaf from the parents. The same seedlings planted in a field trial were sampled at eighteen months and three years: cineole yield increased steadily and at three years closely resembled parent values. A similar result was reported for *E. citriodora*, where it took three years of growth for oil and citronellal content of progeny to attain stable levels (Mascarenhas *et al*. 1987).

Doran (1992) studied the variation in 1,8-cineole concentration in controlled cross progeny of *E. camaldulensis* at regular intervals up to twenty-five months from planting. Correlations between offspring and mid-parent values were poor ($r = 0.23$–$0.55$) until twenty two months, when $r$ reached 0.76. Correlation improved to 0.86 at twenty-five months, although the level of cineole in the progeny of several crosses was still increasing, suggesting that further changes in ranking with age might take place. This study included progeny of a distinctive high-spathulenol chemotype. These non-commercial chemotypes could not be distinguished from the high 1,8-cineole forms until plants were 18–25 months old, an observation which provided another incentive to delay selection until at least twenty-five months.

It appears that the earliest age that trees in eucalypt breeding populations can be reliably ranked for oil traits is about three years. This is still relatively early as general heavy flowering in most species would not occur until some years later.

*Genotype × environment interaction*   The statistical expression, genotype × environment interaction (GEI), is applied when genotypes grow differently relative to each other in different environments. It is estimated by analysing data from two or more trials of the same clones or seedlots and by joint regression analysis to obtain regression coefficients to measure the stability of families across sites for the trait in question (Matheson and Cotterill 1990, Williams and Matheson 1994).

In a study of variation in the two-year growth of extensive provenance-progeny trials of *E. camaldulensis* on three sites in Thailand, Pinyopusarerk *et al*. (1996) reported significant site × family interaction for height, stem diameter and stem volume growth. A marked difference in stability of families across sites was found and provenances from Petford and the Thai land race were more unstable than other provenances, that is they performed very well at good sites but not so well (in relative terms) at poorer sites. This mirrors what has been reported for other tree species such as *Pinus radiata* (Matheson and Raymond 1984). The results have implications for future selection and breeding of *E. camaldulensis* in Thailand. For example, it may be necessary to breed for commercial traits in different environments using a multiple-population breeding option. If a common breeding population is to be adopted it will be necessary to place more emphasis on selecting the seedlots that have shown good general adaptability over several test sites.

In breeding for oils, the extra analytical costs of assessing oil traits is a significant constraint on dividing the breeding population into regional groups, or into sublines. The breeder will be looking for genotypes which are relatively stable over all the proposed planting sites for manipulation in a single breeding population. The critical trials to determine if this is a viable option, ideally using clones on a range of sites, have still to be established for most of the major oil-producing eucalypts.

*Hybridising ability*

Griffin *et al.* (1988) reviewed the occurrence of natural and manipulated inter-specific hybrids within the genus *Eucalyptus*, and confirmed the long-standing hypothesis that within sub-genera there are generally no strong barriers to the production of hybrid seed following cross-pollination. Hybrids may be desirable because they are heterotic or because they combine traits that were not found together in either parental species (Griffin 1989b). Sites which are marginal for pure species have provided, so far, the most successful habitats for use of eucalypt hybrids. For example, hybrids of *E. grandis* with hardier species such as *E. urophylla*, *E. camaldulensis* and *E. tereticornis* are showing great promise on sites in South Africa which, because of drought, are marginal for growing pure *E. grandis* (Van Wyk *et al.* 1989).

Leaf oil composition has sometimes assisted botanists to detect eucalypt hybrids, for example, *E. ovata* × *E. crenulata* (Simmons and Parsons 1976) and *E. alpina* × *E. baxteri* (Whiffin and Ladiges 1992). Pryor and Bryant (1958) showed that the essential oils of interspecific hybrid progeny vary greatly but usually give compositions which are intermediate between those of the parents. Transgressive compounds, which occur in higher proportions than in either parent, have been reported in *E. ovata* × *E. crenulata* (Simmons and Parsons 1976), possibly resulting from enzyme complementation.

To date, there has been little interest in applying hybridisation to the improvement of oil-producing eucalypts. An exception is the interest being shown in Brazil in the hybrid between *E. citriodora* and *E. torelliana* (L. Couto pers. comm.). Here, the aim is to incorporate the better crown shape and density and coppicing ability of *E. torelliana* into the breeding programme. Whether or not this can be achieved by recurrent selection for general combining ability (additive gene effects) in both species or whether it requires the more complex procedures of reciprocal recurrent selection (Nikles 1993, Dieters *et al.* 1995) has still to be determined.

*Ease of vegetative propagation*

In tree breeding, vegetative propagation is commonly used for the establishment of clone banks and clonal seed orchards, and can be utilised for the production population if the species in question lends itself to mass propagation by this means. Clonal forestry generally relies on propagation by stem cuttings rather than micropropagation, which is usually too expensive (Haines 1994). Clonal forestry is now practised with several eucalypt species (e.g. *E. camaldulensis*, *E. globulus*, *E. grandis*, *E. gunnii*, *E. tereticornis*, *E. urophylla* and their hybrids) and work is continuing to master the technique with other commercial species such as *E. nitens*. A summary of the methods employed with different species is provided by Eldridge *et al.* (1993).

Of the important oil-yielding eucalypts, mass propagation by cuttings is currently practised only with *E. camaldulensis*. Doran *et al.* (1992) and Doran and Williams (1994), for example, describe a programme of selection for rooting ability in combination with growth and oil traits within the tropical *E. camaldulensis* provenance of Petford. The use of cuttings is made more attractive in this species by the substantial and quick gains that can be made in yield and oil quality, as these traits are highly variable and at sub-optimal levels in unselected seedling stock. The high overall selection intensity required, due to selection simultaneously for several traits, and particularly rooting ability (only 30 per cent of clones passed the selection criteria for rootability), reinforced the need to start with a broad genetic base.

Micropropagation by tissue culture of genotypes of *E. radiata* selected for oil production in South Africa was recommended by Donald (1991) as a means of improving oil yield in this species. This was an alternative to striking cuttings from coppice shoots which is difficult in this species. However, the expense of micropropagation is unlikely to be justified in the present market. Vegetative propagation as a tool in improving oil yields, therefore, is likely to be practised on only a small scale at present. It does, however, offer an excellent opportunity with easily rooted species to quickly exploit the gains from breeding by employing mass vegetative propagation to establish the production population. The costs and benefits of such a programme need first, of course, to be carefully considered.

*Flowering and seed production*

In their natural environment, most eucalypts are prolific seed producers and it is generally not difficult to produce adequate quantities of seed from open-pollinated orchards, as long as certain conditions are met. Selection of a site where the species will produce abundant seed is essential, and at present can only be done by observing seed production of the species on a range of sites. There are several documented examples in which eucalypt seed orchards have produced very poor seed crops, often when the species is planted on sites favourable for growth within its natural range (Eldridge *et al.* 1993). When planted as an exotic, seed production may occasionally be suppressed. Thus, at least one eucalyptus oil producer in Brazil purchases *E. globulus* seed from Australia because of the species' reluctance to seed locally, and *E. smithii*, too, is shy to seed in Swaziland (J. Coppen pers. comm.). The use of bee-hives to encourage cross-pollination in orchards, and therefore greater seed set, is recommended (Moncur 1991); multiple colonies are better than one as they promote competition between bees.

The earlier an orchard flowers the sooner a breeder can move from the first to the second and subsequent generations. Some species, such as *E. nitens* and *E. regnans*, cannot be relied upon to flower until aged six or more years. Application of a growth regulator which stunts vegetative growth but encourages abundant early flowering is a major breakthrough which speeds genetic improvement of these and many other species (Reid *et al.* 1995).

## Selection criteria and selection indices

The following selection criteria are pertinent to improving oil production in eucalypt species:

1  High leaf biomass, both as a sapling and when coppiced
2  High concentration of oil in the leaves
3  Good coppicing ability
4  Freedom from insect pests and diseases
5  Broad adaptability
6  High oil quality, that is, an oil that is attractive to the market.

The first five criteria are aimed at maximising the oil yield per hectare, while the sixth seeks to produce an oil that is not only acceptable to the market but commands the best possible price. This is a large number of traits to select for simultaneously. Genetic gain in individual traits generally tends to diminish as the number of traits under selection increases. In an index combining two, three, four and five traits, the gain in each trait will be about 70, 60, 50 and 45 per cent, respectively, of the gain expected under selection for a single character alone, assuming traits are uncorrelated (Cotterill and Dean 1990). In practice, therefore, breeders must try to minimise the number of traits under selection. In selection for oil production it may be sufficient to select for stem sectional area (diameter), as this measure is often highly correlated with biomass production and coppicing ability, and oil concentration, while setting limits for oil quality (e.g. for medicinal, cineole-rich oils ⩾70 per cent cineole, <1 per cent α-phellandrene).

The efficiency of selection may be improved by utilising a multi-trait selection index. Smith–Hazel indices are often used and incorporate information on heritabilities, genetic correlations and phenotypic correlations, but the calculations are complex (Cotterill and Dean 1990). If the key traits in the index are assumed to be uncorrelated, then simple indices such as the primary index can work just as well for selection in first-generation breeding populations and are far easier to calculate (Harwood and Aken 1997). A primary index incorporating the key oil traits of stem sectional area and oil concentration might take the form:

$$\text{Index value} = w_1 h_1^2 (P_{SA}/SD_{SA}) + w_2 h_2^2 (P_{OC}/SD_{OC})$$

where $w_1$ and $w_2$ are the economic weights assigned to the two traits; $h_1^2$ and $h_2^2$ are the heritabilities; $P_{SA}/SD_{SA}$ is the individual (block adjusted) phenotypic value for stem sectional area normalised by dividing it by the standard deviation of that trait; and $P_{OC}/SD_{OC}$ is the individual (block adjusted) phenotypic value for oil concentration in the leaves, again normalised by dividing it by the standard deviation of that trait.

## Sampling and analytical considerations when selecting eucalypts for oil traits

To determine the extent of variation in oil traits due to genetic sources it is important to avoid or minimise the confounding influence of non-genetic sources of variation. This is achieved by use of appropriate sampling and analytical techniques. Failure to apply suitable methodological and sampling controls is, unfortunately, a feature of many published studies of essential oils, rendering them of dubious value. The following factors should be considered in any studies of inter- or intra-specific variation in eucalyptus oil.

### *Variation due to ontogeny, age of plant and phenology*

Eucalypts develop up to five distinct morphological types of leaf (viz., cotyledons, seedling leaves, juvenile leaves, intermediate leaves and adult leaves) during their lifetime, each corresponding to a certain ontogenetic stage in their development. Oil concentration in seedling leaves is invariably much lower than that of the other stages, while the comparison between the juvenile and intermediate phases and the adult are inconsistent and appear to depend largely on species (Doran 1991). In some eucalypts (e.g. *E. polybractea*), the harvesting of coppice shoots results in higher oil concentrations than are achieved from mature crowns (Brooker *et al.* 1988). Oil composition may also change with ontogeny. Barton *et al.* (1991) found that juvenile leaves from six-month-old seedlings of *E. kochii* grown in a glasshouse were lower in 1,8-cineole and higher in terpene hydrocarbons, in comparison to their parents. Coppen and Jacovelli (1992) studied oil yields and oil composition of juvenile and intermediate/adult leaves of a commercial sixteen-month-old *E. smithii* seedling crop in Swaziland. In ten separate trees sampled, a consistent pattern of decreasing oil yield in the leaf and decreasing cineole content of the oil was obtained as one progressed up the tree.

The poor juvenile–mature correlation in oil traits has already been discussed. To better ensure the reliable ranking of a plant's oil-production potential, sampling early ontogenetic stages, including early coppice leaves, should be avoided. In progeny trials it is best to wait until plants are at least three years of age before screening and ranking plants for oils.

Although there are no known reports of the effects of flowering on oil traits in eucalypts, evidence from other genera suggests that oils traits are highly variable immediately before and during flowering, so sampling during this period is probably best avoided.

## Variation due to leaf age and position in the crown

Physiological leaf age is commonly denoted by the terms 'young leaf' (growing tips), 'mature leaf' (mean age of about six months) and 'aged leaf' (*ca.* 12–18 months). The contrast in oil concentration between young and mature leaves is often marked. Aged leaves generally yield less oil than recently matured leaves. In *E. camaldulensis*, the concentration of 1,8-cineole increased with increasing leaf age down the branch, reaching the highest levels in fully expanded, non-lignified leaves 3–4 months of age. Thereafter, cineole concentration declined rapidly and reached stable levels with leaf lignification (Doran and Bell 1994). Simmons and Parsons (1987) concluded that the oil concentration at various stages of leaf maturity appears to be determined by a complex pattern of quantitative change in individual or groups of compounds, some remaining constant, some increasing and some decreasing with age. Patterns vary between individuals of the one species on the one site and appear dependent on the genetic constitution of individuals.

Bryant (1950) drew attention to the largely unavoidable, but potentially confounding, influences of the position of the leaf on the tree and the influence of light intensity. Light has been shown to be a key factor influencing oil production in a number of essential oil-producing taxa. Several studies have found that the direction (aspect) faced by sampled branches and the sampling height do not have a significant influence on oil concentration (e.g. Brooker *et al.* 1988, Doran and Brophy 1990), although the findings of Coppen and Jacovelli (1992), noted above, suggest that the age of the trees, or the particular species concerned, may be important.

When sampling leaves for genetic studies, therefore, it is important to limit these sources of variation by attempting to collect only fully expanded and mature leaves of similar physiological age across the entire progeny trial. As an added precaution, shaded parts of the canopy should be avoided.

## Seasonal variation

There is a relatively extensive literature on seasonal variation, no doubt because of the commercial imperative of determining the harvest times that return the greatest yields of oil for a particular species. However, inappropriate sampling procedures which confound seasonal variation with other key sources of variation, such as genotype, leaf age and ontogeny, are an unfortunate feature of much of this literature.

A study of three eucalypt species (*E. delegatensis*, *E. globulus* and *E. nitens*) in which three sources of variation (seasonal, leaf ageing and ontogenetic) were studied at regular intervals over twelve months on labelled shoots of marked trees showed that (i) variation in oil concentration and composition occurred during the active growing season only, (ii) oil concentration increased steadily during the growing season and was at its maximum during autumn, when the proportion of newly matured leaves was at its peak, and (iii) the oils of individual species and provenances differed markedly in the patterns of change in proportions of different compounds throughout the growing season (Haifeng Li 1993).

One generally consistent result from the studies undertaken to date is that genotypic differences outweigh any environmental effects on both oil composition and concentration. Encouragingly for reliability of selection for oil concentration in *E. camaldulensis*, individuals with high oil concentration in one season maintained their superiority over several seasons, despite there being marked shifts in oil concentration over time (Doran and Bell 1994). These authors concluded that, although the season of sampling appears to be of less importance to reliability of rankings than factors such as leaf age, greater reproducibility of results can be attained if fully matured leaves are sampled during the dormant or slower growing period of the year.

## Techniques of oil extraction and chemical and statistical analysis

### Oil extraction

Because of the large number of samples required for reliable estimates of genetic parameters, some workers have replaced steam distillation, the conventional method of extracting volatile oils from plants, by the speedier solvent extraction (Ammon *et al.* 1985b). Both methods have advantages and disadvantages and both are costly in terms of labour and materials. It is advisable to carry out some preliminary work to determine which is the best method for the particular situation before embarking on a large-scale screening programme.

Whenever possible, the moisture content of the leaves being extracted should be determined and oil yields reported on a moisture-free (or 'dry matter') basis. Only in this way can meaningful comparisons be made between different samples, and samples analysed at different times and by different workers. Leaves which are picked and distilled 'fresh' can vary significantly in their moisture content depending on the time of year, time between picking and distillation, conditions of storage, etc., and oil yields reported on a fresh basis, particularly when determined by steam distillation, can only be regarded as indicative. Azeotropic distillation of the water contained in the leaves using a Dean and Stark apparatus is a simple, though time-consuming, method for determining moisture content and preferable to oven drying, which may result in loss of volatiles other than water. Ammon *et al.* (1985a) describe an alternative method, suitable for use with their solvent extraction technique for *Eucalyptus* oil analysis (Ammon *et al.* 1985b).

Whatever method of oil extraction is employed, it should be appreciated that yields determined in the laboratory represent an 'exhaustive' extraction. Yields likely to be obtained on 'fresh' leaf using similar material but in a large-scale, commercial context will inevitably be less than these.

### Chemical analysis

Essential oils are generally mixtures of terpenes, often quite complex. However, their relative volatility means that they can be separated by a technique known as gas chromatography (GC). GC is a standard analytical tool which, when optimised, not only separates the constituents of the mixture but quantifies them as well. In competent hands it is very powerful, especially when linked with other instrumental methods such as mass spectrometry (GC-MS). However, the inexperienced investigator can encounter many problems leading to inaccurate results. Such work is best done by a laboratory specialising in this technique and with experience in essential oil analysis.

### Statistical analysis

Quantitative genetic methods are appropriate for estimating genetic parameters of oil traits. However, since classical methods of quantitative genetics assume that the frequency distribution of a metric trait approximates the normal (Falconer 1989), their application to compositional data is inappropriate when oil components are not independent and not normally distributed

(Birks and Kanowski 1988). Birks and Kanowski (1993) recommend analysis of log ratios of proportional data (i.e. where constituents are reported in terms of relative abundance) in order to remove the constraint of summation to unity and provide an interpretable covariance structure, but this technique does not always overcome the problem of skewed distributions which occur when a eucalypt has different chemical forms. The chemical form contributing to the skewed distribution of the data may need to be identified and deleted from the data set.

## Case study: *Eucalyptus camaldulensis*

An outline of the breeding strategy proposed by Doran (1992) for improving oil production in Petford *E. camaldulensis* is given below. In developing the strategy it was assumed that increased wood production would be the principal breeding objective, with improvements in oil production a secondary aim.

### *Improving growth traits*

The starting point, therefore, is the 'main' strategy directed at improving growth traits. For this, a simple strategy, defined as recurrent selection for general combining ability with open pollination in a single population keeping maternal identity, was recommended. This strategy combines the base, breeding and propagation populations in a single plantation on each site and for each generation. These plantations serve sequentially as progeny tests of trees selected in the previous generation, as a resource for selection and breeding for the next generation, and finally as commercial seed orchards. They also provide estimates of genetic parameters (e.g. heritabilities, genotype × environment interactions, age–age correlations) and may be used to identify material to place in a vegetative propagation programme. Breeding strategies employing this basic approach have been developed for *Acacia*, *Casuarina*, *Eucalyptus*, *Melaleuca* and *Grevillea* species in many countries over the last decade (Harwood 1996) and very useful gains have been achieved. For example, *Eucalyptus grandis* was bred using a similar strategy in Florida where, in four generations over some sixteen years, genetic gains of 159 per cent in stem volume production and a substantial increase in cold tolerance were reported (Meskimen 1983, Eldridge *et al*. 1993). The strategy is shown diagrammatically in Figure 4.7.

### *Improving oil traits*

Controlled pollination, to establish an elite population for oil production and for vegetative propagation, was recommended as a practical means of improving oil traits in parallel with the main aim of improving growth traits.

By four years of age the 'main' breeding population will have been gradually thinned down on the basis of the key selection criteria of health, vigour and stem form. In the programme given in Figure 4.7, which has many individual seedlots under test, we might expect the five best-growing individuals of 350 of the original 400 open-pollinated families to be retained at this stage (i.e. 1750 progenies). These trees would be screened individually for oil concentration and the abundance of 1,8-cineole in their oil. The forty trees assessed as having the best general combining ability (GCA) for oil traits would be crossed at the earliest opportunity following selection using an assortative mating pattern. The controlled-pollinated families generated would be raised in progeny tests and the best individuals in the best families selected by combined index selection, giving oil traits and growth traits equal economic weight. These

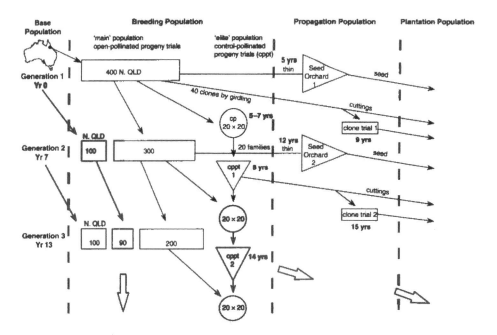

*Figure 4.7* Schematic diagram of the breeding strategy proposed for a tropical provenance (Petford) of *Eucalyptus camaldulensis*. The strategy combines an open-pollinated breeding strategy for growth traits with the establishment of an 'elite' population by controlled pollination for improvement of oil traits, together with clonal testing and deployment (adapted from Doran 1992 and Eldridge 1995).

selections would then be available for mass vegetative propagation. The tendency of the nucleus towards inbreeding would be reduced by including selections from new families in later generations of the 'main' breeding population. In addition, the forty parents selected as the basis for the elite population could be propagated vegetatively through girdling and collection of cuttings from basal shoots (Kijkar 1991). As rooted cuttings, they would be established in clonal tests. Selections that root readily and grow well might be used immediately for clonal oil-producing plantations.

This strategy has a number of advantages over the alternative one of attempting to improve oil traits in the 'main' breeding population. Selection for oils is delayed until the fourth year, when the initial thinning for growth traits in the 'main' population has been completed. At this stage the worst families have been removed and the poorest-growing individuals in the selected families have also been culled (Figure 4.8). The costly chemical analytical work and selection for oil traits is limited, therefore, to those families and individuals that have grown well to half rotation age and produce large quantities of leaf biomass. Four years is also a sufficient age for trees to express their mature oil characteristics and oil-yielding potential.

Climbing to pollinate the ortet at this age could be a problem as the trees will be quite tall (*ca.* 12 m). Scaffolding could be erected around each tree to make the operation safer, or the selected trees could be grafted to a clone bank to make pollination safer and easier. However, this would add at least two years, and possibly three or four years, to the interval between generations (Eldridge 1995).

*Figure 4.8* Seedling seed orchard of *Eucalyptus camaldulensis* at Ratchaburi, Thailand, 5.8 years from planting and after the second thinning (photo: K. Pinyopusarerk).

## Expected gain

The genetic gains per generation expected from mass selection to improve oil traits such as concentration and abundance of 1,8-cineole can be determined thus:

$$\Delta G = i h_i^2 \sigma_p$$

where $i$ is the standardised selection differential and $\sigma_p$ is the phenotypic standard deviation.[1]

One of the main advantages of the elite population strategy for improvement of oils over that of trying to breed for these traits in the 'main' breeding population is that favourable genetic parameters for these characters can be harnessed to give a high rate of gain. The case study given above may be used as an example. The forty trees with the best 1,8-cineole yields in the leaves were selected from the 1750 trees remaining in the 'main' population after first thinning. If these selections were to be simply interbred in a conventional clonal seed orchard using open pollination, it would lead to a production plantation with 39 per cent more 1,8-cineole in its leaves. This estimate is based on the genetic parameters and phenotypic standard deviation for this trait in trials in Zimbabwe reported by Doran and Matheson (1994).

The elite population strategy of assortative crossing of best (GCA) mates and testing in progeny tests, however, can be confidently predicted to lead to clones of an oil-yielding capacity far greater than that of their parents. For example, Doran (1992) reported on the progeny of a high oil-yielding controlled cross in a field trial in Queensland, Australia, which had rapid height

---

[1] Note that this formula applies only to mass selection (i.e. selection among individual trees) without reference to pedigree (i.e. selection within and between families).

growth in combination with 1,8-cineole yields consistently above 4 per cent (w/w, dry matter basis). This was 55 per cent greater than the mid-parent value and about double the oil concentration of 'average' trees. The use of selections like this in clonal plantations of this species would improve oil yields dramatically.

The relatively simple strategy described above suits the circumstances in countries such as Thailand, where there is a major breeding programme of tropical provenances of *E. camaldulensis* (Raymond 1991, Eldridge 1995). This is not to imply that this is the 'best' or 'only' strategy available for use with this species. Each strategy must be tailored to suit the relevant circumstances, with due consideration to the available base material, the biological constraints of the species being bred, and the human, physical and financial constraints on the programme. Eldridge *et al.* (1993) provide examples of other strategies developed by breeders for the improvement of wood production and some other traits (e.g. cold tolerance) in eucalypts.

## Acknowledgements

I would like to thank my colleagues, Alan Brown, Chris Harwood, Carolyn Raymond, Emlyn Williams and Roger Arnold, for providing helpful comments during the development of this chapter. Kron Aken, Maurice McDonald and Sarah Doran assisted with the Figures.

## References

Allard, R.W. (1960) *Principles of Plant Breeding*, Wiley, New York.
Ammon, D.G., Barton, A.F.M., Clarke, D.A. and Tjandra, J. (1985a) Rapid and accurate chemical determination of the water content of plants containing volatile oils. *Analyst*, 110, 917–920.
Ammon, D.G., Barton, A.F.M., Clarke, D.A. and Tjandra, J. (1985b) Rapid and accurate determination of terpenes in the leaves of *Eucalyptus* species. *Analyst*, 110, 921–924.
Barnes, R.D. (1987) The tree breeding programme in Zimbabwe: plan of work for 1987/88. Oxford Forestry Institute, Oxford, UK.
Bartle, J. (1994) New horizons for forestry: tree crops for the wheatbelt. *Newsletter of the Institute of Foresters of Australia Inc.*, 35(2), 4–7.
Barton, A.F.M., Cotterill, P.P. and Brooker, M.I.H. (1991) Heritability of cineole yield in *Eucalyptus kochii*. *Silvae Genetica*, 40, 37–38.
Birks, J.S. and Kanowski, P.J. (1988) Interpretation of the composition of coniferous resin. *Silvae Genetica*, 37, 29–39.
Birks, J.S. and Kanowski, P.J. (1993) Analysis of resin compositional data. *Silvae Genetica*, 42, 340–350.
Borralho, N.M.G., Kanowski, P.J. and Cotterill, P.P. (1992) Genetic control of growth of *Eucalyptus globulus* in Portugal. 1. Genetic and phenotypic parameters. *Silvae Genetica*, 41, 39–45.
Brooker, M.I.H., Barton, A.F.M., Rockel, B.A. and Tjandra, J. (1988) The cineole content and taxonomy of *Eucalyptus kochii* Maiden & Blakely and *E. plenissima* (Gardner) Brooker, with an appendix establishing these two taxa as subspecies. *Aust. J. Bot.*, 36, 119–129.
Bryant, L.H. (1950) Variations in oil yield and oil composition of some species of eucalypts and tea trees. *Forestry Commission (Division of Wood Technology) NSW Technical Notes*, 4, 6–10.
Butcher, P.A., Matheson, A.C. and Slee, M.U. (1996) Potential for genetic improvement of oil production in *Melaleuca alternifolia* and *M. linariifolia*. *New Forests*, 11, 31–51.
Coppen, J.J.W. and Jacovelli, P.A. (1992) Results of a sampling study to determine the variability in leaf oil yield and cineole content in *Eucalyptus smithii* according to height within a tree and between trees. Unpubl. report, Natural Resources Institute, Chatham, UK.
Cotterill, P.P. and Dean, C.A. (1990) *Successful Tree Breeding with Index Selection*, CSIRO, Melbourne.
Cotterill, P.P. and Jackson, N. (1981) Index selection with restrictions in tree breeding. *Silvae Genetica*, 30, 2–3.

Dieters, M.J., Nikles, D.G., Toon, P.G. and Pomroy, P. (1995) Hybrid superiority in forest trees – concepts and applications. In B.M. Potts, N.M.G. Borralho, J.B. Reid, R.N. Cromer, W.N. Tibbits and C.A. Raymond (eds), *Eucalypt Plantations: Improving Fibre Yield and Quality, Proc. CRCTHF-IUFRO Conf.*, Hobart, Australia, February 1995, Cooperative Research Centre for Temperate Hardwood Forestry, Hobart, pp. 152–155.

Donald, D.G.M. (1980) The production of cineole from *Eucalyptus*: a preliminary report. *S. Afr. For. J.*, (114), 64–67.

Donald, D.G.M. (1991) *Eucalyptus* species as an oil source in South Africa. In A.P.G. Schönau (ed.), *Intensive Forestry: The Role of Eucalypts, Proc. IUFRO Symp.*, Durban, September 1991, Vol. 2, Southern African Institute of Forestry, Pretoria, pp. 985–989.

Doran, J.C. (1991) Commercial sources, uses, formation and biology. In D.J. Boland, J.J. Brophy and A.P.N. House (eds), *Eucalyptus Leaf Oils: Use, Chemistry, Distillation and Marketing*, ACIAR/CSIRO, Inkata Press, Melbourne, pp. 11–25.

Doran, J.C. (1992) Variation in and Breeding for Oil Yields in Leaves of *Eucalyptus camaldulensis*, PhD Thesis, Australian National Univ., Canberra.

Doran, J.C., Arnold, R.J. and Walton, S.J. (1998) Variation in first-harvest oil production in *Eucalyptus radiata*. *Aust. For.*, 61, 27–33.

Doran, J.C., Baker, G.R., Murtagh, G.J. and Southwell, I.A. (1997) Improving tea tree yield and quality through breeding and selection. RIRDC Research Paper Series, No. 97/53, Rural Industries Research and Development Corporation, Canberra.

Doran, J.C. and Bell, R.E. (1994) Influence of non-genetic factors on yield of monoterpenes in leaf oils of *Eucalyptus camaldulensis*. *New Forests*, 8, 363–379.

Doran, J.C. and Brophy, J.J. (1990) Tropical red gums – a source of 1,8-cineole-rich eucalyptus oil. *New Forests*, 4, 157–178.

Doran, J.C., Carter, A.S. and Matheson, A.C. (1992) Variation in root strike of Petford *Eucalyptus camaldulensis* clones. In *Mass Production Technology for Genetically Improved Fast Growing Forest Tree Species, Proc. IUFRO Symp.*, Bordeaux, France, September 1992, Vol. 1, pp. 407–414.

Doran, J.C. and Matheson, A.C. (1994) Genetic parameters and expected gains from selection for monoterpene yields in Petford *Eucalyptus camaldulensis*. *New Forests*, 8, 155–167.

Doran, J.C. and Saunders, A.R. (1993) Variation in and Breeding for Essential Oils in *Eucalyptus globulus* subsp. *globulus*. Unpublished report, CSIRO Division of Forestry, Canberra.

Doran, J.C. and Williams, E.R. (1994) Fast-growing *Eucalyptus camaldulensis* clones for foliar-oil production in the tropics. *Commonwealth For. Rev.*, 73, 261–266.

Eldridge, K. (1995) Breeding plan for *Eucalyptus camaldulensis* in Thailand, 1995 revision. Report prepared for CSIRO Division of Forestry, Canberra.

Eldridge, K., Davidson, J., Harwood, C. and van Wyk, G. (1993) *Eucalypt Domestication and Breeding*, Clarendon Press, Oxford, UK.

Eldridge, K.G. and Griffin, A.R. (1983) Selfing effects in *Eucalyptus regnans*. *Silvae Genetica*, 32, 216–221.

Falconer, D.S. (1989) *Introduction to Quantitative Genetics*, 3rd edn, Longman, New York.

Grant, G.D. (1997) Genetic Variation in *Eucalyptus polybractea* and the Potential for Improving Leaf Oil Production, MSc Thesis, Australian National Univ., Canberra.

Griffin, A.R. (1989a) Strategies for the genetic improvement of yield in *Eucalyptus*. In J.S. Pereira and J.J. Landsberg (eds), *Biomass Production by Fast-growing Trees*, Kluwer, Dordrecht, Netherlands, pp. 247–265.

Griffin, A.R. (1989b) Sexual reproduction and tree improvement strategy – with particular reference to eucalypts. In G.L. Gibson, A.R. Griffin and A.C. Matheson (eds), *Breeding Tropical Trees: Population Structure and Genetic Improvement Strategies in Clonal and Seedling Forestry*, Oxford Forestry Institute, UK, and Winrock International, USA, pp. 52–67.

Griffin, A.R., Burgess, I.P. and Wolf, L. (1988) Patterns of natural and manipulated hybridisation in the genus *Eucalyptus* L'Hérit. – a review. *Aust. J. Bot.*, 36, 41–66.

Haifeng Li (1993) Phytochemistry of *Eucalyptus* Species and its Role in Insect-Host-Tree Selection, PhD Thesis, Univ. of Tasmania, Hobart, Australia.

Haines, R. (1994) *Biotechnology and Forest Tree Improvement*, Forestry Paper 118, Food and Agriculture Organization of the United Nations, Rome.

Hanson, W.D. (1963) Heritability. In W.D. Hanson and H.F. Robinson (eds), *Statistical Genetics*, National Academy of Sciences Publication No. 982, National Research Council, Washington DC, USA, pp. 125–140.

Harwood, C.E. (1996) Recent developments in improvement strategy for tropical tree species. In A. Rimbawanto, A.Y.P.B.C. Widyatmoko, H. Suhaendi and T. Furukoshi (eds), *Tropical Plantation Establishment: Improving Productivity Through Genetic Practices, Proc. Internat. Seminar*, Yogyakarta, Indonesia, December 1996, Forest Tree Improvement Research and Development Institute, Yogyakarta, and Japan International Cooperation Agency, Japan, Part II, pp. 1–21.

Harwood, C.E. and Aken, K. (1997) Lecture notes to *Training Course in Statistical and Design Considerations in the Production of High Quality Seed from Planted Trees*, Yogyakarta, March 1997. Internal Report, Australian Tree Seed Centre, CSIRO Forestry and Forest Products, Canberra.

Hellyer, R.O., Lassak, E.V., McKern, H.H.G. and Willis, J.L. (1969) Chemical variation within *Eucalyptus dives*. *Phytochemistry*, 8, 1513–1514.

Hillis, W.E. (1986) By-products of eucalypt wood production. *China–Australia Afforestation Project Technical Communication*, 27, 1–3.

House, A.P.N. and Bell, J.C. (1996) Genetic diversity, mating system and systematic relationships in two red mahoganies, *Eucalyptus pellita* and *E. scias*. *Aust. J. Bot.*, 44, 157–174.

Johnstone, P.C. (1984) Chemosystematics and Morphometric Relationships of Populations of some Peppermint Eucalypts, PhD Thesis, Monash Univ., Melbourne.

Kijkar, S. (1991) *Producing Rooted Cuttings of* Eucalyptus camaldulensis. *Handbook*, ASEAN-Canada Forest Tree Seed Centre, Muak-Lek, Thailand.

Lassak, E.V. (1988) The Australian eucalyptus oil industry, past and present. *Chem. Aust.*, 55, 396–398.

Libby, W.J. (1973) Domestication strategies for forest trees. *Can. J. For. Res.*, 3, 265–275.

Mascarenhas, A.F., Khuspe, S.S., Nadgauda, R.S., Gupta, P.K. and Khan, B.M. (1987) Potential of cell culture in plantation forestry programs. In J.W. Hanover and D.E. Keathley (eds), *Genetic Manipulation of Woody Plants*, Plenum Press, New York and London, pp. 391–412.

Matheson, A.C. (1990) Breeding strategies for MPTs. In N. Glover and N. Adams (eds), *Tree Improvement of Multipurpose Tree Species*, Network Technical Series 2, Winrock International Institute for Agricultural Development, USA, pp. 67–99.

Matheson, A.C. and Cotterill, P.P. (1990) Utility of genotype × environment interactions. *For. Ecol. Manage.*, 30, 159–174.

Matheson, A.C. and Raymond, C.A. (1984) The impact of genotype × environment interactions on Australian *Pinus radiata* breeding programs. *Aust. For. Res.*, 14, 11–25.

Meskimen, G. (1983) Realised gain from breeding *Eucalyptus grandis* in Florida. In R.B. Standiford and F.T. Ledig (eds), *Eucalyptus in California*, Report PSW-69, USDA Forest Service, Berkeley, USA, pp. 121–128.

Moncur, M.W. (1991) Advances in managing seed production in eucalypts, with special reference to *Eucalyptus nitens* and *Eucalyptus globulus*. Paper to *Honey and Tree Seed Farming Conf.*, Melbourne, November 1991, International Tree Crops Institute, Melbourne.

Moran, G.F. (1992) Patterns of genetic diversity in Australian tree species. *New Forests*, 6, 49–66.

Moran, G.F. and Bell, J.C. (1983) *Eucalyptus*. In S.D. Tanksley and T.J. Orton (eds), *Isozymes in Plant Genetics and Breeding*, Part B, Elsevier Science, Amsterdam, pp. 423–441.

Moran, G.F., Bell, J.C. and Griffin, A.R. (1989) Reduction in levels of inbreeding in seed orchards of *Eucalyptus regnans* F. Muell. compared with natural populations. *Silvae Genetica*, 38, 32–36.

Namkoong, G., Barnes, R.D. and Burley, J. (1980) A philosophy of breeding strategy for tropical forest trees. *Tropical Forestry Paper 16*, Commonwealth [now Oxford] Forestry Institute, Oxford, UK.

Nikles, D.G. (1993) Breeding methods for production of interspecific hybrids in clonal selection and mass propagation programs in the tropics and subtropics. In J. Davidson (ed.), *Recent Advances in Mass Clonal Multiplication of Forest Trees for Plantation Programmes, Proc. Regional Symp.*, UNDP/FAO FORTIP (RAS/91/004) Project Field Document, 4, Food and Agriculture Organization of the United Nations, Los Banos, Philippines, pp. 218–253.

Penfold, A.R. and Morrison, F.R. (1927) The occurrence of a number of varieties of *Eucalyptus dives* as determined by chemical analyses of the essential oils. Part 1. *J. Proc. Roy. Soc. N.S.W.*, 61, 54–67.

Penfold, A.R. and Willis, J.L. (1953) Physiological forms of *Eucalyptus citriodora* Hook. *Nature*, 171, 883–884.

Pinyopusarerk, K., Doran, J.C., Williams, E.R. and Wasuwanich, P. (1996) Variation in growth of *Eucalyptus camaldulensis* provenances in Thailand. *For. Ecol. Manage.*, 87, 63–73.

Pryor, L.D. and Bryant, L.H. (1958) Inheritance of oil characters in *Eucalyptus. Proc. Linn. Soc. N.S.W.*, 83, 55–64.

Purnell, R.C. (1986) Early results from provenance/progeny trials of *Eucalyptus nitens* in South Africa. In *Breeding Theory, Progeny Testing and Seed Orchards, Proc. IUFRO Conf.*, Williamsburg, October 1986, North Carolina State Univ., Raleigh, USA, pp. 500–513.

Raymond, C.A. (1988) *Eucalyptus globulus – A Breeding Plan for China*, ACIAR Project 8457, Australian Centre for International Agricultural Research, Canberra.

Raymond, C.A. (1991) *Eucalyptus camaldulensis – A Breeding Plan for Thailand*, ACIAR Project 8808, Australian Centre for International Agricultural Research, Canberra.

Reid, J.B., Hasan, O., Moncur, M.W. and Hetherington, S. (1995) Paclobutrazol as a management tool for tree breeders to promote early and abundant seed production. In B.M. Potts, N.M.G. Borralho, J.B. Reid, R.N. Cromer, W.N. Tibbits and C.A. Raymond (eds), *Eucalypt Plantations: Improving Fibre Yield and Quality, Proc. CRCTHF-IUFRO Conf.*, Hobart, Australia, February 1995, Cooperative Research Centre for Temperate Hardwood Forestry, Hobart, pp. 293–298.

Sedgley, M. and Griffin, A.R. (1989) *Sexual Reproduction of Tree Crops*, Academic Press, London.

Simmons, D. and Parsons, R.F. (1976) Analysis of a hybrid swarm involving *Eucalyptus crenulata* and *E. ovata* using leaf oils and morphology. *Biochem. Syst. Ecol.*, 4, 97–101.

Simmons, D. and Parsons, R.F. (1987) Seasonal variation in the volatile leaf oils of two *Eucalyptus* species. *Biochem. Syst. Ecol.*, 15, 209–215.

Squillace, A.E., Wells, O.O. and Rockwood, D.L. (1980) Inheritance of monoterpene composition in cortical oleoresin of loblolly pine. *Silvae Genetica*, 29, 141–151.

Van Wyk, G., Schönau, A.P.G. and Schön, P.P. (1989) Growth potential and adaptability of young eucalypt hybrids in South Africa. In G.L. Gibson, A.R. Griffin and A.C. Matheson (eds), *Breeding Tropical Trees: Population Structure and Genetic Improvement Strategies in Clonal and Seedling Forestry*, Oxford Forestry Institute, UK, and Winrock International, USA, pp. 325–333.

Volker, P.W., Dean, C.A., Tibbits, W.N. and Ravenwood, I.C. (1990) Genetic parameters and gains expected from selection in *Eucalyptus globulus* in Tasmania. *Silvae Genetica*, 33, 18–21.

Whiffin, T. and Bouchier, A. (1992) Chemical and morphological variation within a population of *Eucalyptus radiata* (Myrtaceae) exhibiting leaf volatile oil chemical forms. *Aust. Syst. Bot.*, 5, 95–107.

Whiffin, T. and Ladiges, P.Y. (1992) Patterns of variation and relationships in the *Eucalyptus alpina–E. baxteri* complex (Myrtaceae) based on leaf volatile oils. *Aust. Syst. Bot.*, 5, 695–709.

Williams, E.R. and Matheson, A.C. (1994) *Experimental Design and Analysis for Use in Tree Improvement*, CSIRO Information Services, Melbourne.

# 5 Eucalyptus chemistry

*Joseph J. Brophy and Ian A. Southwell*

## Introduction

The medicinal and aromatic properties of *Eucalyptus* are normally associated with the steam-volatile components, and so this chapter, with the exception of the last section, is devoted to the chemistry of the essential (or volatile) oils. This then generally restricts discussion to molecules of molecular weight less than 250 amu, which are either terpenoid or aromatic in structure. The ease with which volatile oils can be analysed using gas chromatography (GC) and GC coupled techniques has meant that many investigations have been undertaken and the structures of numerous constituents elucidated. Consequently, the number of volatile compounds reported from *Eucalyptus* far exceeds the number of non-volatile ones. The leaves are the most frequently investigated part of the plant but interesting constituents have also been isolated from the bark and the wood.

Brophy has investigated many new *Eucalyptus* species using GC-MS techniques (see below) and the results, together with a listing of the major components from these and previous studies of eucalyptus oils, have been published in book form (Boland *et al*. 1991).

The bulk of the present chapter comprises a revised and up-dated list of those *Eucalyptus* species for which the volatile oil has been analysed, showing oil yields and the identity and relative abundance of the most significant constituents (Table 5.2). The data for this table, as earlier (Boland *et al*. 1991), are taken mostly from Australian sources, representing trees growing in their natural habitat. Although some non-Australian data are included, these are usually so similar to the analyses of endemic populations that the wider inclusion of such data is considered unnecessary. A few species which do not occur in Australia (e.g. *E. leizhou No. 1*, a natural hybrid which is used as a source of oil in China, and *E. urophylla*) are included in the table. Coppen and Dyer (1993) have published an extensive bibliography of *Eucalyptus* and its leaf oils, indexed by species and country and covering the literature from the 1920s to late 1992, which includes exotic eucalypts.

The taxonomy of such a large and complex genus as *Eucalyptus* is under continual revision. For example, the frequently distilled lemon-scented gum, *E. citriodora*, along with many other species previously placed in the *Eucalyptus* genus, has recently been renamed as a member of the *Corymbia* genus (Hill and Johnson 1995). For the purposes of this chapter, however, this species will still be considered under *Eucalyptus*. At the latest count there were 777 species listed (Wilcox 1997).

## Early investigations

Although the first reported eucalyptus oil distillation is thought to be that of the Sydney Peppermint, *E. piperita*, in 1788, regular steam distillations were not frequent until the 1850s, when Bosisto in Victoria, Australia, produced firstly experimental and then commercial quantities of eucalyptus oil for the European market (Penfold 1935, Penfold and Morrison 1950, McKern 1967, 1968). At that time, organic chemistry was only an emerging science and it was not until later the same century that the principal constituent of *E. globulus* oil was named 'eucalyptol' and its structure established as 1,8-cineole (see later, Figure 5.2, structure 1) (Boland *et al.* 1991). Baker and Smith then systematically investigated oils from other *Eucalyptus* species, isolating some forty different compounds (Baker and Smith 1920, McKern 1968). Several schools of essential oil research were established in New South Wales, Queensland, and South and Western Australia, all publishing data on *Eucalyptus* oils and the chemical structures of their constituents. Most of these results were included in Guenther's series of volumes on essential oils (Guenther 1950) and, later, in Penfold and Willis's treatise *The Eucalypts* (Penfold and Willis 1961).

The advent of gas chromatography and, more recently, techniques coupling gas chromatography with mass spectrometry (GC-MS) and infrared spectroscopy (GC-IR), rapidly accelerated the analysis of volatile oils and the identification of their constituents and this has given rise to an extensive literature on the oils of *Eucalyptus*. Today, this literature continues to grow.

## Sample preparation and analytical techniques

### Sample preparation

As a guide to the commercial viability of production of any eucalyptus oil, laboratory distillations must first be performed to determine yields and chemical composition. Oil yields, usually expressed as a percentage of the leaf weight, are measured on either a dry or fresh weight basis.[1] The highest oil yield recorded in the authors' laboratories was 12.75 per cent from the partly dried leaf of *E. dives*. The equipment of choice for laboratory scale distillation is the Clevenger apparatus (Clevenger 1928) or a modification of it (e.g. Hughes 1970, Whish 1996).

The distillation process is the most time-consuming step in the assessment of essential oil quality. Consequently micro-extraction methods using a suitable organic solvent are frequently used to determine oil yield and quality, especially where large numbers of samples are involved (Ammon *et al.* 1985, Southwell and Stiff 1989, Brophy *et al.* 1989). A room temperature extraction time of 30–40 h can be reduced to one hour if microwave irradiation is first carried out (Southwell *et al.* 1995). Although the product is not as clean as the steam-distilled oil, the extract accurately reflects leaf quality and, with the addition of an internal standard, can also give a reliable estimate of selected constituents of the oil. The use of a GC vial insert (0.1 ml) means that this method can be adapted to micro-analyses (Brophy *et al.* 1989, Chen and Spiro 1995, Spiro and Chen 1995). Consequently this is the method of choice for many laboratories (including our own) for the screening of essential oil-bearing plants, including *Eucalyptus*.

Soxhlet extractions are also used for *Eucalyptus* analysis, as well as simultaneous distillation-extraction procedures like those adapted to the Likens–Nickerson apparatus (Schultz *et al.* 1977). The principles of general essential oil preparation reviewed by researchers such as Koedam (1987) are most applicable to *Eucalyptus* leaf analysis.

---

1 If possible, the moisture content of the distilled leaf should be determined so that oil yields can be expressed on a dry weight basis. This then enables all yields that are determined to be compared on an equal basis. The moisture content of fresh leaf may be as high as 70 per cent but declines rapidly after harvesting. However, even leaf which has been cut and air-dried has a residual moisture content (around 5 per cent) and should not be assumed to be perfectly dry.

Although more recent techniques, including supercritical fluid extraction (Boelens and Boelens 1997), microwave extraction (Craveiro *et al.* 1989) and headspace analysis (Chialva and Gabri 1987) have been used to analyse plant volatiles, *Eucalyptus* has not received the same attention as other species. There is one report of the supercritical fluid extraction of *Eucalyptus* (Milner *et al.* 1997). Other methods of isolation include vacuum distillation and this has been the method of choice in a series of papers by Bignell *et al.* (1994–1998) who report on the oils obtained from almost 200 species of *Eucalyptus*. The procedure has the advantage that it does not lead to the production of artifacts or the degradation of some of the constituents of the oil which can happen with conventional steam distillation, that is, the composition thus determined is closer to the intrinsic composition of the oil within the leaves than that obtained via steam distillation. However, by the same token, the composition may not reflect what is obtained in practice (or could be expected) commercially. Moreover, the results reported by Bignell *et al.* are derived from single trees and may not be representative either of the particular population of which the tree is part or of the wider distribution of the species. Nevertheless, they are reported here (Table 5.2) because in a significant number of cases they are the only analyses reported for the particular *Eucalyptus* species.

## Analytical techniques

Before the advent of gas chromatography most essential oils were analysed by measuring optical rotation, relative density, refractive index and solubility in alcohol. Bench chemistry methods were also used to determine acid value, ester value (before and after esterification), carbonyl value, phenol content and the concentration of some specific isolates (e.g. 1,8-cineole). These methods still have a place today in monographs published by standards organisations such as the International Standardization Organization (ISO) and pharmacopoeias such as the British Pharmacopoeia (BP). With cineole-type eucalyptus oils, the *o*-cresol method of Cocking (1920, 1927) is still used as the bench chemistry method for the determination of cineole. Further information on eucalyptus oil standards is given in Appendix 5; details of how and where the standards can be obtained are given in Appendix 7.

In the main, the identification of individual eucalyptus oil constituents used to be dependent on the isolation of pure compounds, either by fractional distillation under vacuum or using alumina or silica column chromatography. This was then followed by derivatisation and mixed melting point comparison with an authentic derivative of the suspected compound. With modern chromatographic, spectroscopic and X-ray crystallographic techniques, identifications previously taking months or years can now be achieved in hours or days. Examples illustrating the use of these techniques in identifying *Eucalyptus* constituents are abundant in the literature (e.g. Boland *et al.* 1992, Ghisalberti *et al.* 1995, Singh and Etoh 1995). ISO standards now include a chromatographic profile and typical gas chromatographic trace in addition to physicochemical and other data.

The most commonly used technique for the identification of volatile constituents is gas chromatography-mass spectrometry (GC-MS). Using GC-MS a mass spectrum can be obtained for each peak from a chromatogram and compared with a library of reference spectra in order to obtain a spectrum of best fit. The spectra of the common terpenes are in most spectral libraries (e.g. Adams 1995) and a collection of β-triketone mass spectra has recently been published (Brophy *et al.* 1996). Mass spectral data alone are insufficient for unambiguous identification, especially for terpenoids with similar spectra, and should be supported by retention index measurements on two columns of different polarity (Stevens 1996) or additional information (IR, NMR, etc., e.g. Liener 1996). The use of gas chromatography-infrared spectroscopy (GC-IR) is one way of providing such data. The critical factor in all GC techniques is the resolution obtained by the chromatographic column. For example, 1,8-cineole often co-elutes with limonene on non-polar columns and with β-phellandrene on polar columns. Columns of intermediate polarity often give superior resolution of these key *Eucalyptus* constituents (Brophy *et al.* 1989).

Advances in computer technology and software development have encouraged the use of multivariate statistical methods for principal component and cluster analysis of constituents (e.g. Brooker and Lassak 1981, Silvestre *et al.* 1997).

## Infraspecific variation and chemotaxonomy

Early investigations of eucalyptus oils suggested that the composition of an oil was constant within the same species (Baker and Smith 1920) and new taxa were sometimes established solely on chemical evidence (McKern 1968). It was later realised, however, that infraspecific variation could occur. Among many examples are the piperitone and cineole/eudesmol forms of *E. piperita* (McKern 1968) and the piperitone/phellandrene, phellandrene, piperitone/cineole and cineole forms of *E. dives* (Penfold and Morrison 1927a, 1929, Hellyer *et al.* 1969).

The recognition of infraspecific variation led to the introduction of terms such as 'physiological form' or 'chemical variety' and the different types were designated 'Type', 'Variety A', 'Variety B', 'Variety C', etc., in order of their discovery (Hellyer *et al.* 1969). More modern terminology regards these as 'chemotypes' or 'chemovars'. In the case of *E. punctata*, however, sampling seventy-one individual trees showed a wide but continuous variation in the concentration of the major oil components, indicating that there are no grounds for the establishment of separate chemical types (Southwell 1973). This infraspecific variation in oil composition highlights the importance of investigating individual trees and not assuming that the results from single trees are representative of the species generally (Boland *et al.* 1991).

Correlations between the constituents of *Eucalyptus* species and their taxonomic relationship, both within the genus and as part of the Myrtaceae family, have been attempted. Hegnauer, for example, has addressed the issue in his *Chemotaxonomie der Pflanzen* series (Hegnauer 1969, 1990).

With such a large number of species within the eucalypts, it is not surprising that attempts have been made to split the genus *Eucalyptus* into smaller genera. The latest work in this area, by Hill and Johnson (1995), has seen *Corymbia*, previously a sub-genus, raised to full genus status, with approximately 110 species within it. The remaining 700 species (approximately) still belong to the genus *Eucalyptus*, though within this genus there are major and minor groupings. These are shown in the phylogram (Figure 5.1) adapted from Hill and Johnson (1995).

An attempt to discern trends in the essential oils of species within the major groups was carried out by Boland and Brophy (1993) based on analyses of the essential oils of approximately one third of the species. While not conclusive, some trends were observed. *Corymbia* as a genus generally produces poor oil yields and low 1,8-cineole percentages (Boland and Brophy 1993, Brophy *et al.* 1998b). An exception, however, is *C. citriodora* (*Eucalyptus citriodora*) which can produce up to 4.2 per cent of a citronellal-rich (>80 per cent) oil. Significantly more species have now been analysed and the question will be re-examined.

Within the 'Monocalyptus' is a large group of species comprising the majority of section Renantheria which contain significant amounts of *cis*- and *trans*-menth-2-en-1-ol and *cis*- and *trans*-piperitol, as well as, in a significant number of species, piperitone. *E. radiata* and *E. oblonga* occur in this group. Within the eucalypts this is the only place that this occurs. There is also a group of species which contain oils composed principally of β-triketone (e.g. 26, Figure 5.3)

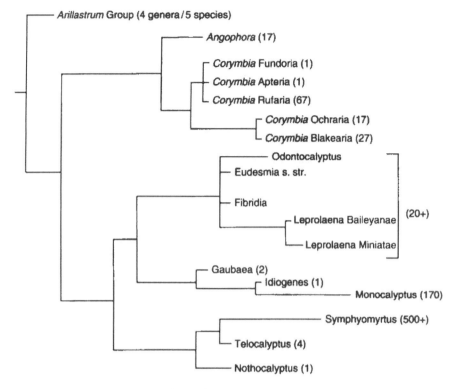

*Figure 5.1* Phylogram of the eucalypts (after Hill and Johnson 1995). Numbers in parentheses refer to the number of species.

and acylphloroglucinol (e.g. 22 and 27, Figure 5.3) derivatives (Brophy *et al.* 1996) in this section, an occurrence not found in the rest of the eucalypts.

'Symphyomyrtus' is the largest group within the eucalypts, containing over 500 species. Within it there are groups of species with high oil yields and 1,8-cineole as the major component. *E. bakeri* and *E. kochii* belong to one section of this group, *E. camaldulensis* to a second, *E. sparsa* and *E. polybractea* to a third, and *E. globulus*, *E. sturgissiana* and *E. smithii* to a fourth.

It was thought (Boland and Brophy 1993) that Western Australian species produced more aromatic compounds than did their eastern Australian relatives, this phenomenon crossing 'group' boundaries. More recent work, however, published in a series of papers by Bignell *et al.* (in which the oil is produced by vacuum distillation rather than steam distillation) has shown that aromatic compounds are present in most of the species examined.

## Commercial utilisation of eucalyptus oils

### Cineole-rich oils

Apart from practical considerations such as oil and biomass yields, and the amenability of the species to appropriate and economic field management, the most important factor in determining

## Cineole-rich oils

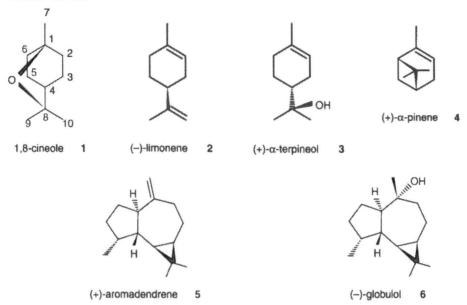

1,8-cineole 1    (−)-limonene 2    (+)-α-terpineol 3    (+)-α-pinene 4

(+)-aromadendrene 5    (−)-globulol 6

## Lemon-scented/perfumery oils

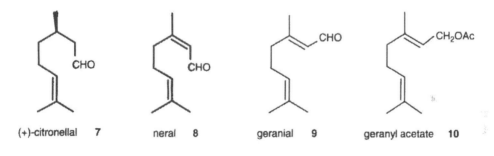

(+)-citronellal 7    neral 8    geranial 9    geranyl acetate 10

## Piperitone/phellandrene-rich oils

## E. olida

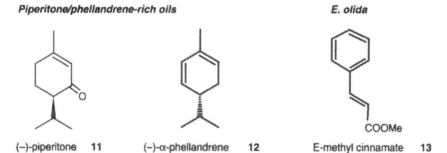

(−)-piperitone 11    (−)-α-phellandrene 12    E-methyl cinnamate 13

*Figure 5.2* Chemical structures of some common constituents of commercial eucalyptus oils.

the commercial value of an oil is its chemical composition. Of the large and diverse range of compounds found in eucalyptus oils (Table 5.2), the most important is the monoterpene ether 1,8-cineole (1, Figure 5.2). It is used for medicinal, flavour and fragrance purposes and has significant biological activity (e.g. mosquito repellency (Klocke *et al.* 1987)). It comprises over 90 per cent of the oil in some specimens of *E. bakeri* (Brophy and Boland 1989), *E. kochii* (Gardner and Watson 1947/48), *E. oblonga* (Baker and Smith 1920), *E. plenissima* (Brooker *et al.* 1988), *E. polybractea* (Boland *et al.* 1991), *E. sparsa* (Brophy and Lassak 1986, Southwell 1987) and *E. sturgissiana* (Boland *et al.* 1991). With the exception of *E. polybractea*, few of these are being exploited as commercial sources of 1,8-cineole.

*E. globulus* remains the chief source of cineole worldwide although the oil contains a lower proportion of it than *E. polybractea* and some other species. However, *E. globulus* is widely grown for its wood and for pulp production and, as a result, the 'waste' leaf is available for production of oil. This and other commercially produced oils with concentrations of cineole less than that in *E. polybractea* (e.g. *E. smithii* and *E. radiata*) are either fractionated to enhance cineole levels to the 70–75 per cent or 80–85 per cent required by monographs published by ISO and various national pharmacopoeias or sold for uses where cineole content is not so critical (e.g. aromatherapy).

The most common constituents co-occurring with 1,8-cineole are limonene (2) and α-terpineol (3), both of which can be derived from the menth-1-en-8-yl cation, the same biogenetic precursor from which cineole is thought to be derived (Croteau 1987, Croteau *et al.* 1994). Other biogenetic pathways then contribute the other monoterpenes, for example, α-pinene (4), sesquiterpenes such as aromadendrene (5), globulol (6) and α-, β- and γ-eudesmol (17–19, Figure 5.3), and aromatic constituents, for example, methyl cinnamate (13).

## Other commercial oils

Although cineole-type (medicinal) eucalyptus oils dominate world production (estimated by Coppen and Hone (1992) at between 2500 and 3100 t in 1991, and by Lawrence (1993) at 3700 t), lemon-scented (perfumery) oils are also produced. The worldwide production of citronellal (7) oils from *E. citriodora* and the citral (i.e. neral+geranial, 8, 9) oils from *E. staigeriana* was estimated at 320 and 70 t respectively for 1984 (Lawrence 1985). A chemotype of *E. citriodora* rich in citronellol (16, Figure 5.3) also exists (Penfold *et al.* 1951). Geranyl acetate (10) has been produced in the past from the leaf and bark of *E. macarthurii* and used as a fragrance constituent (Lassak 1988) although it is not believed to be in current production.

There has been a smaller market for piperitone (11) and α-phellandrene (12)-rich oils (e.g. *E. dives* Type). Piperitone has been used in the synthesis of menthol for flavouring and phellandrene as a fragrance.

The most recent example of a new eucalyptus oil product is that of *E*-methyl cinnamate (13) from *E. olida*. This natural flavouring ingredient was discovered in near pure form in high yield for the first time in *E. olida* in 1985 (Curtis *et al.* 1990) and its production has subsequently been commercialised in Australia, from where it is exported.

## Oils of potential commercial use

Many other eucalyptus oils have the potential to serve as sources of chemical isolates, that is, they contain chemicals with potential commercial application (Figure 5.3).

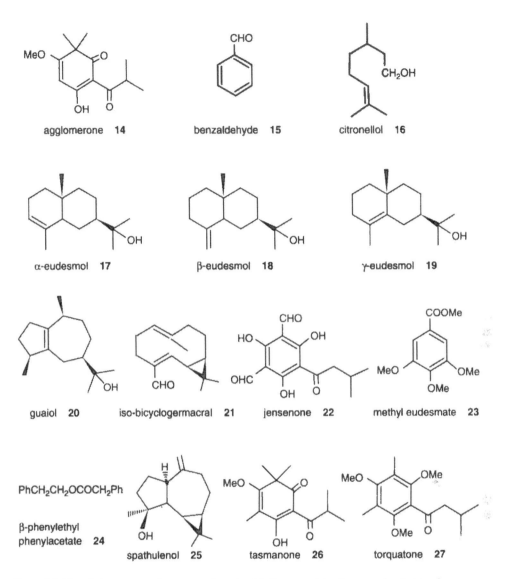

*Figure 5.3* Chemical structures of some constituents with the potential to be sourced as isolates from eucalyptus oils not yet in commercial production.

Food flavour companies have expressed interest in β-phenylethyl phenylacetate (24) which is sometimes the major component of the leaf and bark oil of *E. crenulata* (Lassak and Southwell 1969, Boland *et al.* 1991) and the leaf oil of *E. aggregata* (Hellyer *et al.* 1966, Boland *et al.* 1991). The leaf oil of *E. crenulata* is also rich in methyl eudesmate (methyl 3,4,5-trimethoxybenzoate, 23). Oils from other species of *Eucalyptus* provide excellent sources of the eudesmols (17–19), the ketones agglomerone (14), jensenone (22), tasmanone (26) and torquatone (27), isobicyclogermacral (21), spathulenol (25), benzaldehyde (15) and many other isolates. The best *Eucalyptus* sources of chemicals 14–27 are shown in Table 5.1.

Table 5.1 Constituents with the potential to be sourced as chemical isolates from *Eucalyptus* leaf oils not yet in commercial production

| Isolate[a] | Species | Oil yield[b] (%) | Isolate in oil (%) | Reference |
|---|---|---|---|---|
| Agglomerone (14) | E. agglomerata | 0.9 | 80 | Hellyer (1964, 1968) |
| Benzaldehyde (15) | E. yarraensis | 0.1 | 90 | Lassak and Southwell (1977) |
| Citronellol (16) | E. citriodora | 3.4 | 85 | Penfold et al. (1951) |
| Eudesmols (17–19) | E. oblonga | 0.7 | 97 | Hellyer and McKern (1963) |
| Guaiol (20) | E. maculata | 1.1 | 40 | Lassak and Southwell (1977) |
| Isobicyclogermacral (21) | E. dawsonii | 3.2 | 44 | Boland et al. (1991) |
| Jensenone (22) | E. jensenii | 0.3 | 70 | Boland et al. (1992) |
| Methyl eudesmate (23) | E. crenulata | 0.7 | 47 | Hellyer et al. (1964) |
| β-Phenylethyl phenylacetate (24) | E. aggregata | 1.5 | 91 | Hellyer et al. (1966) |
| Spathulenol (25) | E. rubida | 1.1 | 36 | Boland et al. (1991) |
| Tasmanone (26) | E. cloeziana | 1.9 | 96 | Boland et al. (1991) |
| Torquatone (27) | E. caesia | 0.9 | 50 | Bowyer and Jefferies (1959) |

a Figures in parentheses refer to chemical structures (Figure 5.3).
b Fresh weight basis except for *E. dawsonii* and *E. rubida* (dry weight).

The commercial development of an oil need not be dependent on the major constituents. For example, the grandinol-type β-triketones from *E. grandis* possess powerful root and photosynthetic electron transport inhibition properties (Crow et al. 1971, 1976, Yoshida et al. 1988, Yoneyama et al. 1989) and the minor constituents of *E. citriodora* and *E. camaldulensis* oils have been investigated for mosquito repellent activity (Nishimura et al. 1986, Nishimura 1987, Watanabe et al. 1993).

## Biogenesis of eucalyptus oil constituents

The formation of the mono- and sesquiterpenoids in *Eucalyptus* follows the general principles of isoprenoid biosynthesis (Erman 1985, Croteau 1987). The finer details of these principles advance as the frequency and complexity of labelling experiments increases. For example, recent investigations now suggest that in higher plants isopentenyl pyrophosphate is formed not from mevalonic acid but via the alternative triose phosphate/pyruvate pathway (Lichtenthaler et al. 1997, Eisenreich et al. 1997). These early stages of isoprenoid biosynthesis seem to have been more often studied than the finer details of the coupling and cyclisation steps to form monoterpene, sesquiterpene, diterpene and sesterterpene end products.

The Croteau group, working principally on the monoterpenoids, has isolated cyclases and studied many reactions using isotope labelling and inhibition techniques, principally in mint, pine, marjoram, sage and fennel. From this picture, an extension and refinement of Ruzicka's original 'Biogenetic Isoprene Rule' regarding the head-to-tail coupling of isoprene units has developed (Arigoni et al. 1993). Although specific studies on *Eucalyptus* are rare, principles relating to these other species can be applied to the formation of eucalyptus oil terpenoids. For example, the isolation of cineole synthase from sage (*Salvia officinalis*) led Croteau et al. (1994) to propose alternative stereochemical routes to 1,8-cineole (Figure 5.4) that should also be relevant for *Eucalyptus*.

The non-terpenoids in eucalyptus oils are probably derived via the shikimic acid or phenylalanine–cinnamic acid routes (e.g. methyl eudesmate (23), methyl cinnamate (13),

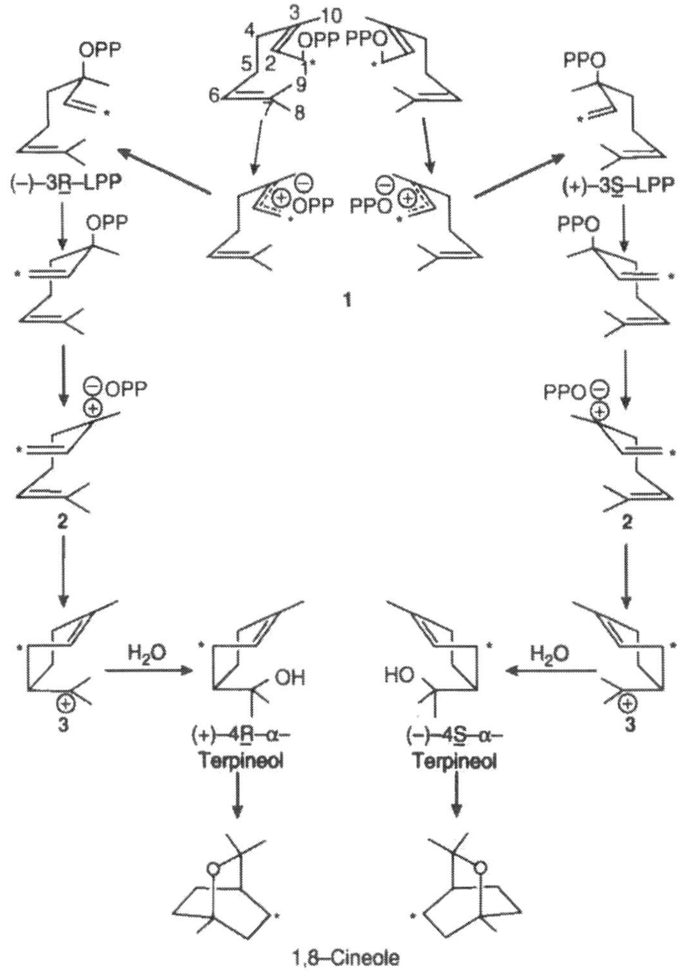

*Figure 5.4* Alternative stereochemical routes from geranyl pyrophosphate to 1,8-cineole. LPP is linalyl pyrophosphate; OPP is the pyrophosphate moiety. Asterisks indicate the two possible positions of labelling from [1-$^3$H]geranyl pyrophosphate (from Croteau *et al.* 1994; used with permission from Academic Press and the author).

β-phenylethyl phenylacetate (24)). However, much more research needs to be done on the formation of the oil components in *Eucalyptus*.

## Metabolism

The bioactivity and chemical ecology of eucalyptus oil are dealt with in detail in other chapters of this book. Some aspects of the chemistry of metabolites are reviewed here.

The metabolism of eucalyptus oil constituents in humans has been poorly studied. Uroterpenol (menth-1-en-7,8-diol) has been isolated from human urine (Wade *et al.* 1966), presumably as a metabolite of dietary limonene which is also a common eucalyptus oil constituent. A review published in 1984 (Santhanakrishnan) on 'Biohydroxylation of Terpenes in Mammals' only reported studies with dogs, rabbits and marsupials (the koala and possum (Southwell *et al.* 1980)). Terpene metabolism generally involved hydroxylation and further oxidation through to the carboxylic acid where primary carbon atoms were involved.

Marsupials, such as the koala, possum and glider, can survive almost exclusively on a diet of *Eucalyptus* leaves and consequently can ingest large amounts (1 g/kg) of eucalyptus oil (Eberhard *et al.* 1975, Foley *et al.* 1987, McLean *et al.* 1993). The concurrent excretion of large amounts of glucuronic acid implies excretion as glucuronide ethers or esters although isolation of these conjugates was not reported (Southwell 1975, Southwell *et al.* 1980).

Metabolites of 1,8-cineole have been investigated from a number of species, ranging from microorganisms through insects to possums, rabbits and rats. Hydroxylation can occur at most carbon atoms but microorganisms, insects, rabbits and rats favour oxidation of the ring, while marsupials favour oxidation of the exposed *gem*-dimethyl group (Flynn and Southwell 1979, Southwell *et al.* 1995). On occasions the hydroxyl groups are oxidised through to ring carbon ketones or to terminal carboxylic acid groups. Insect species feeding on 1,8-cineole-rich leaves metabolise the moiety differently, suggesting a possible insect communication role for the metabolites.

## Non-volatile constituents

The chemistry of the volatile oils is only one facet of *Eucalyptus* secondary metabolite chemistry (Dayal 1988). The genus also contains flavonoids, triterpenes, long chain ketones, glycosides, acylphloroglucinol derivatives and adducts combining more than one of these chemical entities. Typical examples are shown in Figure 5.5, some of which are reviewed in more detail in Chapter 12.

Illustrative of the variety of compounds which occur in *Eucalyptus* are those found in the leaf waxes. These include flavonoids such as eucalyptin (5-hydroxy-7,4'-dimethoxy-6,8-dimethylflavone) (28) (Lamberton 1964, Courtney *et al.* 1983), triterpenes such as ursolic acid (29) (Courtney *et al.* 1983), long-chain ketones such as tritriacontane-16,18-dione (Horn *et al.* 1964) and its 4-hydroxy equivalent (30), a natural antioxidant (Osawa and Namiki 1985), and ketones such as 4,6-dimethoxy-2-hydroxyacetophenone (Courtney *et al.* 1983). None of these is known to be in commercial production from *Eucalyptus*.

On the other hand, the flavone-glycoside rutin (31), (3,5,7,3',4'-pentahydroxy flavone-3-rhamnoglucoside), which occurs in concentrations of 13 per cent and 23 per cent (dried leaf) in *E. macrorhyncha* and *E. youmanii*, respectively, has been produced for medical applications for many years (McKern 1960). Occasional export of dried leaf for such purposes still occurs from Australia (Goodwin pers. comm. 1996).

Ghisalberti (1996), in reviewing bioactive acylphloroglucinol derivatives from *Eucalyptus*, updates and extends an earlier review on volatile β-triketones (Hellyer 1968). Singh *et al.* (1999) have also reviewed non-volatile constituents of eucalypts. Recent investigations have reported new and known triterpenes from *E. globulus* wood (Santos *et al.* 1997) and *E. camaldulensis* leaf (Begum *et al.* 1997, Siddiqui *et al.* 1997). These include unusual cinnamic acid esters of oleanolates (32) and ursolates. In addition, unusual couplings of a jensenone-type acylphloroglucinol to pinane, aromadendrane, guaiane (to give e.g. macrocarpal D, 33) or eudesmane terpenoids give adducts with promising bioactivity (Aukrust and Skattebol 1996, Savina *et al.* 1991, Takasaki *et al.* 1994, Singh and Etoh 1995, Osawa *et al.* 1996, Singh *et al.* 1996).

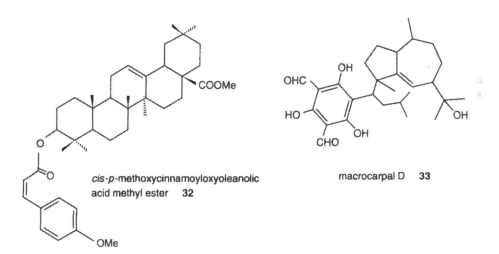

Figure 5.5 Chemical structures of some non-volatile *Eucalyptus* constituents.

Table 5.2 Oil yields and characteristic constituents of the essential oils of *Eucalyptus* species

| Species | Oil yield[a] (%) | Characteristic constituents[b] | Reference |
|---|---|---|---|
| E. abergiana | 0.3 dry | α-pinene (57), β-pinene (4), globulol (13) | Brophy unpubl. |
| E. absita | tr[c] | aromadendrene (8), torquatone (25) | Bignell et al. (1995c) |
| E. acaciiformis | 0.2 | d-α-pinene, geranyl acetate (?), sesquiterpenes | Baker and Smith (1920) |
|  | 1.0–1.4 | 1,8-cineole (66), p-cymene, α-terpineol, globulol, viridiflorol | Boland et al. (1991) |
| E. accedens | 0.9 | d-α-pinene, 1,8-cineole, eudesmol (mixture of isomers) | Baker and Smith (1920) |
|  | 0.9 dry | 1,8-cineole (54), aromadendrene (8), trans-pinocarveol (15) | Bignell et al. (1997g) |
| E. acies | 0.2 dry | 1,8-cineole (34), linalool (16), α-terpineol (15) | Bignell et al. (1997b) |
| E. acmenoides | 0.1 | α-pinene, phellandrene, 1,8-cineole, sesquiterpenes | Baker and Smith (1920) |
|  | 0.9–1.0 | α-pinene (44), 1,8-cineole (43) | Boland et al. (1991) |
| E. acroleuca | 0.7–1.2 dry | α-pinene (57), 1,8-cineole (42), α-terpineol (6) | Brophy unpubl. |
| E. aequioperta | 1.1 dry | α-pinene (10), 1,8-cineole (36), bicyclogermacrene (12) | Bignell et al. (1997e) |
| E. agglomerata | 0.9–1.4 | agglomerone | Unpubl. (1961), Hellyer (1964) |
|  | 0.2 dry | tasmanone (8), agglomerone (3), spathulenol (6) | Bignell et al. (1998) |
|  | 0.4–0.6 | tasmanone, agglomerone, aromatic compounds | Brophy unpubl. |
| E. aggregata (syn. E. rydalensis) | tr-0.3 | d-α-pinene, limonene, 1,8-cineole, p-cymene, methyl eudesmate, β-phenylethyl phenylacetate, sesquiterpenes | Baker and Smith (1920), Hellyer et al. (1966) |
|  | 0.6–1.5 | isoamyl phenylacetate, β-phenylethyl phenylacetate (91) | Boland et al. (1991) |
| E. albens | 0.1 | α-pinene, 1,8-cineole, p-cymene | Baker and Smith (1920) |
|  | tr | globulol (4), spathulenol (8), α-, β-, γ-eudesmol (6 total) | Bignell et al. (1997f) |
| E. albida | 0.4 dry | α-pinene (7), β-pinene (10), bicyclogermacrene (20), α-, β-, γ-eudesmol (10 total) | Bignell et al. (1997d) |
| E. alpina | 0.4 | α-pinene, 1,8-cineole | Baker and Smith (1920) |
| E. amplifolia | 0.9–1.2 | α-pinene, dl-limonene, 1,8-cineole, aromadendrene, α-terpineol, α-eudesmol, β-eudesmol | Hellyer and McKern (1966) |
|  | 0.4 dry | 1,8-cineole (24), aromadendrene (24), globulol (6) | Bignell et al. (1996a) |
| E. amygdalina (syn. E. salicifolia) | 1.6–2.3 | 1,8-cineole | Baker and Smith (1920) |
|  | 1.0 | α-pinene, 1,8-cineole (70), trans-pinocarveol, α-terpineol, terpinyl acetate | Ahmadouch et al. (1985) |

Table 5.2 (Continued)

| Species | Oil yield[a] (%) | Characteristic constituents[b] | Reference |
|---|---|---|---|
| | 2.5–6.3 dry | α-phellandrene (16), 1,8-cineole (15), p-cymene (8), cis- and trans-menth-2-en-1-ol (8 total), piperitone (13) | Li et al. (1995) |
| | 0.2 | terpinen-4-ol (4), piperitone (63), β-eudesmol (4) | Brophy unpubl. |
| E. andrewsii | 1.6 | α-pinene (4), limonene (13), 1,8-cineole (62) | Brophy unpubl. |
| subsp. andrewsii | 1.3 | l-α-phellandrene, p-cymene, piperitone, piperitol (probably trans-) | Baker and Smith (1920), Jones and White (1928) |
| | 0.6 | limonene (4), 1,8-cineole (48), α-, β-, γ-eudesmol (20 total) | Brophy unpubl. |
| subsp. campanulata (syn. E. campanulata) | 0.9 | α-phellandrene, eudesmol (mixture of isomers) | Baker and Smith (1920), Jones and White (1928) |
| | 0.6 | p-cymene (23), menth-2-en-1-ols (13 total), α-, β-, γ-eudesmol (30 total) | Brophy unpubl. |
| E. angophoroides | 0.2 | pinene, l-α-phellandrene, 1,8-cineole, sesquiterpenes | Baker and Smith (1920), |
| | 1.3–1.6 | α-pinene, limonene, 1,8-cineole (70), globulol | Boland et al. (1991) |
| E. angulosa | 0.9–1.2 | d-α-pinene, 1,8-cineole, sesquiterpenes | Baker and Smith (1920) |
| | 0.3 dry | α-pinene (32), 1,8-cineole (14), aromadendrene (11), α-, β-, γ-eudesmol (total 8) | Bignell et al. (1994a) |
| E. angustissima | 3.4 dry | 1,8-cineole (76), α-terpineol (4) | Bignell et al. (1997d) |
| E. annulata | 1.5 dry | α-pinene (36), 1,8-cineole (42), aromadendrene (4) | Bignell et al. (1996e) |
| E. annuliformis | 5.5 dry | α-pinene (23), 1,8-cineole (41), β-eudesmol (6) | Bignell et al. (1994c) |
| E. apiculata | 0.8 | d-α-pinene, 1,8-cineole, l-piperitone, sesquiterpenes | Baker and Smith (1920) |
| E. apodophylla | 0.1–0.2 | α-pinene (37), limonene (31), α-terpineol (6) | Boland et al. (1991) |
| | 0.5 | α-pinene (4), limonene (16), apodophyllone (5), isotorquatone (11), torquatone (11), jensenone (4) | Menut et al. (1999) |
| E. approximans subsp. approximans | 1.8 | p-cymene, cis- and trans-p-menth-2-en-1-ol, piperitone, cis- and trans-piperitol | Lassak and Southwell (1982) |
| subsp. codonocarpa | 2.1 | α- and β-phellandrene, p-cymene, cis- and trans-p-menth-2-en-1-ol, piperitone | Lassak and Southwell (1982) |
| E. aquilina | 1.5 dry | 1,8-cineole (59), α-terpineol (10), bicyclogermacrene (10) | Bignell et al. (1997b) |
| | 1.4 | 1,8-cineole (60), α-terpineol (10), α-, β-, γ-eudesmol (5 total) | Brophy unpubl. |
| E. arachnaea | tr | 1,8-cineole (11), aromadendrene (17), trans-pinocarveol (8), globulol (8), isobicyclogermacral (7) | Bignell et al. (1997c) |

Table 5.2 (Continued)

| Species | Oil yield[a] (%) | Characteristic constituents[b] | Reference |
|---|---|---|---|
| E. archeri | 2.1–2.5 dry | α-pinene (19), limonene (4), 1,8-cineole (60) | Li et al. (1996) |
| E. argophloia | 0.6 | α-pinene (65), limonene, isoamyl isovalerate | Boland et al. (1991) |
| E. argutifolia | 0.9 dry | α-pinene (15), 1,8-cineole (54), β-eudesmol (5) | Bignell et al. (1995a) |
| E. argyphea | 0.3 dry | α-pinene (15), 1,8-cineole (26), trans-pinocarveol (8) | Bignell et al. (1997d) |
| E. aromaphloia | 1.5–4.4 | isovaleraldehyde, α-pinene, 1,8-cineole | Pryor and Willis (1954), Boland et al. (1991) |
| E. aspratilis | 1.5 dry | α-pinene (19), 1,8-cineole (18), γ-terpinene (26) | Bignell et al. (1996e) |
| E. astringens | 0.5–0.6 dry | α-pinene, 1,8-cineole, aromadendrene | Marshall and Watson (1937/38) |
| | 1.7 | α-pinene (36), limonene, 1,8-cineole (43), pinocarveol | Holeman et al. (1987) |
| | 2.9 dry | α-pinene (13), p-cymene (18), β-caryophyllene (15) | Bignell et al. (1996e) |
| | 0.9 | α-pinene (5), 1,8-cineole (57), trans-pinocarveol (20) | Brophy unpubl. |
| E. aureola | 0.4–0.8 | α-pinene (71), guaiol (5), α-, β-, γ-eudesmol (5 total) | Brophy et al. (1998b) |
| E. badjensis | 2.8 | α-pinene, 1,8-cineole (70), α-terpineol, α-, β-, γ-eudesmol | Boland et al. (1991) |
| E. baeuerlenii | 0.4 | α-pinene, 1,8-cineole, eudesmol (mixture of isomers) | Baker and Smith (1920) |
| | 1.0–2.6 dry | limonene, 1,8-cineole (70), α-terpineol, α-, β-, γ-eudesmol | Boland et al. (1991) |
| E. baileyana | 0.8 | d-α-pinene, l-aromadendrene | Baker and Smith (1920) |
| E. bakeri | 1.0 | 1,8-cineole, p-isopropylphenol, phloracetophenone 2,4-dimethylether | Penfold (1927) |
| | 1.8–3.0 | 1,8-cineole (90) | Brophy and Boland (1989) |
| | 2.5 dry | β-phellandrene (2), 1,8-cineole (80), p-cymene (4) | Bignell et al. (1996b) |
| E. balanopelex | 0.7 dry | α-pinene (44), 1,8-cineole (28), α-terpineol (4) | Bignell et al. (1997d) |
| E. balladoniensis subsp. sedens | 0.3 dry | α-pinene (10), aromadendrene (11), α-, β-, γ-eudesmol (42 total) | Bignell et al. (1998) |
| E. bancroftii | 0.5 | l-α-pinene, l-α-phellandrene, 1,8-cineole | Baker and Smith (1920) |
| | 0.9 dry | α-pinene (4), 1,8-cineole (77) | Bignell et al. (1996a) |
| E. banksii | 0.3 | 1,8-cineole | Unpubl. (1961) |
| E. barberi | 2.3 dry | 1,8-cineole (72), α-terpineol (3), α-terpinyl acetate (8) | Li et al. (1996) |
| E. baueriana (syn. E. fletcheri) | 0.3 | α-phellandrene, 1,8-cineole, sesquiterpenes | Baker and Smith (1920) |
| | 1.2–3.3 dry | α-pinene, 1,8-cineole (61), α-terpineol, globulol | Boland et al. (1991) |

Table 5.2 (Continued)

| Species | Oil yield[a] (%) | Characteristic constituents[b] | Reference |
|---|---|---|---|
| | 0.7 dry | p-cymene (17), β-caryophyllene (16), bicyclogermacrene (14), spathulenol (22) | Bignell et al. (1997a) |
| E. baxteri | 0.3 dry | 1,8-cineole (30), α-, β-, γ-eudesmol (33 total) | Bignell et al. (1997b) |
| | 0.8 | 1,8-cineole (30), α-, β-, γ-eudesmol (28 total), aromatic compounds (12) | Brophy unpubl. |
| E. behriana | 0.6 | α-pinene, 1,8-cineole | Baker and Smith (1920) |
| | 0.8–1.1 dry | 1,8-cineole, p-cymene, globulol, β-eudesmol | Boland et al. (1991) |
| | 0.5 dry | 1,8-cineole (36), p-cymene (11), β-eudesmol (25) | Bignell et al. (1995c) |
| E. bensonii | 2.5 | agglomerone (72), tasmanone (15), α-, β-, γ-eudesmol (5 total) | Brophy unpubl. |
| E. benthamii | | | |
| var. benthamii | 1.4 | α-pinene (53), aromadendrene (10), globulol (16) | Boland et al. (1991) |
| var. dorrigoensis | 1.1–1.2 | α-pinene (5), 1,8-cineole (37), p-cymene (10), aromadendrene, globulol | Boland et al. (1991) |
| E. beyeri | | | |
| cineole-rich form | 2.3 | α-pinene, 1,8-cineole (65), globulol, α-, β-, γ-eudesmol | Boland et al. (1991) |
| eudesmol-rich form | 1.8 | globulol (19), α-, β-, γ-eudesmol (total 63) | Boland et al. (1991) |
| E. blakelyi | 2.1 | α- and β-pinene, 1,8-cineole (60), p-cymene | Holeman et al. (1987) |
| | 1.1 dry | limonene (3), 1,8-cineole (67), p-cymene (3) | Bignell et al. (1996a) |
| E. blaxlandii | 0.8 | d-α-pinene, 1,8-cineole | Baker and Smith (1920) |
| | 1.5 | α-pinene (13), 1,8-cineole (11), α-, β-, γ-eudesmol (8 total) | Brophy unpubl. |
| E. bleeseri | 0.4 dry | α-pinene (9), β-pinene (10), bicyclogermacrene (34) | Bignell et al. (1997f) |
| E. bloxsomei | 0.8–1.0 | α-pinene (30), β-pinene (11), guaiol (15), globulol (7) | Brophy et al. (1998b) |
| E. bosistoana | 1.0 | pinene, 1,8-cineole, terpineol, sesquiterpenes, aldehydes | Baker and Smith (1920) |
| | 1.5 | α-pinene (6), β-phellandrene (7), limonene (7), 1,8-cineole (60) | Holeman et al. (1987) |
| E. botryoides | 0.1 | d-α-pinene, sesquiterpenes | Baker and Smith (1920) |
| | 0.5 | α-pinene | Miranda et al. (1981) |
| | 0.5 | α-pinene (13), 1,8-cineole (11), p-cymene (42), α-terpineol (5) | Brophy unpubl. |
| E. brachycalyx | 1.3 dry | α-pinene (8), 1,8-cineole (25), p-cymene (13) | Bignell et al. (1994b) |
| E. brachycorys | 0.8 dry | β-pinene (7), 1,8-cineole (25), p-cymene (9), cryptone (7) | Bignell et al. (1997g) |
| E. brachyphylla | 1.7 dry | α-pinene (18), 1,8-cineole (58), β-eudesmol (4) | Bignell et al. (1996c) |
| E. brassiana | 0.5–0.7 | α-pinene, 1,8-cineole, citronellal | Singh et al. (1991) |

Table 5.2 (Continued)

| Species | Oil yield[a] (%) | Characteristic constituents[b] | Reference |
|---|---|---|---|
| E. brevipes | 0.7 dry | 1,8-cineole (78), α-terpineol (4) | Bignell et al. (1997g) |
| E. brevistylis | 0.8 | conglomerone (20), β-caryophyllene (8), humulene (10), sesquiterpenes (30) | Brophy unpubl. |
| E. bridgesiana (syn. E. stuartiana) | 0.4–0.7 | α-pinene, 1,8-cineole | Baker and Smith (1920) |
| | 0.4–1.0 | α-pinene, limonene, 1,8-cineole (64), α-terpineol | Boland et al. (1991) |
| | 0.5 dry | 1,8-cineole (70), aromadendrene (4) | Bignell et al. (1997f) |
| E. brockwayi | 0.2 dry | α-pinene (21), 1,8-cineole (13), isoamyl isovalerate (10), aromadendrene (14) | Bignell et al. (1996e) |
| E. brookeriana | 1.6–6.9 dry | isovaleraldehyde, α-pinene, limonene, 1,8-cineole, α-terpineol | Brooker and Lassak (1981) |
| | 3.6 dry | α-pinene (18), α-phellandrene (5), 1,8-cineole (53) | Li et al. (1996) |
| E. brownii | 1.9–2.3 | 1,8-cineole (80–90) | Boland et al. 1991 |
| E. bunites | 0.7–1.3 | α-pinene (75), β-pinene (7), α-, β-, γ-eudesmol (5 total) | Brophy et al. (1998b) |
| E. buprestium | 0.2 dry | α-pinene (12), 1,8-cineole (21), linalool (5), α-terpineol (5) | Bignell et al. (1997b) |
| E. burdettiana | tr | α-pinene (11), trans-pinocarveol (13), bicyclogermacrene (8) | Bignell et al. (1996b) |
| E. burgessiana | 1.1–1.5 | α-pinene, α- and β-phellandrene, α-terpinene, p-cymene, terpinen-4-ol, cis-and trans-p-menth-2-en-1-ol, cis-and trans-piperitol, α-, β-, γ-eudesmol | Lassak and Southwell (1982) |
| E. burra-coppinensis | 3.6 dry | α-pinene (13), 1,8-cineole (62), β-eudesmol (5) | Bignell et al. (1994c) |
| E. cadens | 0.5–1.3 | α-pinene, 1,8-cineole (75), trans-pinocarveol | Boland et al. (1991) |
| E. caesia | — | aromadendrene, globulol, torquatone | Bowyer and Jefferies (1959) |
| subsp. caesia | 0.8 dry | α-pinene (14), aromadendrene (27), torquatone (18) | Bignell et al. (1996c) |
| subsp. magna | 0.5 dry | α-pinene (7), aromadendrene (16), torquatone (29) | Bignell et al. (1996c) |
| E. calcareana | 1.4 dry | α-pinene (23), 1,8-cineole (20), aromadendrene (19) | Bignell et al. (1995a) |
| E. calcicola | 0.3 dry | α-pinene (14), limonene (12), 1,8-cineole (32) | Bignell et al. (1997g) |
| | 0.2 | 1,8-cineole (13), α-terpineol (20), geraniol (6), α-, β-, γ-eudesmol (5 total), aromatic compounds (10) | Brophy unpubl. |
| E. caleyi (syn. E. coerulea) | 0.4 | α-pinene, α-phellandrene, 1,8-cineole, eudesmol (mixture of isomers) | Baker and Smith (1920) |
| | 0.1 dry | aromadendrene (8), bicyclogermacrene (34), globulol (9), viridiflorol (8) | Bignell et al. (1997a) |
| | 0.8 | α-pinene (15), limonene (6), 1,8-cineole (53) | Brophy unpubl. |
| E. caliginosa | tr | aromadendrene (8), globulol (5), α-, β-, γ-eudesmol (20 total) | Bignell et al. (1997f) |
| | 0.7–1.5 | α-pinene (30), 1,8-cineole (6), α-, β-, γ-eudesmol (17 total) | Brophy unpubl. |

Table 5.2 (Continued)

| Species | Oil yield[a] (%) | Characteristic constituents[b] | Reference |
|---|---|---|---|
| E. calophylla | 0.3 | d-α-pinene, p-cymene, sesquiterpenes | Baker and Smith (1920) |
|  | tr | α-pinene (10), γ-terpinene (12), E,E-farnesol (21) | Bignell et al. (1996g) |
|  | 0.5 | α-pinene (51), γ-terpinene (10), E,E-farnesol (11) | Brophy unpubl. |
| E. calycogona | 1.0 | d-α-pinene, 1,8-cineole, sesquiterpenes | Baker and Smith (1920) |
|  | — | α-pinene, α-phellandrene, 1,8-cineole, p-cymene, spathulenol, torquatone | Brophy and Lassak unpubl. |
| var. calycogona | 1.4 dry | α-pinene (27), 1,8-cineole (45) | Bignell et al. (1997e) |
| E. camaldulensis (syn. E. rostrata, E. rostrata var. borealis) |  |  |  |
| cineole-rich form | 0.8–1.2 | pinene, 1,8-cineole | Baker and Smith (1920), Gandini (1936) |
| cineole-rich form | 1.2–1.7 | α-pinene, limonene, 1,8-cineole (84), globulol | Doran and Brophy (1990), Boland et al. (1991) |
| cineole-poor form | 0.2–0.4 | phellandrene, 1,8-cineole, p-cymene, cuminal, phellandral | Baker and Smith (1920), Gandini (1936) |
| cineole-poor form | 2.3 | β-pinene, bicyclogermacrene, globulol, viridiflorol | Doran and Brophy (1990), Boland et al. (1991) |
| var. camaldulensis | 0.3 dry | p-cymene (22), cryptone (14), spathulenol (17) | Bignell et al. (1996a) |
| var. obtusa | 1.5 dry | α-pinene (15), 1,8-cineole (62), aromadendrene (3) | Bignell et al. (1996a) |
| E. cameronii | 0.9 | α-pinene (19), 1,8-cineole (20), spathulenol (6), α-, β-, γ-eudesmol (30 total) | Brophy unpubl. |
| E. camfieldii | 0.5–1.0 | tasmanone | Unpubl. (1961), Hellyer et al. (1963) |
| E. campaspe | 0.7–1.2 | α-pinene, 1,8-cineole, p-cymene, aromadendrene, geraniol | Phillips (1923), Watson and Gardner (1944/45) |
|  | 1.9 dry | α-pinene (14), 1,8-cineole (50), aromadendrene (5), torquatone (9) | Bignell et al. (1996d) |
| E. camphora | 1.3 | α-pinene, 1,8-cineole, eudesmol (mixture of isomers) | Baker and Smith (1920) |
| subsp. aquatica | 1.9–4.0 dry | 1,8-cineole (57), α-, β-, γ-eudesmol (33 total) | Boland et al. (1991) |
| subsp. camphora | 2.3–2.6 dry | p-cymene (15), α-, β-, γ-eudesmol (75 total) | Boland et al. (1991) |
| subsp. humeana | 1.6–2.2 dry | α-pinene, limonene, 1,8-cineole (70), p-cymene, terpinyl acetate | Boland et al. (1991) |
| subsp. relicta | 2.6–3.4 dry | α-pinene, limonene, 1,8-cineole (84), α-terpineol | Boland et al. (1991) |
| E. canaliculata | 0.6 dry | p-cymene (27), δ-terpineol (19), cuminal (9) | Bignell et al. (1998) |

Table 5.2 (Continued)

| Species | Oil yield[a] (%) | Characteristic constituents[b] | Reference |
|---|---|---|---|
| *E. capillosa* | | | |
| subsp. *capillosa* | 1.8 dry | α-pinene (9), 1,8-cineole (11), bicyclogermacrene (17), globulol (7) | Bignell et al. (1997c) |
| subsp. *polyclada* | 1.1 dry | α-pinene (6), 1,8-cineole (15), *p*-cymene (10), bicyclogermacrene (12), spathulenol (12) | Bignell et al. (1997c) |
| *E. capitellata* | 0.2 | 1,8-cineole (19), spathulenol (10), α-, β-, γ-eudesmol (20 total) | Brophy unpubl. |
| *E. captiosa* | 0.4 dry | α-pinene (8), 1,8-cineole (38), *trans*-pinocarveol (17), β-eudesmol (7) | Bignell et al. (1997g) |
| *E. carnabyi* | 1.0 dry | α-pinene (28), 1,8-cineole (14), aromadendrene (16) | Bignell et al. (1994c) |
| *E. carnei* | 1.7 dry | α-pinene (33), 1,8-cineole (45), aromadendrene (3) | Bignell et al. (1996c) |
| *E. catenaria* | 0.2–1.2 | α-pinene (80), globulol (2), guaiol (3) | Brophy et al. (1998b) |
| *E. celastroides* | | | |
| subsp. *celastroides* | 2.4 dry | α-pinene (27), 1,8-cineole (37), aromadendrene (11) | Bignell et al. (1997e) |
| subsp. *virella* | 0.6 dry | 1,8-cineole (29), *p*-cymene (14), cryptone (9) | Bignell et al. (1997e) |
| *E. cephalocarpa* | 0.7–1.8 dry | α-pinene, limonene, 1,8-cineole (85) | Boland et al. (1991) |
| *E. ceratocorys* | 1.2 dry | α-pinene (26), 1,8-cineole (36), aromadendrene (10) | Bignell et al. (1994a) |
| *E. chapmaniana* | 2.4–2.7 | 1,8-cineole (60), γ-terpinene, α-terpineol, α-, β-, γ-eudesmol | Boland et al. (1991) |
| *E. chlorophylla* | 0.5–0.6 | limonene, 1,8-cineole (80), α-terpineol, globulol | Boland et al. (1991) |
| *E. cinerea* (syn. *E. stuartiana* var. *cordata*) | 1.1–1.2 | α-pinene, 1,8-cineole, butyraldehyde, valeraldehyde | Baker and Smith (1920) |
| | 2.8–3.5 | α-pinene, 1,8-cineole (78) | Boland et al. (1991) |
| *E. citriodora* | | | |
| 'Type' | 0.5–4.2 | citronellal (>65) | Baker and Smith (1920), Boland et al. (1991), Bignell et al. (1997f) |
| 'Var. A' | 0.8–2.0 | citronellol and its acetate, citronellal | Penfold and Morrison (1948), Penfold et al. (1951) |
| 'Intermediate form' | 0.1–1.4 | citronellal (20–50), guaiol | Penfold et al. (1948), Penfold et al. (1950), Harris and McKern (1950), Unpubl. (1961) |
| 'Hydrocarbon form' | 0.2–0.5 | citronellal (<10), hydrocarbons | Penfold et al. (1950), Unpubl. (1961) |
| *E. cladocalyx* (syn. *E. corynocalyx*) | 0.1 | *d*-α-pinene, 1,8-cineole | Baker and Smith (1920) |
| | tr | *p*-cymene (5), caryophyllene oxide (14), spathulenol (10), β-eudesmol (4) | Bignell et al. (1996g) |
| *E. clarksoniana* | 0.2–0.5 | α- and β-pinene, bicyclogermacrene, globulol | Boland et al. (1991) |

Table 5.2 (Continued)

| Species | Oil yield[a] (%) | Characteristic constituents[b] | Reference |
|---|---|---|---|
| E. clelandii | — | α-pinene, 1,8-cineole, *trans*-pinocarveol, β-eudesmol, torquatone | Brophy and Lassak unpubl. |
|  | 1.8 dry | α-pinene (19), 1,8-cineole (63), *trans*-pinocarveol (4) | Bignell et al. (1995a) |
| E. clivicola | 0.2 dry | α-pinene (13), 1,8-cineole (21), aromadendrene (11), bicyclogermacrene (12) | Bignell et al. (1997c) |
| E. cloeziana | 0.1 dry | α-pinene (16), caryophyllene oxide (8) | Bignell et al. (1997f) |
| pinene-rich form | 0.2–0.6 | α-pinene (78), β-pinene, limonene, α-terpineol, globulol | Brophy and Boland (1990), Boland et al. (1991) |
| tasmanone-rich form | 1.5–1.9 | tasmanone (96) | Brophy and Boland (1990), Boland et al. (1991) |
| E. cneorifolia | 2.0 | α- and β-phellandrene, 1,8-cineole, *p*-cymene, *l*-cryptone | Baker and Smith (1920), Berry (1947) |
|  | 0.9 dry | 1,8-cineole (40), *p*-cymene (14), cryptone (9) | Bignell et al. (1997d) |
| E. coccifera | 0.6 | *l*-α-phellandrene, sesquiterpenes | Baker and Smith (1920) |
|  | 0.9–2.5 dry | α-phellandrene (9), *p*-cymene (10), *cis*- and *trans*-menth-2-en-1-ol (10 total), α-, β-, γ-eudesmol (32 total) | Li et al. (1995) |
|  | 0.6 | 1,8-cineole (20), globulol (8), α-, β-, γ-eudesmol (13 total) | Brophy unpubl. |
| E. concinna | 1.7–1.8 dry | α-pinene, 1,8-cineole, aromadendrene, geraniol | Marshall and Watson (1939/40) |
|  | 4.2 dry | α-pinene (23), β-pinene (8), 1,8-cineole (14) | Bignell et al. (1994b) |
| E. conferruminata | 0.7 dry | α-pinene (23), 1,8-cineole (22), β-eudesmol (4) | Bignell et al. (1996b) |
| E. conglobata | 0.1 dry | α-pinene (8), *p*-cymene (16), β-caryophyllene (10), spathulenol (24) | Bignell et al. (1995a) |
| E. conglomerata | 0.1–0.2 | *l*-α-phellandrene, conglomerone, sesquiterpenes | Lahey and Jones (1939) |
|  | 0.6 | β-caryophyllene (8), globulol (12), E,E-farnesol (9), conglomerone (11) | Brophy unpubl. |
| E. conica | 0.6 | α-pinene, 1,8-cineole, sesquiterpenes | Baker and Smith (1920) |
|  | 0.9 dry | 1,8-cineole (15), *p*-cymene (12), bicyclogermacrene (8), spathulenol (17) | Bignell et al. (1997a) |
| E. consideniana | 1.2 | α-pinene, α-phellandrene, 1,8-cineole, eudesmol (mixture of isomers) | Baker and Smith (1920) |
|  | 2.4 | 1,8-cineole (50), *p*-cymene (10), α-terpineol (8) | Brophy unpubl. |
| E. coolabah subsp. *coolabah* | tr | *p*-cymene (6), aromadendrene (14), palustrol (14), globulol (9) | Bignell et al. (1997a) |

Table 5.2 (Continued)

| Species | Oil yield[a] (%) | Characteristic constituents[b] | Reference |
|---|---|---|---|
| subsp. *microtheca* | tr | bicyclogermacrene (53), δ-cadinene (4) | Bignell *et al.* (1997a) |
| E. *cooperiana* | 2.4 dry | α-pinene (25), 1,8-cineole (16), aromadendrene (19) | Bignell *et al.* (1996e) |
| E. *cordata* | 2.3 | *d*-α-pinene, 1,8-cineole | Baker and Smith (1920) |
|  | 4.4–6.4 dry | α-pinene (16), 1,8-cineole (64), α-terpineol (4), α-terpinyl acetate (6) | Li *et al.* (1996) |
| E. *cornuta* | 1.2 | *d*-α-pinene, 1,8-cineole | Baker and Smith (1920) |
|  | 0.4 dry | 1,8-cineole (62), *p*-cymene (5), *trans*-pinocarveol (7) | Bignell *et al.* (1996b) |
| E. *coronata* | 0.8 dry | 1,8-cineole (44), α-, β-, γ-eudesmol (25 total) | Bignell *et al.* (1997b) |
| E. *corrugata* | 2.2 dry | 1,8-cineole (12), *p*-cymene (17), cryptone (12) | Bignell *et al.* (1994b) |
| E. *cosmophylla* | 0.6 | α-pinene, 1,8-cineole | Baker and Smith (1920) |
|  | 0.2 dry | 1,8-cineole (51), *trans*-pinocarveol (5) | Bignell *et al.* (1996g) |
| E. *crebra* | 0.2 | α-pinene, *l*-α-phellandrene, 1,8-cineole, sesquiterpenes | Baker and Smith (1920) |
| cineole-rich form | 0.5–0.9 | α-pinene (9), 1,8-cineole (66), globulol | Boland *et al.* (1991) |
| pinene-rich form | 0.5–0.9 | α-pinene (26), β-pinene (30), limonene, aromadendrene, globulol | Boland *et al.* (1991) |
|  | 0.5 dry | 1,8-cineole (37), aromadendrene (9), spathulenol (6) | Bignell *et al.* (1997a) |
| E. *crenulata* | 0.9 | isovaleraldehyde, α-pinene, γ-terpinene, *p*-cymene, methyl eudesmate, terpinolene, terpinen-4-ol | Hellyer *et al.* (1964) |
|  | 0.7–0.9 | γ-terpinene, *p*-cymene, terpinen-4-ol, β-phenylethyl phenylacetate (35), phenylacetic acid esters | Boland *et al.* (1991) |
| bark | 1.5 | isovaleraldehyde, α- and β-pinene, α-terpinene, *p*-cymene, methyl eudesmate, β-phenylethyl phenylacetate | Lassak and Southwell (1969) |
| E. *cretata* | 1.9 dry | α-pinene (15), 1,8-cineole (19), *p*-cymene (16), aromadendrene (7) | Bignell *et al.* (1995a) |
| E. *croajingolensis* |  | α-pinene (14), 1,8-cineole (30), α-terpineol (30), terpinen-4-ol (4) | Brophy and Doran unpubl. |
| E. *crucis* |  |  |  |
| subsp. *crucis* | 2.2 dry | α-pinene (16), 1,8-cineole (62), β-eudesmol (5) | Bignell *et al.* (1996c) |
| subsp. *lanceolata* | 3.8 dry | α-pinene (8), 1,8-cineole (62), β-eudesmol (9) | Bignell *et al.* (1997g) |
| E. *cullenii* | 0.4–1.8 dry | 1,8-cineole (40), *p*-cymene (38) | Brophy unpubl. |
| E. *cunninghamii* (syn. E. *rupicola*) | 1.9–2.8 | α-pinene, α- and β-phellandrene, *p*-cymene, piperitone | Lassak and Southwell (1982) |
| E. *cuprea* | 1.3 dry | α-pinene (13), 1,8-cineole (49), α-terpineol (6) | Bignell *et al.* (1995c) |

Table 5.2 (Continued)

| Species | Oil yield[a] (%) | Characteristic constituents[b] | Reference |
|---|---|---|---|
| E. curtipes | 0.1 dry | α-pinene (6), globulol (12), spathulenol (33) | Brophy et al. (1998b) |
| E. curtisii | 0.1 | β-pinene (17), E-β-ocimene (11), globulol (9) | Brophy et al. (1998a) |
| E. cyanophylla | 2.1 dry | α-pinene (16), 1,8-cineole (14), aromadendrene (31), globulol (9) | Bignell et al. (1995a) |
| E. cyclostoma | 2.0 dry | α-pinene (31), 1,8-cineole (34), β-eudesmol (3) | Bignell et al. (1997e) |
| E. cylindriflora | 1.5 dry | α-pinene (20), 1,8-cineole (34), trans-pinocarveol (12) | Bignell et al. (1996c) |
| E. cylindrocarpa | 1.9 dry | α-pinene (22), β-pinene (9), 1,8-cineole (17), p-cymene (10) | Bignell et al. (1997e) |
| E. cypellocarpa | 1.9–2.6 | 1,8-cineole (64), β-eudesmol (20) | Boland et al. (1991) |
| E. dalrympleana | 0.5 | α-pinene, 1,8-cineole | Baker and Smith (1920) |
|  | 0.7–1.1 | α-pinene, limonene, 1,8-cineole (78), p-cymene | Boland et al. (1991) |
|  | 0.8 dry | p-cymene (20), globulol (11), viridiflorol (8), spathulenol (32) | Li et al. (1996) |
| subsp. heptantha | 0.1 | globulol (8), viridiflorol (8), spathulenol (25) | Brophy unpubl. |
| E. dawsonii | 0.8 | l-α-phellandrene, sesquiterpenes | Baker and Smith (1920) |
| eudesmol-rich form | 4.2 dry | globulol (16), α-, β-, γ-eudesmol (total 61) | Boland et al. (1991) |
| isobicyclo germacral-rich form | 2.2–3.2 dry | spathulenol (9), isobicyclogermacral (44), farnesol (7), sesquiterpene alcohol (14) | Boland et al. (1991) |
|  | 0.7 dry | alloaromadendrene (8), isobicyclogermacral (28) | Bignell et al. (1998) |
| E. dealbata | 0.9 | α-pinene, 1,8-cineole, sesquiterpenes | Baker and Smith (1920) |
|  | 1.7 dry | α-pinene (2), 1,8-cineole (81) | Bignell et al. (1996a) |
| E. deanei | 0.6 | α-pinene, 1,8-cineole, p-cymene, cuminal, phellandral, cryptone, sesquiterpenes | Baker and Smith (1920), Mus. A. and S. unpubl. |
| E. decipiens | tr | 1,8-cineole (9), trans-pinocarveol (15), isobicyclogermacral (12) | Bignell et al. (1997d) |
| E. decorticans | 0.3 | leptospermone, flavesone | Bick et al. (1965), Hellyer (1968) |
|  | 0.1–0.2 | β-pinene (28), 1,8-cineole (10), α-terpineol (6), globulol (12) | Brophy unpubl. |
| E. decurva | 0.2 dry | α-pinene (4), 1,8-cineole (44), aromadendrene (4) | Bignell et al. (1996d) |
| E. deglupta (syn. E. naudiniana) | 0.2 dry | isovaleraldehyde, α-pinene, α-phellandrene, p-cymene, ocimene, nerolidol | Webb et al. (1956) |
|  | 0.1 | nerolidol (66) | Martínez et al. (1986) |
| E. delegatensis (syn. E. gigantea) subspp. delegatensis, tasmaniensis | 1.8–3.9 | l-α- and β-phellandrene, p-cymene, piperitone, cis- and trans-p-menth-2-en-1-ol, trans-piperitol, 4-phenyl-2-butanone, methyl cinnamate, α-, β-, γ-eudesmol | Baker and Smith (1920), Weston (1984) |

Table 5.2 (Continued)

| Species | Oil yield[a] (%) | Characteristic constituents[b] | Reference |
|---|---|---|---|
|  | 2.6 dry | $l$-α- and β-phellandrene, $p$-cymene, piperitone, cis- and trans-$p$-menth-2-en-1-ol, trans-piperitol, methyl cinnamate, α-, β-, γ-eudesmol | Boland et al. (1991) |
| seedling leaves | tr-0.1 | α- and β-phellandrene, α-terpinene, $p$-cymene, trans-piperitol, 4-phenyl-2-butanone, methyl cinnamate, α-, β-, γ-eudesmol | Boland et al. (1982) |
| subsp. tasmaniensis | 3.2–5.6 dry | α-phellandrene (16), β-phellandrene (11), $p$-cymene (11), cis- and trans-menth-2-en-1-ol (20 total) | Li et al. (1995) |
| E. dendromorpha | 0.3–2.1 dry | α- and β-phellandrene, α-terpinene, $p$-cymene, cis- and trans-$p$-menth-2-en-1-ol, piperitone | Lassak and Southwell (1982) |
| E. densa subsp. improcera | 1.1 dry | α-pinene (19), 1,8-cineole (33), bicyclogermacrene (10) | Bignell et al. (1997c) |
| E. denticulata | 0.7 dry | 1,8-cineole (11), γ-terpinene (22), $p$-cymene (30), α-terpineol (3) | Li et al. (1994) |
| E. desmondensis | 1.1 dry | α-pinene (12), 1,8-cineole (48), aromadendrene (10) | Bignell et al. (1997c) |
| E. desquamata | 3.1 dry | α-pinene (3), 1,8-cineole (79), $p$-cymene (3) | Bignell et al. (1995c) |
| E. deuaensis | 6–7 dry | α-pinene, limonene, 1,8-cineole, $p$-cymene, terpinolene, α-terpineol, piperitone, α-, β-, γ-eudesmol | Boland et al. (1986) |
| E. dielsii | 4.1 dry | α-pinene (50), 1,8-cineole (34) | Bignell et al. (1996c) |
| E. dimorpha | 0.5 | α-pinene (52), globulol (6), α-, β-, γ-eudesmol (10 total) | Brophy et al. (1998b) |
| E. diptera | 1.7 dry | α-pinene (25), 1,8-cineole (20), aromadendrene (12) | Bignell et al. (1996d) |
| E. dissimulata | 1.0 dry | α-pinene (9), 1,8-cineole (30), aromadendrene (12), alloaromadendrene (12) | Bignell et al. (1997d) |
| E. distans | 1.5–1.9 | 1,8-cineole (69), $p$-cymene (25) | Boland et al. (1991) |
|  | 1.6 dry | 1,8-cineole (37), $p$-cymene (15), bicyclogermacrene (7) | Bignell et al. (1997a) |
| E. diversicolor | 0.8–1.1 | $d$-α-pinene, 1,8-cineole, $d$-α-terpineol, butyl butyrate | Baker and Smith (1920) |
|  | tr | 1,8-cineole (18), trans-pinocarveol (18), α-terpineol (11) | Bignell et al. (1996g) |
| E. diversifolia (syn. | 0.4 | $l$-limonene (?), 1,8-cineole, sesquiterpenes | Baker and Smith (1920) |
| E. santalifolia) | 0.6 dry | α-pinene (17), limonene (22), 1,8-cineole (38) | Bignell et al. (1997b) |
| E. dives | 4.7–6.0 dry | α-phellandrene (20), piperitone (53) | Boland et al. (1991), Bignell et al. (1998) |
| 'Type' | 3.0–4.0 | $l$-α-phellandrene (20–30), $l$-piperitone, (40–56) | Baker and Smith (1920), Hellyer et al. (1969) |
| 'Var. A' | 1.5–5.1 | $l$-α-phellandrene (60–80), $l$-piperitone (2–8) | Penfold and Morrison (1927a), (1929) |

Table 5.2 (Continued)

| Species | Oil yield[a] (%) | Characteristic constituents[b] | Reference |
|---|---|---|---|
| 'Var. B' | 2.9–3.9 | *l*-α-phellandrene, 1,8-cineole (25–40), piperitone (12–18) | Penfold and Morrison (1927a) |
| 'Var. C' | 3.0–4.0 | 1,8-cineole (68–75), terpineol, geraniol, citral | Baker and Smith (1920), Penfold and Morrison (1927a) |
| *E. dolichorhyncha* | 1.3 dry | α-pinene (11), 1,8-cineole (32), aromadendrene (12) | Bignell *et al.* (1994a) |
| *E. doratoxylon* | tr | 1,8-cineole (8), bicyclogermacrene (29), spathulenol (8) | Bignell *et al.* (1996d) |
|  | 0.1 | α-pinene (6), bicyclogermacrene (15), globulol (8) | Brophy unpubl. |
| *E. drepanophylla* | tr | α-pinene (28), 1,8-cineole (15), β-eudesmol (5), torquatone (6) | Bignell *et al.* (1997a) |
| *E. drummondii* | 2.9 dry | α-pinene (24), 1,8-cineole (43), globulol (2) | Bignell *et al.* (1994c) |
| *E. dumosa* | 1.0 | *d*-α-pinene, 1,8-cineole, sesquiterpenes | Baker and Smith (1920), Penfold and Morrison (1951) |
|  | 1.5 dry | *p*-cymene (16), bicyclogermacrene (15), spathulenol (21) | Bignell *et al.* (1995a) |
| *E. dundasii* | 0.2 | *d*-α-pinene, 1,8-cineole, aromadendrene, esters | Marshall and Watson (1934/35) |
|  | 0.5 dry | α-pinene (35), 1,8-cineole (35), aromadendrene (6) | Bignell *et al.* (1996e) |
| *E. dunnii* | 0.9–2.1 | α-pinene, 1,8-cineole (44), aromadendrene, α-terpineol, globulol | Boland *et al.* (1991) |
|  | 0.6 dry | α-pinene (7), 1,8-cineole (32), aromadendrene (16), *trans*-pinocarveol (7) | Bignell *et al.* (1997f) |
| *E. dura* | tr–0.1 | α-pinene (15), 1,8-cineole (15), spathulenol (3) | Brophy unpubl. |
|  | 0.1 | β-phellandrene (16), 1,8-cineole (23) | Doimo *et al.* (1999) |
| *E. dwyeri* | 1.6–2.0 dry | 1,8-cineole, *p*-cymene, isobicyclogermacrene, spathulenol, sesquiterpene alcohols | Boland *et al.* (1991) |
|  | 2.3 dry | α-pinene (9), 1,8-cineole (68), *trans*-pinocarveol (3) | Bignell *et al.* (1996a) |
| *E. ebbanoensis* | 1.8 dry | α-pinene (26), aromadendrene (15), bicyclogermacrene (21) | Bignell *et al.* (1996f) |
| *E. effusa* |  |  |  |
| subsp. *effusa* | 0.3 dry | α-pinene (8), 1,8-cineole (31), β-caryophyllene (8) | Bignell *et al.* (1996g) |
| subsp. *exsul* | 1.3 dry | α-pinene (4), 1,8-cineole (71), *trans*-pinocarveol (7) | Bignell *et al.* (1996g) |
| *E. elaeophloia* | 0.2 | 1,8-cineole (8), *p*-cymene (20), spathulenol (31) | Brophy unpubl. |
| *E. elata* (syn. *E. andreana*, *E. numerosa*) | 6.5 dry | α-phellandrene, *p*-cymene, piperitone, *cis*- and *trans*-menth-2-en-1-ol, *cis*- and *trans*-piperitol | Boland *et al.* (1991) |

Table 5.2 (Continued)

| Species | Oil yield[a] (%) | Characteristic constituents[b] | Reference |
|---|---|---|---|
| 'Type' (as E. radiata) | 1.0–3.0 | l-α-phellandrene, l-piperitone (5–10), trans-piperitol (10–15) | Baker and Smith (1920), Penfold and Morrison (1932) |
| 'Var. A' (syn. E. lindleyana var. stenophylla) | 1.0–3.0 | l-α-phellandrene, l-piperitone (40–55), trans-piperitol | Baker and Smith (1920), Penfold and Morrison (1932) |
| 'Var. B' (syn. E. lindleyana) | 1.7–2.7 | l-α-phellandrene, 1,8-cineole (12–15), l-piperitone (20–30), trans-piperitol | Baker and Smith (1920), Penfold and Morrison (1932) |
| E. erectifolia | 0.3 dry | α-pinene (11), 1,8-cineole (53), α-terpineol (5) | Bignell et al. (1997b) |
| E. eremicola | 3.1 dry | α-pinene (11), β-pinene (21), p-cymene (10), trans-pinocarveol (6) | Bignell et al. (1995b) |
| E. eremophila | 1.8 dry | d-α-pinene, 1,8-cineole, aldehydes, aromadendrene, phenols, alcohols, geraniol (?), eudesmols (?), esters | Marshall and Watson (1936/37) |
| subsp. eremophila | 3.0 dry | α-pinene (12), 1,8-cineole (51), trans-pinocarveol (8) | Bignell et al. (1996d) |
| E. erythrandra | 1.4 dry | α-pinene (35), 1,8-cineole (32), aromadendrene (8) | Bignell et al. (1994a) |
| E. erythrocorys | 0.5 dry | α-terpineol (24), $C_{10}$ diol ? (24) | Bignell et al. (1996f) |
| E. erythronema | 2.5 dry | 1,8-cineole, geraniol | Watson (1941/42) |
| var. erythronema | 2.4 dry | α-pinene (10), 1,8-cineole (57), trans-pinocarveol (10) | Bignell et al. (1996c) |
| var. marginata | 1.7 dry | α-pinene (8), 1,8-cineole (10), p-cymene (13), spathulenol (20) | Bignell et al. (1996c) |
| E. erythrophloia | 0.6 dry | α-pinene (50), β-pinene (11), globulol (5) | Brophy unpubl. |
| E. eudesmioides | | | |
| subsp. eudesmioides | 1.3 dry | α-pinene (10), 1,8-cineole (53), globulol (3) | Bignell et al. (1996f) |
| E. eugenioides (syn. E. acervula, E. wilkinsoniana) | 0.2–1.4 | d- and l-α-pinene, phellandrene, 1,8-cineole, geraniol, geranyl acetate, sesquiterpenes | Baker and Smith (1920), Mus. A. and S. unpubl. |
| | 0.2 dry | α-pinene (23), α-, β-, γ-eudesmol (18 total) | Bignell et al. (1997f), Brophy unpubl. |
| E. ewartiana | 1.5 dry | α-pinene (5), 1,8-cineole (75), α-terpineol (5) | Bignell et al. (1996c) |
| E. exilipes | 0.5–0.9 | α-pinene (14), limonene (16), 1,8-cineole (8), α-, β-, γ-eudesmol (18 total) | Brophy unpubl. |
| E. exilis | 0.1 dry | α-pinene (7), 1,8-cineole (43), α-terpineol (5) | Bignell et al. (1997b) |
| E. eximia | 0.5–0.8 | d-α-pinene, eudesmol (mixture of isomers), sesquiterpenes | Baker and Smith (1920) |
| | 0.8–1.0 | α-pinene (33), globulol (7), α-, β-, γ-eudesmol (35 total) | Brophy et al. (1998b) |
| E. exserta | 0.8 | d-α-pinene, 1,8-cineole, sesquiterpenes, aldehydes | Baker and Smith (1920) |
| | 0.4–1.2 | α-pinene, 1,8-cineole, globulol | Boland et al. 1991 |

Table 5.2 (Continued)

| Species | Oil yield[a] (%) | Characteristic constituents[b] | Reference |
|---|---|---|---|
| | 0.4 dry | α-pinene (5), 1,8-cineole (74), trans-pinocarveol (3) | Bignell et al. (1996a) |
| E. falcata | tr | 1,8-cineole (11), aromadendrene (11), trans-pinocarveol (29) | Bignell et al. (1997d) |
| E. famelica | 0.5 dry | 1,8-cineole (86) | Bignell et al. (1997e) |
| E. fasciculosa | tr | 1,8-cineole | Baker and Smith (1920) |
| | 0.3 dry | 1,8-cineole (58), p-cymene (5), α-terpineol (5) | Bignell et al. (1995c) |
| E. fastigata | 0.1–2.4 | d-α-pinene, α-phellandrene, 1,8-cineole | Baker and Smith (1920) |
| | 0.5–1.3 | α-, β-, γ-eudesmol (up to 70 total) | Brophy unpubl. |
| E. fibrosa | 0.3–0.5 | α-pinene (7), limonene (5), 1,8-cineole (55) | Brophy unpubl. |
| subsp. fibrosa | 0.1 dry | 1,8-cineole (37), aromadendrene (12), globulol (5) | Bignell et al. (1997a) |
| E. ficifolia | 0.2 | α-thujene (5), α-pinene (66), β-pinene (8), γ-terpinene (14), p-cymene (8) | Briggs and Bartley (1970) |
| | 0.2–0.5 | α-pinene (36), farnesol (17) | Boland et al. (1991) |
| | tr | bicyclogermacrene (43), spathulenol (3) | Bignell et al. (1996g) |
| E. flavida | 1.6 dry | α-pinene (19), 1,8-cineole (62) | Bignell et al. (1997c) |
| E. flindersii | 0.8 dry | α-pinene (12), 1,8-cineole (49), γ-terpinene (3) | Bignell et al. (1996a) |
| E. flocktoniae | 1.8 | α-pinene, 1,8-cineole, aromadendrene, alcohols | Watson (1934/35) |
| | 2.9 dry | α-pinene (31), 1,8-cineole (40), aromadendrene (7) | Bignell et al. (1995b) |
| E. foecunda (misidentified as E. uncinata by Baker and Smith) | 1.4 | d-pinene, 1,8-cineole, p-cymene, aldehydes, sesquiterpenes | Baker and Smith (1920) |
| | 0.9 dry | 1,8-cineole (62), β-eudesmol (5) | Bignell et al. (1997d) |
| E. formanii | 1.9 dry | α-pinene (17), 1,8-cineole (57), trans-pinocarveol (7) | Bignell et al. (1997d) |
| E. forrestiana | 3.3 dry | α-pinene (31), 1,8-cineole (17), aromadendrene (7), bicyclogermacrene (12) | Bignell et al. (1994a) |
| E. fraseri | 2.7 dry | α-pinene (26), 1,8-cineole (30), aromadendrene (11) | Bignell et al. (1995a) |
| E. fraxinoides | 1.0 | l-α-phellandrene, 1,8-cineole, piperitone, α-, β-, γ-eudesmol | Baker and Smith (1920), Lassak and Southwell (1982) |
| E. froggattii | — | leptospermone | Hellyer (1968) |
| E. fusiformis | 0.1–0.2 dry | α-pinene, pinocarvone, aromadendrene, trans-pinocarveol, α-terpineol, borneol, spathulenol | Boland et al. (1987) |
| E. gamophylla | 2.6 dry | 1,8-cineole (6), bicyclogermacrene (47), spathulenol (9) | Bignell et al. (1996f) |
| E. gardneri subsp. gardneri | 1.3 dry | 1,8-cineole (6), spathulenol (10), isobicyclogermacral (41) | Bignell et al. (1997c) |
| subsp. ravensthorpensis | 1.5 dry | p-cymene (7), β-caryophyllene (11), isobicyclogermacral (33) | Bignell et al. (1997c) |

Table 5.2 (Continued)

| Species | Oil yield[a] (%) | Characteristic constituents[b] | Reference |
|---|---|---|---|
| E. georgei | 1.8 dry | α-pinene (5), γ-terpinene (26), p-cymene (39), aromadendrene (7) | Bignell et al. (1995a) |
| E. gillenii | 0.6 dry | α-pinene (8), 1,8-cineole (51), aromadendrene (11) | Bignell et al. (1996a) |
| E. gillii | 3.2 dry | α-pinene (21), 1,8-cineole (10), β-caryophyllene (21), bicyclogermacrene (13) | Bignell et al. (1995b) |
| E. gittinsii | 0.3 dry | 1,8-cineole (47), aromadendrene (8), globulol (4) | Bignell et al. (1996f) |
| E. glaucescens | 2.1–2.6 | α-pinene, 1,8-cineole (49), isoamyl isovalerate, aromadendrene, globulol | Boland et al. (1991) |
| E. globoidea (syn. E. yangoura) | 0.2–1.1 | 1,8-cineole (40), p-cymene (15), α-, β-, γ-eudesmol (10 total) | Brophy unpubl. |
| E. globulus | | | |
| subsp. bicostata (syn. E. bicostata) | 1.1–2.4 dry, 1.0–2.0 | d-α-pinene, 1,8-cineole, eudesmol (mixture of isomers) | Penfold and Morrison (1930), Boland et al. (1991) |
| | 0.8 dry | 1,8-cineole (69), aromadendrene (6) | Bignell et al. (1996f) |
| subsp. globulus | 0.5–1.5 | isovaleraldehyde, d-α-pinene, 1,8-cineole, l-pinocarveol, α-terpineol, globulol, sesquiterpenes | Baker and Smith (1920), Gildemeister and Hoffmann (1961) |
| | 1.4–2.4 | α-pinene (11), 1,8-cineole (69) | Boland et al. (1991) |
| | 2.8–4.0 dry | α-pinene (18), 1,8-cineole (55), globulol (5) | Li et al. (1996) |
| | 1.8 dry | α-pinene (13), 1,8-cineole (41), aromadendrene (12) | Bignell et al. (1996f) |
| fruits | ? | 1,8-cineole, aromadendrene, alloaromadendrene, globulol | Nishimura et al. (1979) |
| fruits | 2.1 | 1,8-cineole (73), 27 other compounds incl. methyl eugenol, eugenol, carvacrol | Baslas and Saxena (1984) |
| subsp. maidenii (syn. E. maidenii) | 1.0–2.8 | isovaleraldehyde, α-pinene, 1,8-cineole, aromadendrene, borneol, isoamyl alcohol, eudesmol (mixture of isomers) | Baker and Smith (1920), Rutowski and Winogradowa (1927), Boland et al. (1991) |
| | 1.6 dry | 1,8-cineole (59), aromadendrene (12) | Bignell et al. (1996f) |
| subsp. pseudoglobulus (syn. E. pseudoglobulus, E. stjohnii) | 4.0–5.6 dry | α-pinene, 1,8-cineole (70), β-eudesmol | Boland et al. (1991) |
| | 1.6 dry | α-pinene (12), 1,8-cineole (52) | Bignell et al. (1996f) |
| E. glomerosa | 1.6–2.2 dry | 1,8-cineole (62), pinocarvone (5), trans-pinocarveol (10) | Bignell et al. (1996g), (1997g) |
| E. gomphocephala | tr–0.1 | α-pinene, α-phellandrene | Baker and Smith (1920), Miranda et al. (1981) |
| | tr | trans-pinocarveol (7), carvone (9), globulol (6), β-eudesmol (4) | Bignell et al. (1996g) |
| E. gongylocarpa | 1.9 dry | α-pinene (47), 1,8-cineole (33) | Bignell et al. (1996f) |
| | — | α-pinene, 1,8-cineole, trans-pinocarveol, α-terpineol, globulol | Brophy and Lassak unpubl. |

Table 5.2 (Continued)

| Species | Oil yield[a] (%) | Characteristic constituents[b] | Reference |
|---|---|---|---|
| bark [as *E. eudesmioides*] | 1.0 | 1,8-cineole, globulol | Blumann *et al.* (1953) |
| *E. goniantha* subsp. | 1.2 | *d*-α-pinene, 1,8-cineole, aromadendrene | Marshall and Watson (1934/35) |
| *goniantha* | 0.4 dry | α-pinene (37), 1,8-cineole (20), *trans*-pinocarveol (12) | Bignell *et al.* (1997d) |
| *E. goniocalyx* (syn. *E. elaeophora*, *E. cambagei*) | 1.0–2.5 | *d*-α-pinene, 1,8-cineole, eudesmol (mixture of isomers), aldehydes, sesquiterpenes | Baker and Smith (1920) |
|  | 0.4–0.9 | limonene, 1,8-cineole (78), α-terpineol | Boland *et al.* (1991) |
| *E. gracilis* | 0.9 | *d*-α-pinene, 1,8-cineole, *p*-cymene, aldehydes | Baker and Smith (1920) |
|  | 0.6 dry | *p*-cymene (11), cryptone (10) | Bignell *et al.* (1997e) |
| *E. grandifolia* | 0.2 dry | α-pinene (74), spathulenol (2) | Brophy unpubl. |
| *E. grandis* | 0.2–0.6 | α-pinene (25), 1,8-cineole (6), spathulenol, leptospermone (26), flavesone (12), isoflavesone (3) | Boland *et al.* (1991) |
| [as *E. saligna* var. *pallidivalvis*] | 0.1–0.3 | *d*-α-pinene, 1,8-cineole, esters, alcohols | Baker and Smith (1920) |
| *E. granitica* | 0.4 | 1,8-cineole (82) | Brophy unpubl. |
|  | 0.5–1.3 dry | α-pinene (18), 1,8-cineole (60), α-terpineol (2) | Brophy unpubl. |
| *E. gregsoniana* | 0.8 | 1,8-cineole (18), α-, β-, γ-eudesmol (50 total) | Brophy unpubl. |
| *E. griffithsii* | 0.3 dry | α-pinene (8), 1,8-cineole (26), *trans*-pinocarveol (11) | Bignell *et al.* (1994b) |
| *E. grossa* | 2.5 dry | α-pinene (37), 1,8-cineole (36), aromadendrene (5) | Bignell *et al.* (1996e) |
|  | — | α-pinene, myrcene, 1,8-cineole, globulol, viridiflorol, spathulenol | Brophy and Lassak unpubl. |
| *E. guilfoylei* | tr dry | viridiflorene (30), myrtenol (6), torquatone (7) | Bignell *et al.* (1997g) |
| *E. gummifera* (syn. *E. corymbosa*) | 0.1 | α-pinene, sesquiterpenes | Baker and Smith (1920) |
|  | tr | β-caryophyllene (6), bicyclogermacrene (34), torquatone (6) | Bignell *et al.* (1997f) |
|  | 0.4 | β-pinene (32), globulol (14) | Brophy unpubl. |
| *E. gunnii* | 0.7 | *d*-α-pinene, *l*-α-phellandrene, 1,8-cineole | Baker and Smith (1920) |
|  | 0.5–0.9 | α-phellandrene, *p*-cymene, globulol, bicyclogermacrene, spathulenol | Boland *et al.* (1991) |
|  | 0.8–3.4 dry | α-pinene (16), 1,8-cineole (38), *p*-cymene (7), spathulenol (10) | Li *et al.* (1996) |
| *E. haemastoma* | 0.3–0.5 | *d*-α-pinene, 1,8-cineole, eudesmol (mixture of isomers) | Penfold and Morrison (1927b), Mus. A. and S. unpubl. |
| *E. haematoxylon* | 0.2 dry | α-pinene (17), γ-terpinene (16), *E,E*-farnesol (28) | Bignell *et al.* (1996g) |
| *E. halophila* | 0.6 dry | α-pinene (17), β-pinene (16), 1,8-cineole (12), aromadendrene (13) | Bignell *et al.* (1996e) |

Table 5.2 (Continued)

| Species | Oil yield[a] (%) | Characteristic constituents[b] | Reference |
|---|---|---|---|
| E. hamersleyana | tr | bicyclogermacrene (38), calamenene (7) | Bignell et al. (1997f) |
| E. hebetifolia | 0.3 dry | α-pinene (22), 1,8-cineole (31), aromadendrene (8) | Bignell et al. (1997c) |
| E. henryi | 0.9–1.3 dry | 1,8-cineole (27), δ-cadinene (5), cadinols/muurolols (22 total) | Brophy unpubl. |
| E. histophylla | 1.5 dry | α-pinene (22), 1,8-cineole (43), aromadendrene (5) | Bignell et al. (1997c) |
| E. horistes (syn. E. oleosa var. borealis) | 2.1–4.7 dry | 1,8-cineole | Gardner and Watson (1947/48) |
| | 3.4 dry | 1,8-cineole (88), p-cymene (1) | Bignell et al. (1995b) |
| E. imlayensis | 0.3–0.4 | α-pinene, 1,8-cineole, α-terpineol, spathulenol | Boland et al. (1991) |
| E. incerata | 2.8 dry | α-pinene (7), bicyclogermacrene (66) | Bignell et al. (1998) |
| E. incrassata (syn. E. costata) | 2.8 dry | α-pinene (18), 1,8-cineole (30), aromadendrene (13) | Bignell et al. (1994a) |
| E. indurata | 0.5 dry | 1,8-cineole (22), trans-pinocarveol (14), spathulenol (6) | Bignell et al. (1997d) |
| E. insularis | tr | α-pinene (20), β-pinene (16), 1,8-cineole (15) | Bignell et al. (1997b) |
| E. intermedia | 0.1 | d-α-pinene, aldehydes, sesquiterpenes | Baker and Smith (1920) |
| | 0.2–0.5 | α-pinene (41), β-pinene (26), α-, β-, γ-eudesmol | Boland et al. (1991) |
| E. intertexta (syn. E. intertexta var. fruticosa) | 0.2 | α-pinene, 1,8-cineole | Baker and Smith (1920) |
| | 1.5 dry, 0.4–1.6 | α-pinene, 1,8-cineole, p-cymene, trans-pinocarveol, globulol | Brophy and Lassak (1986), Boland et al. (1991) |
| | 1.1 dry | α-pinene (18), limonene (4), 1,8-cineole (68) | Bignell et al. (1995c) |
| 'E. irbyi' [hybrid, E. dalrympleana subsp. dalrympleana × E. gunnii subsp. gunnii] | 0.2 | α-pinene, α-phellandrene, 1,8-cineole, sesquiterpenes | Baker and Smith (1920) |
| E. jacksonii | 1.0 dry | jacksonone (60), cryptone (6) | Bignell et al. (1997b) |
| E. jensenii | 0.1–0.3 | jensenone (70) | Boland et al. (1991) |
| E. johnsoniana | 0.8 dry | 1,8-cineole (56), spathulenol (3) | Bignell et al. (1997b) |
| E. johnstonii (syn. E. muelleri) | 1.3 | d-α-pinene, 1,8-cineole | Baker and Smith (1920) |
| | 0.5–1.8 | α-pinene, limonene, 1,8-cineole, γ-terpinene, α-terpinyl acetate, α-, β-, γ- eudesmol | Boland et al. (1991) |
| | 4.6–5.0 dry | α-pinene (19), limonene (6), 1,8-cineole (62) | Li et al. (1996) |
| E. jutsonii | 2.3 dry | α-pinene (10), β-pinene (9), 1,8-cineole (63) | Bignell et al. (1996b) |
| E. kartzoffiana | 1.5–2.2 | α-pinene, 1,8-cineole (72), α-terpineol, globulol | Boland et al. (1991) |

Table 5.2 (Continued)

| Species | Oil yield[a] (%) | Characteristic constituents[b] | Reference |
|---|---|---|---|
| E. kessellii | 1.1 dry | α-pinene (19), 1,8-cineole (16), β-caryophyllene (11), aromadendrene (11) | Bignell et al. (1997d) |
| E. kingsmillii | 1.3 dry | α-pinene (23), 1,8-cineole (31), bicyclogermacrene (8) | Bignell et al. (1994c) |
| subsp. alatissima | 2.1 dry | 1,8-cineole (55), trans-pinocarveol (13) | Bignell et al. (1997g) |
| 'E. kirtoniana' [hybrid, E. robusta × E. tereticornis] | 0.3 | limonene (?), citral, sesquiterpenes | Baker and Smith (1920) |
| E. kitsoniana | 2.1 | α-pinene, dipentene, 1,8-cineole, aromadendrene, sesquiterpene alcohols | Hellyer and McKern (1966) |
| E. kochii | | | |
| subsp. kochii (syn. E. oleosa | 2.5–3.5 dry | α-pinene, 1,8-cineole, p-cymene | Gardner and Watson (1947/48), Brooker et al. (1988) |
| var. kochii) | 3.0 dry | α-pinene (1), 1,8-cineole (82), cryptone (1) | Bignell et al. (1995b) |
| subsp. plenissima (syn. E. plenissima) | 2.2–8.6 dry | α-pinene, 1,8-cineole, p-cymene | Gardner and Watson (1947/48), Brooker et al. (1988) |
| E. kondininensis | 3.5 dry | α-pinene (28), 1,8-cineole (50), γ-terpinene (4) | Bignell et al. (1995a) |
| E. kruseana | 1.0 dry | α-pinene (22), 1,8-cineole (51), β-eudesmol (6) | Bignell et al. (1996c) |
| E. kumarlensis | 0.7 dry | 1,8-cineole (56), pinocarvone (6), trans-pinocarveol (13) | Bignell et al. (1997d) |
| E. kybeanensis | 1.7 | p-cymene (12), α-, β-, γ-eudesmol (53 total) | Brophy unpubl. |
| E. laevopinea | 0.6 | l-α-pinene, 1,8-cineole | Baker and Smith (1920) |
| | tr | 1,8-cineole (7), α-, β-, γ-eudesmol (25 total) | Bignell et al. (1997f) |
| | 1.3 | α-pinene (20), 1,8-cineole (18), α-, β-, γ-eudesmol (40 total) | Brophy unpubl. |
| | 1.1 | α-pinene (60), 1,8-cineole (7), α-, β-, γ-eudesmol (10 total) | Brophy unpubl. |
| E. lane-poolei | 1.0 dry | α-pinene (24), 1,8-cineole (51), trans-pinocarveol (4) | Bignell et al. (1994c) |
| E. lansdowneana | | | |
| subsp. albopurpurea | 0.5 dry | α-pinene (9), β-phellandrene (9), bicyclogermacrene (32) | Bignell et al. (1995c) |
| subsp. lansdowneana | 1.3 dry | α-pinene (27), 1,8-cineole (39), trans-pinocarveol (2) | Bignell et al. (1995c) |
| E. largiflorens (syn. E. bicolor) | 0.9 | pinene, 1,8-cineole, aldehydes | Baker and Smith (1920) |
| | 0.9 dry | α-pinene (4), 1,8-cineole (44), aromadendrene (18) | Bignell et al. (1995c) |

Table 5.2 (Continued)

| Species | Oil yield[a] (%) | Characteristic constituents[b] | Reference |
|---|---|---|---|
| 'E. laseroni' [hybrid, E. caliginosa × E. stellulata] | 0.4 | α-pinene, α-phellandrene, 1,8-cineole, sesquiterpenes | Baker and Smith (1920) |
| E. latens | 1.6 dry | α-pinene (9), 1,8-cineole (40), p-cymene (23) | Bignell et al. (1997d) |
| E. lateritica | 0.9 dry | 1,8-cineole (24), tasmanone (37), lateriticone (14) | Bignell et al. (1997b) |
| E. latifolia | tr | α-pinene (12), aromadendrene (15), globulol (7) | Bignell et al. (1997f) |
| E. lehmannii | 0.9 | d-α-pinene, 1,8-cineole, aldehydes | Baker and Smith (1920) |
|  | 2.1 dry | α-pinene (27), 1,8-cineole (32), γ-terpinene (15) | Bignell et al. (1996b) |
| E. leichhardtii | 0.4–0.9 | α-pinene (84), 3-phenylpropanal (5) | Brophy et al. (1998b) |
| E. leizhou No. 1 [natural hybrid of E. exserta in China] | — | α-pinene (13), 1,8-cineole (37), p-cymene (4), trans-pinocarveol (4) | Chen et al. (1983) |
| E. lenziana | tr | bicyclogermacrene (10), δ-cadinene (20), α-, β-, γ-eudesmol (10 total) | Bignell et al. (1997f) |
| E. leptocalyx | 0.8 dry | α-pinene (20), 1,8-cineole (38), trans-pinocarveol (9) | Bignell et al. (1996b) |
| E. leptoloma | 1.1–1.2 | α-pinene (88), 3-phenylpropanal (5) | Brophy et al. (1998b) |
| E. leptophleba | 0.4–0.6 | α- and β-pinene, spathulenol, caryophyllene oxide | Boland et al. (1991) |
|  | tr | p-cymene (13), bicyclogermacrene (8), spathulenol (8) | Bignell et al. (1997a) |
| E. leptophylla | 0.8 dry | 1,8-cineole (66), aromadendrene (5), trans-pinocarveol (5) | Bignell et al. (1997d) |
| E. leptopoda | 1.3 | α-pinene, 1,8-cineole, aromadendrene, geraniol, aldehydes | Marshall and Watson (1936/37) |
|  | 2.0 dry | α-pinene (15), limonene (2), 1,8-cineole (72) | Bignell et al. (1994c) |
| subsp. elevata | 2.1 dry | 1,8-cineole (77), trans-pinocarveol (6) | Bignell et al. (1996g), (1997g) |
| E. lesouefii | 5.1 dry | α-pinene (16), 1,8-cineole (17), aromadendrene (10), bicyclogermacrene (18) | Bignell et al. (1995a) |
| leaf, bud, twig | 1.4, 0.8, 0.3 | α-pinene, 1,8-cineole, torquatone | Brophy et al. unpubl. |
| E. leucophloia | 1.2–1.7 | 1,8-cineole, p-cymene, globulol, viridiflorol, spathulenol | Boland et al. (1991) |
| E. leucoxylon | 0.8–2.5 | d-α-pinene, 1,8-cineole, aromadendrene | Baker and Smith (1920), Penfold and Morrison (1951) |
| subsp. leucoxylon | 1.1 dry | α-pinene (7), aromadendrene (36), bicyclogermacrene (5) | Bignell et al. (1995c) |
| subsp. petiolaris | 2.1 dry | α-pinene (26), 1,8-cineole (28), aromadendrene (18) | Bignell et al. (1995c) |
| E. ligulata | 0.1 dry | α-pinene (7), 1,8-cineole (49), geraniol (5) | Bignell et al. (1997b) |

Table 5.2 (Continued)

| Species | Oil yield[a] (%) | Characteristic constituents[b] | Reference |
|---|---|---|---|
| | 0.7–1.0 | α-pinene (24), 1,8-cineole (37), geraniol (3) | Brophy unpubl. |
| E. ligustrina | 0.1 | geraniol, geranyl acetate, eudesmol (mixture of isomers), sesquiterpenes | Baker and Smith (1920) |
| | 0.1–0.5 | 1,8-cineole (18), α-, β-, γ-eudesmol (50 total) | Brophy unpubl. |
| E. livida | 1.8 dry | α-pinene (8), 1,8-cineole (18), aromadendrene (9), bicyclogermacrene (25) | Bignell et al. (1997c) |
| E. longicornis (syn. E. oleosa var. longicornis) | 1.2 | α-pinene, 1,8-cineole, sesquiterpenes | Baker and Smith (1920) |
| E. longifolia | 0.6 | α-pinene, 1,8-cineole, sesquiterpenes | Baker and Smith (1920) |
| | 1.1 | α-pinene (59), β-pinene (16), 1,8-cineole (5) | Ndou and von Wandruszka (1985) |
| | 0.6 dry | 1,8-cineole (26), p-cymene (19), aromadendrene (14), palustrol (7) | Bignell et al. (1996g) |
| E. loxophleba | 2.4 | α-pinene, 4-methylpent-2-yl acetate, 1,8-cineole (67), trans-pinocarveol, α-terpineol | Boland et al. (1991) |
| subsp. gratiae | 3.0 dry | α-pinene (10), 1,8-cineole (72) | Bignell et al. (1997g) |
| subsp. lissophloia | 1.9 dry | 4-methylpent-2-yl acetate (4), 1,8-cineole (63), trans-pinocarveol (11) | Bignell et al. (1997g) |
| subsp. loxophleba | 1.6 dry | 4-methylpent-2-yl acetate (14), 1,8-cineole (25), aromadendrene (32) | Bignell et al. (1997g) |
| E. lucasii | 0.7 dry | 1,8-cineole (50), aromadendrene (8), trans-pinocarveol (6) | Bignell et al. (1997a) |
| E. lucens | — | α-pinene, 1,8-cineole, camphor, pinocarvone, trans-pinocarveol | Brophy and Lassak (1986) |
| E. luehmanniana | 0.4 | 1,8-cineole (41), α-terpineol (10), α-, β-, γ-eudesmol (20 total) | Brophy unpubl. |
| E. macarthurii | 0.2–1.1 | d-α-pinene, myrcene, 1,8-cineole, p-cymene, linalool, geranyl acetate, geraniol, carissone, eudesmol (mixture of isomers) | Baker and Smith (1920), McQuillin and Parrack (1956), Agarwal et al. (1970), Boland et al. (1991) |
| bark | 0.1–0.4 | d-α-pinene, geranyl acetate, geraniol | Baker and Smith (1920), Neybergh (1953) |
| E. macrandra | 0.2 dry | α-pinene (10), 1,8-cineole (31), aromadendrene (8) | Bignell et al. (1996e) |
| subsp. olivacea | tr | 1,8-cineole (19), aromadendrene (20), alloaromadendrene (4) | Bignell et al. (1996e) |
| E. macrocarpa | | | |
| subsp. elachantha | 0.7 dry | α-pinene (13), aromadendrene (10), bicyclogermacrene (25), | Bignell et al. (1994c) |
| subsp. macrocarpa | 2.5 dry | α-pinene (21), 1,8-cineole (15), aromadendrene (21) | Bignell et al. (1994c) |
| E. macrorhyncha | | | |
| subsp. cannonii | 1.0 | globulol (20), viridiflorol (14), spathulenol (9), α-, β-, γ-eudesmol (17 total) | Brophy unpubl. |

Table 5.2 (Continued)

| Species | Oil yield$^a$ (%) | Characteristic constituents$^b$ | Reference |
|---|---|---|---|
| subsp. macrorhyncha | 0.3 | α-pinene, α-phellandrene, 1,8-cineole, eudesmol (mixture of isomers) | Baker and Smith (1920), Brophy et al. (1982) |
| E. maculata | 0.5 dry | α-pinene (30), globulol (8), α-, β-, γ-eudesmol (18 total) | Bignell et al. (1997f) |
| | 0.7 | limonene (5), 1,8-cineole (55), δ-cadinene (4) | Brophy unpubl. |
| NSW form | 0.2–0.8 | d-α-pinene, d-limonene, dipentene, 1,8-cineole, cadinene, cadinol, sesquiterpenes | Baker and Smith (1920), McKern et al. (1954) |
| Queensland form | 0.7–1.1 dry | d-α-pinene, guaiol, sesquiterpenes | McKern et al. (1954), Unpubl. (1961) |
| E. malacoxylon | 1.1–1.2 | α-pinene, limonene, 1,8-cineole (69), globulol | Boland et al. (1991) |
| E. mannensis | 1.5 dry | α- and β-pinene, limonene, 1,8-cineole, p-cymene | Brophy and Lassak (1986) |
| | 2.5 dry | α- and β-pinene, 1,8-cineole (86) | Bignell et al. (1996b) |
| E. mannifera | 0.7 | α-pinene (6), aromadendrene (17), globulol (30) | Brophy unpubl. |
| subsp. gullickii (syn. E. gullickii) | 0.4 | α-pinene, 1,8-cineole, eudesmol (mixture of isomers) | Baker and Smith (1920) |
| subsp. maculosa (syn. E. maculosa) | 1.1 | α-pinene, 1,8-cineole, eudesmol (mixture of isomers) | Baker and Smith (1920) |
| | 1.2–2.0 | limonene (12), 1,8-cineole (68), α-terpineol (7) | Boland et al. (1991) |
| subsp. praecox (syn. E. lactea) | 0.6 | α-pinene, 1,8-cineole, p-cymene | Baker and Smith (1920) |
| E. marginata | 0.2 | α-pinene, p-cymene, sesquiterpenes, aldehydes | Baker and Smith (1920) |
| | 0.8–1.3 | limonene (11), 1,8-cineole (14), p-cymene (13), menthyl acetate (4) | Brophy unpubl. |
| subsp. marginata | 0.5 dry | p-cymene (10), cryptone (9), spathulenol (12) | Bignell et al. (1997b) |
| subsp. thalassica | tr | p-cymene (13), cryptone (8) | Bignell et al. (1997b) |
| 'E. marsdenii' [probably a hybrid] | 0.7 | sesquiterpenes | Baker and Smith (1920) |
| E. mckieana | — | agglomerone | Hellyer (1968) |
| E. megacarpa | 0.5 | l-α-pinene, α-terpinene, l-limonene, 1,8-cineole | Baker and Smith (1920) |
| | tr dry | 1,8-cineole (11), terpinolene (10) | Bignell et al. (1997g) |
| E. megacornuta | tr | α-pinene (46), aromadendrene (9) | Bignell et al. (1996b) |
| E. melanoleuca | 0.2–0.4 | α-pinene (36), 1,8-cineole (32), trans-pinocarveol (3), β-eudesmol (2) | Brophy unpubl. |
| E. melanophitra | 0.8 dry | α-pinene (16), 1,8-cineole (15), γ-terpinene (27), bicyclogermacrene (7) | Bignell et al. (1997c) |
| E. melanophloia | 0.1 | α-pinene, α-phellandrene, 1,8-cineole, p-cymene, eudesmol (mixture of isomers), sesquiterpenes | Baker and Smith (1920) |
| | 0.6–0.9 | α-pinene (40), β-phellandrene, 1,8-cineole, α-terpineol, globulol | Boland et al. (1991) |

Table 5.2 (Continued)

| Species | Oil yield[a] (%) | Characteristic constituents[b] | Reference |
|---|---|---|---|
| | 0.2 dry | α-phellandrene (5), bicyclogermacrene (30), globulol (8) | Bignell et al. (1995c) |
| E. melanoxylon | 2.3 dry | α-pinene (11), 1,8-cineole (26), p-cymene (11) | Bignell et al. (1994b) |
| E. melliodora | 0.9, 2.3–3.1 dry | α-pinene, α-phellandrene, 1,8-cineole | Baker and Smith (1920), Boland et al. (1991) |
| | 0.1 dry | 1,8-cineole (43), β-caryophyllene (5) | Bignell et al. (1997a) |
| E. merrickiae | 1.5 dry | α-pinene (2), 1,8-cineole (25), aromadendrene (7), β-eudesmol (6) | Bignell et al. (1996e) |
| E. michaeliana | — | leptospermone | Hellyer (1968) |
| E. micranthera | 0.5 dry | α-pinene (20), a $C_6$ acetate (27), aromadendrene (5) | Bignell et al. (1997d) |
| E. microcarpa (syn. E. woollsiana) | 0.4–0.5 | α-pinene, 1,8-cineole, p-cymene, phellandral, cuminal | Baker and Smith (1920) |
| | tr dry | trans-pinocarveol (8), δ-terpineol (4), torquatone (6) | Bignell et al. (1995c) |
| E. microcorys | 0.7 | isovaleraldehyde, d-α-pinene, 1,8-cineole, borneol and its esters | Jones and Lahey (1938) |
| | 0.1 dry | α-pinene (12), 1,8-cineole (24), trans-pinocarveol (11) | Bignell et al. (1997f) |
| | 0.8 | α-pinene (27), 1,8-cineole (57), α-terpineol (5) | Brophy unpubl. |
| E. microtheca | 0.5 | α-pinene, l-α-phellandrene, sesquiterpenes | Baker and Smith (1920) |
| | 0.3–0.4 | 1,8-cineole, p-cymene, globulol | Boland et al. (1991) |
| E. miniata | 1.0 dry | α-pinene (85) | Bignell et al. (1998) |
| | 0.2–1.2 dry | α-pinene (26), 1,8-cineole (17), isobaeckeol, homoisobaeckeol and their methyl ethers, apodophyllone, isotorquatone, torquatone | Brophy unpubl. |
| E. mitchelliana | 3.3–3.7 | 1,8-cineole (14), p-cymene (6), α-, β-, γ-eudesmol (43 total) | Brophy unpubl. |
| E. moluccana (syn. E. hemiphloia) | 0.4–0.9 | 1,8-cineole, p-cymene, l-cryptone, l-phellandral, cuminal, | Baker and Smith (1920), Penfold (1922) |
| | 0.7 | α-pinene, limonene, 1,8-cineole (80), p-cymene, cryptone, α-terpineol | De Riscala and Retamar (1981) |
| | 0.1 dry | p-cymene (6), cryptone (10), spathulenol (7) | Bignell et al. (1997a) |
| E. moorei | 0.4 | 1,8-cineole (26), α-, β-, γ-eudesmol (40 total) | Brophy unpubl. |
| var. latiuscula | 1.1 | elemol (6), α-, β-, γ-eudesmol (70 total) | Brophy unpubl. |
| var. moorei | 0.8 | α-pinene, 1,8-cineole, eudesmol (mixture of isomers) | Baker and Smith (1920) |
| E. morrisbyi | 0.9 | α-pinene, 1,8-cineole (64), α-terpineol, α-terpinyl acetate | Boland et al. (1991) |
| | 1.2–4.6 dry | α-pinene (8), limonene (6), 1,8-cineole (67), α-terpineol (6) | Li et al. (1996) |
| E. morrisii | 1.7 | α-pinene, 1,8-cineole | Baker and Smith (1920) |

Table 5.2 (Continued)

| Species | Oil yield[a] (%) | Characteristic constituents[b] | Reference |
|---|---|---|---|
| E. muelleriana (syn. E. dextropinea) | 0.9 | d-α-pinene, geranyl acetate | Smith (1905), Baker and Smith (1920) |
| E. multicaulis | 0.8 | p-cymene (6), α-, β-, γ-eudesmol (46 total), E,E-farnesol (7) | Brophy unpubl. |
| E. myriadena | 1.5 dry | 1,8-cineole (71), trans-pinocarveol (9) | Bignell et al. (1997e) |
| E. neglecta | 0.8–1.0 | 1,8-cineole (68), globulol | Boland et al. (1991) |
| 'E. nepeanensis' [probably a hybrid] | 0.5 | α-pinene, 1,8-cineole, sesquiterpenes | Baker and Smith (1920) |
| E. nesophila | 0.6 dry | α-pinene (17), globulol (68), spathulenol (3) | Brophy unpubl. |
| E. newbeyi | tr | aromadendrene (4), trans-pinocarveol (6), globulol (10) | Bignell et al. (1996b) |
| E. nicholii | 1.7–2.3 | 1,8-cineole (84) | Boland et al. (1991) |
| E. nigra (includes E. phaeotricha sensu Boland et al. (1991)) | 0.1 | l-α-pinene, 1,8-cineole, leptospermone | Baker and Smith (1920), Hellyer (1968) |
| | 1.0–1.9 | α-pinene, linalool, α-, β-, γ-eudesmol | Boland et al. (1991) |
| E. nitens | 1.0 | α-pinene, limonene, 1,8-cineole, cis-ocimene, p-cymene, isoamyl isovalerate, α-terpineol, leptospermone | Hellyer (1968), Franich (1986) |
| | 0.1–0.6 | α-pinene (17), 1,8-cineole (30), sesquiterpenes | Boland et al. (1991) |
| E. nitida (syn. E. amygdalina var. nitida, E. radiata 'Var. D') | 1.6 | l-α-phellandrene, 1,8-cineole, eudesmol (mixture of isomers) | Baker and Smith (1920) |
| | 2.2–5.6 dry | α-phellandrene (16), 1,8-cineole (14), p-cymene (13), cis- and trans-menth-2-en-1-ol (9 total), α-, β-, γ-eudesmol (17 total) | Li et al. (1995) |
| | 1.7 | α-phellandrene (11), 1,8-cineole (10), p-cymene (14), menth-2-en-1-ols (12 total), α-, β-, γ-eudesmol (17 total) | Brophy unpubl. |
| E. normantonensis | 1.2 dry | α-pinene, α-terpinene, limonene, 1,8-cineole, β-trans-ocimene, p-cymene, trans-pinocarveol | Brophy and Lassak (1986) |
| | 0.6 | α-pinene (28), limonene (8), 1,8-cineole (57) | Boland et al. (1991) |
| | 1.0 | 1,8-cineole (43), γ-terpinene (23), p-cymene (23) | Boland et al. (1991) |
| | 0.4 dry | 1,8-cineole (30), p-cymene (9), spathulenol (11) | Bignell et al. (1997a) |
| E. nortonii | 0.7 | p-cymene, globulol, spathulenol | Boland et al. (1991) |
| E. notabilis | — | leptospermone | Hellyer (1968) |
| E. nova-anglica | 0.5–2.1 | d-α-pinene, aromadendrene, sesquiterpenes | Baker and Smith (1920), Unpubl. (1961) |
| | 1.2 dry | aromadendrene (22), α-, β-, γ-eudesmol (30 total) | Bignell et al. (1998) |

Table 5.2 (Continued)

| Species | Oil yield[a] (%) | Characteristic constituents[b] | Reference |
|---|---|---|---|
| eudesmol-rich form | 2.0 | globulol, viridiflorol, spathulenol (total 15), α-, β-, γ-eudesmol) (total 50) | Boland et al. (1991) |
| nerolidol-rich form | 2.7 | nerolidol (78), globulol (10) | Boland et al. (1991) |
| aromadendrene-rich form | 1.0–2.2 | aromadendrene (40), globulol (30) | Boland et al. (1991) |
| E. nutans | 0.9 dry | α-pinene (10), 1,8-cineole (14), aromadendrene (12), bicyclogermacrene (18) | Bignell et al. (1996d) |
| E. obliqua | 0.7 | l-α-phellandrene, 1,8-cineole, p-cymene, aldehydes | Baker and Smith (1920) |
| | 3.8 | β-phellandrene, p-cymene, piperitone | Boland et al. (1991) |
| | 2.2–5.4 dry | β-phellandrene (7), p-cymene (20), cis- and trans-menth-2-en-1-ol (16 total), piperitone (17) | Li et al. (1995) |
| | tr | piperitone (15), bicyclogermacrene (20), spathulenol (7) | Bignell et al. (1997b) |
| E. oblonga (syn. E. sparsifolia) | 0.2–1.7 | 1,8-cineole, eudesmol (mixture of isomers), leptospermone, agglomerone, tasmanone | Baker and Smith (1920), Hellyer (1968) |
| | 1.8 | p-cymene (15), α-, β-, γ-eudesmol (60 total) | Brophy unpubl. |
| E. obtusiflora | 0–1.3 | 1,8-cineole | Lassak and Southwell (1982), Boland et al. (1991) |
| | 3.7 dry | α-pinene (7), 1,8-cineole (16), bicyclogermacrene (30), spathulenol (10) | Bignell et al. (1995a) |
| E. occidentalis | 1.0 | d-α-pinene, 1,8-cineole, aldehydes | Baker and Smith (1920) |
| var. occidentalis | 1.4 dry | α-pinene (19), β-caryophyllene (6), bicyclogermacrene (29) | Bignell et al. (1996e) |
| E. ochrophloia | 4.3 dry | α- and β-pinene, limonene, 1,8-cineole, p-cymene, α-terpineol, citronellol, globulol, α-, β-, γ-eudesmol | Brophy and Lassak (1986) |
| | 0.1 dry | verbenone (9), globulol (5), spathulenol (6) | Bignell et al. (1995c) |
| E. odorata | 1.9 | 1,8-cineole, aldehydes | Baker and Smith (1920) |
| | — | α-pinene, 1,8-cineole (83), p-cymene | Carmo and Frazão (1985) |
| | 0.9 dry | 1,8-cineole (17), p-cymene (12), cryptone (9) | Bignell et al. (1995c) |
| E. oldfieldii | 2.9 dry | α-pinene (24), 1,8-cineole (34), p-cymene (12) | Bignell et al. (1994c) |
| E. oleosa (syn. E. oleosa var. obtusa) | 1.1–2.1 | α-pinene, 1,8-cineole, aromadendrene, geraniol, aldehydes | Baker and Smith (1920), Marshall and Watson (1936/37) |
| | 4.6 dry | α-pinene (25), limonene (4), 1,8-cineole (52) | Bignell et al. (1995b) |
| E. olida | 2.0–6.0 | E-methyl cinnamate (98), β-trans-ocimene (2) | Curtis et al. (1990), Smale et al. (2000) |

Table 5.2 (Continued)

| Species | Oil yield[a] (%) | Characteristic constituents[b] | Reference |
|---|---|---|---|
| E. olsenii | 1.8–2.2 | 1,8-cineole (29), p-cymene (25), menth-2-en-ols (17 total), α-, β-, γ-eudesmol (8 total) | Brophy unpubl. |
| E. orbifolia | 1.7 dry | isovaleraldehyde, limonene, 1,8-cineole, p-cymene | Brophy and Lassak (1986) |
|  | tr | α-pinene (7), 1,8-cineole (66), α-terpineol (6) | Bignell et al. (1996c) |
| E. oreades | 1.2 | l-α-phellandrene, eudesmol (mixture of isomers), sesquiterpenes | Baker and Smith (1920) |
|  | 0.4–0.5 | β-phellandrene, 1,8-cineole, p-cymene, piperitone, α-, β-, γ-eudesmol | Boland et al. (1991) |
| E. orgadophila | 0.2–0.8 | α- and β-pinene, β-phellandrene, α-terpineol, sesquiterpenes | Boland et al. (1991) |
| E. ornata | 0.5 dry | α-pinene (12), 1,8-cineole (21), aromadendrene (7), trans-pinocarveol (17) | Bignell et al. (1997d) |
| E. orophila | — | limonene (8), 1,8-cineole (16), terpinyl acetate (22) | Pryor et al. (1995) |
| E. ovata (syn. E. paludosa; E. acervula of some authors) | 0.2–1.3 dry | α- and β-pinene, myrcene, α-phellandrene, α-terpinene, limonene, 1,8-cineole, p-cymene, terpinolene, linalool, terpinen-4-ol, α-terpineol, piperitone, geraniol, cuminal, globulol, viridiflorol, eudesmol (mixture of isomers) | Baker and Smith (1920), Unpubl. (1961), Brooker and Lassak (1981) |
|  | 0.3–3.9 dry | α-pinene (12), 1,8-cineole (23), linalool (13), nerolidol (6) | Li et al. (1996) |
|  | tr dry | terpinen-4-ol (28), globulol (12) | Bignell et al. (1998) |
| E. ovularis | 1.0 dry | α-pinene (11), 1,8-cineole (45), trans-pinocarveol (9) | Bignell et al. (1997e) |
| E. oxymitra | 0.3 dry | α-pinene (16), 1,8-cineole (46), trans-pinocarveol (9) | Bignell et al. (1994c) |
| E. pachycalyx | 0.6–1.0 | α-pinene, 1,8-cineole | Boland et al. (1991) |
| E. pachyloma | 0.7 dry | α-pinene (10), 1,8-cineole (53) | Bignell et al. (1997b) |
| E. pachyphylla | 0.6 dry | α-pinene, limonene, 1,8-cineole, γ-terpinene, β-trans-ocimene, p-cymene, trans-pinocarveol, α-terpineol, globulol | Brophy and Lassak (1986) |
|  | 0.3 dry | α-pinene (5), 1,8-cineole (59), aromadendrene (7) | Bignell et al. (1994c) |
| E. paliformis | 2.9 | α-phellandrene, 1,8-cineole, p-cymene, cis- and trans-p-menth-2-en-1-ol, terpinen-4-ol, cis- and trans-piperitol, piperitone | Lassak and Southwell (1982) |
|  | 0.6 | p-cymene (30), menth-2-en-1-ols (12 total), piperitone (30) | Brophy unpubl. |
| E. paludicola | 0.1 | α-pinene, 1,8-cineole, esters, alcohols | Baker and Smith (1920) |
|  | 0.2 | α-pinene (5), 1,8-cineole (44), p-cymene (6), piperitone (6) | Ndou et al. (1985) |

Table 5.2 (Continued)

| Species | Oil yield[a] (%) | Characteristic constituents[b] | Reference |
|---|---|---|---|
| | 0.2 dry | 1,8-cineole (29), aromadendrene (12), α-terpineol (10) | Bignell et al. (1996g) |
| | 0.4 | α-pinene (10), 1,8-cineole (55), trans-pinocarveol (10) | Brophy unpubl. |
| E. papuana | 0.3–0.7 dry | papuanone (70) | Boland and Brophy (1993) |
| E. papuana s. lat. | 0.0 | | Brophy unpubl. |
| E. papuana vel aff. | tr | aromadendrene (60), bicyclogermacrene (2), spathulenol (30) | Brophy unpubl. |
| E. paracolpica | 0.1 dry | α-pinene (20), β-pinene (6), limonene (19), globulol (4) | Brophy unpubl. |
| E. parramattensis | 0.6 | α-pinene, 1,8-cineole, aldehydes | Baker and Smith (1920) |
| | 1.0 dry | 1,8-cineole (72), aromadendrene (6) | Bignell et al. (1996a) |
| E. parvifolia | — | α-pinene, 1,8-cineole, eudesmols (mixture of isomers), aldehydes | Baker and Smith (1920) |
| | 1.3 | 1,8-cineole (81), α-terpineol | Boland et al. (1991) |
| E. patellaris | 0.3 dry | 1,8-cineole (22), aromadendrene (33), verbenone (10) | Bignell et al. (1997a) |
| E. patens | tr | limonene (31), 1,8-cineole (7), torquatone (8) | Bignell et al. (1997b) |
| E. pauciflora (syn. E. coriacea, E. phlebophylla) | 0.6–1.5 | l-α-pinene, l-α-phellandrene, 1,8-cineole, piperitone, eudesmol (mixture of isomers), may exist in chemical varieties | Baker and Smith (1920) |
| | — | myrcene, 1,8-cineole, elemol | Yatagai and Takashi (1983) |
| | 1.8–3.4 dry | α-pinene (11), α-phellandrene (6), 1,8-cineole (9), α-, β-, γ-eudesmol (40 total) | Li et al. (1995) |
| | 0.6–2.7 | spathulenol (4), α-, β-, γ-eudesmol (71 total) | Brophy unpubl. |
| subsp. debeuzevillei | 2.1 | α-, β-, γ-eudesmol (60 total), elemol (16) | Brophy unpubl. |
| subsp. niphophila | 1.4 dry | cis- and trans-menth-2-en-1-ol (30 total), trans-piperitol (20) | Bignell et al. (1998) |
| (syn. E. niphophila) | 0.4 | piperitone (6), elemol (9), α-, β-, γ-eudesmol (55) | Brophy unpubl. |
| subsp. pauciflora | 0.5 dry | piperitone (14), α-, β-, γ-eudesmol (42 total) | Bignell et al. (1998) |
| E. peeneri | 5.8 dry | α-pinene (21), β-pinene (6), 1,8-cineole (32) | Bignell et al. (1995b) |
| E. pellita | 0.1 | α-pinene, 1,8-cineole, globulol | Boland et al. (1991) |
| | 0.1 dry | 1,8-cineole (25), trans-pinocarveol (14), α-, β-, γ-eudesmol (10 total) | Bignell et al. (1997f) |
| | 0.3 | α-pinene (77), globulol (3) | Brophy et al. (1998b) |
| E. pendens | tr | bicyclogermacrene (22), spathulenol (8), torquatone (6) | Bignell et al. (1997b) |

Table 5.2 (Continued)

| Species | Oil yield[a] (%) | Characteristic constituents[b] | Reference |
|---|---|---|---|
| E. perangusta | tr | 1,8-cineole (17), bicyclogermacrene (30), spathulenol (4) | Bignell et al. (1997d) |
| E. percostata | 3.9 dry | α-pinene (12), p-cymene (12), bicyclogermacrene (21), spathulenol (14) | Bignell et al. (1995a) |
| E. perriniana | 1.1 | α-pinene, 1,8-cineole, butyl butyrate, aldehydes | Baker and Smith (1920) |
|  | 2.8 dry | 1,8-cineole (86) | Boland et al. (1991) |
|  | 3.2 dry | α-pinene (11), limonene (5), 1,8-cineole (66), α-terpineol (2) | Li et al. (1996) |
| E. petalophylla | 0.3–1.0 | α-pinene (24), γ-terpinene (9), aromadendrene (5), globulol (15) | Brophy et al. (1998b) |
| E. petraea | 0.7 dry | α-pinene (11), 1,8-cineole (57) | Bignell et al. (1997a) |
| E. phaenophylla | 1.7 dry | α-pinene (8), 1,8-cineole (7), aromadendrene (7), bicyclogermacrene (21), spathulenol (15) | Bignell et al. (1997c) |
| subsp. interjacens | 0.4 dry | α-pinene (8), p-cymene (15), spathulenol (30) | Bignell et al. (1997g) |
| E. phenax (syn. E. 'anceps') | 0.7–0.8 | d-α-pinene, 1,8-cineole | Berry and Swanson (1942) |
|  | 0.5 dry | α-pinene (14), 1,8-cineole (30), aromadendrene (18), trans-pinocarveol (11) | Bignell et al. (1995a) |
| E. pileata | 3.7 dry | α-pinene (24), 1,8-cineole (20), bicyclogermacrene (16) | Bignell et al. (1995a) |
| sp. aff. pileata | 2.1 dry | α-pinene (26), 1,8-cineole (29), aromadendrene (12) | Bignell et al. (1997e) |
| E. pilularis | 0.1 | d-α-pinene, l-α-phellandrene, 1,8-cineole, eudesmol (mixture of isomers), sesquiterpenes | Baker and Smith (1920) |
|  | 0.5–0.8 | α-phellandrene (8), p-cymene (30), spathulenol (6), isobicyclogermacrene (15) | Brophy unpubl. |
| E. pimpiniana | 1.9 dry | α-pinene (39), 1,8-cineole (18), aromadendrene (7) | Bignell et al. (1994b) |
| E. piperita |  |  |  |
| 'Var. A' | 0.8 | α-pinene, α-phellandrene, 1,8-cineole, piperitone (5), eudesmol (mixture of isomers) | Baker and Smith (1920), Penfold and Morrison (1924) |
| 'Type' | 2.3 | l-α-phellandrene, l-piperitone (40–50) | Penfold and Morrison (1924) |
| subsp. piperita | 1.5 | α-phellandrene (8), 1,8-cineole (25), α-, β-, γ-eudesmol (20 total) | Brophy unpubl. |
| subsp. urceolaris | 0.6 | α-phellandrene (18), 1,8-cineole (23), p-cymene (17), terpinen-4-ol (6), α-, β-, γ-eudesmol (10 total) | Brophy unpubl. |
| E. planchoniana | tr | α-phellandrene, sesquiterpenes | Baker and Smith (1920) |
|  | tr–0.3 | 1,8-cineole (47), spathulenol (5), α-, β-, γ-eudesmol (20 total) | Brophy unpubl. |

Table 5.2 (Continued)

| Species | Oil yield[a] (%) | Characteristic constituents[b] | Reference |
|---|---|---|---|
| E. platycorys | 4.0 dry | α-pinene (20), 1,8-cineole (44), β-eudesmol (4) | Bignell et al. (1996b) |
| E. platyphylla | 0.6–0.8 | α-pinene (75), limonene (11) | Boland et al. (1991) |
| E. platypus | 0.8 | d-α-pinene, 1,8-cineole, aldehydes, sesquiterpenes | Baker and Smith (1920) |
| var. heterophylla | 0.6–1.2 | α-pinene (36), 1,8-cineole (15), aromadendrene (6), globulol (13) | Brophy unpubl. |
|  | 0.5 dry | α-pinene (13), 1,8-cineole (14), aromadendrene (19), bicyclogermacrene (7) | Bignell et al. (1996d) |
| var. platypus | 2.1 dry | α-pinene (15), 1,8-cineole (24), aromadendrene (11), bicyclogermacrene (11) | Bignell et al. (1996d) |
| E. pluricaulis |  |  |  |
| subsp. pluricaulis | 0.3 dry | 1,8-cineole (28), β-caryophyllene (8), aromadendrene, bicyclogermacrene (8), isobicyclogermacral (5) | Bignell et al. (1997c) |
| subsp. porphyrea | 2.5 dry | α-pinene (33), 1,8-cineole (21), aromadendrene (15), globulol (5) | Bignell et al. (1997c) |
| E. polyanthemos (syn. E. ovalifolia, | 0.3–0.8 | α-pinene, l-α-phellandrene, 1,8-cineole, sesquiterpenes | Baker and Smith (1920) |
| E. ovalifolia var. lanceolata) | 0.8–2.6 dry | 1,8-cineole (60), viridiflorene, globulol | Boland et al. (1991) |
| E. polybractea (syn. | 1.2–2.5 | 1,8-cineole, p-cymene, cuminal, phellandral, cryptone | Baker and Smith (1920) |
| E. fruticetorum) | 0.7–5.0 | 1,8-cineole (92) | Boland et al. (1991) |
| E. populnea (syn. | 0.8 | α-pinene, 1,8-cineole, sesquiterpenes | Baker and Smith (1920) |
| E. populifolia) | 0.4 dry | 1,8-cineole (61), aromadendrene (13), alloaromadendrene (3) | Bignell et al. (1995c) |
| E. porosa | 2.1 dry | α-pinene (10), 1,8-cineole (17), p-cymene (12) | Bignell et al. (1995c) |
| E. praetermissa | 0.8 dry | α-pinene (27), 1,8-cineole (34), bicyclogermacrene (10) | Bignell et al. (1997c) |
| E. preissiana | 0.6 dry | 1,8-cineole (54), α-, β-, γ-eudesmol (12 total) | Bignell et al. (1997b) |
| subsp. lobata | 0.3 dry | 1,8-cineole (38), viridiflorene (10) | Bignell et al. (1997g) |
| E. propinqua | 0.3 | α-pinene, 1,8-cineole, aldehydes | Baker and Smith (1920) |
|  | 1.0–1.8 | α-pinene, 1,8-cineole, monoterpene alcohols | Boland et al. (1991) |
| E. sp. aff. propinqua | 1.3–1.7 | α-pinene, 1,8-cineole, p-cymene | Boland et al. (1991) |
| E. pruiniramis | 3.1 dry | α-pinene (23), 1,8-cineole (35), aromadendrene (14) | Bignell et al. (1997g) |
| E. pruinosa | tr–0.4 | bicyclogermacrene, δ-cadinene, globulol, spathulenol | Boland et al. (1991) |
|  | tr | 1,8-cineole (63), trans-pinocarveol (5), carvone (9) | Bignell et al. (1997a) |
| E. pterocarpa | 2.5 dry | α-pinene (14), 1,8-cineole (17), aromadendrene (27) | Bignell et al. (1995a) |

Table 5.2 (Continued)

| Species | Oil yield[a] (%) | Characteristic constituents[b] | Reference |
| --- | --- | --- | --- |
| E. pulchella (syn. E. linearis) | 1.1–1.5 | l-α-phellandrene, 1,8-cineole, piperitone, eudesmol (mixture of isomers) | Baker and Smith (1920) |
|  | 4.2–5.6 dry | α-phellandrene (6), 1,8-cineole (47), aromatics (8) | Li et al. (1995) |
|  | 2.7 | 1,8-cineole (46), apodophyllone (18) | Brophy unpubl. |
| E. pulverulenta | 2.2–4.8 dry | isovaleraldehyde, α-pinene, limonene, 1,8-cineole, α-terpineol, α-terpinyl acetate | Baker and Smith (1920), Brophy et al. (1985), Boland et al. (1991) |
| E. pumila | 1.6 | d-α-pinene, 1,8-cineole, aldehydes | Baker and Smith (1920) |
|  | 0.5 dry | 1,8-cineole (72) | Bignell et al. (1998) |
| E. punctata | 0.2–2.3 | α- and β-pinene, myrcene, α-terpinene, limonene, 1,8-cineole, p-cymene, menthyl acetate, terpinen-4-ol, cryptone, α-terpineol, geraniol, cuminal, phellandral | Baker and Smith (1920), Southwell (1973) |
| E. pyriformis | 1.0–1.1 | d-α-pinene, 1,8-cineole, aromadendrene, eudesmol (mixture of isomers), aldehydes | Marshall and Watson (1937/38) |
|  | 1.5 dry | α-pinene (20), 1,8-cineole (39), aromadendrene (5) | Bignell et al. (1994c) |
| E. quadrangulata | 0.7 | d-α-pinene, 1,8-cineole | Baker and Smith (1920) |
| cineole-rich form | 1.6–2.1 | α-pinene (25), 1,8-cineole (55) | Boland et al. (1991) |
| γ-terpinene-rich form | 1.2 | α-pinene, 1,8-cineole (25), γ-terpinene (19), p-cymene, globulol | Boland et al. (1991) |
| eudesmol-rich form | 1.5 | p-cymene (33), α-, β-, γ-eudesmol (total 40) | Boland et al. (1991) |
| E. quadrans | 0.8 dry | 1,8-cineole (43), trans-pinocarveol (20) | Bignell et al. (1997e) |
| E. quadricostata | 0.3–0.5 | α-pinene (21), β-pinene (15), 1,8-cineole (10), α-terpineol (11), globulol (5) | Brophy unpubl. |
| E. radiata | 6.3 dry | α-phellandrene (26), 1,8-cineole (12), cis- and trans-menth-2-en-1-ol (12 total), terpinen-4-ol (4), piperitone (10) | Li et al. (1995) |
| subsp. radiata 'Type' (syn. E. australiana, E. radiata var. australiana) | 3.0–3.5 | α-pinene, 1,8-cineole, α-terpineol, geraniol, citral | Baker and Smith (1920), Penfold and Morrison (1935) |
|  | 1.7–2.0 dry | 1,8-cineole (70) | Boland et al. (1991) |
|  | 1.5–2.3 | α-terpinene, α-phellandrene, γ-terpinene, terpinen-4-ol | Penfold and Morrison (1935) |
| 'Var. B' (syn. E. phellandra) | 3.0–4.5 | l-α-phellandrene, 1,8-cineole, terpineol, citral | Baker and Smith (1920), Penfold and Morrison (1935) |
| 'Var. C' (syn. E. australiana 'Var. C') | 0.4–3.9 | l-α-phellandrene, p-cymene, l-piperitone, l-piperitol | Penfold and Morrison (1940) |

Table 5.2 (Continued)

| Species | Oil yield[a] (%) | Characteristic constituents[b] | Reference |
|---|---|---|---|
| subsp. *radiata* | 0.4 dry | 1,8-cineole (65) | Bignell *et al.* (1998) |
|  | 3.4 dry | α-phellandrene (8), γ-terpinene (12), terpinen-4-ol (25) | Bignell *et al.* (1998) |
| subsp. *sejuncta* | 0.3–0.8 | α-pinene (80) | Brophy unpubl. |
| 'E. *rariflora*' [hybrid, E. *crebra* × E. *populnea*] | 2.0–2.9 dry | *l*-α- and β-pinene, car-4-ene, α-phellandrene, 1,8-cineole, *p*-cymene, aromadendrene, dehydroangustione | Penfold *et al.* (1930) |
| E. *raveretiana* | tr–0.1 | α-pinene (84), campholenic aldehyde | Boland *et al.* (1991) |
| E. *ravida* (syn. E. *salubris* var. *glauca*) | 1.6 dry | α-pinene (14), 1,8-cineole (35), torquatone (8) | Bignell *et al.* (1996d) |
|  | — | α-pinene, 1,8-cineole, *trans*-pinocarveol, β-eudesmol, torquatone | Brophy and Lassak unpubl. |
| E. *recta* | tr dry | α-pinene (11), 1,8-cineole (15), *trans*-pinocarveol (8), bicyclogermacrene (16) | Bignell *et al.* (1998) |
| E. *redacta* | 0.5 dry | 1,8-cineole (43), *trans*-pinocarveol (11), bicyclogermacrene (6) | Bignell *et al.* (1997g) |
| E. *redunca* | 1.2 | α-pinene, 1,8-cineole, sesquiterpenes | Baker and Smith (1920) |
|  | 0.1 dry | α-pinene (14), 1,8-cineole (19), aromadendrene (6), bicyclogermacrene (16) | Bignell *et al.* (1997c) |
| E. *regnans* | 0.8 | *l*-α-phellandrene, eudesmol (mixture of isomers) | Baker and Smith (1920) |
|  | 3.2 dry | α-phellandrene, *p*-cymene, *cis*- and *trans*-menth-2-en-1-ol, elemol, α-, β-, γ-eudesmol | Boland *et al.* (1991) |
|  | 3.0–5.8 dry | α-phellandrene (6), *p*-cymene (5), α-, β-, γ-eudesmol (43 total), aromatics (7) | Li *et al.* (1995) |
| E. *remota* | 0.7–2.0 | eudesmol (mixture of isomers) | Unpubl. (1961) |
|  | tr | α-, β-, γ-eudesmol (50 total), isobicyclogermacral (5) | Bignell *et al.* (1997b) |
| E. *resinifera* (syn. E. *hemilampra*) | 0.4 | α-pinene, 1,8-cineole, valeraldehyde, sesquiterpenes | Baker and Smith (1920) |
|  | 0.7 | 1,8-cineole (72) | Martínez *et al.* (1986) |
|  | tr | 1,8-cineole (13), *trans*-pinocarveol (11), spathulenol (14) | Bignell *et al.* (1997f) |
| E. *rhodantha* | 0.9 dry | α-pinene (28), 1,8-cineole (33), *trans*-pinocarveol (3) | Bignell *et al.* (1994c) |
| E. *rhombica* | 0.3–0.5 | 1,8-cineole (34), globulol (15), spathulenol (5) | Brophy unpubl. |
| E. *rigens* | 2.0 dry | 1,8-cineole (69), *trans*-pinocarveol (11) | Bignell *et al.* (1997e) |
| E. *rigidula* | 3.3 dry | α-pinene (25), 1,8-cineole (7), aromadendrene (19), bicyclogermacrene (12) | Bignell *et al.* (1997d) |
| E. *risdonii* | 1.4 | *l*-α-phellandrene, 1,8-cineole, amyl acetate | Baker and Smith (1920) |
|  | 1.3–2.3 dry | 1,8-cineole (60), α-terpineol (6), α-terpinyl acetate (6) | Li *et al.* (1995) |

Table 5.2 (Continued)

| Species | Oil yield[a] (%) | Characteristic constituents[b] | Reference |
|---|---|---|---|
| | 0.3 | 1,8-cineole (61), α-terpineol + terpinyl acetate (15 total) | Brophy unpubl. |
| E. robertsonii (syn. E. radiata subsp. robertsonii) | 1.3 dry | 1,8-cineole (61), α-terpineol (15) | Bignell et al. (1998) |
| | 1.4 | 1,8-cineole (65), linalool (5), α-terpineol (13) | Brophy unpubl. |
| E. robusta | 0.2 | α-pinene, α-phellandrene, 1,8-cineole, sesquiterpenes | Baker and Smith (1920) |
| | 1.7 | 1,8-cineole, p-cymene, linalool, terpinen-4-ol, α-terpineol, piperitone, citronellyl acetate | Dayal and Maheshwari (1985) |
| | tr | α-pinene (13), trans-pinocarveol (27) | Bignell et al. (1997f) |
| E. rodwayi | 0.5 | α-pinene, 1,8-cineole | Baker and Smith (1920) |
| cineole-rich form | 2.3–3.3 dry | α-pinene (18), 1,8-cineole (60), α-terpineol (3) | Li et al. (1996) |
| | 1.6–2.2 | α-pinene (20), 1,8-cineole (63) | Boland et al. (1991) |
| phellandrene-rich form | 2.1 | α-phellandrene (43), 1,8-cineole (10), p-cymene (8), globulol (9) | Boland et al. (1991) |
| E. rossii | 0.6 | α-pinene, 1,8-cineole, eudesmol (mixture of isomers) | Baker and Smith (1920) |
| | 1.6 | 1,8-cineole (34), α-, β-, γ-eudesmol (34 total) | Brophy unpubl. |
| E. rubida | 0.1 | α-pinene, α-phellandrene, 1,8-cineole, sesquiterpenes | Baker and Smith (1920) |
| | 0.6–1.1 | p-cymene, spathulenol (36) | Boland et al. (1991) |
| | 1.1–3.9 dry | α-pinene (11), 1,8-cineole (45), α-terpineol (6), spathulenol (6) | Li et al. (1996) |
| E. rubiginosa | 0.2 | α-pinene (47), β-pinene (23) | Brophy et al. (1998a) |
| E. rudderi | 0.3 | p-cymene, aldehydes | Baker and Smith (1920) |
| E. rudis | 1.2 | d-α-pinene, 1,8-cineole, aldehydes | Baker and Smith (1920) |
| | 0.7 dry | α-phellandrene (3), alloaromadendrene (4), bicyclogermacrene (67) | Bignell et al. (1996a) |
| E. rugosa | 0.9 dry | α-pinene (9), 1,8-cineole (31), aromadendrene (9) | Bignell et al. (1994b) |
| E. saligna | 0.1 | α-pinene, p-cymene | Baker and Smith (1920) |
| | 0.3–0.5 | α-pinene (73), campholenic aldehyde, α-terpineol | Boland et al. (1991) |
| E. salmonophloia | 1.4–3.6 | α-pinene, 1,8-cineole, aromadendrene | Watson (1935/36) |
| | 1.6 | piperitone (42) | Martínez et al. (1986) |
| | 2.7 dry | 1,8-cineole (10), p-cymene (17), cryptone (11) | Bignell et al. (1996e) |
| E. salubris | 1.4 | α-pinene, 1,8-cineole, p-cymene, aldehydes | Baker and Smith (1920) |
| | 1.5 dry | α-pinene (16), 1,8-cineole (49), aromadendrene (5) | Bignell et al. (1996d) |
| | — | α-pinene, 1,8-cineole, p-cymene, pinocarvone, trans-pinocarveol | Brophy and Lassak unpubl. |

Table 5.2 (Continued)

| Species | Oil yield[a] (%) | Characteristic constituents[b] | Reference |
|---|---|---|---|
| E. sargentii subsp. sargentii | 1.6 dry | α-pinene (12), 1,8-cineole (49), trans-pinocarveol (5) | Bignell et al. (1996e) |
| E. saxatilis | 3.5–5.1 dry | 1,8-cineole (71), sesquiterpenes | Boland et al. (1991) |
| E. scabrida | 0.3 | α-pinene (23), globulol (7), guaiol (11), α-, β-, γ-eudesmol (8 total) | Brophy et al. (1998b) |
| E. scias | 0.2–1.8 | α-pinene (18), 1,8-cineole (43), α-terpineol (12) | Doran et al. (1995) |
| E. sclerophylla | 0.7 | 1,8-cineole (14), globulol (15), viridiflorol (7), spathulenol (28) | Brophy unpubl. |
| E. scoparia | 1.2–1.5 | α-pinene, 1,8-cineole (71), globulol | Boland et al. (1991) |
| E. scyphocalyx | 2.9 dry | α-pinene (30), 1,8-cineole (38), aromadendrene (10) | Bignell et al. (1996b) |
| E. seeana | 0.8 | α-pinene, 1,8-cineole | Baker and Smith (1920) |
|  | 0.4 dry | 1,8-cineole (10), p-cymene (25), palustrol (9) | Bignell et al. (1996a) |
| E. semiglobosa | tr | α-pinene (36), 1,8-cineole (21), aromadendrene (6), bicyclogermacrene (15) | Bignell et al. (1997d) |
| E. semota | 0.8 dry | 1,8-cineole (61), p-cymene (3) | Bignell et al. (1997g) |
| E. sepulcralis | 0.4 dry | α-pinene (8), 1,8-cineole (62), α-terpineol (5) | Bignell et al. (1997b) |
| E. serpentinicola | 1.5 | 1,8-cineole (37), α-, β-, γ-eudesmol (34 total) | Brophy unpubl. |
| E. sessilis | 2.0 dry | α-pinene (12), 1,8-cineole (61), aromadendrene (5) | Bignell et al. (1994c) |
| E. setosa | tr–0.1 dry | aromadendrene (13), globulol (40), β-eudesmol (5) | Brophy unpubl. |
| E. sheathiana | 4.6 dry | α-pinene (15), limonene (4), 1,8-cineole (57) | Bignell et al. (1995a) |
| E. shirleyi | 0.6–0.7 | α- and β-pinene, bicyclogermacrene, δ-cadinene, globulol | Boland et al. (1991) |
| E. siderophloia | 0.1 | pinene, phellandrene | Baker and Smith (1920) |
|  | 0.2 | α-pinene (66), α-phellandrene, trans-pinocarveol, α-terpineol | Boland et al. (1987) |
| E. sideroxylon | 1.5–2.5 | α-pinene, 1,8-cineole, sesquiterpenes | Baker and Smith (1920) |
|  | 2.6–3.2 | 1,8-cineole, α-terpinyl acetate, globulol, α-, β-, γ-eudesmol | Boland et al. (1991) |
|  | 0.4 dry | α-pinene (14), 1,8-cineole (60), bicyclogermacrene (5) | Bignell et al. (1997a) |
| E. sieberi (syn. E. sieberiana) | 0.5 | l-α-phellandrene, 1,8-cineole, piperitone | Baker and Smith (1920) |
|  | 0.4 | β-pinene (24), car-3-ene (15), β-phellandrene (28), 1,8-cineole (14) | Arora et al. (1971) |
|  | 3.3–3.9 dry | α-phellandrene (7), β-phellandrene (6), 1,8-cineole (8), α-, β-, γ-eudesmol (35 total) | Li et al. (1995) |
|  | 1.0–1.3 | 1,8-cineole (26), p-cymene (27), α-, β-, γ-eudesmol (20 total) | Brophy unpubl. |

Table 5.2 (Continued)

| Species | Oil yield[a] (%) | Characteristic constituents[b] | Reference |
|---|---|---|---|
| E. signata | 0.0–1.1 | β-phellandrene (26), p-cymene (19), α-terpineol (12), spathulenol (7) | Brophy unpubl. |
| E. smithii | 1.1–2.2 | isovaleraldehyde, α-pinene, 1,8-cineole, eudesmol (mixture of isomers) | Baker and Smith (1920), Mus. A. and S. unpubl. |
|  | 2.4–3.0 | α-pinene, 1,8-cineole (81), globulol, β-eudesmol | Boland et al. (1991) |
|  | 0.7 dry | 1,8-cineole (78) | Bignell et al. (1998) |
| E. socialis | 1.0 dry | α-pinene (25), 1,8-cineole (41), trans-pinocarveol (7) | Bignell et al. (1995b) |
|  | 2.2 dry | α-pinene (9), β-pinene (13), p-cymene (21), spathulenol (10) | Bignell et al. (1996g) |
| E. sparsa | 2.6 dry | α-pinene, limonene, 1,8-cineole, p-cymene | Brophy and Lassak (1986) |
|  | 5.1 dry | α-pinene (5), 1,8-cineole (15), bicyclogermacrene (64) | Bignell et al. (1995c) |
| E. spathulata | 1.4 | d-α-pinene, 1,8-cineole, aromadendrene, aldehydes | Phillips (1923) |
| subsp. grandiflora | 1.3 dry | α-pinene (33), 1,8-cineole (38), aromadendrene (3) | Bignell et al. (1996d) |
| subsp. spathulata | 2.1 dry | α-pinene (20), 1,8-cineole (44), aromadendrene (9) | Bignell et al. (1996d) |
| E. sphaerocarpa | 1.3–1.4 | α-phellandrene, 1,8-cineole, p-cymene | Boland et al. (1991) |
| E. squamosa | 0.7 | isovaleraldehyde, α-pinene, 1,8-cineole | Baker and Smith (1920), Mus. A. and S. unpubl. |
| E. staeri | tr | calamenene (5), globulol (4), spathulenol (14), torquatone (7) | Bignell et al. (1997b) |
| E. staigeriana | 1.2–1.5 | l-limonene, citral, geranyl acetate, geraniol, sesquiterpenes | Baker and Smith (1920) |
|  | 2.9–3.4 | limonene, β-phellandrene, methyl geranate, geranial, geranyl acetate | Boland et al. (1991) |
|  | 2.3 dry | 1,8-cineole (24), terpinolene (10), δ-terpineol (12), methyl geranate (7), geranial (19), geraniol (7) | Bignell et al. (1997a) |
| E. stannicola | 0.1–0.5 | spathulenol (10), α-, β-, γ-eudesmol (60 total) | Brophy unpubl. |
| E. steedmanii | 1.2 dry | α-pinene (21), 1,8-cineole (32), aromadendrene (15) | Bignell et al. (1996d) |
| E. stellulata | 0.3 | l-α-phellandrene, 1,8-cineole, sesquiterpenes | Baker and Smith (1920) |
|  | 0.6–1.1 | β-phellandrene (12), p-cymene (27), α-, β-, γ-eudesmol (10 total) | Brophy unpubl. |
| E. stenostoma | 1.7–2.4 | α-pinene, α- and β-phellandrene, α-terpinene, 1,8-cineole, p-cymene, cis- and trans-p-menth-2-en-1-ol, terpinen-4-ol, cis- and trans-piperitol, α-terpineol, eudesmol (all three isomers) | Lassak and Southwell (1982) |
|  | 0.5 | p-cymene (29), piperitone (38) | Brophy unpubl. |
| E. stoatei | 2.6 dry | α-pinene (13), 1,8-cineole (34), aromadendrene (12) | Bignell et al. (1994a) |

Table 5.2 (Continued)

| Species | Oil yield[a] (%) | Characteristic constituents[b] | Reference |
|---|---|---|---|
| E. stowardii | 1.4 dry | α-pinene (47), 1,8-cineole (19) | Bignell et al. (1996e) |
| E. striaticalyx | 4.4 dry | α-pinene (29), 1,8-cineole (25), β-eudesmol (4) | Bignell et al. (1995a) |
| subsp. beadellii | 0.7 dry | α-pinene (15), 1,8-cineole (16), aromadendrene (17), trans-pinocarveol (22) | Bignell et al. (1997e) |
| subsp. canescens | 3.7 dry | α-pinene (13), 1,8-cineole (13), aromadendrene (30), globulol (7), spathulenol (8) | Bignell et al. (1997e) |
| subsp. gypsophila | 0.8 dry | 1,8-cineole (56), trans-pinocarveol (17), β-eudesmol (8) | Bignell et al. (1997e) |
| subsp. striaticalyx | 1.7 dry | 1,8-cineole (78), p-cymene (5) | Bignell et al. (1997e) |
| E. stricklandii | 2.8 dry | α-pinene (20), 1,8-cineole (14), torquatone (21) | Bignell et al. (1996c) |
|  | — | α-pinene, 1,8-cineole, torquatone, sesquiterpene alcohols | Brophy and Lassak unpubl. |
| E. stricta | 0.5 | α-pinene, 1,8-cineole, piperitone, eudesmol (mixture of isomers), aldehydes | Baker and Smith (1920), Mus. A. and S. unpubl. |
| E. sturgissiana | 1.1–2.5 | 1,8-cineole (90) | Boland et al. (1991) |
| E. subangusta |  |  |  |
| subsp. cerina | 3.1 dry | α-pinene (9), α-phellandrene (16), 1,8-cineole (6), β-caryophyllene (11), bicyclogermacrene (42) | Bignell et al. (1997g) |
| subsp. pusilla | 1.3 dry | α-pinene (18), 1,8-cineole (57), β-eudesmol (6) | Bignell et al. (1997g) |
| subsp. subangusta | 3.3 dry | α-pinene (27), 1,8-cineole (35), bicyclogermacrene (8) | Bignell et al. (1997g) |
| E. subcrenulata | 2.5–4.6 dry | α-pinene, 1,8-cineole (62), α-terpineol | Boland et al. (1991) |
|  | 1.1–4.1 dry | α-pinene (15), 1,8-cineole (48), p-cymene (4), spathulenol (6) | Li et al. (1996) |
| E. suberea | 1.4 dry | tasmanone (94) | Bignell et al. (1997b) |
| E. sublucida | 1.9 dry | α-pinene (31), 1,8-cineole (16), trans-pinocarveol (12) | Bignell et al. (1996g) |
| E. subtilior | 0.6 | 1,8-cineole (58), α-, β-, γ-eudesmol (10 total) | Brophy unpubl. |
| E. subtilis | 2.7 dry | α-pinene (7), 1,8-cineole (58), aromadendrene (9) | Bignell et al. (1997c) |
| E. suffulgens | 0.3–0.9 | α-pinene (20), β-pinene (38), α-terpineol (12), α-, β-, γ-eudesmol (5 total) | Brophy unpubl. |
| E. suggrandis | 1.1 dry | α-pinene (27), 1,8-cineole (26), carvone (19) | Bignell et al. (1998) |
| E. synandra | 2.4 dry | α-pinene (16), β-pinene (2), 1,8-cineole (71) | Bignell et al. (1994c) |
| 'E. taeniola' [hybrid, E. amygdalina E. sieberi] | 0.7 | l-α-phellandrene, eudesmol (mixture of isomers), sesquiterpenes | Baker and Smith (1920) |
| E. talyuberlup | tr | α-pinene (39), limonene (4), p-cymene (6) | Bignell et al. (1996b) |

Table 5.2 (Continued)

| Species | Oil yield[a] (%) | Characteristic constituents[b] | Reference |
|---|---|---|---|
| E. tectifica | 0.4–1.0 | 1,8-cineole (65), p-cymene, aromadendrene | Boland et al. (1991) |
| E. tenera | 1.8 dry | 1,8-cineole (62), trans-pinocarveol (11) | Bignell et al. (1998) |
| E. tenuipes | 0.4 | α-pinene (5), β-pinene (31), α-terpineol (11), globulol (5) | Brophy et al. (1998a) |
| E. tenuiramis | 3.1–5.7 dry | α-phellandrene (11), 1,8-cineole (35), cis- and trans-menth-2-en-1-ol (10 total) | Li et al. (1995) |
| E. tenuis | 1.3 dry | α-pinene (15), 1,8-cineole (24), p-cymene (21), trans-pinocarveol (11) | Bignell et al. (1995a) |
| E. terebra | 0.6 dry | α-pinene (12), 1,8-cineole (45), trans-pinocarveol (13) | Bignell et al. (1996d) |
| E. tereticornis ('var. cineolifera' is a cineole-rich form) | 0.5–0.9 | α-pinene, phellandrene, 1,8-cineole, p-cymene, cuminal | Baker and Smith (1920), Shiva et al. (1984) |
| | 0.9–1.4 | α- and β-pinene, 1,8-cineole, spathulenol, α-, β-, γ-eudesmol, globulol | Boland et al. (1991) |
| | 0.6 dry | p-cymene (28), cryptone (15), caryophyllene oxide (9) | Bignell et al. (1996a) |
| E. tessellaris | 0.2 | α-pinene, 1,8-cineole, aromadendrene | Baker and Smith (1920) |
| | tr–0.1 | β-pinene, limonene, spathulenol | Boland et al. (1991) |
| | tr | α-pinene (5), aromadendrene (10), globulol (5), torquatone (5) | Bignell et al. (1997f) |
| E. tetragona | 0.5 dry | d-α-pinene, l-α-phellandrene, 1,8-cineole | Watson (1935/36) |
| | 0.1 dry | aromadendrene (13), bicyclogermacrene (11), α-, β-, γ-eudesmol (18 total) | Bignell et al. (1996f) |
| | 0.4–0.6 | α-pinene (13), globulol (12), α-, β-, γ-eudesmol (38 total) | Brophy unpubl. |
| E. tetraptera | 2.1 dry | α-pinene (21), 1,8-cineole (24), aromadendrene (13) | Bignell et al. (1994a) |
| | 0.2 | α-pinene (10), 1,8-cineole (44), globulol (7), α-, β-, γ-eudesmol (5 total) | Brophy unpubl. |
| E. tetrodonta | 0.2 | α-pinene (55), α-terpineol (9), globulol (3) | Brophy unpubl. |
| E. thozetiana | tr | 1,8-cineole (21), verbenone (7), torquatone (3) | Bignell et al. (1997a) |
| E. tindaliae | 0.2 | α-pinene (35), α-, β-, γ-eudesmol (30 total) | Brophy unpubl. |
| 'E. tinghaensis' [hybrid, E. caliginosa × E. mckieana] | 0.4–0.9 | | Unpubl. (1961) |
| E. tintinnans | 0.1 dry | α-pinene (64), limonene (5), bicyclogermacrene (8) | Bignell et al. (1997f) |
| E. todtiana | tr | β-pinene (11), 1,8-cineole (9), viridiflorene (8) | Bignell et al. (1997b) |

Table 5.2 (Continued)

| Species | Oil yield[a] (%) | Characteristic constituents[b] | Reference |
|---|---|---|---|
| | tr-0.1 | 1,8-cineole (36), globulol (6), spathulenol (11), α-, β-, γ-eudesmol (10 total) | Brophy unpubl. |
| E. torelliana | 0.3 | α- and β-pinene, p-cymene, ocimene, aromadendrene | Webb et al. (1956) |
| | 0.3 dry | α-pinene (54), β-phellandrene (9), β-caryophyllene (8) | Brophy unpubl. |
| E. torquata | — | α-pinene, 1,8-cineole, eudesmol (mixture of isomers), torquatone | Bowyer and Jefferies (1959) |
| | 3.6 dry | α-pinene (19), β-eudesmol (10), torquatone (41) | Bignell et al. (1994b) |
| E. trachyphloia | 0.2 | d-α-pinene, aromadendrene, aldehydes | Baker and Smith (1920) |
| | 0.2 dry | bicyclogermacrene (54), globulol (8) | Bignell et al. (1997f) |
| E. transcontinentalis (syn. E. oleosa var. glauca) | 3.9 dry | α-pinene (26), 1,8-cineole (47), aromadendrene (7) | Bignell et al. (1995b) |
| E. triflora | 0.7–0.8 | α-pinene, α- and β-phellandrene, p-cymene, cis- and trans-p-menth-2-en-1-ol, terpinen-4-ol, cis- and trans-piperitol, piperitone | Lassak and Southwell (1982) |
| | 0.3–0.5 | β-pinene (6), p-cymene (29), cis- and trans-menth-2-en-1-ol (12 total), piperitone (29) | Brophy unpubl. |
| E. trivalvis | 1.0 dry | α-pinene (10), 1,8-cineole (52), aromadendrene (5) | Bignell et al. (1997g) |
| E. tumida | 0.6 dry | α-pinene (31), 1,8-cineole (20), aromadendrene (10) | Bignell et al. (1997c) |
| E. umbonata | tr | α-, β-, γ-eudesmol (20 total), torquatone (37) | Bignell et al. (1997f) |
| E. umbra | 0.6 | d-α-pinene, 1,8-cineole, sesquiterpenes | Baker and Smith (1920) |
| | 0.7 | α-pinene, 1,4-cineole, 1,8-cineole | Miranda et al. (1981) |
| | 1.5 | α-pinene (37), 1,8-cineole (30) | Brophy unpubl. |
| subsp. carnea (syn. E. carnea) | 0.2 | d-α-pinene, 1,8-cineole, terpinyl acetate | Baker and Smith (1920) |
| E. umbrawarrensis | 0.4–0.6 | α-pinene, β-pinene (41), α-terpineol, globulol, spathulenol | Boland et al. (1991) |
| E. uncinata | 0.6 dry | 1,8-cineole (47), pinocarvone (8), aromadendrene (6), trans-pinocarveol (21) | Bignell et al. (1997d) |
| 'E. unialata' [hybrid, E. globulus × E. viminalis] | 0.9 | d-α-pinene, 1,8-cineole, sesquiterpenes | Baker and Smith (1920) |
| E. urnigera | 1.1 | d-α-pinene, 1,8-cineole | Baker and Smith (1920) |
| | 1.4–5.6 dry | α-pinene (16), 1,8-cineole (56), α-terpineol (3), α-terpinyl acetate (4) | Li et al. (1996) |
| E. urophylla | 0.2 | α-pinene (7), p-cymene (75) | Singh et al. (1988) |

Table 5.2 (Continued)

| Species | Oil yield[a] (%) | Characteristic constituents[b] | Reference |
|---|---|---|---|
| | 0.6–1.4 | α-pinene (20), β-pinene (5), 1,8-cineole (40) | Doran et al. (1995) |
| E. varia | | | |
| subsp. salsuginosa | 0.2 dry | α-pinene (8), 1,8-cineole (12), bicyclogermacrene (31), spathulenol (5), α-, β-, γ-eudesmol (8 total) | Bignell et al. (1997c) |
| subsp. varia | 0.6 dry | β-caryophyllene (9), bicyclogermacrene (76), spathulenol (3) | Bignell et al. (1997c) |
| E. vegrandis | 1.9 dry | α-pinene (35), 1,8-cineole (34) | Bignell et al. (1998) |
| E. vernicosa | 0.8 | d-α-pinene, 1,8-cineole, aldehydes | Baker and Smith (1920) |
| | 2.2–2.8 dry | α-pinene (18), 1,8-cineole (62), α-terpineol (3) | Li et al. (1996) |
| E. vicina | 1.2 dry | 1,8-cineole (64), trans-pinocarveol (9), aromadendrene (3) | Bignell et al. (1996a) |
| E. viminalis | 0.5 | d-α-pinene, α-phellandrene, 1,8-cineole, sesquiterpenes | Baker and Smith (1920) |
| | 1.1–1.6 | α-pinene, 1,8-cineole (64), globulol | Boland et al. (1991) |
| | 2.2–3.4 dry | α-pinene (9), limonene (5), 1,8-cineole (50), globulol (7) | Li et al. (1996) |
| 'Var. A' | 0.8 | α-pinene, 1,8-cineole, benzaldehyde, sesquiterpenes | Baker and Smith (1920) |
| E. virens | 0.1–0.3 | α-pinene (80), bicyclogermacrene (4) | Brophy unpubl. |
| 'E. virgata' [possibly a hybrid, E. luehmanniana × E. obtusiflora] | 0.3–0.6 | l-α-phellandrene, 1,8-cineole, piperitone, eudesmol (mixture of isomers), sesquiterpenes | Baker and Smith (1920) |
| E. viridis | 1.5 | α-pinene, 1,8-cineole | Penfold and Morrison (1951) |
| | 2.1 | α- and β-pinene, 1,8-cineole (93), p-cymene, cryptone, α-terpineol, geraniol, isopulegol | Ghanim and Jayaraman (1978) |
| | 0.1 dry | α-pinene (5), 1,8-cineole (44), aromadendrene (18) | Bignell et al. (1995c) |
| 'E. vitrea' [hybrid, E. pauciflora subsp. pauciflora × E. radiata subsp. radiata] | 1.5 | l-α-phellandrene, 1,8-cineole, sesquiterpenes | Baker and Smith (1920) |
| E. wandoo | tr | α-pinene (8), 1,8-cineole (24), γ-terpinene (10), p-cymene (18) | Bignell et al. (1997c) |
| p-cymene-rich form | 0.7 | p-cymene (65) | Brophy unpubl. |
| p-cymene-poor form | 1.8 | 1,8-cineole (28), γ-terpinene (12), p-cymene (10), α-, β-, γ-eudesmol (20 total) | Brophy unpubl. |
| E. watsoniana subsp. capillata | 0.1–0.6 | α-pinene (38), bicyclogermacrene (9), globulol (8), α-, β-, γ-eudesmol (10 total) | Brophy et al. (1998b) |

Table 5.2 (Continued)

| Species | Oil yield[a] (%) | Characteristic constituents[b] | Reference |
|---|---|---|---|
| subsp. watsoniana | tr 0.1–0.4 | bicyclogermacrene (52), globulol (4) α-pinene (5), β-pinene (26), bicyclogermacrene (8), globulol (15) | Bignell et al. (1997f) Brophy et al. (1998b) |
| E. websteriana | 1.4 dry | α-pinene (16), 1,8-cineole (62), β-eudesmol (3) | Bignell et al. (1996c) |
| E. wetarensis | — | α-pinene (5), α-phellandrene (14), γ-terpinene (15), p-cymene (20) | Pryor et al. (1995) |
| E. whitei | 0.1–1.9 | nerolidol (55), α-terpineol (8) | Brophy unpubl. |
| E. wilcoxii | 1.4–2.3 dry | 1,8-cineole, γ-terpinene, α-, β-, γ-eudesmol | Boland et al. (1991), Boland and Kleinig (1983) |
| E. williamsiana | 0.6–1.0 | α-, β-, γ-eudesmol (70 total) | Brophy unpubl. |
| E. willisii | — | 1,8-cineole (74), α-terpineol (4), α-, β-, γ-eudesmol (4 total) | Brophy and Doran unpubl. |
| E. woodwardii | 2.6 dry | α-pinene (28), 1,8-cineole (17), aromadendrene (11), torquatone (12) | Bignell et al. (1995a) |
|  | — | α-pinene, 1,8-cineole, α-, β-, γ-eudesmol, torquatone | Brophy and Lassak unpubl. |
| E. wyolensis | 1.4 dry | α-pinene (7), 1,8-cineole (12), spathulenol (15) | Bignell et al. (1996g) |
| E. xanthonema | 0.7 dry | α-pinene (20), 1,8-cineole (15), aromadendrene (14), bicyclogermacrene (18) | Bignell et al. (1997c) |
| E. yalatensis | tr dry | α-pinene (8), β-pinene (10), 1,8-cineole (14), torquatone (4) | Bignell et al. (1995b) |
| E. yarraensis | 0.1–0.6 | benzaldehyde | Lassak and Southwell (1977) |
|  | 0.1 | benzaldehyde (84), nerolidol, globulol | Boland et al. (1991) |
|  | tr dry | terpinen-4-ol (17), piperitone (11), spathulenol (11) | Bignell et al. (1998) |
| E. yilgarnensis | 0.8 dry | 1,8-cineole (12), p-cymene (14), cryptone (10) | Bignell et al. (1997e) |
| E. youmanii | 1.3 | α-pinene, myrcene, α-phellandrene, limonene, 1,8-cineole, p-cymene, terpinen-4-ol, α-terpineol, α-, β-, γ-eudesmol | Brophy et al. (1982) |
|  | tr | 1,8-cineole (12), alloaromadendrene (11), spathulenol (10) | Bignell et al. (1997f) |
| E. youngiana | 3.1 dry | α-pinene (25), 1,8-cineole (47), trans-pinocarveol (3) | Bignell et al. (1994c) |
| E. yumbarrana | 1.4 dry | α-pinene (29), 1,8-cineole (37), trans-pinocarveol (8) | Bignell et al. (1995b) |
| E. zopherophloia | 1.7 dry | α-pinene (9), 1,8-cineole (72) | Bignell et al. (1997g) |

a Based on fresh weight of leaf except where 'dry' indicates dry weight basis.
b Figures in parentheses refer to the percentage abundance of the constituent in the oil.
c Indicates trace (<0.05%).

## References

Adams, R.P. (1995) *Identification of Essential Oil Components by Gas Chromatography-Mass Spectroscopy*, Allured, Carol Stream, Illinois, USA.

Agarwal, S.G., Vashist, V.N. and Atal, C.K. (1970) The essential oil of Jammu grown *Eucalyptus macarthurii*. *Flavour Ind.*, 1, 625–626.

Ahmadouch, A., Bellakdar, J., Berrada, M., Denier, C. and Pinel, R. (1985) Chemical analysis of the essential oils of five species of *Eucalyptus* acclimated in Morocco (in French). *Fitoterapia*, 56, 209–220.

Ammon, D.G., Barton, A.F.M., Clarke, D.A. and Tjandra, J. (1985) Rapid and accurate determination of terpenes in the leaves of *Eucalyptus* species. *Analyst*, 110, 921–924.

Arigoni, D., Cane, D.E., Shim, J.H., Croteau, R. and Wagschal, K. (1993) Monoterpene cyclization mechanisms and the use of natural abundance deuterium NMR – Short cut or primrose path? *Phytochemistry*, 32, 623–631.

Arora, S.K., Agarwal, S.G., Vashist, V.N. and Madan, C.L. (1971) The essential oil of *Eucalyptus sieberiana* F. Muell. raised in Jammu. *Indian Perfumer*, 15(Part I), 16–18.

Aukrust, I.R. and Skattebol, L. (1996) The synthesis of (−)-robustadial A from *Eucalyptus robusta* and some analogues. *Acta Chem. Scand.*, 50, 132–140.

Baker, R.T. and Smith, H.G. (1920) *A Research on the Eucalypts, Especially in Regard to Their Essential Oils*, 2nd edn, NSW Government Printer, Sydney.

Baslas, R.K. and Saxena, S. (1984) Chemical examination of essential oil from the fruits of *Eucalyptus globulus* Labill. *Herba Hungarica*, 23(3), 21–23.

Begum, S., Farhat, B.S. and Siddiqui, S. (1997) Triterpenoids from the leaves of *Eucalyptus camaldulensis* var. *obtusa*. *J. Nat. Prod.*, 60, 20–23.

Berry, P.A. (1947) The seasonal variation of the essential oil from the growing tips of *E. cneorifolia* with special reference to the occurrence of cymene. *J. Aust. Chem. Inst.*, 14, 176–200.

Berry, P.A. and Swanson, T.B. (1942) A note on the essential oil of *Eucalyptus conglobata* var. *anceps*. *J. Proc. Roy. Soc. N.S.W.*, 76, 53–54.

Bick, I.R.C., Blackman, A.J., Hellyer, R.O. and Horn, D.H.S. (1965) The isolation and structure of flavesone. *J. Chem. Soc.*, 3690–3693.

Bignell, C.M., Dunlop, P.J., Brophy, J.J. and Jackson, J.F. (1994a) Volatile leaf oils of some south-western and southern Australian species of the genus *Eucalyptus*. Part I – Subgenus Symphyomyrtus, Section Dumaria, Series Incrassatae. *Flavour Fragr. J.*, 9, 113–117.

Bignell, C.M., Dunlop, P.J., Brophy, J.J. and Jackson, J.F. (1994b) Volatile leaf oils of some south-western and southern Australian species of the genus *Eucalyptus*. Part II – Subgenus Symphyomyrtus, Section Dumaria, Series Torquatae. *Flavour Fragr. J.*, 9, 167–171.

Bignell, C.M., Dunlop, P.J., Brophy, J.J. and Jackson, J.F. (1994c) Volatile leaf oils of some south-western and southern Australian species of the genus *Eucalyptus*. Part III – Subgenus Symphyomyrtus, Section Bisectaria, Series Macrocarpae. *Flavour Fragr. J.*, 9, 309–313.

Bignell, C.M., Dunlop, P.J., Brophy, J.J. and Jackson, J.F. (1995a) Volatile leaf oils of some south-western and southern Australian species of the genus *Eucalyptus*. Part IV – Subgenus Symphyomyrtus, Section Dumaria, Series Dumosae. *Flavour Fragr. J.*, 10, 85–91.

Bignell, C.M., Dunlop, P.J., Brophy, J.J. and Jackson, J.F. (1995b) Volatile leaf oils of some south-western and southern Australian species of the genus *Eucalyptus*. Part V – Subgenus Symphyomyrtus, Section Bisectaria, Series Oleosae. *Flavour Fragr. J.*, 10, 313–317.

Bignell, C.M., Dunlop, P.J., Brophy, J.J. and Jackson, J.F. (1995c) Volatile leaf oils of some south-western and southern Australian species of the genus *Eucalyptus*. Part VI – Subgenus Symphyomyrtus, Section Adnataria. *Flavour Fragr. J.*, 10, 359–364.

Bignell, C.M., Dunlop, P.J., Brophy, J.J. and Jackson, J.F. (1996a) Volatile leaf oils of some south-western and southern Australian species of the genus *Eucalyptus*. Part VII – Subgenus Symphyomyrtus, Section Exsertaria. *Flavour Fragr. J.*, 11, 35–41.

Bignell, C.M., Dunlop, P.J., Brophy, J.J. and Jackson, J.F. (1996b) Volatile leaf oils of some south-western and southern Australian species of the genus *Eucalyptus*. Part VIII – Subgenus Symphyomyrtus,

(a) Section Bisectaria, Series Cornutae and Series Bakeranae and (b) Section Dumaria, unpublished Series Furfuraceae group. *Flavour Fragr. J.*, 11, 43–47.

Bignell, C.M., Dunlop, P.J., Brophy, J.J. and Jackson, J.F. (1996c) Volatile leaf oils of some south-western and southern Australian species of the genus *Eucalyptus*. Part IX – Subgenus Symphyomyrtus, Section Bisectaria, (a) Series Elongatae, (b) unpublished Series Stricklandiae, (c) Series Kruseanae and (d) Series Orbifoliae. *Flavour Fragr. J.*, 11, 95–100.

Bignell, C.M., Dunlop, P.J., Brophy, J.J. and Jackson, J.F. (1996d) Volatile leaf oils of some south-western and southern Australian species of the genus *Eucalyptus*. Part X – Subgenus Symphyomyrtus, Section Bisectaria, (a) unpublished Series Erectae, (b) Series Contortae and (c) Series Decurvae. *Flavour Fragr. J.*, 11, 101–106.

Bignell, C.M., Dunlop, P.J., Brophy, J.J. and Jackson, J.F. (1996e) Volatile leaf oils of some south-western and southern Australian species of the genus *Eucalyptus*. Part XI – Subgenus Symphyomyrtus. A – Section Bisectaria, (a) Series Occidentales, (b) unpublished Series Annulatae, (c) Series Micromembranae, (d) Series Obliquae, (e) Series Dundasianae, (f) Series Cooperianae, (g) Series Halophilae, (h) Series Salmonophloiae and (i) Series Pubescentes. B – Section Dumaria, (a) Series Merrickianae. *Flavour Fragr. J.*, 11, 107–112.

Bignell, C.M., Dunlop, P.J., Brophy, J.J. and Jackson, J.F. (1996f) Volatile leaf oils of some south-western and southern Australian species of the genus *Eucalyptus*. Part XII. A – Subgenus Eudesmia. B – Subgenus Symphyomyrtus, (a) Section Exsertaria, (b) Series Globulares. *Flavour Fragr. J.*, 11, 145–151.

Bignell, C.M., Dunlop, P.J., Brophy, J.J. and Jackson, J.F. (1996g) Volatile leaf oils of some south-western and southern Australian species of the genus *Eucalyptus* (Series I). Part XIII – (a) Series Subulatae, (b) Series Curviptera, (c) Series Contortae, (d) Series Incognitae, (e) Series Terminaliptera, (f) Series Inclusae, (g) Series Microcorythae and (h) Series Cornutae. *Flavour Fragr. J.*, 11, 339–347.

Bignell, C.M., Dunlop, P.J., Brophy, J.J. and Jackson, J.F. (1997a) Volatile leaf oils of some Queensland and northern Australian species of the genus *Eucalyptus* (Series II). Part I – Subgenus Symphyomyrtus, Section Adnataria, (a) Series Oliganthae, (b) Series Ochrophloiae, (c) Series Moluccanae, (d) Series Polyanthemae, (e) Series Paniculatae, (f) Series Melliodorae and (g) Series Porantheroideae. *Flavour Fragr. J.*, 12, 19–27.

Bignell, C.M., Dunlop, P.J., Brophy, J.J. and Fookes, C.J.R. (1997b) Volatile leaf oils of some south-western and southern Australian species of the genus *Eucalyptus* (Series I). Part XIV – Subgenus Monocalyptus. *Flavour Fragr. J.*, 12, 177–183.

Bignell, C.M., Dunlop, P.J. and Brophy, J.J. (1997c) Volatile leaf oils of some south-western and southern Australian species of the genus *Eucalyptus* (Series I). Part XV – Subgenus Symphyomyrtus, Section Bisectaria, Series Levispermae. *Flavour Fragr. J.*, 12, 185–193.

Bignell, C.M., Dunlop, P.J. and Brophy, J.J. (1997d) Volatile leaf oils of some south-western and southern Australian species of the genus *Eucalyptus* (Series I). Part XVI – Subgenus Symphyomyrtus, Section Bisectaria, Series Cneorifoliae, Series Porantherae and Series Falcatae. *Flavour Fragr. J.*, 12, 261–267.

Bignell, C.M., Dunlop, P.J. and Brophy, J.J. (1997e) Volatile leaf oils of some south-western and southern Australian species of the genus *Eucalyptus* (Series I). Part XVII – Subgenus Symphyomyrtus, (i) Section Bisectaria, Series Calycogonae and (ii) Section Dumaria, Series Dumosae, Series Rigentes and Series Ovulares. *Flavour Fragr. J.*, 12, 269–275.

Bignell, C.M., Dunlop, P.J. and Brophy, J.J. (1997f) Volatile leaf oils of some Queensland and northern Australian species of the genus *Eucalyptus* (Series II). Part II – Subgenera (a) Blakella, (b) Corymbia, (c) unnamed, (d) Idiogenes, (e) Monocalyptus and (f) Symphyomyrtus. *Flavour Fragr. J.*, 12, 277–284.

Bignell, C.M., Dunlop, P.J. and Brophy, J.J. (1997g) Volatile leaf oils of some south-western and southern Australian species of the genus *Eucalyptus* (Series I). Part XVIII. A – Subgenus Monocalyptus. B – Subgenus Symphyomyrtus, (i) Section Guilfoyleanae, (ii) Section Bisectaria, Series Accedentes, Series Occidentales, Series Levispermae, Series Loxophlebae, Series Macrocarpae, Series Orbifoliae, Series Calycogonae, (iii) Section Dumaria, Series Incrassatae and Series Ovulares. *Flavour Fragr. J.*, 12, 423–432.

Bignell, C.M., Dunlop, P.J. and Brophy, J.J. (1998) Volatile leaf oils of some south-western and southern Australian species of the genus *Eucalyptus* (Series I). Part XIX. *Flavour Fragr. J.*, 13, 131–139.

Blumann, A., Michael, N. and White, D.E. (1953) The chemistry of Western Australian plants. VIII. The essential oil of *Eucalyptus eudesmioides* bark. *J. Chem. Soc.*, 788–789.

Boelens, M.H. and Boelens, H. (1997) Differences in chemical and sensory properties of orange flower and rose oils obtained from hydrodistillation and from supercritical $CO_2$ extraction. *Perfum. Flavor.*, 22(3), 31–35.

Boland, D.J. and Brophy, J.J. (1993) Essential oils of the eucalypts and related genera. In R. Teranishi, R.G. Buttery and H. Sugisawa (eds), *Bioactive Volatiles from Plants, ACS Symposium Series 525*, American Chemical Society, Washington DC, pp. 72–87.

Boland, D.J., Brophy, J.J., Flynn, T.M. and Lassak, E.V. (1982) Volatile leaf oils of *Eucalyptus delegatensis* seedlings. *Phytochemistry*, 21, 2467–2469.

Boland, D.J., Brophy, J.J. and Fookes, C.J.R. (1992) Jensenone, a ketone from *Eucalyptus jensenii*. *Phytochemistry*, 31, 2178–2179.

Boland, D.J., Brophy, J.J. and House, A.P.N. (eds) (1991) *Eucalyptus Leaf Oils: Use, Chemistry, Distillation and Marketing*, ACIAR/CSIRO, Inkata Press, Melbourne.

Boland, D.J., Gilmour, P.M. and Brophy, J.J. (1986) *Eucalyptus deuaensis* (Myrtaceae), a new species of mallee from Deua National Park, south-eastern New South Wales. *Brunonia*, 9, 105–112.

Boland, D.J. and Kleinig, D.A. (1983) *Eucalyptus wilcoxii* (Myrtaceae), a new species from south-eastern New South Wales. *Brunonia*, 6, 241–250.

Boland, D.J., Kleinig, D.A. and Brophy, J.J. (1987) *Eucalyptus fusiformis* (Myrtaceae), a new species of ironbark (in the informal E. series Paniculatae Pryor et Johnson) from north-eastern New South Wales. *Brunonia*, 10, 201–209.

Bowyer, R.C. and Jefferies, P.R. (1959) Studies in plant chemistry. 1. The essential oils of *Eucalyptus caesia* Benth. and *E. torquata* Luehm. and the structure of torquatone. *Aust. J. Chem.*, 12, 442–446.

Briggs, L.H. and Bartley, J.P. (1970) Constituents of the essential oil of *Eucalyptus ficifolia*. *Aust. J. Chem.*, 23, 1499.

Brooker, M.I.H., Barton, A.F.M., Rockel, B.A. and Tjandra, J. (1988) The cineole content and taxonomy of *Eucalyptus kochii* Maiden & Blakely and *E. plenissima* (Gardner) Brooker, with an appendix establishing these two taxa as subspecies. *Aust. J. Bot.*, 36, 119–129.

Brooker, M.I.H. and Lassak, E.V. (1981) The volatile leaf oils of *Eucalyptus ovata* Labill. and *E. brookerana* A.M. Gray (Myrtaceae). *Aust. J. Bot.*, 29, 605–615.

Brophy, J.J. and Boland, D.J. (1989) Leaf essential oil of *Eucalyptus bakeri*. In D.J. Boland (ed.), *Trees for the Tropics*, ACIAR, Canberra, pp. 205–207.

Brophy, J.J. and Boland, D.J. (1990) Leaf essential oil of two chemotypes of *Eucalyptus cloeziana*. F. Muell. *J. Essent. Oil Res.*, 2, 87–90.

Brophy, J.J., Davies, N.W., Southwell, I.A., Stiff, I.A. and Williams, L.R. (1989) Gas chromatographic quality control for oil of *Melaleuca* terpinen-4-ol type (Australian tea tree). *J. Agric. Food Chem.*, 37, 1330–1335.

Brophy, J.J., Forster, P.I. and Goldsack, R.J. (1998a) The essential oils of three unusual eucalypts: *Eucalyptus curtisii*, *E. rubiginosa* and *E. tenuipes* (Myrtaceae). *Flavour Fragr. J.*, 13, 87–89.

Brophy, J.J., Forster, P.I., Goldsack, R.J. and Hibbert, D.B. (1998b) The essential oils of the yellow bloodwood eucalypts (*Corymbia*, Section *Ochraria*, Myrtaceae). *Biochem. Syst. Ecol.*, 26, 239–250.

Brophy, J.J., Goldsack, R.J., Forster, P.I., Clarkson, J.R. and Fookes, C.J.R. (1996) Mass spectra of some β-triketones from Australian Myrtaceae. *J. Essent. Oil Res.*, 8, 465–470.

Brophy, J.J and Lassak, E.V. (1986) The volatile leaf oils of some central Australian species of *Eucalyptus*. *J. Proc. Roy. Soc. N.S.W.*, 119, 103–108.

Brophy, J.J., Lassak, E.V. and Toia, R.F. (1985) The steam volatile leaf oil of *Eucalyptus pulverulenta*. *Planta Medica*, 51, 170–171.

Brophy, J.J., Lassak, E.V., Win, S. and Toia, R.F. (1982) The volatile leaf oils of *Eucalyptus youmanii* and *E. macrorhyncha*. *J. Sci. Soc. Thailand*, 8, 137–145.

Carmo, M.M. and Frazão, S. (1985) Studies on the composition of essential oils of 'cineolic eucalyptus'. In A. Baerheim and J.J.C. Scheffer (eds), *Essential Oils and Aromatic Plants*, Martinus Nijhoff/Dr W. Junk Publishers, Dordrecht, pp. 163–166.

Chen, S.S. and Spiro, M. (1995) Kinetics of microwave extraction of rosemary leaves in hexane, ethanol and a hexane + ethanol mixture. *Flavour Fragr. J.*, 10, 101–112.

Chen, Y., Yang, L., Li, S. and Jiang, Z. (1983) Study on the chemical components of essential oil from the leaves of *Eucalyptus* spp. (in Chinese; English summary). *Chem. Ind. For. Prods.*, 3(2), 16–31.

Chialva, F. and Gabri, G. (1987) Headspace *versus* classical analysis. In P. Sandra and C. Bicchi (eds), *Capillary Gas Chromatography in Essential Oil Analysis*, Huethig, Heidelberg, pp. 123–154.

Clevenger, J.F. (1928) Apparatus for the determination of volatile oils. *J. Am. Pharm. Assoc.*, 17, 345–349.

Cocking, T.A. (1920) A new method for the estimation of cineole in *Eucalyptus* oil. *Pharm. J.*, 105, 81–83.

Cocking, T. (1927) Estimation of cineole. *Perfum. Essent. Oil Rec.*, 18, 254–257.

Coppen, J.J.W. and Dyer, L.R. (1993) *Eucalyptus and its Leaf oils: An Indexed Bibliography*, Natural Resources Institute, Chatham, UK.

Coppen, J.J.W. and Hone, G.A. (1992) *Eucalyptus Oils. A Review of Production and Markets*, Bulletin 56, Natural Resources Institute, Chatham, UK.

Courtney, J.L., Lassak, E.V. and Speirs, G.B. (1983) Leaf wax constituents of some myrtaceous species. *Phytochemistry*, 22, 947–949.

Craveiro, A.A., Matos, F.J.A., Alencar, J.W. and Plumel, M.M. (1989) Microwave oven extraction of an essential oil. *Flavour Fragr. J.*, 4, 43–44.

Croteau, R. (1987) Biosynthesis and catabolism of monoterpenoids. *Chem. Rev.*, 87, 929–954.

Croteau, R., Alonso, W.R., Koepp, A.E. and Johnson, M.A. (1994) Biosynthesis of monoterpenes: partial purification, characterization, and mechanism of action of 1,8-cineole synthase. *Arch. Biochem. Biophys.*, 309, 184–192.

Crow, W.D., Nicholls, W. and Sterns, M. (1971) Root inhibitors in *Eucalyptus grandis*: naturally occurring derivatives of the 2,3-dioxabicyclo[4.4.0]decane system. *Tetrahedron Letts.*, 18, 1353–1356.

Crow, W.D., Osawa, T., Platz, K.M. and Sutherland, D.M. (1976) Root inhibitors in *Eucalyptus grandis*. II. Synthesis of inhibitors and origin of peroxide linkage. *Aust. J. Chem.*, 29, 2525–2531.

Curtis, A., Southwell, I.A. and Stiff, I.A. (1990) *Eucalyptus*, a new source of E-methyl cinnamate. *J. Essent. Oil Res.*, 2, 105–110.

Dayal, R. (1988) Phytochemical examination of eucalypts – a review. *J. Sci. Ind. Res.*, 47, 215–220.

Dayal, R. and Maheshwari, M.L. (1985) Terpenoids of the essential oil of *Eucalyptus*. *Indian Forester*, 111, 1076–1079.

De Riscala, E. and Retamar, J.A. (1981) Essential oil of *Eucalyptus moluccana* (*E. hemiphloia*) (in Spanish). *Essenze Derivati Agrumari*, 51, 263–270.

Doimo, L., Fletcher, R.J., D'Arcy, B.R. and Bird, L. (1999) A new essential oil from *Eucalyptus dura* L.A.S. Johnson & K.D. Hill. *J. Essent. Oil Res.*, 11, 149–150.

Doran, J.C. and Brophy, J.J. (1990) Tropical red gums – a source of 1,8-cineole-rich *Eucalyptus* oil. *New Forests*, 4, 157–178.

Doran, J.C., Williams, E.R. and Brophy, J.J. (1995) Patterns of variation in the seedling leaf oils of *Eucalyptus urophylla*, *E. pellita* and *E. scias*. *Aust. J. Bot.*, 43, 327–336.

Eberhard, I.H., McNamara, J., Pearse, R.J. and Southwell, I.A. (1975) Ingestion and excretion of *Eucalyptus punctata* D.C. and its essential oil by the koala, *Phascolarctos cinereus* (Goldfuss). *Aust. J. Zool.*, 23, 169–179.

Eisenreich, W., Sagner, S., Zenk, M.H. and Bacher, A. (1997) Monoterpenoid essential oils are not of mevalonoid origin. *Tetrahedron Letts.*, 38, 3889–3892.

Erman, W.F. (1985) Chemistry of the monoterpenes. In P.G. Gassman (ed.), *Studies in Organic Chemistry*, Vol. 11A and B, Marcel Dekker, New York.

Flynn, T.M. and Southwell, I.A. (1979) 1,3-Dimethyl-2-oxabicyclo [2.2.2]-octane-3-methanol and 1,3-dimethyl-2-oxabicyclo [2.2.2]-octane-3-carboxylic acid, urinary metabolites of 1,8-cineole. *Aust. J. Chem.*, 32, 2093–2095.

Foley, W.J., Lassak, E.V. and Brophy, J.J. (1987) Digestion and absorption of *Eucalyptus* essential oils in greater glider (*Petauroides volans*) and brushtail possum (*Trichosurus vulpecula*). *J. Chem. Ecol.*, 13, 2115–2130.

Franich, R.A. (1986) Essential oil composition of juvenile leaves from coppiced *Eucalyptus nitens*. *Phytochemistry*, 25, 245–246.

Gandini, A. (1936) The analysis of the essence of *Eucalyptus rostrata*. *Ann. Chim.*, 26, 344–351.

Gardner, C.A. and Watson, E.M. (1947/48) The Western Australian varieties of *Eucalyptus oleosa* and their essential oils. *J. Proc. Roy. Soc. W. Aust.*, 34, 73–86.

Ghanim, A. and Jayaraman, I. (1978) Essential oil of Jodhpur grown *Eucalyptus viridis*. *Ind. J. Pharm. Sci.*, 41, 80–81.

Ghisalberti, E.L. (1996) Bioactive acylphloroglucinol derivatives from *Eucalyptus* species. *Phytochemistry*, 41, 7–22.

Ghisalberti, E.L., Skelton, B.W. and White, A.H. (1995) Structural study of torquatone, an acylphloroglucinol derivative from *Eucalyptus* sp. *Aust. J. Chem.*, 48, 1771–1774.

Gildemeister, E. and Hoffmann, F. (1961) *Die Atherischen Ole*, Vol. 6, Akademie-Verlag, Berlin, p. 196.

Guenther, E. (ed.) (1950) The Eucalyptus oils. In *The Essential Oils*, Vol. 4, Van Nostrand, New York, pp. 437–525.

Harris, C.M. and McKern, H.H.G. (1950) A note on the occurrence of guaiol in some essential oils of *Eucalyptus citriodora* Hook. *Researches on Essential Oils of the Australian Flora*, Museum of Applied Arts and Sciences, Sydney, 2, 15.

Hegnauer, R. (1969) *Chemotaxonomie der Pflanzen*, Vol. 5, Birkhauser Verlag, Basel, Switzerland, pp. 163–195, 439–441, 457.

Hegnauer, R. (1990) *Chemotaxonomie der Pflanzen*, Vol. 9, Birkhauser Verlag, Basel, Switzerland, pp. 116–132.

Hellyer, R.O. (1964) The structure of agglomerone, a new β-triketone. *Aust. J. Chem.*, 17, 1418–1422.

Hellyer, R.O. (1968) The occurrence of β-triketones in the steam-volatile oils of some myrtaceous Australian plants. *Aust. J. Chem.*, 21, 2825–2828.

Hellyer, R.O., Bick, I.R.C., Nicholls, R.G. and Rottendorf, H. (1963) The structure of tasmanone. *Aust. J. Chem.*, 16, 703–708.

Hellyer, R.O., Keyzer, H. and McKern, H.H.G. (1964) The volatile oils of the genus *Eucalyptus* (family Myrtaceae). III. The leaf oil of *E. crenulata* Blakely and Beuzeville. *Aust. J. Chem.*, 17, 283–285.

Hellyer, R.O., Lassak, E.V. and McKern, H.H.G. (1966) The volatile oils of the genus *Eucalyptus* (family Myrtaceae). V. The leaf oil of *E. aggregata* Deane and Maiden. *Aust. J. Chem.*, 19, 1765–1767.

Hellyer, R.O., Lassak, E.V., McKern, H.H.G. and Willis, J.L. (1969) Chemical variation within *Eucalyptus dives*. *Phytochemistry*, 8, 1513–1514.

Hellyer, R.O. and McKern, H.H.G. (1963) The volatile oils of the genus *Eucalyptus* (family Myrtaceae). II. The leaf oils of *E. oblonga* DC. (syn. *E. sparsifolia* Blakely) and *E. mitchelliana* Cambage. *Aust. J. Chem.*, 16, 515–519.

Hellyer, R.O. and McKern, H.H.G. (1966) The volatile oils of the genus *Eucalyptus* (family Myrtaceae). IV. The leaf oils of *E. amplifolia* Naudin and *E. kitsoniana* Maiden. *Aust. J. Chem.*, 19, 1541–1543.

Hill, K.D. and Johnson, L.A.S. (1995) Systematic studies in the eucalypts. 7. A revision of the bloodwoods, genus *Corymbia* (Myrtaceae). *Telopea*, 6, 185–504.

Holeman, M., Rombourg, M., Fechtal, M., Gorrichon, J.P. and Lassaigne, G. (1987) *Eucalyptus astringens* Maiden, *Eucalyptus blakelyi* Maiden and *Eucalyptus bosistoana* F. Muell.: the same chemotype (in French). *Plantes Médicinales et Phytothérapie*, 4, 311–316.

Horn, D.H.S., Kranz, Z.H. and Lamberton, J.A. (1964) The composition of *Eucalyptus* and some other leaf waxes. *Aust. J. Chem.*, 17, 464–476.

Hughes, A. (1970) A modified receiver for heavier than water essential oils. *Chem. Ind.*, 1536.

Jones, T.G.H. and Lahey, F.N. (1938) Essential oils from the Queensland flora. Part XVI. *Eucalyptus microcorys*. *Proc. Roy. Soc. Qld*, 50, 43–45.

Jones, T.G.H. and White, M. (1928) The essential oil of *Eucalyptus andrewsi* from Queensland. *Proc. Roy. Soc. Qld*, 40, 132–133.

Klocke, J.A., Darlington, M.V. and Balandrin, M.F. (1987) 1,8-Cineole (eucalyptol), a mosquito feeding and ovipositional repellent from volatile oil of *Hemizonia fitchii* (Asteraceae). *J. Chem. Ecol.*, 13, 2131–2141.

Koedam, A. (1987) Some aspects of essential oil preparation. In P. Sandra and C. Bicchi (eds), *Capillary Gas Chromatography in Essential Oil Analysis*, Huethig, Heidelberg, pp. 13–27.

Lahey, F.N. and Jones, T.G.H. (1939) Essential oils of the Queensland flora. XIV. *Eucalyptus conglomerata*. *Proc. Roy. Soc. Qld*, 50, 10–13.

Lamberton, J.A. (1964) The occurrence of 5-hydroxy-7,4'-dimethoxy-6-methylflavone in *Eucalyptus* waxes. *Aust. J. Chem.*, 17, 692–696.

Lassak, E.V. (1988) The Australian eucalyptus oil industry, past and present. *Chem. Aust.*, 55, 396–398.

Lassak, E.V. and Southwell, I.A. (1969) The bark oil of *Eucalyptus crenulata*. *Phytochemistry*, 8, 667–668.

Lassak, E.V. and Southwell, I.A. (1977) Essential oil isolates from the Australian flora. *Int. Flav. Food Addit.*, 8, 126–132.

Lassak, E.V. and Southwell, I.A. (1982) The steam volatile leaf oils of some species of *Eucalyptus* Subseries Strictinae. *Phytochemistry*, 21, 2257–2261.

Lawrence, B.M. (1985) A review of the world production of essential oils (1984). *Perfum. Flavor.*, 10(5), 2–16.

Lawrence, B.M. (1993) A planning scheme to evaluate new aromatic plants for the flavour and fragrance industries. In J. Janick and J.E. Simon (eds), *New Crops*, John Wiley, New York, pp. 620–627.

Li, H., Madden, J.L. and Davies, N.W. (1994) Variation in leaf oil of *Eucalyptus nitens* and *E. denticulata*. *Biochem. Syst. Ecol.*, 22, 631–640.

Li, H., Madden, J.L. and Potts, B.M. (1995) Variation in volatile leaf oils of the Tasmanian *Eucalyptus* species. I. Subgenus *Monocalyptus*. *Biochem. Syst. Ecol.*, 23, 299–318.

Li, H., Madden, J.L. and Potts, B.M. (1996) Variation in volatile leaf oils of the Tasmanian *Eucalyptus* species. II. Subgenus *Symphyomyrtus*. *Biochem. Syst. Ecol.*, 24, 547–569.

Lichtenthaler, H.K., Schwender, J., Disch, A. and Rohmer, M. (1997) Biosynthesis of isoprenoids in higher plant chloroplasts proceeds via a mevalonate independent pathway. *FEBS Letts.*, 400, 271–274.

Liener, I.E. (1996) Editorial. *J. Agric. Food Chem.*, 44, 1.

Marshall, G.E. and Watson, E.M. (1934/35) The essential oils of the Western Australian eucalypts. Part II. The oils of *E. kesselli* and *E. dundasi*. *J. Proc. Roy. Soc. W. Aust.*, 21, 107–111.

Marshall, G.E. and Watson, E.M. (1936/37) The essential oils of the Western Australian eucalypts. Part IV. The oils of *E. oleosa* F.v.M., *E. eremophila* Maiden and *E. leptopoda* Benth. *J. Proc. Roy. Soc. W. Aust.*, 23, 1–5.

Marshall, G.E. and Watson, E.M. (1937/38) The essential oils of the Western Australian eucalypts. Part V. The oils of *E. astringens* Maiden and *E. pyriformis* Turez. *J. Proc. Roy. Soc. W. Aust.*, 24, 65–68.

Marshall, G.E. and Watson, E.M. (1939/40) The essential oils of the Western Australian eucalypts. Part VI. The oil of *E. concinna* Maiden et Blakely. *J. Proc. Roy. Soc. W. Aust.*, 26, 15–16.

Martínez, M.M., Guimerás, J.L.P., Hernández, J.M., Zayas, J.R.P., Días, M.J.Q. and Montejo, L. (1986) Preliminary study of the essential oils of *Eucalyptus* species introduced into Topes de Collantes (in Spanish). *Rev. Cub. Farm.*, 20, 159–168.

McKern, H.H.G. (1960) The natural plant products industry of Australia. *J. Proc. Aust. Chem. Inst.*, 27, 295–308.

McKern, H.H.G. (1967) Volatile plant oils in Australia. *Aust. Nat. History*, 15, 352–354.

McKern, H.H.G. (1968) Research into the volatile oils of the Australian flora, 1788–1967. In Council of the Royal Society of New South Wales (ed.), *A Century of Scientific Progress*, Royal Society of NSW, Sydney, pp. 310–331.

McKern, H.H.G., Spies, M.C. and Willis, J.L. (1954) The essential oil of *Eucalyptus maculata* Hooker. Part I. *J. Proc. Roy. Soc. N.S.W.*, 88, 15–21.

McLean, S., Foley, W.J., Davies, N.W., Brandon, S., Duo, L. and Blackman, A.J. (1993) Metabolic fate of dietary terpenes from *Eucalyptus radiata* in common ringtail possum (*Pseudocheirus peregrinus*). *J. Chem. Ecol.*, 19, 1625–1643.

McQuillin, F.J. and Parrack, J.D. (1956) The isomeric eudesmols and their association with carissone in *Eucalyptus macarthurii*. *J. Chem. Soc.*, 2973–2978.

Menut, C., Bessiere, J.M., Samate, A.D., Millogo-Rasolodimby, J. and Nacro, M. (1999) Apodophyllone and isotorquatone, two arenic ketones from *Eucalyptus apodophylla*. *Phytochemistry*, 51, 975–978.

Milner, C.P., Trengrove, R.D. and Dunlop, P.J. (1997) Supercritical $CO_2$ extraction of the essential oils of eucalypts: a comparison with other methods. In H.-F. Linskens and J.F. Jackson (eds), *Modern Methods of Plant Analysis*, Vol. 19, *Plant Volatile Analyis*, Springer-Verlag, Berlin, pp. 141–158.

Miranda, M., Zayas, J.R.P. and Henriques, R.D. (1981) Study of the principal components of 19 species of *Eucalyptus* acclimated in Cuba (in Spanish). *Rev. Cub. Farm.*, 15, 106–114.

Ndou, T.T. and von Wandruszka, R.M.A. (1985) Essential oils of South African *Eucalyptus* species (Myrtaceae). *S. Afr. J. Chem.*, 39, 95–100.

Neybergh, A.G. (1953) Some essential oil plants from the east of the colony. *Bull. Agric. Congo Belge*, 44, 1–33.

Nishimura, H. (ed.) (1987) *Eucalyptus as Biochemical Resources in the Future*, Uchida Rokakuho, Tokyo, p. 123.

Nishimura, H., Fukazawa, Y., Mizutani, J., Calvin, M. and Paton, D.M. (1979) Essential oils of *Eucalyptus* as reproductive biomass. In *Koen Yoshishu-Koryo, Terupen oyobi Seiyu Kagaku ni kansuru Toronkai, 23rd*, Chem. Soc. Japan, Tokyo, pp. 195–197.

Nishimura, H., Mizutani, J., Umino, T. and Kurihara, T. (1986) New repellents against mosquitoes, p-menthane-3,8-diols in *Eucalyptus citriodora* and related compounds. *6th Internat. Congr. Pesticide Chem.*, Ottawa, 1986, Paper 2D/E-07.

Osawa, K., Yasuda, H., Morita, H., Takeya, K. and Itokawa, H. (1996) Macrocarpals H, I, J from the leaves of *Eucalyptus globulus*. *J. Nat. Prod.*, 59, 823–827.

Osawa, T. and Namiki, M. (1985) Natural antioxidants isolated from *Eucalyptus* leaf waxes. *J. Agric. Food Chem.*, 33, 777–780.

Penfold, A.R. (1922) A critical examination of aromatic aldehydes occurring in certain *Eucalyptus* oils. *J. Chem. Soc.*, 266–269.

Penfold, A.R. (1927) The essential oil of *Eucalyptus bakeri* Maiden. *J. Proc. Roy. Soc. N.S.W.*, 61, 179–189.

Penfold, A.R. (1935) The development of our knowledge concerning the essential oils of the eucalypts. *Aust. J. Pharm.*, 16(1), 29–32.

Penfold, A.R. and Morrison, F.R. (1924) Notes on *Eucalyptus piperita* and its essential oils, with special reference to their piperitone content. Part I. *J. Proc. Roy. Soc. N.S.W.*, 58, 124–127.

Penfold, A.R. and Morrison, F.R. (1927a) The occurrence of a number of varieties of *Eucalyptus dives* as determined by chemical analyses of the essential oils. Part I. *J. Proc. Roy. Soc. N.S.W.*, 61, 54–67.

Penfold, A.R. and Morrison, F.R. (1927b) The essential oils of *Eucalyptus micrantha* and *E. haemastoma*. Part I. *J. Proc. Roy. Soc. N.S.W.*, 61, 267–278.

Penfold, A.R. and Morrison, F.R. (1929) The occurrence of a number of varieties of *Eucalyptus dives* as determined by chemical analyses of the essential oils. Part III. *J. Proc. Roy. Soc. N.S.W.*, 63, 79–84.

Penfold, A.R. and Morrison, F.R. (1930) Notes of the essential oils from some cultivated eucalypts. Part II. *J. Proc. Roy. Soc. N.S.W.*, 64, 210–223.

Penfold, A.R. and Morrison, F.R. (1932) The occurrence of a number of varieties of *Eucalyptus radiata* (*E. numerosa*) as determined by chemical analyses of the essential oils. Part I. *J. Proc. Roy. Soc. N.S.W.*, 66, 181–193.

Penfold, A.R. and Morrison, F.R. (1935) The essential oils of *Eucalyptus australiana* and its physiological forms. Part I. *J. Proc. Roy. Soc. N.S.W.*, 69, 111–122.

Penfold, A.R. and Morrison, F.R. (1940) The essential oils of *Eucalyptus australiana* and its physiological forms. Part III. *J. Proc. Roy. Soc. N.S.W.*, 74, 277–282.

Penfold, A.R. and Morrison, F.R. (1948) The occurrence of a physiological form of *Eucalyptus citriodora* Hooker. *Aust. J. Sci.*, 11, 29.

Penfold, A.R. and Morrison, F.R. (1950) Australian eucalyptus oils. In E. Guenther (ed.), *The Essential Oils*, Vol. 4, Van Nostrand, New York, pp. 437–482.

Penfold, A.R. and Morrison, F.R. (1951) *Commercial Eucalyptus Oils*, Museum of Applied Arts and Sciences Bulletin No. 2 (5th edn), Government Printer, Sydney.

Penfold, A.R., Morrison, F.R. and McKern, H.H.G. (1948) Studies in the Myrtaceae and their essential oils. Part II. Some sources of error in the study of plant populations: *Eucalyptus citriodora* Hook. *Researches on Essential Oils of the Australian Flora*, Museum of Applied Arts and Sciences, Sydney, 1, 8–11.

Penfold, A.R., Morrison, F.R., McKern, H.H.G. and Willis, J. (1950) Studies in the physiological forms of the Myrtaceae. Part 5. *Eucalyptus citriodora* Hook. and the incidence of its physiological forms. *Researches on Essential Oils of the Australian Flora*, Museum of Applied Arts and Sciences, Sydney, 2, 12–14.

Penfold, A.R., Morrison, F.R., Willis, J.L., McKern, H.H.G. and Spies, M.C. (1951) The essential oil of a physiological form of *Eucalyptus citriodora* Hook. *J. Proc. Roy. Soc. N.S.W.*, 85, 120–122.

Penfold, A.R., Radcliff, C.B. and Short, F.W. (1930) The essential oil of *Eucalyptus rariflora* Bailey. *J. Proc. Roy. Soc. N.S.W.*, 64, 101–114.

Penfold, A.R. and Willis, J.L. (1961) *The Eucalypts*, Leonard Hill, London.

Phillips, L.W. (1923) The essential oils of some Western Australian plants. *J. Proc. Roy. Soc. W. Aust.*, 9, 107–119.

Pryor, L.D., Williams, E.R. and Gunn, B.V. (1995) A morphometric analysis of *Eucalyptus urophylla* and related taxa with descriptions of two new species. *Aust. Syst. Bot.*, 8, 57–70.

Pryor, L.D. and Willis, J.H. (1954) A new Victorian (and South Australian) eucalypt. *Vict. Nat.*, 71, 125–129.

Rutowski, B.N. and Winogradowa, I.W. (1927) The components of Caucasian *Eucalyptus* oils. *Trans. Sci. Chem. Pharm. Inst. (Moscow)*, 39–68.

Santhanakrishnan, T.S. (1984) Biohydroxylation of terpenes in mammals. *Tetrahedron*, 40, 3597–3609.

Santos, G.G., Alves, J.C.N., Rodilla, J.M.L., Duante, A.P., Lithgow, A.M. and Urones, J.G. (1997) Terpenoids and other constituents of *Eucalyptus globulus*. *Phytochemistry*, 44, 1309–1312.

Savina, A.A., Zakharov, V.F. and Tsybul'ko, N.S. (1991) Structure of euvimal-1, a new phenol aldehyde from the leaves of *Eucalyptus viminalis*. *Chem. Nat. Compd.*, 27, 696–701.

Schultz, T.H., Flath, R.A., Mon, T.R., Eggling, S.B. and Teranishi, R. (1977) Isolation of volatile components from a model system. *J. Agric. Food Chem.*, 25, 446–449.

Shiva, M.P., Paliwal, G.S., Chandra, K. and Mathur, M. (1984) Pinene rich essential oil from *Eucalyptus tereticornis* leaves from Tarai and Bhabar areas of Uttar Pradesh. *Indian Forester*, 110, 23–27.

Siddiqui, B.S., Farhat, S. and Siddiqui, S. (1997) Eucalyptic acid and eucalyptolic acid, two new triterpenes from the leaves of *Eucalyptus camaldulensis*. *Planta Medica*, 63, 47–50.

Silvestre, A.J.D., Cavaleiro, J.A.S., Delmond, B., Filliatre, C. and Bourgeois, G. (1997) Analysis of the variation of the essential oil composition of *Eucalyptus globulus* Labill. from Portugal using multivariate statistical analysis. *Industrial Crops Prods.*, 6, 27–33.

Singh, A.K., Brophy, J.J. and Gupta, K.C. (1988) The essential oil of *Eucalyptus urophylla*: a rich source of *p*-cymene. *Indian Perfumer*, 32, 201–204.

Singh, A.K., Gupta, K.C. and Brophy, J.J. (1991) Chemical constituents of the leaf essential oil of *Eucalyptus brassiana*. *J. Essent. Oil Res.*, 3, 45–47.

Singh, A.K., Khare, M. and Kumar, S. (1999) Non-volatile constituents of eucalypts: a review on chemistry and biological activities. *J. Med. Arom. Plant Sci.*, 21, 375–407.

Singh, I.P. and Etoh, H. (1995) New macrocarpal-am-1 from *Eucalyptus amplifolia*. *Biosci. Biotech. Biochem.*, 59, 2330–2332.

Singh, I.P., Takahashi, K. and Etoh, H. (1996) Potent attachment-inhibiting and promoting substances for the blue mussel, *Mytilis edulis galloprovincialis*, from two species of *Eucalyptus*. *Biosci. Biotech. Biochem.*, 60, 1522–1523.

Smale, P.E., Nelson, M.A., Porter, N.G. and Hay, A.J. (2000) Essential oil of *Eucalyptus olida* L. Johnson and K. Hill. 1: Variability of yield and composition in foliage from a seedling population. *J. Essent. Oil Res.*, 12, 569–574.

Smith, H.G. (1905) The refractive indices, with other data, of the oils of 118 species of *Eucalyptus*. *J. Proc. Roy. Soc. N.S.W.*, 39, 39–47.

Southwell, I.A. (1973) Variation in the leaf oil of *Eucalyptus punctata*. *Phytochemistry*, 12, 1341–1343.

Southwell, I.A. (1975) Essential oil metabolism in the koala. III. Novel urinary monoterpenoid lactones. *Tetrahedron Letts.*, (24), 1885–1888.

Southwell, I.A. (1987) Essential oil isolates from the Australian flora. *Flavour Fragr. J.*, 2, 21–27.

Southwell, I.A., Flynn, T.M. and Degabriele, R. (1980) Metabolism of α- and β-pinene, *p*-cymene and 1,8-cineole in the brushtail possum, *Trichosurus vulpecula*. *Xenobiotica*, 10, 17–23.

Southwell, I.A., Maddox, C.D.A. and Zalucki, M.P. (1995) Metabolism of 1,8-cineole in tea tree (*Melaleuca alternifolia* and *M. linariifolia*) by pyrgo beetle (*Paropsisterna tigrina*). *J. Chem. Ecol.*, 21, 439–453.

Southwell, I.A. and Stiff, I.A. (1989) Ontogenetical changes in monoterpenoids of *Melaleuca alternifolia* leaf. *Phytochemistry*, 28, 1047–1051.

Spiro, M. and Chen, S.S. (1995) Kinetics of isothermal and microwave extraction of essential oil constituents of peppermint leaves into several solvent systems. *Flavour Fragr. J.*, 10, 259–272.

Stevens, R. (1996) Editorial. *Flavour Fragr. J.*, 11, 1.

Takasaki, M., Konoshima, T., Kozuka, M., Ito, K., Crow, W.D. and Paton, D.M. (1994) Euglobal-In-1, a new euglobal from *Eucalyptus incrassata. Chem. Pharm. Bull.*, 42, 2113–2116.

Unpublished records, Museum of Applied Arts and Sciences, Sydney. Referred to in Penfold, A.R. and Willis, J.L. (1961) *The Eucalypts*, Leonard Hill, London.

Wade, A.P., Wilkinson, G.S., Dean, F.M. and Price, A.W. (1966) The isolation, characterization and structure of uroterpenol, a monoterpene from human urine. *Biochem. J.*, 101, 727–734.

Watanabe, K., Shono, Y., Kakimizu, A., Okada, A., Matsuo, M., Satoh, A. and Nishimura, H. (1993) New mosquito repellent from *Eucalyptus camaldulensis. J. Agric. Food Chem.*, 41, 2164–2166.

Watson, E.M. (1934/35) The essential oils of the Western Australian eucalypts. Part I. The oil of *E. flocktoniae* Maiden. *J. Proc. Roy. Soc. W. Aust.*, 21, 101–105.

Watson, E.M. (1935/36) The essential oils of the Western Australian eucalypts. Part III. The oils of *E. salmonophloia* F.v.M. and *E. tetragona* F.v.M. *J. Proc. Roy. Soc. W. Aust.*, 22, 113–118.

Watson, E.M. (1941/42) The essential oils of the Western Australian eucalypts. Part VII. The oil of *E. erythronema* Turcz. *J. Proc. Roy. Soc. W. Aust.*, 28, 247–249.

Watson, E.M. and Gardner, C.A. (1944/45) The essential oils of the Western Australian eucalypts. Part VIII. The oils of *E. campaspe* S. Moore and *E. kochii* Maiden et Blakely. *J. Proc. Roy. Soc. W. Aust.*, 31, 33–36.

Webb, L.J., Sutherland, M.D. and Wells, J.W. (1956) Quoted in Penfold, A.R. and Willis, J.L. (1961) *The Eucalypts*, Leonard Hill, London, p. 489.

Weston, R. (1984) Composition of essential oil from leaves of *Eucalyptus delegatensis. Phytochemistry*, 23, 1943–1945.

Whish, J.P.M. (1996) A flexible distillation system for the isolation of essential oils. *J. Essent. Oil Res.*, 8, 405–410.

Wilcox, M.D. (1997) *A Catalogue of the Eucalypts*, Groom Poyry, Auckland.

Yatagai, M. and Takahashi, T. (1983) An approach to biomass utilization. II. Components of *Eucalyptus* leaf oils. *Mokuzai Gakkaishi*, 29, 396–399.

Yoneyama, K., Asami, T., Crow, W.D., Takahashi, N. and Yoshida, S. (1989) Photosynthetic electron transport inhibition by phlorophenone derivatives. *Agric. Biol. Chem.*, 53, 471–475.

Yoshida, S., Asami,T., Kawano, T., Yoneyama, K., Crow, W.D., Paton, D.M. and Takahashi, N. (1988) Photosynthetic inhibitors in *Eucalyptus grandis. Phytochemistry*, 27, 1943–1946.

# 6 Distillation of eucalyptus leaf oils
## Theory and practice

*E.F.K. Denny*

## Theory

### The distilling operation

Several species of *Eucalyptus* yield oils which are valuable in commerce. The oil glands are secreted deeply in the leaves, well below the epidermal cuticle and other cells which together form the surface layers of the foliage. These oils can be recovered from the herb by steam distillation and are then referred to as essential oils because, being the products of distillation they are, by definition, essences.

When trees grown for timber are felled, the 'waste' leaves and small branches may be gathered by hand and loaded onto carts or trucks for transport to the distillery. In plantations developed primarily for essential oil production the trees are smaller and may be harvested by relatively harsh machinery. This usually damages the foliage and exposes some of the oil ducts, resulting in some loss of oil by evaporation. If the harvester loads the cut material directly into a trailer which also serves as the processing container, as is the case in Australia, any such losses are not significant. Otherwise, unloading and rehandling of this machine-cut eucalypt may result in an appreciable loss of oil.

At the distillery, the hand-cut material must be loaded manually and tramped down tightly into big vats, called stills. On the other hand, the mechanically harvested foliage is received ready packed for distillation in trailers which are specially designed to act efficiently as stills. In both cases, a steam-tight lid, often incorporating the outlet pipe, is fitted to the top of the still. A flow of steam, either from water boiled in the bottom of the still itself or, preferably, piped from a separate boiler, is then introduced below the plant material so that it percolates upwards through the charge.

Starting at the bottom, the steam condenses on all the herb surfaces and, surrendering its latent heat, raises the temperature of successive layers to boiling point. When the appropriate temperature reaches the top of the still, any oil that is exposed on the herb surfaces will start to boil away. The on-coming steam will then drive a mixture of steam and oil vapour off the top of the charge and from there it is led through a condenser. In the liquid state, the oil and water are virtually immiscible and separate spontaneously. The oil floats on top of the water and is easily removed for bulk storage and subsequent sale. This operation continues until, for all practical purposes, the oil is exhausted from the herb. The spent charge is then removed from the still and replaced by fresh plant material.

It is many centuries since aromatic herbs were first packed into vats and steam was passed through the charge to extract the volatile oils. But despite its antiquity, certain aspects of this hydrodistillation process have not been well understood and it is much abused. Modern equipment has made little fundamental advance on the primitive cooking methods of the past. The principle involved in recovering the volatile oils by vaporising them in the presence of steam is well known. But this vaporisation demands the transfer of heat from the steam to the oil and it is this function which is gravely misunderstood. Any improvement in the efficiency and economy of the industry depends on a better appreciation of this aspect.

## Conditions for distilling the oil

For the steam to boil away any oil that is exposed on the herb surface and gather up the vapour, certain conditions are required. Firstly, since vapours contain more heat energy per gram than their parent liquids, the oil cannot be turned into a vapour unless heat is applied to it. If the steam is to do this, and it is the only thing that can, it must be at a higher temperature than the liquid oil. The amount of heat required to vaporise one gram of liquid, without raising the vapour's temperature above that of the liquid, is a characteristic of the particular compound and is termed its *latent heat* of vaporisation. Conversely, if steam or any other vapour is condensed back to liquid, its energy content is reduced and the characteristic quantity of latent heat is given out.

Secondly, the conditions for boiling depend upon certain properties of vapours. It is not practical to discuss these in isolation from other aspects of vapour behaviour which will be important later on and so they are now considered below.

Liquids continually emit moving vapour molecules from their surfaces. These impinge on their surroundings and exert a characteristic *vapour pressure* on them. This pressure increases with rising temperature. If a closed liquid–vapour system is at a constant temperature, equal numbers of molecules are continuously leaving the liquid surface and returning to it. In this *equilibrium* condition the number of molecules present in any unit volume of the vapour space, their *concentration*, is exactly that required to exert the vapour pressure that is characteristic of the compound at the prevailing temperature. A vapour having this concentration and exerting this pressure is termed a *saturated vapour*.

All vapours in immediate contact with their parent liquid will be saturated vapours. It follows that if heat is applied to a vapour space or its volume is expanded, extra molecules must vaporise from the liquid surface to maintain the saturated vapour pressure in the new conditions. But if the vapour is no longer in contact with its liquid the extra molecules are simply not available. The pressure that can be exerted by the heated or expanded vapour will then be less than would be exerted by a saturated vapour of the same kind, at the same temperature. Alternatively, its temperature is higher than that of the saturated vapour exerting the same pressure. A vapour in this condition is said to be *superheated*.

If a superheated vapour comes into contact with its parent liquid, the latter will instantly evaporate to ensure that saturated equilibrium conditions are restored. Since the superheated vapour will usually provide the heat to support this evaporation, its own temperature will be reduced. It is very important for eucalypt distillers to understand that when superheated steam enters moist surroundings, it will immediately take up sufficient water molecules as vapour to revert to being saturated steam at a lower temperature.

If heat is applied to a liquid, its temperature will not rise beyond the point where its vapour pressure becomes equal to the surrounding pressure, whatever that may be. Continued application of heat will merely convert liquid into vapour at the rate set by the liquid's requirement of latent heat for vaporisation. The liquid is then said to *boil* and the temperature at which this occurs is called the *boiling point* under the prevailing pressure.

A liquid's vapour pressure will attain equality with a low surrounding pressure at a lower temperature than it would require for equalising a high surrounding pressure. If a table is prepared showing a liquid's vapour pressure at different temperatures, the latter are automatically the boiling points when the surrounding pressure is the same as the indicated vapour pressure. Such a table will show that even a very small increase in the pressure required of a vapour is inevitably associated with the corresponding rise in its temperature, and *vice versa*.

If the vapour-generating vessel is open to the atmosphere, the liquid will boil when its vapour pressure becomes equal to the atmospheric pressure. In average conditions at sea level the pressure in the atmosphere will support a column of mercury 760 mm tall, so-called standard atmospheric pressure or 1 atm abs (atmosphere absolute). Alternative expressions for this standard atmospheric pressure are 14.7 psi, 1.033 kg/cm$^2$ or 101.3 kPa. These are examples of the many other scales by which pressures are measured in commercial practice. But in all cases, ordinary industrial pressure gauges show the difference between atmospheric pressure and that inside the vessel to which they are attached. To obtain the absolute pressure, the ambient atmospheric pressure must be added to the gauge reading.

*Steam*

Steam is the vapour of pure water. Reference books give tables showing the properties of dry saturated steam at different temperatures. They show the total heat of steam, which is the amount required to raise the water from 0°C to boiling point under the given pressure, plus the latent heat required to vaporise the water. It may be given in calories, joules or British thermal units (Btu).[1] Steam generated under 8 atm abs, equal to 7 atm gauge, has a total heat some 3.8 per cent greater than the same mass of steam under atmospheric pressure. So, if steam is generated under high pressure and allowed to expand under a lower pressure as it enters an essential oil still, the surplus heat must be accounted for.

In practice, all saturated steam carries microscopic liquid particles in the form of cloud. Only superheated steam will carry none at all. Good commercial boilers generate steam consisting of about 97 per cent by weight dry saturated steam and 3 per cent of liquid cloud, the so-called 'wetness fraction'. Assume this steam emerges from the distillery boiler under 7 atm of gauge pressure and passes along efficiently insulated piping to the bottom of a still working at virtual atmospheric pressure. As it expands on entering the still, this steam will give out enough heat to vaporise all its own wetness fraction plus about 1.5 per cent of its own weight in further water which, importantly, will be taken from the herb surfaces inside the still.

The amount of steam for distillation is stated in terms of its rate of flow in mass per minute for each square metre of charge cross-section area traversed by the moving vapours. For *Eucalyptus*, a good flow is around 3 kg/min/m$^2$. Faster flows process more herb per hour and consume more steam per kg of oil recovered. But as rates of steam displacement decline below 2 kg/min/m$^2$, they increasingly invite loss of oil to internal reflux, an aspect which is discussed further below. Eucalypt distillers need their boilers and stills to be matched so that the steam's rate of flow cannot fall below 2 kg/min/m$^2$ of charge (= still) cross-section area. For this purpose, boilers may be taken as generating 1.6 kg of steam at 100°C and 1 atm of absolute pressure for each kW of their proven capacity. But at the generating pressures likely to be required, one should not rely on more than 1.2 kg of steam per hour per rated kW. The old boiler ratings in horsepower (HP) are misleading and should be ignored. They are not convertible on the basis of 1 HP = 0.746 kW.

---

[1] One calorie will raise the temperature of 1 g of water by 1°C and is equal to 4.18 J of energy. One Btu will raise a 1 pound mass of water 1°F.

*Mixed vapours*

If two immiscible liquids are in equilibrium with the same vapour space, each will contribute its own characteristic vapour pressure to that space as though the other were not there. The mixture will boil at the liquids' points of contact when the sum of their two vapour pressures becomes equal to the surrounding pressure. This enables the distiller to vaporise these natural oils in the presence of steam without even approaching the temperatures at which they would boil alone. For example, 1,8-cineole, an important component of many eucalyptus oils, exerts a vapour pressure of 67.3 mm mercury (Hg) at 97.42°C. At the same temperature steam exerts 692.7 mm Hg pressure. The sum of the vapour pressures is then 760 mm Hg and equal to that in the atmosphere. So cineole can be boiled away from the eucalyptus leaf in the presence of steam under atmospheric pressure at 97.42°C, whereas it would boil alone at nearer 177°C and might start to decompose. The hydrodistillation of all essential oils depends on bringing the two immiscible liquids into contact at a temperature very close to the boiling point of the water. Then the addition of even a very small amount of vapour pressure from the oil will cause the mixture's total pressure to equalise the surrounding pressure. The mixture then boils away from the herb surface at the rate that the latent heat to support this evaporation can be supplied.

For vapour pressure, VP, and molecular weight, MW, the ratio of oil to water in the ideal saturated mixed vapour is given by:

$$\frac{\text{mass}_{oil}}{\text{mass}_{water}} = \frac{VP_{oil} \times MW_{oil}}{VP_{water} \times MW_{water}} \tag{1}$$

Since the molecular weights are predetermined, the vapour pressures control both the temperature at which the oil and water boil together and the composition of the mixed vapour rising off the herb. Although these proportions are useful for calculation purposes, it is quite impossible for them to exist in the final distillate issuing from the still. It is also relevant that, although all vapour pressures increase with any given rise in temperature, that of the oil increases by a substantially greater factor than does that of water. This phenomenon originally led to a false theory from which many eucalyptus distilleries still suffer. But, for other reasons, it is still important for the higher boiling oils, including perhaps *E. dives* (Type), whose oil exerts much lower vapour pressures than does cineole.

*The transference of heat*

For the oil to vaporise inside a field still it must receive latent heat from condensing steam. This can happen only if the steam is at a higher temperature than the liquid oil. At the oil's point of evaporation, the concentration of its vapour and the proportion of the total ambient pressure it exerts will be at its maximum. Consequently, the pressure remaining to be exerted by the steam is minimal and its temperature is the lowest that can produce a boiling mixture of oil and water under the prevailing pressure.

When the oil vapour is dispersed through the steam in the general vapour space its concentration, and the share of the total pressure it can exert, are very much reduced. The share of the total pressure remaining to be exerted by the steam is correspondingly increased. So, in the general vapour space, the steam's temperature cannot fall below that at which it must exert the total ambient pressure, less only a meagre contribution from the diluted oil vapour. Inevitably, this temperature is higher than that at the point of evaporation on the herb, where the oil vapour is saturated and exerts more of the pressure.

This difference in temperature between the general vapour space and the oil's point of evaporation is the vital principle of the distillation process. It is the temperature gradient that makes possible the transference of heat to vaporise the oil, and its magnitude governs the rate at which the oil receives its latent heat and boils away from the herb surface. In short, the factors which govern the delivery of the oil's latent heat also control the efficiency of its distillation from the herb.

## How theory affects practice

### Feedback and processing time

Clearly, the oil can evaporate only as fast as its latent heat of vaporisation can be applied to it. Then, for every oil and set of conditions, there must be a balance point for the maximum obtainable oil content of the general distillate vapour. Any further enrichment of the vapour would lower its temperature and reduce the magnitude of the gradient. This would reduce the rate at which heat is delivered to the oil and retard its rate of evaporation until the normal balance situation was restored. This is a perfect example of an automatic control system of a type frequently called '*feedback*'.

If the maximum yield of oil is to be achieved with the minimum consumption of fuel and time, as much of the distillate as possible must be made to pass with the maximum ratio of oil to water that feedback will permit. This phenomenon affects not only the running costs per kg of oil, but also bears on the next subject for discussion, the time taken to process each charge.

The time that a charge must occupy the still falls into two parts. The *heating time* is the period from the first admission of steam to the still until it breaks through the top of the charge and the distillate is flowing normally. This depends on the availability of steam, the rise from day temperature to boiling point, and the mass of the charge. For calculating this time the mean specific heat of eucalyptus herb is about 0.8 kcal/kg and the latent heat surrendered by condensing steam is taken as 540 kcal/kg. The *extraction time* runs from the start of the normal flow of distillate to the point when the herb is exhausted of commercially recoverable oil, the commercial 'end-point'.

### How the oil comes to the herb surface

Initially the oil is secreted deeply inside the eucalyptus leaves. By the end of the heating period, hot water has condensed on all the leaf surfaces and penetrated their cutaneous layers. This water sets up a diffusion process by which the oil seeks to equalise its concentration throughout each microsystem. As oil diffusing to the surface is removed by passing steam it is replaced from the region of greater oil presence inside the leaf. This movement of the oil to the outside of the leaf continues at a rate which is proportional to the factor by which the oil's concentration inside the herb exceeds that on the surface.

### Relation between charge height and extraction time

If the height of the charge is increased, the number of points the steam will pass from which it may gather oil is also increased. As the vapours rise through the charge, they continually condense on, and revaporise from, each successive herb layer as they come to it. They tend to be enriched by nearly equal amounts of oil from each unit layer of herb.

So the amount of surface oil continually in transit at the top of the still tends to increase proportionately with charge height. This travelling oil adds to the surface concentration on the top layer of herb, and the factor by which the oil's concentration inside the herb exceeds that on the surface at that level is reduced by an equal amount for each successive layer from which the steam has already gathered oil. The diffusion rate declines and the extraction time for the oil in the top layer, which is that for the charge as a whole, increases accordingly; that is, by some factor of charge height.

The steam's ability to deliver heat to vaporise the oil has been seen to be proportional to the magnitude of the temperature gradient leading heat from the general vapour space to the liquid oil on the leaf. Since this gradient declines as the vapour space is enriched with oil from every herb layer below the top of the charge, the rate that the oil receives heat and boils away from that level must also decline proportionately with charge height.

For practical, commercial purposes, a given steam flow may be seen as requiring a basic time, $t$, to exhaust the oil from typical glands on the top layer of herb. The limiting factors extend this time by an increment, $\delta$, for each unit of charge height. The total extraction time, $T$ (minutes), for a charge of any height, $H$, is then given by:

$$T = t + (H \cdot \delta t) \qquad (2)$$

The parameters $t$ and $\delta t$ are determined from two simultaneous equations in this form which are obtained from very accurately timed distillations of two charges having different heights.

Once the steam has traversed about 75 cm of a typical charge of eucalyptus leaf, it will have gathered enough oil for the limiting factors of feedback and diffusion to take effect. Then, all the vapours from above this putative level will have a constant ratio of oil to water and it will be the maximum which feedback will permit. All oil that started in herb above this 75 cm level in the still will be extracted in the richest distillate that nature will allow. To maximise the efficiency this offers, stills and their charges should be as tall as conveniently practical for rapid handling. Both this supposed 75 cm level and formula 2 are simplifications involving a minor systematic error which increases with diminishing charge heights. But it becomes insignificant with charges more than 1.25 m tall. Most commercial distilleries work with charge heights between 1.5 and 2.0 m.

Another very important effect of feedback is that the amount of steam required to pass to complete a distillation depends on the quantity of oil to be recovered and is independent of the actual mass of herb in the still. If a charge contains so much barren plant material that its overall content of oil per kg is only one-third of average, its distillation will consume the same amount of steam as a normal charge of the same species, having the same diameter but only one-third the height, processed under similar conditions.

### Typical distillation curve for Eucalyptus polybractea

Figure 6.1 plots the oil delivered against the water passed for *E. polybractea*, and shows how the early part of the distillate from a charge of commercial height has a nearly constant ratio of oil to water. Feedback and other factors limit the oil content of distillate vapours after they have traversed a certain amount of herb. This characteristic applies to most other subcutaneously secreted oils, although the level in the charge at which the limiting factors intervene will vary considerably. Nevertheless, it enables engineers to calculate variations in extraction time due to variations in the total oil content in the still.

### Comparison of species' yields and extraction times

Estimates of the extraction times for distillation of *Eucalyptus* species other than *E. polybractea* can be made by first finding the time for *E. polybractea* in the new still and conditions, and then

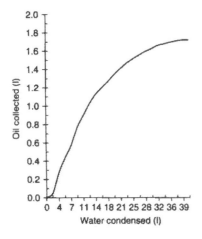

*Figure 6.1* Distillation curve for *E. polybractea*.

*Table 6.1* Relative extraction times for *Eucalyptus* species other than *E. polybractea*

| Species | Oil yield (%)[a] | Relative extraction time |
|---|---|---|
| E. polybractea | 1.35 | 1.00 |
| E. globulus | 0.90 | 0.67 |
| E. smithii | 1.30 | 0.96 |
| E. citriodora | 1.125 | 1.51 |
| E. dives (Type) | 2.70 | 2.00 |

a  Fresh weight basis, v/w.

applying the appropriate time factor taken from Table 6.1. For calculation purposes, variations in the oil content of the herb may be dealt with by using 'virtual' charge heights. The oil content of a herb layer, 1 cm thick and 1 m² in area, will be known for the herb sample which was test distilled to determine the parameters $t$ and $\delta t$ in formula 2. This then becomes the standard oil content for use with those parameters. When calculating the extraction time to be expected for a charge of similar herb in any other still, its expected total oil yield is divided by its cross-section area to give the oil content of a column 1 m² in cross section. The column's oil content is then divided by the standard oil content figure to get a virtual height $H$, for use in formula 2.

The figures used for the cineole-type *E. polybractea* serve as an example. Using direct steam generated under 2 atm gauge in a separate boiler, the test material was distilled under atmospheric pressure to 95 per cent of its estimated virtual exhaustion oil content, the economic endpoint. Then $t$ was 18.27 min and $\delta t$ was 0.411 min when distillate flow was 1.36 l/min/m² and herb oil content 37.3 ml per layer 1 cm thick and 1 m² in area. With charge heights and distillate flows (see below) duly adjusted, the parameters used in formula 2 will give accurate extraction times for charges of *E. polybractea*. They may also be used to obtain guidance on the extraction times for cineole-type oils from other species of *Eucalyptus* such as *E. globulus* and *E. smithii*.

## Relation between steam speed and extraction time

It is practical to treat the valuable oxygenated compounds as comprising most of the oil. Due to their greater affinity for water they diffuse to the surface ahead of the minor hydrocarbon components. Their rate of diffusion is proportional to the factor by which their concentration inside the herb exceeds that on the surface. If the oxygenated bodies vaporised immediately on reaching the leaf surface, as in fact the monocyclic hydrocarbons do, the concentration there would always be nil. Faster flows of steam would make no difference to the relative concentrations and could not speed up oil recovery. But it is known that the distillation of subcutaneous oils is accelerated by faster flows of steam. This implies that, under any given flow, there must be an appropriate quantity of oil on the herb surface whose concentration sets an orderly limit on the rate of diffusion. By quicker removal of this oil, faster steam reduces its surface concentration. Since the oil's concentration inside the herb is not immediately affected, it will now exceed that on the surface by a larger factor and diffusion will accelerate accordingly.

When the oil diffuses to the herb surface it forms numerous small circular patches of homogeneous oil. It can be shown that the oil will vaporise only from the perimeters of these patches where it is in contact with the water that condensed to raise the temperature. Also for comparison with the oil's concentration inside the herb, the area of the circular spot of diffused oil is proportional to its concentration on the herb surface.

From the foregoing it can be shown that increasing the rate of steam flow by a factor of $X^3$ will, ideally, speed up oil recovery by a factor of only $X^2$. But, in practice, fluids do not increase their rates of flow along restricted channels in the full proportion of the accelerating force. A lag factor, $R$, must be introduced which may be expressed as its effect on the speed of oil recovery. The resistance to diffusion increases, and the numerical value of $R$ declines, with any faster rate of steam flow. The formula relating different speeds of steam flow to the speeds with which oil will be recovered is then given by

$$Z = RY^{2.3} \tag{3}$$

where $Z$ is the factor of change in speed of oil recovery against the clock; $Y$ the factor of change in steam speed, with speed expressed in kg/min/m² of charge top area and $R$ the diffusion lag factor appropriate to factor of change in steam speed, $Y$.

The new extraction time, $T$, due to a factor of change in steam speed, $Y$, may be derived from the original extraction time, $t$, using the relation

$$T = \frac{t}{RY^{2.3}} \tag{4}$$

The new amount of steam, $W$, required to pass as distillate due to changing the steam speed by the factor $Y$, is compared with the original amount required, $w$, by

$$W = \frac{w}{R} \cdot Y^{1.3} \tag{5}$$

## The lag factor for Eucalyptus

Figure 6.2 relates the steam speed's factor of change, $Y$, to the diffusion lag factor, $R$, and applies strictly for changing from a base condensate flow of 1.36 l/min/m² of charge top area. However,

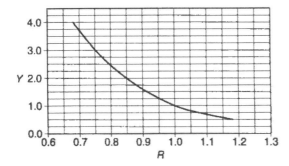

*Figure 6.2* Curve showing empirical relationship between Y and R for *Eucalyptus*.

in practice it appears to give usable values for $R$ when $Y$ happens to be a multiple of some other base rate of flow. The graph may be used to help convert any original extraction time to what it would have been at the graph's nominated base flow rate. If this is used as $t$ in formula 4, and for deriving a new value for $Y$, then the associated value for $R$ will theoretically be free from systematic error.

### Herb surfaces and wet steam

Because water is a very poor conductor of heat, steam can impart its latent heat to the liquid oil only by condensing onto the water surfaces in immediate close contact with the perimeter of the surface oil patch. On absorptive herb surfaces, the oil and water intermingle by capillary action all along their contact interface. This greatly increases both the active heat transfer area and, naturally, the oil's proportion of the distillate. With those herbs which rely on their absorptive surfaces to create an adequate heat transfer area, the distillation fails when the surfaces become saturated and are no longer absorptive.

Compared with herbs which are naturally absorptive or have been wilted in the sun to get the same effect, the absorptive capacity of eucalyptus leaf is very limited. It may well be nearly saturated by the water condensed to heat the charge, yet the oil's rate of vaporisation is not impaired as that of surface-born oils like mint and lavender would be. The surface oil patches of eucalyptus are homogeneous and virtually circular. They are also minute, and so numerous that even without intermingling at their perimeters, the aggregate length of oil–water interface is enough to create an adequate heat transfer area, into which steam can usefully condense. Eucalyptus distillations differ from some others in that vaporisation of the oil is not impaired if the herb surface is nearly saturated by steam that has condensed. The inefficiencies and losses are due to the steam's wetness fraction.

### Internal reflux

If the steam is very wet, amounts of cloud that are excessive for eucalyptus distillations will lodge on the leaf surfaces in the still. This deposited moisture is added to the steam which originally condensed to raise the temperature and it may be further reinforced by fluid from collapsing aqueous plant cells. This moisture soon floods the nearly non-absorptive leaves to the point where they can no longer hold it. A downward flow of liquids is established, dripping and gathering pace and volume from one leaf to the next, which washes oil to the bottom of the still. The loss of oil is evidenced by an excess of discoloured water in the bottom of the still at the end of each distillation. Field tests have shown well-established distilleries losing 30 per cent of the herb's recoverable oil to this cause.

With satellite boiler steam generated under less than 2 atm gauge, a distillate flow of 3 kg/min/m$^2$ will create an upward wind through the charge which opposes this reflux and reduces, but does not eliminate, the loss. In wetter conditions a faster flow of steam may help, but more of it will be used per kg of oil recovered. As noted earlier when discussing steam requirements, losses of oil to reflux become increasingly severe as flow rates decline progressively below 2 kg/min/m$^2$. At all flow rates, losses increase as the steam's wetness fraction rises. The distiller can gauge his loss from the amount of dirty water gathering in the bottom of his still. It should not be there – but it usually is.

*The hydrophylic effect*

If a cloud particle strikes an exposed oil surface it will roll up a coating of oil which it can carry upward through the charge. Coated cloud particles which lodge on herb surfaces higher up the still, and are then vaporised, accelerate the extraction of the oil. So relatively rich distillates and apparently short extraction times are achieved by many of the very primitive operators distilling *Eucalyptus* and tea trees in the Australian 'bush', because they use the very wet steam which results from boiling the water in the bottom of the still.

Many oil-coated cloud particles fail to lodge on herb surfaces further up the charge. They escape from the still in the liquid state and are too small to be caught by any normal separator. Their oil coatings are lost in the discarded water. Evidently, the amount of oil lost to this hydrophylic effect is proportional to the number of cloud particles reaching the top of the still and the area of oil held exposed to them while they pass.

Due to lodgement on each successive layer of herb, the residue of the steam's original wetness fraction which reaches the top of the still declines as a function of charge height. Conversely, the time that the oil area remains exposed at that level increases as a function of charge height. Tests with charges of practical commercial height show that these effects can compensate each other, so that the amount of oil lost to the hydrophylic effect is proportional only to the initial wetness fraction of the steam and the cross-section area of the charge. For each case, this loss of oil tends to be a significant fixed quantity, the same per unit area for all practical charge heights. So taller stills give better yields of oil.

*E. polybractea* seldom assays an oil content under 1.75 per cent by weight (fresh basis). When it is packed to a density of 250 kg/m$^3$ in commercial stills 1.5 m in diameter by 1.7 m tall, and processed with steam generated under some 2 atm of gauge pressure, most field distillers would be satisfied with a return of 11.5 kg of oil. Closer inspection shows that the apparent fixed loss of this oil to the hydrophylic effect is 1.5 kg/m$^2$ of charge area, representing over 20 per cent of the herb's recoverable oil at commercial charge heights.

Comparing equal flows of naturally wet steam, lower generating pressures in the boiler give shorter extraction times for equal returns of oil. Steam from the lowest pressure expands and dries out least on entering the still. Compared with steam from higher pressure, it increases the number of coated cloud particles carrying oil up the charge, as well as those escaping from the still. Since this accelerates both the speed with which oil is removed from the herb and the rate that it is lost to the hydrophylic effect, it shortens extraction time and enriches the distillate without raising the yield. However, it is not feasible to exploit the speed of this hydrophylic translocation of the oil because with all stills of practical commercial height the 20–25 per cent loss of the available oil persists.

## Comparing steam generated under low and high pressures

Consider a typical example of a common low-pressure boiler which delivers 300 kg of heating steam and then passes 1000 kg of steam through a 2.5 t trailer-load of herb. It recovers only 40 kg of oil when the charge is assayed to contain 50 kg. If the steam's wetness fraction at the still is as low as an unlikely 3 per cent, almost 40 kg of cloud will have been carried into the still. It is not difficult to show that each kg of cloud particles could carry off more than the 250 g of oil needed to explain the loss.

Now let the boiler deliver steam to the still under 7 atm gauge. On expanding into the atmospheric pressure still, this steam may take up almost 1.5 per cent of its own weight in water from the herb, instead of importing 3 per cent of cloud particles. It follows that, even if there were no reduction in the requirement of steam, which is contrary to all theory and experience, the amount of cloud available to promote reflux or to carry off oil will be some 58 kg less than with the lower pressure steam. A very big 'state of the art' distillery processing tea tree (*Melaleuca*), for which the requirements are similar to *Eucalyptus*, adopted these parameters and reduced the consumption of steam per kg of oil to two-thirds of normal, while improving the quality and markedly increasing the yield of oil per tonne of herb.

## Distillation under gauge pressure

Measurable gauge pressure in the still is helpful for recovering very high-boiling oils like vetiver. However, all the oils from eucalyptus are moderately volatile and do not justify the higher operating and capital costs that pressure distillation incurs. It can be shown (Denny 1991) that raising the operating pressure from 1 atm abs to 2, enriches the distillate by a factor of 1.5 and only two-thirds as much steam must pass to extract a given yield of oil. Pressure also increases the temperature gradient, but not by enough to matter with the eucalypts.

# Practice – the equipment

## The stills

All heat-losing surfaces of stills should be insulated with rockwool, or equivalent, about 50 mm thick.

## Laboratory stills

Laboratory steam distillations using stills only 20 or 30 cm tall are usually unhelpful. They are no guide to herb parameters and are misleading as to herb quality and yield. If the condensate water is returned to the still and the process is continued to total exhaustion, comparisons of yield only may be made between multiple tests. With very limited quantities of herb, better indications of yield and quality are given by 'water' distillations to exhaustion in a glass Clevenger apparatus. With suitable plant materials, comminuted if necessary, an accurate image of commercial operation can be obtained with 'drainpipe' stills only 20 cm in diameter, provided they are about 1.3 m tall. In the laboratory it may be possible to avoid the hydrophylic effect by desiccating the steam for the drainpipe still.

## Pilot still with small steam evaporator

This still (Figure 6.3) is popular for small-scale operation and is capable of passing 45 l of distillate per hour at atmospheric pressure. It is made from two galvanised 200-l oil drums and so is

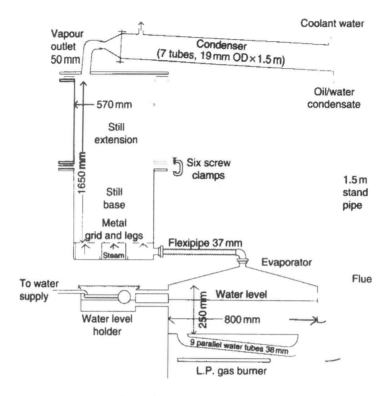

*Figure 6.3* Pilot still with small steam evaporator.

not expensive; even the evaporator can be put together in any normal farm workshop. The condenser, however, may need to be fabricated professionally. The still lid should be stainless steel but it does not require difficult cutting or welding. This still will hold over 100 kg of fresh herb and the evaporator will give a steam displacement rate very close to 3 kg/min/m$^2$ with a suitable gas burner. However, with this wet steam and the virtually non-absorptive eucalyptus leaf, the 20–25 per cent loss of oil to the hydrophylic effect is inevitable. But the scheme is very good for absorptive herbs.

*Water and steam distillation with under-charge kettle*

This system (Figure 6.4) is usually directly fired by wood fuel under a cylindrical still with a plain, flat bottom, rather than by oil as indicated. The former was a common method of distilling eucalyptus oil, but it is not a good one. The steam's wetness fraction is imponderably high and causes a strong reflux flow. The thermal efficiency of the circular flat plate kettle is scarcely 15 per cent, and solid fuel cannot possibly generate an upward flow of steam through a charge of equal diameter sufficient to reduce the reflux. There is a massive hydrophylic translocation of oil, which accelerates both its extraction and its loss. This gives the primitive still an illusion of efficiency, with extraction times that seem short for its slow rate of flow, and distillates so rich that it passes barely two-thirds of the amount of steam required by many, supposedly more advanced, distilleries using small low-pressure boilers to recover the same amount of oil. One can see this causing the competing claims of superior efficiency to protagonists of each system when, in fact, both methods were often equally inefficient.

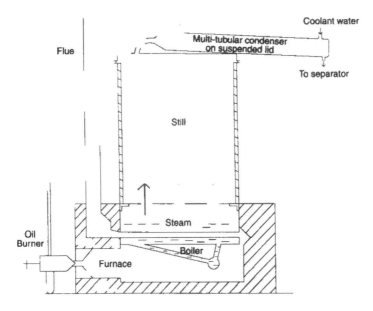

*Figure 6.4* Water and steam distillation with under-charge kettle.

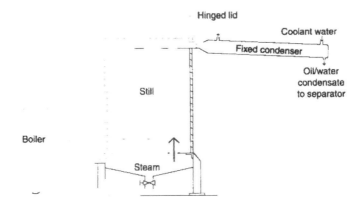

*Figure 6.5* Direct steam distillation with a separate boiler.

If the kettle for 'water and steam' distillation has the 'Babcock' tubes shown in Figure 6.4, and the furnace is fired by a diesel burner, the overall thermal efficiency can approach 80 per cent if well designed. The faster boiling rate reduces reflux but the steam is still wet enough to promote serious hydrophylic loss when processing the non-absorptive eucalyptus leaf. However, the system is very suitable for naturally absorptive and partially dried herbs.

*Direct steam distillation with a separate boiler*

With steam generated from a separate boiler (Figure 6.5) and under different gauge pressures up to about 2 atm, the ratio of the amount of oil returned, to the amount lost, in any given still usually varies relatively little. With the lowest generating pressures the distillate ratio of oil to

*Figure 6.6* Packing still with comminuted *E. globulus* leaf and twig, Portugal (photo: J. Coppen).

water approaches that of the water and steam system. But the direct steam's potential for faster distillate flows can save time. On the other hand, if the steam is generated under some 7 atm gauge, and expands to atmospheric pressure as it enters the still, it superheats sufficiently to ensure that no wetness fraction at all is carried to the herb. Its residual surplus heat is then still enough to dry out any additional condensation due to thermal inefficiency. At the end of the distillation the space beneath the charge is perfectly dry, showing that reflux is eliminated, and greatly improved yields suggest that losses due to the hydrophylic effect have been minimised.

If the boiler must work at low pressure, it might be worth testing comminution of the material with a chopper as it is being loaded into the still. It then packs more densely and traps more of the coated cloud particles. An example of this in Portugal, where *E. globulus* is distilled, is shown in Figure 6.6.

*Trailer stills*

These are large boxes filled directly during mechanical harvesting in the field and in the context of eucalyptus distillation are found only in Australia. They eliminate all manual methods of filling and emptying the stills and were developed in the early 1970s by GR Davis Pty Ltd as a means of reducing high labour costs. An example of one such trailer still is shown in Chapter 7 (Figure 7.2).

They are subject to all the normal principles of distillation and are usually large enough to hold at least 2.5 t of herb. On arrival at the distillery, a lid with a central vapour outlet and flexible hose (which connects to the condenser) is lowered over the trailer and fixed securely to it. Ideally, a typical cross-section area of $8\,m^2$ would demand the output of an 800 kW boiler. Thermal insulation of the trailer's vertical walls saves fuel, but the herb close to the wall is such a small part of the whole charge that the loss of some of its oil is not obvious enough to cause alarm. In practice, operators ignore the accumulation of dirty water in the bottom of the trailer still. Insulation of the trailer lid gives maximum steam flow, but causes unacceptable back pressure when air is being expelled at the start of each run if the condenser is a single tube type tapering to a restricted final outlet. Unlike steam, air does not collapse in volume on entering the condenser. If some of the early vapours condense on the cool lid it reduces their speed and less pressure is required to expel air at the slower rate. An array of parallel, floor level, sparge pipes joined to a common header is commonly used to admit steam to the still. Pressure in the header should be maintained close to 1 atm gauge. Alternatively, the steam may enter a pressure equalising space under a perforated false floor carrying the herb. The perforations must be fine enough to avoid sticks from gripping in them which would prevent the load sliding off easily when tipped for final discharge. At the end of the distillation the lid is removed and the trailer towed away to be replaced by a new one.

## Handling systems

Distilling essential oils involves getting the fresh herb to the distillery and may also entail moving the exhausted bulk away from it. These are major activities involving both capital and operating costs. Frequently, inadequate provision is made for these during the planning stage and this leads to administrative and financial difficulties later on. In some cases, spent herb from eucalyptus distillations is used to fuel the boiler or is composted on-site for subsequent sale, and it does not incur the same sort of removal costs.

In Australia, trailer stills with power tipping systems for discharging spent loads automatically solve the transport problem. There will be at least one more stand for trailers at the distillery than the number distilling at any one time. This allows for receptions and removals in the idle bay without disrupting distillation. At change time the steam is merely switched from the exhausted trailer to the new one. So the system is a genuine continuous operation.

Orthodox cylindrical stills can use 'cartridges'. These are stout bins with mesh bottoms which may be filled by the harvester in the field or by hand at a point near the stills. They can be moved on roller beds to a point under a gantry where an overhead trolley hoist picks them up. They may be liners for normal stills or they may constitute the still themselves. In either case, their bottom perimeters will rest on a steam seal such as 'piano note' cross-section neoprene, carried on a flange sealed to the still wall 150 mm above its floor. Where the cartridge is a liner for an orthodox still, the lid closure and insulation are normal, but the height of lift and gantry may be inconvenient. If the cartridge is merely lowered onto a simple still base just above floor level, the lift required is much reduced and the gantry simplified. The still lids are then a common fit to all cartridge tops. Then the cartridge must either carry its own insulation, which may thus be exposed to damage, or each base must have doors lined with rockwool which close tightly against the still cylinder. The latter is the best scheme, but not so easy to install. Loosely fitting doors turn the cartridge wall into an effective air condenser with disastrous results. When carried on the gantry, the cartridges are suspended by ropes from the hoist's swingletree to pivots on the cylinder wall. The latter are designed to facilitate tipping the exhausted herb at the end of the gantry. A tractor with a front end blade then pushes it to a compost heap. The cartridge system facilitates transportation and also permits such rapid replacement of exhausted charges that it is virtually a continuous operation.

In older systems, the still is filled manually and the herb forms a 'pudding' on a grid that can be lifted from the still for discharge. The still is necessarily out of action while it is being filled and the grid legs cause wasteful steam leaks through the charge.

On both theoretical and practical grounds, essential oil distillers should be very careful of so-called 'continuous distillation systems'. They usually involve a moving column of herb traversed by a counter current of steam. They do not suit eucalyptus distillation.

### The flow diagram

Figure 6.7 depicts the general arrangement of water and steam piping for a modern eucalypt distillery. It is necessary to treat the boiler feed water to minimise salts and oxygen that can harm the boiler tubes. Reputable boiler manufacturers will give advice as to the particular treatment required. All practical rainwater should be collected and used for the boiler. The discharged condensate water can also be used if it does not create foam or undesirable volatile products of decomposition. It is free of salts and, unlike the petroleum hydrocarbons, the traces of essential oil in the distillate water have tested harmless to boiler tubes.

Most of the energy consumed by a distillery is lost in the heat discharged in the condenser's cooling water. Schemes for saving some of this by using it as preheated boiler feed water are more academic theory than practical measures. The best method of preheating the feed water is to hold it in a water jacket round the furnace flue. Some economy may be effected by using spent herb as boiler fuel. However, this requires major capital outlay for a Dutch oven furnace which is justified only where other fuel is not readily available. It is usually better economy to compost the waste material and return it to the fields.

The piping must allow for the distillery to use water at twenty times the amount actually boiled away per day. This provides for the condenser coolant water to be turned off when distillate is not running during changing and heating periods. Design of the condenser will normally allow for a cooling water flow equal to fifteen times the planned distillate flow. Provision for emergencies requires that double this rate be available at the condenser coolant inlet under a pressure of at least 3 m head.

### The condenser

The single pass, multi-tube, vapour-in-tube condenser is immensely superior to all others for essential oil distilleries. It has the following advantages not available from single-tube, spiral

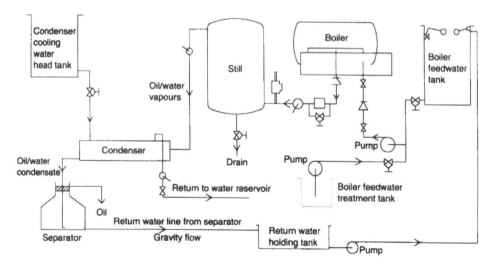

*Figure 6.7* Flow diagram showing layout and fluid transfers in a modern distillery.

coil condensers. It is nearly five times as efficient in the use of cooling water per unit of heat exchange area and takes up much less space. It can be instantly adjusted to deliver the distillate at any chosen temperature for separating oil and water. It readily releases air at the start of each distillation without unacceptable back pressure. It is easy to clean the inside of the tubes and access is easy for cleaning the outside of the tubes, if required. An example of a multi-tubular condenser being used for eucalyptus oil distillation is shown in Figure 6.8. Note that it is here being used in the vertical position but it can equally be used horizontally.

Consideration of the factors affecting condenser design are given elsewhere (Denny 1991), but the sample specifications given below are amenable to linear interpolation and so will cover a wide range of operating conditions. They apply for the following: connecting pipe from the still is $\leqslant 3$ m long with only one 90° bend; coolant water enters at $\leqslant 25°C$ and flows at fifteen times the distillate flow rate; bundles of stainless steel tubes 20 mm (3/4") outside diameter (the best size) and about 1 mm wall thickness, arranged in an equilateral triangle pattern with centres $1\frac{1}{2}$ tube diameters apart; coolant baffles not more than $1\frac{1}{2}$ shell diameters apart; condensate cooled to about 22°C above the temperature of the entering coolant water. Note that the best separation of oil from water occurs at 45–50°C, not lower as is commonly supposed.

The relation between rate of condensate flow and tube numbers is linear. A guide to the number of tubes required to handle satisfactorily a range of condensate flow rates is given in Table 6.2.

*Figure 6.8* Multi-tubular condenser being used for eucalyptus oil distillation, Australia (photo: J. Coppen).

Table 6.2 Recommended parameters for multi-tube condensers of the single pass, vapour-in-tube type

| Rate of condensate flow (kg/hr) | Diameter of connecting pipe (mm) | Number of tubes (20 mm diameter ×2 m long) | Number of tubes (20 mm diameter ×3 m long) |
|---|---|---|---|
| 200 | 75 | 20 | — |
| 500 | 125 | 50 | 35 |
| 1000 | 125 | 100 | 69 |
| 1500 | 175 | — | 104 |

*Ancillary equipment*

Figure 6.9 shows a convenient arrangement for the condenser, separator and end-point indicator. The condenser is used in the near horizontal position, which permits the separator and other equipment to be used at a convenient bench height. The horizontal condenser uses only an insignificant amount more water than when installed in the vertical position but the extra convenience is, by contrast, very significant.

*Separator*

It is a mistake to have the condensate delivered to the separator at a low temperature in the belief that less oil will be lost to solubility than in warmer water. Most of the loss to solubility occurs in the condenser near 100°C and it is very difficult to demonstrate any commercially significant increase in a distilled oil's solubility in water raised from 20°C to 60°C. A higher temperature increases the difference in density between eucalyptus oils and water, thus increasing the force promoting separation. At 45°C the forces due to the water's viscosity, which resist separation, are only 60 per cent of what they are at 20°C. At 45°C, small globules of eucalyptus oil rise through water twice as fast as they do at 25°C. The inner cylinder of the separator holds the first two minutes of full distillate flow, because it will normally be cooler than the water remaining in the separator from the preceding distillation, and, if it were not held until all temperatures had stabilised, it would go straight to the bottom and out of the water discharge pipe with its very rich content of oil.

The separator operates continuously. The central core, at the bottom of which is an inverted funnel, is inserted down the neck and inner cylinder and provides a primary stage separation with minimum turbulence. To ensure that delivered oil holds minimal suspended water, 6–8 cm depth of oil should be retained in the neck. The cross-section area of the outer annulus is calculated to ensure that the water travels downward towards the outlet slightly more slowly than the oil particles rise through it. The cineole-type oils rise at 10 mm/min at 45°C, which is faster than those of *E. dives* (Type) (6.4 mm/min) and *E. citriodora*. Inevitably, a properly designed separator is much larger than most distillers are used to. If the oil particles travel more slowly than the water they will be carried away and lost.

Recommended dimensions of separators for eucalyptus oils of the cineole and *E. dives* types, for a number of different distillate flow rates and at 45°C, are given in Table 6.3.

*End-point indicator*

Taking small samples of distillate in a graduated cylinder is an unreliable way to discern when the distillation should be stopped. An indicator of the type shown in Figure 6.9 for oils such as

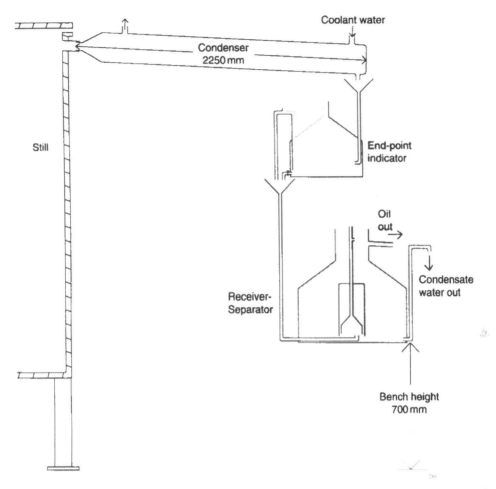

*Figure* 6.9 Arrangement of ancillary equipment.

*Table* 6.3 Recommended dimensions of separators for eucalyptus oils

| Rate of distillate flow (l/min) | Height of both cylinders (cm) | Diameter of inner cylinder and neck (cm) | Diameter of outer cylinder (cm) | |
|---|---|---|---|---|
| | | | Cineole types | E. dives/others |
| 1 | 30 | 9 | 37 | 45 |
| 3 | 30 | 16 | 64 | 79 |
| 6 | 30 | 22 | 90 | 111 |
| 9 | 50 | 22 | 109 | 136 |
| 12 | 50 | 24 | 132 | 156 |
| 15 | 50 | 27 | 140 | 175 |
| 18 | 50 | 30 | 154 | 192 |

eucalyptus, which are less dense than water, with a central, graduated glass column, is strongly recommended. The main body of the vessel has a capacity of about two minutes of distillate flow. During normal running, the plug cock at the bottom of the indicator is open and distillate flows directly into the separator inlet funnel. To measure the rate at which oil is being produced the plug cock is closed. When the oil–water mixture rises to the level of the upper outlet the supernatant oil appears in the graduated glass column; the rate at which the depth of oil increases indicates the rate at which it is being produced.

## Acknowledgements

Simon Coppen is thanked for preparation of Figures 6.3, 6.4, 6.5, 6.7 and 6.9 from the author's original drawings.

## References

Denny, E.F.K. (1991) *Field Distillation for Herbaceous Oils*, 2nd edn, Denny, McKenzie Associates, PO Box 42, Lilydale, Tasmania 7268, Australia.

# Part 2

# Cultivation and production of eucalypts around the world

With special reference to the leaf oils

# 7 Cultivation and production of eucalypts in Australia

## With special reference to the leaf oils

*Geoffrey R. Davis*

## History of oil production[1]

### The early days

Eucalyptus oil was distilled as early as 1788, the first year of white settlement in Australia, when Governor Phillip sent a sample to Sir Joseph Banks, and in 1790, John White, the Surgeon-General, despatched a quart of oil of *Eucalyptus piperita* to England. However, it was not until 1852 that a still was set up for the commercial distillation of eucalyptus oil. Joseph Bosisto, who had emigrated from England four years earlier, established his still on Dandenong Creek, about 40 km southeast of Melbourne, with the encouragement of Ferdinand von Mueller, then Government Botanist in Victoria.

In those early days, before much was known about the chemical composition of eucalyptus oils, several different species were worked, giving oils of varying composition. Stills for the production of oil were established in Gippsland, just east of Melbourne, and in Tasmania. The oil was used as a disinfectant, as a solvent, and for its therapeutic value. For the distiller of the oil, the critical factor was the yield of oil rather than its composition and, initially, species containing a high percentage of oil were favoured.

The search for high-yielding species saw the commercial activity move to western Victoria, particularly in the southern Mallee region, and then east towards central Victoria. At the same time stills were established in Tasmania and South Australia, particularly on Kangaroo Island.

### Early research

While the commercial distillers were seeking better sources of the oil, scientific investigation began to make significant headway only during the last two decades of the nineteenth century. For the industry to develop it was essential to know the chemical composition of the different oils, whether the composition was constant within a species, and the factors, if any, which affected the yield and composition. At this time, most oil production was in Victoria, but the main research was carried out at the Technological Museum in Sydney. This institution, later named the Museum of Applied Arts & Sciences, was set up in 1880 to, *inter alia*, 'investigate the economics of the natural products of Australia, and of New South Wales in particular, and to make this information available to the public' (Grolier Society 1965). Although it covered a wide field, a lot of its research was on the essential oils of the many plants unique to Australia.

---

1 A more detailed history of the industry is given by Shiel (1985).

Its reputation grew and the Museum soon became the foremost authority on Australian essential oils, particularly the eucalyptus oils.

During the Museum's ninety-nine year life, the combination of a botanist and a chemist as the research leaders was of great value to the essential oil industry. The initial work of the eminent botanist J.H. Maiden was followed, in turn, by that of R.T. Baker and H.G. Smith, A.R. Penfold and F.R. Morrison, and J.L. Willis and H.H.G. McKern, who, with their more recent successors R.O. Hellyer, E.V. Lassak and I.A. Southwell, have provided the scientific basis necessary to develop the industry. The early work established the chemical composition of the oils, within the limits of chemical knowledge and methods then available. Penfold, together with his co-workers, was particularly prolific throughout the 1920s, 1930s and 1940s (Coppen and Dyer 1993). The composition of the oil from a given species appeared, then, to be relatively constant. While the oil yield was also characteristic of the species, it could vary markedly between individual trees within a natural population.

However, further research showed that within some species pronounced chemical variation could exist. The plants which exhibit such variation, first termed 'physiological forms' by their discoverers, Penfold and Morrison, are now usually referred to as chemical variants, chemotypes or chemovars. Marked variation in oil composition may exist between populations or between individual trees within a population. An early example of this phenomenon within *Eucalyptus* was found to be that of *E. dives*. Some populations ('Type') produce oil commercially valued for its 40–50 per cent piperitone content; other populations ('Variety C') produce oil rich in cineole, also commercially valuable. Yet other variant populations, of no present commercial importance, are known: 'Variety A' yields an oil consisting chiefly of hydrocarbons, whilst 'Variety B' yields oils of variable composition, made up of constituents found in the other three forms.[2] However, for practical commercial purposes, populations of trees yielding oils of the desired composition have now been well defined and the oils distilled from them vary in their properties only within narrow limits. Recent work indicates that genetically determined quantitative variation (often to an extreme degree), rather than qualitative variation, accounts for these chemical forms.

## Oil-producing species

Of the hundreds of species of eucalyptus, most produce an oil, but few have oil of commercial value. To be of such value, the quantity of oil in the leaves must be at least 1 per cent of the fresh weight of leaf and the chemical composition must be of interest to the market. Apart from one or two speciality oils, such as those from *E. olida* (see below) and *E. staigeriana* (produced in Brazil), there are only three types which presently meet these criteria: oils rich in cineole, piperitone and citronellal. In Australia, the citronellal type has only been produced to a very small extent.

In the early stages of the industry, the species worked near Melbourne were those which produced the cineole-type oil, although the cineole content varied anywhere between 30 and 70 per cent. 'Phellandrene' (mainly $\alpha$-phellandrene) was a common constituent. It is not easy to be sure which species were worked because at that time the taxonomy of the genus was not well established. By the fourth quarter of the nineteenth century *E. globulus* oil, containing 60–70 per cent cineole, was being produced in southern Victoria and Tasmania. Following the first inclusion of

---

2 The terms 'Type', 'Variety A', 'Variety B', etc., were introduced in the 1920s and applied chronologically within a species as each distinct type of oil was enumerated. It is preferable, nowadays, to avoid using these terms and to state the particular chemical variant by name.

eucalyptus oil in the British Pharmacopoeia (BP) in 1885, compliance with the specification required increasingly higher levels of cineole, and by 1924 the minimum cineole content was 70 per cent, the level at which it is today. This made *E. globulus* more popular since the cineole content could be increased to 70 per cent or more by simple rectification and the phellandrene content was negligible. The BP allowed only a very low level of phellandrene. It still does, although there appears to be no good reason for this.

Towards the end of the nineteenth century the main part of the industry was in the central western part of Victoria, where availability of good-yielding, high-cineole, phellandrene-free oil attracted distillers. Although oil was extracted from several species, including *E. viridis* and *E. sideroxylon*, and elsewhere from *E. radiata* and *E. robertsonii* and others, the industry in the central and western part of Victoria was firmly based on *E. polybractea*. Oil was also produced from the same, hardy species in the mallee country of the western plains of New South Wales. The oil is high in cineole, 78–88 per cent, and phellandrene-free; it is also produced in reasonable yields. *E. polybractea* has a remarkable ability to coppice after harvest and grows on land which is of little use for anything else. In other parts of New South Wales, and parts of Victoria, cineole variants of *E. radiata* and *E. dives* were the major source of the medicinal type of eucalyptus oil, particularly the 70 per cent cineole grade which conformed to the several national pharmacopoeias which listed eucalyptus oil.

In the early part of the twentieth century, oils from various types of *E. radiata* and *E. dives* were produced on the south coast of New South Wales and on the mountains further west. In addition to pharmaceutical applications, much of the production, particularly that from *E. dives* and the phellandrene variant of *E. radiata* (then known as *E. phellandra*), was used for mineral flotation and in disinfectants.

## Commercial production

Although hundreds of distillers have been in business during the 140-odd years that eucalyptus oil has been produced in Australia, there have been just a few major ones in the industry. Joseph Bosisto, the initial force in the eucalyptus oil industry, continued to be involved in it until his death in 1898. The following year, J. Bosisto & Company Pty Ltd was constituted, and operated until 1951 when it became a subsidiary of Drug Houses of Australia (DHA). DHA was taken over by Slater Walker in 1968. In 1974, Peter Abbott purchased the eucalyptus oil section of DHA and the name Felton Grimwade & Bickford, the present name of the company. This company also owns the name Bosisto & Co., which is still in use today, and the Bosisto Parrot Brand name.

Early on, F.H. Faulding & Co. became involved in eucalyptus oil production, particularly in Victoria and South Australia and, to some extent, New South Wales. This company has maintained its connection until the present day. In 1880, the Tasmanian Eucalyptus Oil Company started in Tasmania. It moved its operations to Melbourne in about 1920 and remained a major buyer until 1947. W.K. Burnside Pty Ltd were one of the main buyers and leaseholders of land for oil production from the 1920s to the 1960s. This company also operated in New South Wales.

In 1912, Fred Webb of Braidwood, a gold fossicker, distilled oil from the narrow-leaved peppermint, *E. radiata* (phellandrene variant). He went to Sydney and while there ordered a suit from Mr A.J. Bedwell, a tailor who carried out a lot of country order business. The suit, when made, was sent to Braidwood and to Mr Bedwell's consternation payment was made by the arrival of a small drum of eucalyptus oil. Bedwell not only succeeded in selling the drum of oil but soon sought more. By 1919 he gave up his tailoring business and became the major force in eucalyptus oil in New South Wales until selling out to Plaimar Ltd of Perth in 1950.

A.J. Bedwell Pty Ltd continued producing as a Plaimar subsidiary until 1971, when its eucalyptus oil interests were acquired by G.R. Davis Pty Ltd.

The eucalyptus oil industry continued to be centred in Victoria until about 1950, producing mainly cineole-type oils. It had provided work for many men and a lot of oil was produced, but the industry was not stable and was profitable for the producers for short periods only. For most of the time it was subsistence farming. The main problems faced by the industry were fluctuating markets, due mainly to competition from other countries, replacement of the oil with alternative, cheaper products – particularly in the case of mineral flotation – and, in recent years, the stubborn refusal of all Victorian state governments to allow adequate tenure of land.

In the early post-war years there was a world shortage of menthol. Plaimar Ltd had succeeded in earlier years in producing *l*-menthol from *l*-piperitone, the major compound of *E. dives* (Type) oil. Several other manufacturers in Australia, Europe and USA were able to produce liquid menthol from this source and the demand for piperitone-rich *E. dives* oil increased substantially. The main stands of *E. dives* (Type) are in southern New South Wales and, as a result, much of the industry moved there. The production of cineole-rich oils also increased in New South Wales, while continuing in Victoria. The demand for locally produced *E. dives* (Type) oil lasted, with fluctuations, for about twenty years, until other countries, particularly South Africa, were able to produce low-cost oil. More importantly, production of menthol from alternative sources (natural menthol from *Mentha arvensis* and synthetic menthol from turpentine) could supply the world's needs and the piperitone route from *E. dives* became increasingly less economic.

## *Early methods of production*

For the first hundred years of the industry, almost all eucalyptus oil was produced in very simple bush stills by steam distillation at atmospheric pressure. The typical bush still consisted of a simple tank, usually mild steel, into which leaf and terminal branchlets were stacked on a grid about 15–30 cm from the bottom. Water below the grid was heated by a fire directly below the tank. The steam so generated passed up through the leaf to an outlet just below, or sometimes through, the lid. The resulting mixture of hot oil and water vapours was led to a condenser – often simply a long pipe passing through a dam or stream – and the condensate then passed to a separator where the top layer of oil was removed from the oil–water mixture. In some cases steam was produced in a separate boiler and injected into the tank holding the leaf. This was necessarily so when wooden vats, rather than steel tanks, were used to hold the leaf. These simple stills were used because they were easy to construct and operate, were low-cost (second-hand tanks were often used), and could easily be set up and dismantled, an advantage when they were used in areas where there was insufficient leaf available to support a large or permanent operation. There were, of course, some much larger and more sophisticated plants operating in the early stages of the industry, where large natural stands of the required species were available. These plants were usually, but not always, established and owned by one of the major companies referred to earlier. It is worth noting that the oil produced in the simplest direct-fired still is not inferior in quality to that produced in the most sophisticated apparatus.

It is fortunate that in stands of commercial oil-bearing species the required trees usually dominate the stand. Furthermore, where a single species is worked, the composition of the oil is similar throughout the stand, although the yield of oil might differ substantially from tree to tree. There are some exceptions to this but, generally, once a stand of trees with good quality oil is found, the producer can harvest it with confidence.

All the commercial oil-bearing eucalypts coppice well and, providing the interval between harvests is not too short so that there is no decline of vigour in the production of foliage, they can be worked on a sustained yield basis. Coppicing was practised from the early days of oil production. The only tools required were an axe for felling the tree and a heavy knife for removing the leaves and terminal branchlets. In the case of the mallee species, that is, the multi-stemmed trees commonly found in the drier inland areas, the tree is normally cut at, or close, to the ground using a billhook or curved knife. Regrowth comes from the lignotuber just below the ground. In the early years of the industry this arduous work was done mostly by itinerant workers, often as an alternative to prospecting for gold. Few were able to make a good living because of the low value of the oil. Nevertheless, production of oil gave a start to many of the post-war migrants. When farms were being developed on land where oil-bearing trees occurred, production of eucalyptus oil provided a cash return before the traditional agricultural crops matured. New farmers were prepared to sell oil cheaply just to get a cash income.

## *Competition with other producers*

Ever since the industry first started in Australia other countries have planted commercial oil-bearing eucalypts and for almost a century have been able to offer oil on the world market. The main early competitors were Spain and Portugal, where oil was produced, mainly from *E. globulus*, as a by-product of the timber industry. In Australia, despite the advantage of low costs for much of its history, high production costs in the 1950s, and for some time after, made it uncompetitive in many world markets, including the eucalyptus oil market. The advantages of producing oil as a by-product, rather than the sole product, proximity to the main markets, and a low wage structure, enabled first the Iberian countries, and then China, to put oil on the market at prices which Australia could not meet. South Africa, too, began to supply Australia with piperitone-rich oil from *E. dives* (as was noted earlier) and medicinal (cineole-rich) oil from *E. smithii*. It became apparent by the 1960s that if the industry were to survive, mechanisation of production and development of superior trees in plantations was essential.

It is unfortunate that Australia has been unable to retain the larger part of this uniquely Australian product. However, very rapid growth of *Eucalyptus* in other countries, probably due to the absence of natural predators, together with the factors noted above, meant that not enough money could be generated in the industry, either to make it profitable for producers or to provide the capital needed to develop the industry. The value of the product on the world market, except for quite brief periods, has not been enough to cover production costs in this country. Much of the land cleared of oil-bearing trees for planting traditional agricultural crops is actually better suited to growing trees for oil. Except for very small areas this calls for plantation development, which is costly and beyond the financial capacity of most producers. Furthermore, as noted earlier, in Victoria, where larger producers could have financed such development, short-sighted Government policy has frustrated such moves by its refusal to recognise the need for reasonable term land tenure.

## Modern methods of harvesting and production

On the plain country, attempts were made as early as 1950 to mechanise harvesting and, although not the complete solution, they did demonstrate that cutting the tree using powered tools was possible without detriment to its health.

The problem of mechanising harvesting of natural stands is that the land is usually uneven and contains rocks, stumps, logs and holes. There are also, initially, other plants which, if harvested, might contain products which would be co-distilled with the eucalyptus oil. Clearing the site of unwanted obstacles is therefore the first task, and in many areas this is a large and costly job. Once it is possible to operate normal agricultural machinery on the land to be harvested the aim is to reduce the manual work to a minimum. The traditional technique of cutting the coppice with a knife, laying the cut material in heaps, loading the heaps on to a vehicle, transporting the leaf to the still and loading the still, has all to be done, if possible, mechanically.

The idea of bringing the still to the area to be harvested was thus conceived and, after many trials, a machine was developed which was strong enough to cut the mallee coppice at ground level and elevate the material into the mobile still towed behind. Partial mechanisation was achieved by several distillers in the late 1960s, but the first effective, fully mechanical, harvesting system was set up in the mallee country near West Wyalong, New South Wales, in about 1971. Since then, virtually all the industry situated on the plains has been mechanised. A typical area of cleaned natural stands of *E. polybractea*, ready for mechanical harvesting, is shown in Figure 7.1.

In the field, a forage harvester is towed by a tractor, and a series of rotating 'hammers' with cutting edges slice though the shrubby plants at ground level. The chopped pieces are blown up and into the separate 3-t capacity box-shaped 'bin' trailed behind (Figure 7.2). It takes approximately one hour to fill a bin and two to two and a half bin loads to harvest one hectare of land. When harvesting areas near to the distillery, pairs of bins are towed to and from the site by tractor. If the distances involved preclude this then each pair is transported by low loader. One man is able to undertake the harvesting operation at each site on his own. The system has also been adopted for other essential oil production, such as tea tree oil (from *Melaleuca*).

Whatever the raw material, the principles of harvesting and handling, and the subsequent distillation, remain the same. On arrival back at the distillery, a suspended lid is lowered onto

*Figure 7.1* Area of regularly harvested natural stands of *Eucalyptus polybractea*, West Wyalong, New South Wales. Regrowth shown is eighteen months old and ready for harvesting (photo: J. Coppen, courtesy of G.R. Davis Pty Ltd).

*Figure 7.2* Trailer bin, which also functions as a still, packed during mechanical harvesting of *Eucalyptus polybractea*, West Wyalong, New South Wales (photo: J. Coppen, courtesy of G.R. Davis Pty Ltd).

the bin by block and tackle and securely fastened. Steam from a boiler is passed upwards through the charge of plant material and, on distillation, the oil/water vapours are led through a duct in the lid to a multi-tubular condenser and receiver (Figure 7.3). No lagging or double skinning of the bin is found to be necessary, ambient temperatures being sufficiently high to prevent reflux of vapours within the bin during distillation. Once distillation commences, it is complete in approximately one hour. Distillation at atmospheric pressure is effective and there is no advantage in distilling at either reduced or increased pressure. The whole operation can be carried out in a fairly compact, enclosed area, designed to accommodate two bins at a time (Figure 7.4). While one is being distilled the other, with spent leaf, can be removed and replaced by another with fresh leaf. After distillation, the extracted leaf is returned to an area which has recently been harvested and pulled or, in some cases, tipped out of the still. In this way it serves as a much-valued mulch. The leaf so distilled is untouched by hand throughout the entire process.

Although eucalyptus oil is readily distilled from the foliage in quite simple apparatus, some improvement in oil yields has been achieved by using the cohobation technique. Here, the distillate water, after removal of the non-dissolved oil, is re-introduced to the system. The only additional water required is that needed to replace water lost by evaporation and residual water lost when the wet charge of extracted leaf is removed. The water so circulated becomes saturated with oil and does not dissolve any more of it. Cineole-type oils have quite low solubility in water, even hot water, and this technique was seldom used with simple bush stills. However, with larger stills which operate most of the year, the increase in yield with cohobation is significant. A further advantage of cohobation is conservation of water; *E. polybractea* is a dryland species where water is usually limited.

*Figure 7.3* Trailer bin ready for distillation of *Eucalyptus polybractea*. Photo shows flexible duct which leads oil/water vapours from outlet at lid to multi-tubular condenser (photo: J. Coppen, courtesy of G.R. Davis Pty Ltd).

*Figure 7.4* Distillery of G.R. Davis Pty Ltd showing space either side of the boiler for placement of trailer bins. Boiler uses dried spent leaf from previous distillations (photo: J. Coppen).

In the mountain country where *E. dives*, *E. radiata* and several other species are worked, the axe gave way to a chain saw for felling the tree in the 1950s, but it was not until the late 1970s that the idea of a mobile still of the type used on the plains was tried. A still of reasonable size mounted on a truck is difficult to operate in the steep mountain country, and is suitable only in the less steep areas. After felling the tree with a chain saw, or the coppice growth with a heavy knife, the branches are fed manually into a chipper which elevates the chipped material into the still mounted on the truck. The truck then returns to the distillery. Although still requiring some manual handling, this partial mechanisation has reduced production costs significantly where it can be used.

To be able to produce *E. dives* (Type) oil competitively calls for complete mechanisation and this, in turn, requires plantation establishment. Although in some years up to 500 t of oil was produced from natural stands, the market was never stable enough to support the high capital cost of establishing plantations in Australia.

Mechanised harvesting of natural stands does tend to expose the harvested area to soil erosion, both by wind and water. In the period immediately following harvest there is no ground cover and soil blows and washes away rapidly. Various methods of alleviating this problem have been tried. Sowing traditional agricultural cereal crops after harvesting the leaves is sometimes practised. Establishment of a pasture, where this is possible, is effective, providing the stock are not allowed to graze until the tree regrowth is advanced. Light cultivation across the slope is also practised. This has the effect of collecting water, so reducing run-off and therefore erosion, and, of course, stimulating growth. A very effective method of controlling erosion and building soil fertility is to return the extracted leaf to the land whence it came. Considerable expense is involved in this, but in so doing, the regrowth of the trees is stimulated, in addition to controlling erosion. Increased yield of leaf, and therefore oil, eventually offsets the cost of this operation.

## Cultivation for oil production

The ready availability of oil-bearing eucalypts in Australian forests, and the low value and fluctuating demand for the oil, made plantation development unattractive during the first century of production. However, as the nation developed, vast areas of oil-bearing forests were cleared for agriculture, other good areas were incorporated into national parks and reserves of various types, and much private property became unavailable. Mechanisation, so necessary to allow competitive production, was difficult or often impossible on much of the available hill country forests. It became obvious that plantation establishment was essential if the industry was to progress.

The feasibility of plantation establishment became apparent when markets became reasonably assured and as mechanical planting and harvesting techniques were developed. The mechanical harvesting machines were tried and improved on upgraded natural stands and the planting machines were developed by modification of existing planters.

The advantages of plantation production, as distinct from natural forest production, are summarised below:

1. Suitable land on which to grow the required species of *Eucalyptus* can usually be found. Some species grow better on land other than that on which they occur naturally.
2. Alternatively, species can be selected which are most suited to the land which is available – on an existing farm, for example.
3. Land can be chosen on which machinery can be safely used.
4. A single species can be planted if desired and at a density which can be optimised. This is usually much greater than in natural forests, where in most cases other species are present (sometimes to a greater extent than the one required).

5  Insects and other predators can be controlled more easily.
6  In dry country, which is usually the case with Australian production, irrigation is sometimes possible.
7  The compact nature of stands allows short hauls to the still-house after harvesting.
8  Breeding programmes to develop trees with superior growth rate, higher oil yields or better quality oil can be established.

In Australia, the eucalyptus oil industry is moving more towards plantation production. Plantations already established are living up to expectations and one such plantation (*E. polybractea*) is shown in Figure 7.5.

*Aspects of establishment*

*Land preparation*

Land preparation is important. In most cases it is desirable to deep rip the row to be planted – as deep as the available machinery can manage – to allow water and root penetration. Just before planting, the rough surface is reduced to a smoother one on which a planting machine will work effectively. A rotary hoe is suitable for this purpose. For *E. polybractea* one pass is usually sufficient since, in the type of soil in which it thrives, further hoeing tends to break down the soil structure. In dry country, planting in a groove is preferable to planting on flat land or on a mound. In this way any surface water that becomes available is concentrated near the plant.

Layout of the plantation is also important. In particular, it is necessary to ensure that machinery can be used between rows of trees and that the density of planting is adequate to give sufficient

*Figure 7.5* *Eucalyptus polybractea* plantation of G.R. Davis Pty Ltd, West Wyalong, New South Wales (photo: G. Davis).

biomass, though not so dense as to cause suppression of some trees or excess stress in very dry conditions. In very broad terms, and subject to much variation according to actual site, 5000 trees/ha is the likely optimum density for *E. polybractea*.

## Planting

Direct sowing of seed is not practised because of the very fragile nature of the *E. polybractea* plant soon after germination. Even in good conditions, losses are too great to allow adequate establishment. Seedlings are therefore raised in a nursery to a stage where they can stand field conditions. In spring and summer, seeds take about three weeks to germinate (unless climate-controlled greenhouses are used) and the seedlings are ready to plant out after a further six weeks. In autumn and winter, at least twice this time is required. Provided water is available, seedlings can be planted in spring and summer so that the trees are established before the first frosts of winter.

## Weeding

Control of weeds in the early stages is essential. After the trees are well established, and form a canopy, weeds are suppressed. However, since this is a perennial crop, weeds must be controlled after each harvest, until either the canopy is formed or the season for vigorous weed growth is past.

In plantations with straight rows, 'cultivation' of a strip approximately 0.5 m either side of the trees is effective in controlling weeds. The 2 m space between cultivated strips is best left with plants on it to prevent soil erosion. Except when the trees are quite young, and therefore fragile, sheep can be grazed on the natural or sown pasture between the rows. There is likely to be some damage to trees when stock graze but in most cases this is offset by the value of the grazing. Before planting, a pre-emergence herbicide such as Goal™ is very effective. Cultivation in the early stages of seedling growth and coppice regrowth, followed by sheep grazing as the trees develop, can control weeds until there is sufficient canopy to minimise weed competition.

## Pests and diseases

*E. polybractea* is not usually subject to serious insect attack, although a sawfly (*Perga dorsalis*) and a case moth (*Hyalarcta huebneri*) both cause some damage. When the tree is concentrated as a single species in plantations, or in dense natural stands from which other species have been removed, this damage can, under certain conditions, be considerable. Occasional, serious infestation tends to occur when leaf density is high, but rarely in the early regrowth stage. Damage due to infestation can spread very rapidly, destroying large areas of leaf within a few days if not checked. Although the trees are defoliated and the crop is lost, the tree is not killed, and it shoots again from the lignotuber. The case moth is difficult to destroy by chemical spraying as it is protected by its case, except at night when it feeds. However, a very effective control method is to harvest the area of *E. polybractea* under attack.

The greatest hazard which *E. polybractea* has to face is prolonged drought conditions. The tree has adapted to this and survives even extreme droughts by not growing. So while almost a whole crop might be lost in such conditions, the stock is not lost and it grows again when rain eventually falls.

*E. polybractea* is subject to root fungus attack when grown outside its natural habitat, particularly in wetter areas.

*Table 7.1* Commercial oil yields of *Eucalyptus* species in Australia

|  | Commercial yield (% w/w, fresh basis) | Interval between harvests (years) |
|---|---|---|
| E. polybractea[a] | 1.0–1.5 | 1.5–2.0 |
| E. radiata, cineole variant | 2.5–3.5 | 2.5–3.0 |
| E. dives, cineole variant | 2.5–4.0 | 2.5–3.0 |
| E. dives, piperitone variant | 3.0–4.5 | 3.0–4.0 |
| E. viridis[b] | 0.3–0.6 | 1.0–1.5 |
| E. globulus[b] | 0.6–1.1 | 2.0–3.0 |

a The most important species.
b Nowadays only occasionally distilled separately.

## Factors which affect biomass and oil yields

It is not sensible to state quantities of biomass or oil yield per unit area for natural forest stands, from which most Australian-produced oil originates. The number of trees per hectare varies greatly, as does their vigour and the yield of oil, depending as they do on soil, climate and species.

There is a general consistency within a species of yield of oil per unit weight of leaf, and even greater consistency of chemical composition. For commercial purposes, the yield of oil of any species is usually expressed in percentage terms as the weight of oil distilled from freshly harvested leaf material (i.e. per cent w/w, fresh basis). Leaf material is that part of the plant which is put into the still. In the case of trees in which the branches are chopped off and then the leaves stripped from the branches, for example, *E. dives* and *E. radiata*, the material that goes into the still consists of leaves and terminal branchlets only. In the case of trees cut at ground level by hook or by machine, for example, *E. polybractea* and *E. viridis*, the whole of the aerial part of the plant is put into the still. On this basis, yields of the main commercially produced oils are shown in Table 7.1.

The figures in Table 7.1 are general and subject to considerable variation. There is variation from district to district and even more variation with season. The cause of this rise and fall in the yield of oil is not clearly understood, although a number of factors are involved. In the case of *E. polybractea*, which is a dry-country species, the quantity of biomass produced per hectare increases with higher rainfall, while the yield of oil in the leaf increases in dry periods. This is partially explained by the lower moisture content of the leaf in dry times. In order to avoid the problems of variable moisture content, and the inclusion of variable amounts of woody material in the charge of leaf which is distilled, laboratory distillation – in which the moisture content of the leaf is known (by determination) and only leaf (including petiole) is used – is the only satisfactory means of determining oil yields. For the commercial producer, however, such methods are time-consuming and expensive, and may yield results which cannot immediately be related to practice.

It must be noted, too, that the frequency of harvest varies with species and, to some extent, with district. The interval between harvests is indicated in Table 7.1. Although in years of favourable weather the trees can be harvested at shorter intervals, more frequent harvesting will ultimately reduce the vigour of the tree. Trees which have not been harvested too frequently have a long production life and give a sustained yield of oil. Some *E. polybractea* trees have been harvested at approximately eighteen-month intervals for up to eighty years without loss of vigour.

As in natural stands, the biomass and oil yields of plantations vary greatly. Factors which affect these yields include:

1 Seed provenance
2 Soil and nutrient properties

3  Water supply
4  Weather
5  Weeds            ⎫ Discussed
6  Pests and diseases ⎭ earlier

*Seed provenance*

In nature, although there is some consistency of leaf production and oil yield within a stand, there can be variation, sometimes quite marked, between different geographical sites (provenances) due to genetic differences. In addition, some individual trees within a stand may have a particularly high or low vigour and/or oil content. Selection of seed from known, high quality trees or provenances or, better still, vegetative reproduction, is therefore desirable in order to maximise the returns from plantation development. To date, while vegetative propagation has been achieved, it is difficult and much more expensive than planting seedlings. Selection of superior trees, striking cuttings and developing a seed orchard from them remains the most effective means of establishing high-quality plantations.

*Soils*

Natural stands of *E. polybractea*, the main source of oil in Australia, are restricted to a small area on the central western plains of New South Wales and, disjunctively, to a larger area in central and western Victoria. The tree grows well on hard, dry country but less well on more fertile, friable soils. Considerable research has confirmed that the best results for biomass production are obtained in the natural stand areas. Here, *E. polybractea* is often the dominant tree and sometimes occupies the area almost exclusively in terms of eucalypts. Although the tree thrives on the low ridges of the mallee-type country it grows best in the wide, very shallow valleys between the ridges. The plantations established to date are in the natural stand areas.

Addition of conventional fertilisers has very little effect on the growth rate of *E. polybractea* when it is grown in its natural habitat. Trials have shown that there is no useful response to potassium or phosphorous. Indeed, growth can be retarded by heavy application of these elements. Milthorpe *et al*. (1994) reported a poor response to either phosphorous or nitrogen. Substantial application of nitrogen does tend to increase growth, but mainly of stem, with little increase in leaf biomass. Some trials of trace elements have also shown little promise. Other than returning extracted leaf to the harvested areas, therefore, application of fertiliser is not beneficial.

*Water supply*

For the dryland eucalypts, water is a major factor in determining biomass and oil yields. If irrigation is feasible, biomass can be considerably increased if it is employed at the most appropriate time in the growth cycle. However, water for irrigation is seldom available in the mallee-growing areas and it is necessary to resort to maximising retention of rainwater on plantations. Spreading the extracted leaf on the land from which it was harvested is probably the most effective means of conserving moisture. This probably also returns some of the nutrients to the trees.

*Weather*

Without doubt, the major factor in biomass yield is the weather. Although irrigation reduces its effect to some extent, biomass nevertheless increases dramatically in times of good rain, particularly in springs which are followed by hot summers with sufficient follow-up rain. In mallee species, biomass is low at the first harvest after planting, but it increases in subsequent harvests as the coppice growth from the lignotuber produces several stems rather than just the original, single one. Excluding the first harvest, biomass yields are likely to range from 2.5 to 15 t, and oil yields from 30 to 150 kg, per hectare. While this very great variation is due to a combination of the factors already mentioned, the major factor remains the weather and the climatic conditions which prevail during the period of growth between harvests.

In a 5-ha trial plot of *E. polybractea* harvested annually from 1981 to 1990, the yield of oil per hectare varied from 56 to 80 kg (G. Davis unpubl.). The highest yields were in years of higher-than-average rainfall, while drought years resulted in lower oil yields. The lowest yield, 56 kg/ha, was in the final year of the trial; biomass was also low, despite reasonable weather conditions, and this was attributed to a decline in vigour of the trees due to too short a time interval between harvests. In its natural habitat of an average, but variable, annual rainfall of about 400 mm, *E. polybractea* is normally harvested at eighteen-month intervals. When the interval between harvests in the trial was increased from twelve to eighteen months, and then to twenty-one months, the oil yields were 112 kg (equivalent to 75 kg/year) and 140 kg (80 kg/year) per hectare, respectively, that is, at the higher end of the annual range quoted earlier. However, the confounding effect of rainfall means that there is not necessarily a simple inverse relationship between annual yield and frequency of harvesting.

While the variation in oil yield in the trial has been substantial, variation in natural stands is much greater and caused by such factors as differing soil properties, tree provenance, topography (particularly in relation to water run-off) and weed and pest infestation.

## Post-harvest aspects of oil production

### Oil storage

After distillation, the layer of oil which is separated from the oil-water condensate remains wet. If the oil is to be sold in crude form, without further rectification, it is therefore desirable to dry it and for this, anhydrous calcium chloride or anhydrous sodium sulphate can be employed. The oil can then be stored, preferably in galvanised steel drums or in high-density plastic ones. Some plastics are affected by cineole-type oils. Although cineole-type oils do not oxidise rapidly in contact with air, it is recommended that soon after distillation the oil be stored in airtight drums in the shade, that is, away from excessive heat and with light excluded.

### Further processing

After drying, the crude oil is usually colourless or pale yellow. In this form it can be held for several years without deterioration, although it will eventually discolour. However, oil is often refined by carrying out a second, separate, dry distillation after the initial distillation from the leaf. This rectification, if undertaken, is carried out primarily to adjust the cineole content to that required by the various standards or by the buyers, but it also removes any water and colour. It also removes any low-boiling components of the oil, which in some oils have an unpleasant odour. If necessary, high-boiling fractions can also be removed, although this calls for a high temperature towards the end of the rectification process, which is undesirable. Rectification is best carried out at reduced pressure. Rectification at moderate temperatures can also be achieved by passing steam through the oil; this removes colour and unwanted odours, but the oil then has to be dried. A further advantage of rectification is that it has a unifying effect, producing more consistency in composition for different batches of oil, which most buyers require.

It is standard practice to rectify oil distilled from *E. globulus* because the crude oil rarely has a sufficiently high cineole content to conform to the normal standards. *E. smithii* oil, imported into Australia from Africa, is usually rectified to reduce the aldehyde content; isovaleraldehyde, the offending chemical, is obnoxious, even in small concentrations. Although the most commonly produced Australian oils, those from *E. polybractea* and *E. radiata* (Type), do not need to be rectified, in practice they are because of the demand for consistency of composition. We are in an age where natural oils are expected to fit man-made standards, rather than standards being established to accommodate natural oils. While this view tends to hold sway, some buyers consider that rectification has the effect of reducing the individual character of the different oils by removing some of the compounds which give them their unique aroma, and so reduces, rather than enhances, the quality. *E. radiata*, for example, yields an oil with much more aromatic character than, say, *E. globulus*.

There is a demand for eucalyptol, the trade name for virtually pure 1,8-cineole, and this is readily obtained from *E. polybractea*, where the cineole content of the crude oil is high to start with (from levels of around 82 per cent to as high as 90 per cent). However, limonene, which is usually present to the extent of a few percent in the crude oil, has a boiling point very close to that of 1,8-cineole, and this prevents eucalyptol being obtained solely by distillation without a great deal of effort and expense. A two-stage procedure is therefore adopted, making use of differences in freezing point as well as boiling point. Initial rectification is performed by distilling the oil under vacuum and, after first removing the low-boiling components, collecting the major cineole-rich fraction. This is then transferred to a cold store at around $-30°C$ for up to 24 h. The resulting mixture contains mainly frozen cineole; limonene, and some other minor components, remain liquid. The mixture is centrifuged to remove the liquid portion and the frozen part is collected and warmed to furnish eucalyptol. Eucalyptol is hygroscopic and is usually returned to the still for a final, careful redistillation. The end-product is 99 per cent or more pure.

## Standards

For more than 100 years there has been a standard to which eucalyptus oil must conform in order to be used for medicinal purposes. In 1914, the British Pharmacopoeia required a minimum of 55 per cent cineole and a very low content of phellandrene and aldehydes. The latter two compounds were restricted by limit tests, without the actual percentages permitted being stated. The main aldehyde in the cineole-type eucalyptus oils is isovaleraldehyde, already referred to above. As well as being unpleasant itself, short exposure to air induces production of its oxidation product, isovaleric acid, in sufficient amounts to give a rancid odour. Later, national pharmacopoeias, including the British Pharmacopoeia, stipulated 70 per cent as the minimum level of cineole, and specific ranges for a number of physical constants. At present, oil is sold mainly in accordance with national pharmacopoeias or other national and international standards, all of which are similar.

Revised Australian standards for cineole-rich oils have recently been published (SA 1998), replacing those of 1977. In addition to physico-chemical data for oils with cineole contents within the defined ranges (70–75 and 80–85 per cent) there are chromatographic data and information on flash points (see also Appendix 5). Full details are available from Standards Australia. There is an Australian standard for *E. citriodora* oil (AS 2116) but it dates to about the time that the oil was last produced in Australia, approximately thirty years ago; it is unlikely that *E. citriodora* oil will be produced again in Australia.

Adulteration of the cineole-type eucalyptus oils is extremely uncommon. The general use of chromatography in analysis makes detection of adulterants relatively easy and the present low price and ready availability of the oil provides little financial incentive for adulteration.

End uses

*Cineole-type oils*

In the early days of white settlement in Australia many natural products were tried for their general household and medicinal utility. Cineole-type eucalyptus oil was highly valued and used for many purposes. A list of tried and tested uses of the oil, provided by Felton Grimwade & Bickford Pty Ltd, is shown below and indicates the diversity of its applications.

| | | |
|---|---|---|
| Children's colds | Sauna | Carpet shampoo |
| Head colds and influenza | Wool wash | Bathroom cleaner |
| Mouthwash | Washing work clothes and | Linoleum cleaner |
| Scalp massage | nappies | Leather cleaner |
| Muscular aches and pains | Dog wash | Plastic and vinyl cleaner |
| Rubbing or training oil | Sticking plaster removal | Paint brush cleaner and |
| Cuts and abrasions | Removal of chewing gum, | softening hardened paint |
| Insect bites | paint, ball-point ink marks | Telephone cleanser |
| Insect repellent | Removal of tar marks on | Toilet cleanser |
| Hand and skin cleaner | motor vehicles | Vaporiser-humidifier |
| Bath and foot bath additive | Spot and stain remover | Penetrating oil |

Many of the household uses listed above have been exploited commercially and numerous products containing eucalyptus oil are marketed (Chapter 16). The oil is also used in perfumery, in flavouring, and in confectionery for its therapeutic value as well as its flavouring properties.

In more recent years, a boost to the use of eucalyptus oil has been achieved with the development and marketing of eucalyptus wool wash. Although this has been used in Australia for many years, and was usually made up in the household as required, its use was small until it was taken up by some of the major manufacturers of cleaning products around 1985 and marketed in a ready-to-use form. The oil in the wool wash serves three purposes: it acts as a solvent for removing grease, it imparts a fresh clean odour to the garment being washed, and it softens the texture of the wool.

*Other oils*

Eucalyptus oils of the non-cineole type were used in large quantities earlier this century for mineral flotation (*E. dives* Type and *E. radiata* phellandrene variant) and for the manufacture of menthol (*E. dives* Type) and, to a small extent, thymol. Oil is not used for these purposes now. The production of the so-called perfumery oils has never developed in Australia and although all the oil-bearing species occur naturally, the yield of oil from *E. citriodora*, *E. staigeriana* and *E. macarthurii* is too low to enable commercial exploitation to be profitable. Only *E. citriodora* oil (and to a much lesser extent *E. staigeriana* oil) has been produced on a commercial scale in other countries, and this only where production costs are low.

Recently, Australian scientists discovered a eucalypt, originally designated *Eucalyptus* sp. nov. aff. *campanulata* but now classified as *E. olida*, which yields an oil rich in E-methyl cinnamate (Curtis *et al.* 1990). As a result, natural methyl cinnamate is now produced and marketed. However, the species is confined to a remote part of New South Wales, which is not easily accessible and which has very limited stands, so it is now grown in plantations. Unfortunately, the market for natural methyl cinnamate is very small and, at present, further development of plantations of *E. olida* is not justified.

## Production of extracts other than oil from *Eucalyptus*

Several extractive products other than oil have been obtained commercially from eucalypts in Australia, although the ones described below are now largely of historical interest only.

One other outlet for the 'waste', spent leaf arising from oil production is worth mentioning. Apart from its use as a mulch and a slow-acting fertiliser by the oil producer himself, as has already been mentioned, it has established a market as a domestic garden mulch through retail outlets. However, its low value and large bulk, and consequent high transport cost, tends to confine its sale to producers fairly close to a city or large town.

### Tannin extract

Although it is no longer produced, tannin used to be extracted from the wood of *E. wandoo*, a Western Australian species of *Eucalyptus*. Industrial Extracts Ltd, a member of the Plaimar group of companies, extracted and concentrated tannin for about forty years until the late 1960s. The material was used in tanning leather, particularly heavy leather. It was also used as an agent for treating boiler-water, where it acted as an oxygen scavenger, preventing scaling by forming soluble salts in preference to insoluble calcium ones, and forming a preservative iron tannate film on the tubes of the boiler. It was also used as a mud loss preventative in oil drilling. Many thousands of tonnes of these particular types of phenolic products were exported.

### Rutin

Another phenolic compound, the flavone glucoside rutin, occurs in the leaves of a number of eucalypts and was extracted on a commercial scale, particularly from *E. macrorhyncha* and *E. youmanii*, during the 1950s and 1960s (Humphreys 1964). Rutin is used in the treatment of capillary fragility, particularly varicose veins, haemorrhoids and frostbite. However, the economics of rutin extraction from *Eucalyptus* could not compete with that of rutin obtained from *Sophora japonica* from China and Australian production ceased around 1970.

### Eucalyptus kino

The astringent exudate produced from the trunk of many eucalypts is commonly, but erroneously, known as 'gum'. It was used by early settlers as an astringent and also, later, in the wine trade. Kino was exported to Europe, particularly France, until about 1975 when difficulties with its collection, and a declining market, led to a cessation of production. Kino from *E. camaldulensis* was the most popular.

## Research and development

The Australian eucalyptus oil industry has been well served by the research institutes. Almost from the commencement of the industry, the Technological Museum in Sydney (later the Museum of Applied Arts & Sciences) worked on the essential oils of the native flora, particularly eucalypts. Their research on the botany and chemistry assisted the industry to develop. Other aspects, such as selection of the most suitable land for growing oil-bearing species such as *E. polybractea*, were also investigated, particularly by B.E.J. Small of the Museum of Applied Arts & Sciences. Some of this research has been continued at the New South Wales Department of Agriculture. Milthorpe *et al.* (1994), for example, have conducted research into growth rates and oil yields of *E. polybractea* under dryland and irrigation conditions. Optimum planting densities for *E. polybractea* and *E. kochii* have also been investigated (Milthorpe *et al*. 1998).

Currently, much work is focused on the selection and breeding of elite trees for oil production. Dr M.U. Slee of the Forestry Department of the Australian National University, Canberra, is working on a project with *E. polybractea*. In Western Australia, work at Murdoch University, Agriculture WA and the Department of Conservation and Land Management (CALM) is seeking to produce elite trees for oil production from *E. kochii* subsp. *kochii*, *E. kochii* subsp. *plenissima*, *E. horistes*, *E. angustissima* and *E. loxophleba* subsp. *lissophloia*, all Western Australian species, as well as *E. polybractea*.

The Western Australian research is part of a large project aimed at reversing land degradation in the wheatbelt. Loss of tree and shrub cover, sometimes exacerbated by the use of flood irrigation, has led to a rising water table and salt-rich soils. By planting eucalypts it is hoped to reduce groundwater recharge and thus control the salinity (Eastham *et al*. 1993). In choosing good oil-bearing species it is expected that oil production, as well as residue utilisation (to produce, for example, activated carbon and energy), will eventually defray the cost of planting and bring in a cash return. All the species have the mallee habit, well-suited to short production cycles and large-scale mechanised harvesting operations. Placing large quantities of oil on a traditional market, already over-supplied, will be difficult and the project aims, instead, to penetrate the market for industrial solvents, particularly as a natural, environmentally friendly substitute for trichloroethane (Bartle *et al*. 1996, Bartle 1999). If successful, the size of the scheme (of the order of 0.5 million ha of oil mallees) will result in very considerable quantities of cineole-type oil being produced. However, as the project has not yet reached the first large-scale harvest stage the production economics cannot be confirmed and its viability remains uncertain.

Commercial producers themselves have also spent much time investigating the many aspects of plantation development including land selection, land preparation, seed selection, raising of seedlings, planting and harvesting techniques, weed and insect control, soil and water conservation, and irrigation. A very good technique for raising seedlings of *E. polybractea*, using a potting mix derived mainly from eucalyptus leaves, was developed by Constance Davis of G.R. Davis Pty Ltd. This, together with an effective method of selecting high quality seed, has been used to establish large plantations on the western plains of New South Wales. Mrs Davis, in conjunction with the Department of Forestry, Australian National University, has also developed an effective technique for the vegetative propagation of *E. polybractea*.

*Research needs*

Australia, being the home of the eucalypt, has all the oil-bearing species at its disposal. It therefore has the potential to be able to choose the best ones in terms of suitability for planting in a particular part of the country and in terms of oil yield and quality. The future of the industry depends on continued research to develop such high-yielding, high quality stock, as well as even more efficient methods of production. Further work on vegetative reproduction will enable this to proceed rapidly.

While, obviously, there is still much to be learnt about the botany and chemistry of the eucalypts, and about the practical aspects of field management such as growing, harvesting and distilling the oil, the major task facing researchers today is, without doubt, to find new uses for eucalyptus oil. At present, world demand for this oil is easily met by the supply. Furthermore, there are a number of countries not now contributing to the market which could produce oil if increased demand were to develop. A new, large use is required if the industry is to develop further. Research at Murdoch University into the use of eucalyptus oil as an industrial solvent has already been mentioned but the very low cost of the solvents to be replaced is likely to make this a difficult task.

Some years ago, work in Japan and Australia indicated that cineole, or high-cineole eucalyptus oil, could be used advantageously with ethanol as an automobile fuel (e.g. Takeda 1982, Barton and Tjandra 1989). Small amounts of the oil in ethanol, or a petrol–ethanol mix, improve the effectiveness of the fuel. Further research on this should be done, although, even if it is shown that there is an advantage in incorporating some cineole in the mix, it is unlikely to benefit the eucalyptus oil industry unless there is a shortage of petrol.

Chemical research has demonstrated the preponderance in some oils of chemicals other than the usual ones such as cineole or piperitone, and this offers the possibility of the oils being exploited as sources of chemical isolates, providing, of course, that the chemical in question is of commercial interest. The extraction of E-methyl cinnamate from *E. olida* is a recent example of this, and *E. loxophleba*, which is rich in 4-methyl-2-pentyl acetate (Grayling and Knox 1991), is currently being evaluated for its commercial prospects. However, by their very nature, such speciality oils have limited markets in volume terms.

## References

Bartle, J.R. (1999) Why oil mallee? In *Oil Mallee Profitable Landcare, Proc. Oil Mallee Association Seminar*, Perth, Western Australia, March 1999, pp. 4–10.

Bartle, J.R., Campbell, C. and White, G. (1996) Can trees reverse land degradation? In *Farm Forestry and Plantations: Investing in Future Wood Supply, Proc. Australian Forest Growers Conf.*, Mount Gambier, Australia, September 1996, pp. 68–75.

Barton, A.F.M. and Tjandra, J. (1989) Eucalyptus oil as a cosolvent in water–ethanol–gasoline mixtures. *Fuel*, 68(1), 11–17.

Coppen, J.J.W. and Dyer, L.R. (1993) *Eucalyptus and its Leaf Oils. An Indexed Bibliography*, Natural Resources Institute, Chatham, UK.

Curtis, A., Southwell, I.A. and Stiff, L.A. (1990) *Eucalyptus*, a new source of E-methyl cinnamate. *J. Essent. Oil Res.*, 2, 105–110.

Eastham, J., Scott, P.R., Steckis, R.A., Barton, A.F.M., Hunter, L.J. and Sudmeyer, R.J. (1993) Survival, growth and productivity of tree species under evaluation for agroforestry to control salinity in the Western Australian wheatbelt. *Agrofor. Syst.*, 21, 223–237.

Grayling, P.M. and Knox, J.R. (1991) (R)-4-methyl-2-pentyl acetate from *Eucalyptus loxophleba*. *J. Nat. Prod.*, 54, 295–297.

Grolier Society (1965) *The Australian Encyclopedia*, Vol. 3, The Grolier Society of Australia, Sydney, pp. 409–411.

Humphreys, F.R. (1964) The occurrence and industrial production of rutin in southeastern Australia. *Econ. Bot.*, 18, 195–253.

Milthorpe, P.L., Brooker, M.I.H., Slee, A. and Nicol, H.I. (1998) Optimum planting densities for the production of eucalyptus oil from blue mallee (*Eucalyptus polybractea*) and oil mallee (*E. kochii*). *Industrial Crops Prods.*, 8, 219–227.

Milthorpe, P.L., Hillan, J.M. and Nicol, H.I. (1994) The effect of time of harvest, fertilizer and irrigation on dry matter and oil production of blue mallee. *Industrial Crops Prods.*, 3, 165–173.

SA (1998) *Oil of Australian Eucalyptus. Part 1: 70–75 Percent Cineole*, AS 2113.1-1998, and *Part 2: 80–85 Percent Cineole*, AS 2113.2-1998, Standards Australia, Strathfield, Australia.

Shiel, D. (1985) *Eucalyptus – The Essence of Australia*, Queensbury Hill Press, Melbourne.

Takeda, S. (1982) Studies of eucalyptus oil and its application to internal combustion engine. *Pan-Pacific Synfuels Conf.*, 2, 498–508.

# 8 Cultivation and production of eucalypts in the People's Republic of China

## With special reference to the leaf oils

*Shaoxiong Chen*

## Eucalypts as plantation trees in China

### History of their introduction and cultivation

Eucalypts were first introduced into China from Italy in 1890 (Qi 1990). They were initially planted for ornamental and screening purposes, mainly on a small scale in gardens, schools, colleges and scenic spots, along roads and around villages. Some very old trees, up to 80 years old, still exist in southern China and some indication of their size can be seen from Table 8.1.

A number of distinct periods can be recognised in eucalypt domestication in China (Liu *et al.* 1996). Prior to 1950 eucalypts were planted in gardens and for amenity or general purpose use. Predominant species were *Eucalyptus camaldulensis*, *E. citriodora*, *E. exserta*, *E. globulus*, *E. robusta*, *E. rudis* and *E. tereticornis* but little is known of their provenance origin. Establishment of eucalypt plantations on a large scale only began in the 1950s, using species readily available within China. However, severe frost damage and widespread failures occurred in sub-tropical areas due to the use of inappropriate frost-sensitive material. Plantation species were primarily the same as those used prior to 1950, with growth rates averaging up to $7.5\,m^3\,ha^{-1}\,yr^{-1}$ (Wu *et al.* 1994).

*Table 8.1* Documented data on old eucalypts in China

| Species | Age (years) | Height (m) | Dbh (cm) | No. of trees | Location (province) |
| --- | --- | --- | --- | --- | --- |
| E. amplifolia | 62 | 20 | 45 | 1 | Guangdong |
| E. botryoides | 62 | 25 | 70 | 1 | Guangdong |
| E. camaldulensis | 50–76 | 30–39 | 71–150 | 7 | Fujian, Guangdong |
| E. citriodora | 50–70 | 30–40 | 75–101 | 6 | Guangdong, Fujian, Guangxi, Yunnan |
| E. dichromophloia | 70 | 30 | 88 | 1 | Guangdong |
| E. exserta | 50 | 20–30 | 75–80 | 2 | Guangdong |
| E. globulus | 50–75 | 30–50 | 110–161 | 3 | Yunnan |
| E. globulus subsp. bicostata | 50 | 20 | 100 | 1 | Sichuan |
| E. maculata | 70 | 30 | 65 | 1 | Guangdong |
| E. microcorys | 62 | 20 | 34 | 1 | Guangdong |
| E. paniculata | 60–62 | 20–30 | 50–76 | 2 | Guangdong |
| E. robusta | 58–66 | 25–30 | 84–98 | 4 | Guangdong, Fujian |
| E. rudis | 80 | 37 | 154 | 1 | Fujian |
| E. saligna | 65 | 30 | 68 | 1 | Guangdong |
| E. tereticornis | 60–70 | 20–35 | 40–97 | 6 | Guangdong, Jiangxi |

The first eucalypt forest farm, Yuexi Forest Farm, was set up in Zhanjiang, Guangdong Province, by the Southern China Bureau of Agriculture in Leizhou Peninsula, Guangdong Province, in 1954 (Wu 1991). The main species planted were *E. exserta* and *E. citriodora*.

During the 1960s the total plantation area increased slowly towards 300,000 ha with increments of $8-10\,m^3\,ha^{-1}\,yr^{-1}$ as improved land races and better cultural techniques were developed. The second eucalypt forest farm, Guinan Forest Farm, was founded in Guangxi Zhuang Autonomous Region in 1963. *E. exserta* and *E. citriodora* were again the main species. In the 1970s small-scale, rudimentary breeding programmes commenced with species and provenance testing and the testing of land races. The plantation area increased to about 400,000 ha and increments of $10-12\,m^3\,ha^{-1}\,yr^{-1}$ became commonplace.

The period from 1981 onwards saw the greatest progress in eucalypt domestication in China (Bai 1994). Establishment of eucalypt plantations increased substantially, with an average of approximately 30,000 ha being planted each year. The rate has been accelerating since 1990 and by 1995 had reached $50,000\,ha\,yr^{-1}$, based on a survey by the China Eucalypt Research Centre (CERC).

## Present day plantations

More than 300 species of *Eucalyptus* have been introduced to China in the past century (Wang and Brooker 1991), but only about a third of these have survived, of which ten species are used in large-scale plantations: *E. camaldulensis*, *E. citriodora*, *E. exserta*, *E. globulus*, *E. globulus* subsp. *maidenii*, *E. grandis*, *E. urophylla*, the hybrid between *E. urophylla* and *E. grandis*, '*E. 12ABL*', and a local land-race called '*E. leizhou No. 1*'. Several of these are used as sources of oil in China. The taxon *E. 12ABL* is a land race developed from the seed introduced to China in the 1970s from the Congo and reported to originate from a single tree in Madagascar (Eldridge *et al.* 1993). *E. leizhou No. 1* is a natural hybrid of *E. exserta* (Qi 1990). A number of other species, including *E. nitens*, *E. smithii*, *E. tereticornis*, *E. dunnii* and *E. saligna* have recently shown good potential in trials, although plantation areas of these species are relatively limited to date (Liu *et al.* 1996). *E. smithii* is discussed later in connection with its oil.

According to estimates made by CERC in 1996, the area of eucalypt plantations now exceeds 0.92 million ha. Details of the main areas, together with intended end uses (including oil production), are given in Table 8.2.

The plantations are distributed over 17 provinces: Guangdong, Guangxi, Hainan, Fujian, Yunnan, Sichuan, Guizhou, Hunan, Hubei, Jiangxi, Zhejiang, Jiangsu, Shanghai, Anhui, Shenxi, Gangsu and Taiwan (Wu 1991). The distribution extends from latitude 18° to 33°N, longitude 100 to 125°E, and altitude up to 2000 m (Qi 1990). About 90 per cent of the plantation area is concentrated in Guangdong, Guangxi, Hainan and Yunnan provinces, south of latitude 27°N. By the year 2000, the Chinese eucalypt plantation area was forecast to be more than 1.3 million ha (Midgley and Pinyopusarerk 1996), with 15–20 per cent of the annual planting stock being derived from clonal seed orchard seed or from vegetatively multiplied clones.

Because of their rapid growth, ability to coppice, multiple uses and adaptability, eucalypts are widely planted in the countryside, often by small farmers who use part of the harvest themselves and part as a source of cash income. Some of the wood is used for timber and some for fuelwood. Recently cropped leaves from roadside areas of *E. citriodora* grown for timber, and awaiting distillation, are shown in Figure 8.1. Increasing use of improved planting and management methods has enabled short rotations of 8–10 years to be shortened still further.

Apart from oil production and extraction of tannins (Table 8.2), non-wood uses include extraction of phytohormones from the leaves of some species for horticultural applications and

*Table 8.2* Areas under plantation and utilisation of main eucalypt species in China (1996)

| Province | Main species | Area (ha) | Utilisation |
|---|---|---|---|
| Guangdong | E. camaldulensis, E. citriodora, E. exserta, E. leizhou No. 1, E. 12ABL, E. urophylla, E. urophylla × E. grandis | 400,000 | Wood chips, plywood, particle board, fibre board, charcoal, oil, tannins |
| Hainan | E. citriodora, E. exserta, E. 12ABL | 187,000 | Wood chips, plywood, particle board, fibre board, oil |
| Guangxi | E. citriodora, E. exserta, E. leizhou No. 1, E. urophylla, E. urophylla × E. grandis | 170,000 | Wood chips, plywood, particle board, fibre board, oil |
| Yunnan | E. globulus, E. globulus subsp. maidenii | 147,000 | Oil, plywood, particle board, fibre board |
| Sichuan | E. globulus, E. globulus subsp. maidenii, E. robusta | 13,000 | Oil, firewood |
| Guizhou | E. globulus, E. globulus subsp. maidenii | 3000 | Oil, firewood |
| Hunan | E. camaldulensis | 2000 | Firewood |

*Figure 8.1* Eucalyptus citriodora leaves awaiting distillation, Leizhou Peninsula, Guangdong Province (photo: S. Chen).

honey production from bees that feed exclusively on *Eucalyptus* flowers (Zheng 1988). Again, for some families, such activities are a much valued means of income generation. As multipurpose trees, some eucalypts are used in agroforestry. On Hainan Island, for example, *E. exserta* and *E. citriodora* are interplanted with pineapple and in southwest China *E. globulus* is planted alongside tobacco and sweet potato.

*Table 8.3* Eucalyptus species used for 'four-around' planting

| Species | Province |
| --- | --- |
| E. camaldulensis, E. citriodora, E. exserta, E. robusta, E. rudis | Guangdong, Guangxi, Fujian |
| E. camaldulensis, E. globulus, E. globulus subsp. maidenii | Yunnan |
| E. botryoides, E. camaldulensis, E. robusta | Sichuan |
| E. camaldulensis, E. tereticornis | Jiangxi |
| E. camaldulensis, E. robusta | Zhejiang, Hubei, Hunan |
| E. camaldulensis, E. cinerea, E. gunnii | Shanghai |

### 'Four-around' plantings

The term 'four-around' plantings refers to the trees which are planted as shelterbelts along roads and ditches, and around villages and houses (Qi 1990). The main species used in this way are listed in Table 8.3 and include some oil-bearing ones.

Plantings amount to over 1.8 billion trees (Wu 1991), of which over half are in Yunnan and Sichuan provinces. However, eucalypts are not naturally good shelterbelt trees because of their branch shedding habit which permits the wind to blow under the crowns of single rows of trees. It is therefore necessary to plant them as several rows. The trees do not cast a heavy overhead shade but provide a good side shade throughout the year. Another reason for planting eucalypts alongside roads and railways is to reduce noise.

## Silviculture

Although some research is now being focused on the dual-purpose management of eucalypts for oil and wood production (see below), most Chinese eucalyptus oil is produced from 'waste' leaf resulting from thinning or felling of trees grown primarily for timber, pulp or other wood use. Several of the important plantation species (*E. citriodora*, *E. exserta*, *E. globulus* and *E. globulus* subsp. *maidenii*) also happen to be useful sources of oil and the leaves can be collected for distillation whenever it is worthwhile to do so.

In some cases, eucalypt plantations are coppiced and harvested again at intervals of 4–7 years. Timber from these crops is mostly small in diameter with a low value per unit volume. Substantial added value is only obtained when it is processed into paper and other reconstituted products.

Establishing a productive plantation requires good seedlings of suitable species, adequate site preparation, sound planting methods, effective weed control and a satisfactory nutritional status in the soil.

### Nursery techniques

Seed is usually collected from seed orchard and seed production areas. Sometimes it is obtained from phenotypically outstanding trees. Nursery techniques have changed significantly in recent years and simplicity and cheapness have given way to more efficient and productive methods. The advantages of container planting using plastic tubes and pots have been recognised, namely, lower cost, greater survival rate (due to the greater volume of soil in which the roots of the young

Table 8.4 Examples of container sizes and standards of *Eucalyptus* seedlings

| Species | Tube size (cm) | Length of seedling stage (days) | Height (cm) | Diameter at base (mm) | Survival (%) |
|---|---|---|---|---|---|
| E. leizhou No. 1 | 14 × 3.5 | 60–80 | 15–30 | 1.5–3.0 | 95 |
| E. citriodora | 14 × 3.5 | 60–80 | 15–30 | 1.5–3.0 | 95 |
| E. globulus | 11 × 7 | 70–90 | 20–25 | >3.0 | 80 |
| E. globulus subsp. maidenii | 11 × 7 | 70–90 | 20–25 | >3.0 | 80 |

seedlings can grow) and avoidance of damage to seedlings through careless picking out. This permits the planting operation to be carried out over a reasonably long season. Seed is normally sown to a soil bed. When the seedlings are in the second or third leaf-pair stages they are transferred to containers.

The size of the containers is important. The more room in the container, the more medium can be used, and this gives the seedlings better conditions for growth in the nursery and early stages after planting. However, it is then heavier for transporting and handling. The medium used depends very much on what is available locally. Two examples are 53 per cent burned soil + 41 per cent organic matter + 6 per cent phosphorous (applicable to seedlings) and 75 per cent deep yellow soil + 25 per cent peat soil (cuttings). Examples of container sizes and standards for seedlings of four *Eucalyptus* species are shown in Table 8.4. The information is taken from the Leizhou Forestry Bureau and Yunnan Forestry Academy.

To protect the seedlings and cuttings from excessive sunlight and storm rains, different types of shade are used for raising them. The shade density is about 50–70 per cent. Many nurseries also require some protection from the wind, especially in coastal regions. Artificial screens are sometimes erected at right angles to the prevailing winds or live hedges are used. Facilities for watering are essential in the nursery and for large ones this may be done automatically. Watering twice a day, morning and evening, is desirable.

*Establishment*

When the planting area has an old crop of trees and shrubs on it which have no value, they need to be cleared before ploughing can begin. In Guangdong and Guangxi, tractors are used to assist in this task but in Yunnan, where the planting is in mountainous areas, the work is usually done manually. When the area to be cleared for planting is an earlier eucalypt plantation the old stumps are dug out or killed in order to avoid any coppicing. There is a very considerable volume of stump wood below the ground in these areas and it is recovered by the local people for use as fuel for cooking or for making charcoal.

Land preparation is a very important step in planting eucalypts. In Guangdong, Guangxi and Hainan, ripping, ploughing and disc harrowing of the site are generally carried out. The depth of ploughing is normally 30 cm. Planting holes are then prepared, usually 30 × 30 × 30 cm in size. However, in Yunnan, where the sites are stony or very steep, complete cultivation is impracticable. Normal practice is to prepare a soil ditch, 60 × 60 cm (Zhang, Su and Dai, unpubl. 1996), and then make a planting hole at least 50 × 50 × 40 cm, preferably larger. This has proved to be very good for *E. globulus* and *E. globulus* subsp. *maidenii* growth.

A wide range of initial spacings has been used, depending on the planting site and the intended end-use for the eucalypts. In general, poor sites have wider spacings and good sites closer spacings. Some examples are given in Table 8.5.

Table 8.5 Spacings used for four species in the main eucalypt provinces

| Species | Initial spacing (m) | Stocking density (stems/ha) | Utilisation | Province |
| --- | --- | --- | --- | --- |
| E. leizhou No. 1 | 1 × 1 × 3 or 1 × 1.3 × 2.7 | 5000 | Pulpwood | Guangdong |
|  | 1.5 × 4 or 1.5 × 3 | 1667 or 2222 | Pulpwood | Guangxi |
| E. citriodora | 1 × 3 | 3333 | Pulpwood | Guangdong |
|  | 1.5 × 4 or 1.5 × 3 | 1667 or 2222 | Sawlogs, oil | Guangxi |
| E. globulus | 1 × 2 | 5000 | Oil | Yunnan |
|  | 1.5 × 2.5 or 1 × 4 | 2667 or 2500 | Pulpwood, sawlogs | Yunnan |
| E. globulus subsp. maidenii | 1 × 2 | 5000 | Oil | Yunnan |
|  | 1.5 × 2.5 or 1 × 4 | 2667 or 2500 | Pulpwood, sawlogs | Yunnan |

Table 8.6 Fertiliser regimes used in the main eucalypt provinces

| Province | Fertiliser regime |
| --- | --- |
| Guangdong | 1.5–2.5 kg organic matter +75 g NPK (1:2:2) per plant basal dressing; 50 g NPK in 2nd and 3rd years |
| Guangxi | 200 kg NPK (2:1:1) per ha basal dressing; 150 kg NPK in 2nd and 3rd years |
| Yunnan | 5 kg organic matter +5–10 g B per plant basal dressing; 50 g N in 2nd year |

In order to promote uniform growth and facilitate mechanical tending and harvesting, the lines of trees should be regular and straight and the spaces between them even. This is achieved by use of a planting chain which has markers along it at the same intervals as the required spacing. The planting holes are indicated on the ground by lime. Planting is usually done in the wet season so that the plants can take full advantage of the rains in summer rainfall regions. The seedling should be planted firmly in the ground without air space around or below its roots, and the field soil should be in contact with the roots. Plastic containers are removed before planting.

In most sites, young eucalypts respond quickly and generously to fertilisation. It is common practice to apply fertiliser a few days before planting (basal dressing) and twice more thereafter. The amount of fertiliser used and the balance between the different elements is determined by weighing the increased yield against the cost of the fertiliser and its application on a particular type of soil. Examples are given in Table 8.6.

Control of weed growth is greatly helped by good site preparation but there are likely to be many weed seeds in the soil and other seed will blow in from outside. Before the closure of the canopy, therefore, weeds are controlled by disc harrowing, hand pulling or hoeing.

### Management of coppice crops

The first-rotation crop is normally felled at 5–10 years. The way in which the felling is carried out, the type of equipment used, and the period during which it is done are all key factors in the survival and well-being of the plantations through successive coppicing. Experience in Guangxi has shown that better results are obtained with the use of chain-saws than axes, while in Guangdong axes have been reported to give better results than the bow-saw and the two-man crosscut saw. The felling period should be planned to avoid dry periods; in Guangdong province it is from October to March.

The height at which the cut is made is important. If the stool is too high the chances of survival are lower but if the cut is at ground level then the bark is loosened. The recommended height is about 5 cm. The cut should be made as smooth as possible and slanted so as to facilitate water run-off. The accumulation of water on the stool increases the risk of fungal attack.

In most cases a number of shoots develop on each stool, although these are likely to be reduced to six or seven by mutual competition. When the coppice shoots are about 6 months old they are thinned to two or three stems per stool, the main selection criteria being form, wind firmness and vigour. The final number of stems per hectare retained in the thinned coppice crop is not less than the original stocking.

*Tree improvement*

Bai (1992, 1994) has summarised the problems of eucalypt improvement in southern China and discussed issues relating to breeding strategies, use of hybrids (see also Wu *et al*. 1996), clonal forestry, etc. However, during the last two decades considerable progress has been made in eucalypt domestication and eucalypt improvement in China, much of it with Australian assistance (Anon. 1989, Midgley and Yang 1992, Brown 1994). New base/breeding populations have been assembled, clonal seed orchards established, inter- and intra-specific controlled pollination techniques developed and vegetative propagation techniques improved. Species and provenance testing has continued and traits currently of major importance in selection work include growth, wind firmness, disease resistance, cold tolerance and wood quality. Research aimed at identifying high-yielding oil species and developing improved germplasm for oil production is discussed in more detail below.

*Pests and diseases*

At present, our information on the pests and diseases of *Eucalyptus* in China (as well as those which affect other tree crops) is limited. Pests of *Eucalyptus* are not too serious but the main ones are listed in Table 8.7 (Gu and Chen 1988).

The diseases usually occur in the nursery and in the roots of the plant and can sometimes cause serious damage in plantations. The main diseases of eucalypts in China are listed in Table 8.8 (Gong and Ke 1988).

*Table 8.7* Main pests of eucalypts in China

| Pest | Host |
|---|---|
| *Carea subtilis* (Lepidoptera: Noctuidae) | *E. citriodora* |
| *Dappula tertius* (Lepidoptera: Psychidae) | *E. citriodora* |
| *Acanthoecia laminati*[a] (Lepidoptera: Psychidae) | *E. citriodora* |
| *Pelochrista* sp. (Lepidoptera: Tortrichidae) | Leaf of all kinds of eucalypts |
| *Tarbinskiellus portentosus*[b] (Orthoptera: Gryllidae) | All kinds of eucalypts |
| *Adoretus sinicus* (Coleoptera: Scarabaeidae) | *E. camaldulensis, E. 12ABL* |
| *Anomala cupripes* (Coleoptera: Scarabaeidae) | *E. camaldulensis, E. citriodora* |
| *Anomala corpulenta, A. chamaeleon* (Coleoptera: Scarabaeidae) | Leaf/root of all kinds of eucalypts |

a Syn. *Chalia laminati*.
b Syn. *Brachytrupes portentosus*.

*Table 8.8* Main nursery and root diseases of eucalypts in China

| Pathogen | Host |
| --- | --- |
| Nursery diseases | |
| *Fusarium* sp. | *E. citriodora*, *E. gunnii*, *E. robusta*, *E. leizhou* No. 1, *E. leptophleba* |
| *Macrophomina phaseolina* | *E. citriodora*, *E. exserta*, *E. globulus*, *E. globulus* subsp. *bicostata*, *E. gunnii*, *E. pauciflora*, *E. robusta* |
| *Sclerotium rolfsii* | *E. citriodora*, *E. exserta* |
| *Botrytis cinerea* | *E. saligna* |
| Root diseases | |
| *Pseudomonas solanacearum* | *E. grandis*, *E. saligna*, *E. urophylla*, *E. urophylla* × *E. grandis* |
| *Agrobacterium tumefaciens* | *E. citriodora*, *E. exserta* |
| *Verticillium albo-atrum* | *E. citriodora*, *E. exserta* |
| *Ganoderma* sp. | *E. exserta*, *E. robusta* |
| *Fomes* sp. | *E. robusta* |

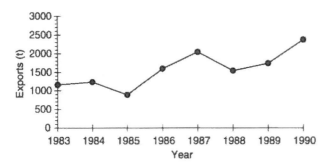

*Figure 8.2* Exports of eucalyptus oil from China, 1983–1990.

## Eucalyptus oil production

### Brief history of the industry

The eucalyptus oil industry first developed in 1953 in Guangzhou, Guangdong Province (Song 1991). This immediately became an important channel for cash income in rural areas of Guangdong and Guangxi. In the 1960s, the first large commercial eucalyptus oil operations in China started in Leizhou Peninsula, Guangdong Province (Qi 1990). By the 1970s, annual production of eucalyptus oil was about 3000 t, of which 30 per cent was exported. In the 1980s these figures increased to 4000 t, of which 50 per cent was exported. During the 1990s, production remained the same but the proportion which was exported rose to around 80 per cent. Exports of eucalyptus oil for the period 1983–90 are shown in Figure 8.2 (Zhang 1997). China is the world's largest producer and exporter of both medicinal (mainly *E. globulus*) and perfumery (*E. citriodora*) oils.

In the last 10–15 years, Yunnan has become the focus of essential oil production in China. The topography and range of climates found there gives rise to a wide variety of native flora, many of which yield essential oils, and enables some species of exotic plants such as eucalypts to flourish (Cu 1988). With a rich and diverse raw material base to support it, primary production of essential oils has gradually progressed to value-added processing such as fractionation and the production of 'absolutes', 'concretes' and other types of extract. Research centres aimed at developing 'in-house' capabilities in processing and formulating essential oils, including eucalyptus oil, have been established, particularly in and around Kunming.

Typically, eucalyptus oil distilled by small farmers from *E. globulus* using fairly rudimentary equipment is either sold to local shops or traders or transported to Kunming for redistillation through a fractionating column. This removes both the lower boiling monoterpenes and, after collection of the desired cineole-rich oil, the high-boiling residue. The latter contains a complex mixture of oxygenated monoterpenes and sesquiterpenes and has, itself, the potential for further exploitation (Zhao *et al*. 1997).

### Eucalyptus *species used for oil production*

The principal species used for oil production in China are listed in Table 8.9, together with oil yields and an indication of typical abundances of the main constituents of the oils (Zhang 1997). Of the medicinal type, *E. leizhou No. 1* produces an oil of relatively poor quality (low 1,8-cineole) and in low yields but its widespread occurrence accounts for its use as a source of oil. *E. exserta* also yields an oil with a relatively low cineole content (Chen *et al*. 1983, Zhang 1997). *E. globulus* is the source of most Chinese eucalyptus oil. *E. citriodora* and *E. dives* are the sole sources of perfumery and industrial oils, respectively.

Biomass and oil yields on a per hectare basis for four of the species given in Table 8.9 are shown in Table 8.10 (Qi 1990). The data represent typical trees managed for commercial production. *E. globulus* is the best of the medicinal oil species. Research has demonstrated some seasonal variation in oil yield within the leaves (Table 8.11), although for the five species reported by Song (1991) the pattern is not consistent.

*Table 8.9* Principal *Eucalyptus* species exploited for leaf oil in China

| Species | Oil yield[a] (%) | Main constituent (relative abundance, %) | Province |
|---|---|---|---|
| Medicinal oils | | | |
| E. globulus | 0.4–1.7 | 1,8-Cineole (60–75) | Yunnan, Sichuan, Guizhou |
| E. globulus subsp. maidenii | 1.5–2.3 | 1,8-Cineole (75) | Yunnan, Sichuan, Guizhou |
| E. smithii | 2.4–3.0 | 1,8-Cineole (75–84) | Yunnan |
| E. exserta | 0.6–0.8 | 1,8-Cineole (35–40) | Guangdong, Guangxi, Hainan |
| E. leizhou No. 1 | 0.4 | 1,8-Cineole (20–38) | Guangdong |
| Perfumery oils | | | |
| E. citriodora | 1.5–1.7 | Citronellal (80–90) | Guangdong, Guangxi, Hainan, Fujian |
| Industrial oils | | | |
| E. dives | 4.7 | Piperitone (52) $\alpha$-Phellandrene (20) | Yunnan, Fujian |

a  Fresh weight basis (averages of different sites).

Table 8.10 Biomass and oil yields of four *Eucalyptus* species

| Species | Harvest (age, yr) | Biomass yield[a] ($t\,ha^{-1}\,yr^{-1}$) | Oil yield ($kg\,ha^{-1}\,yr^{-1}$) |
|---|---|---|---|
| E. globulus | Thinning (6–8) | 30–60 | 450–900 |
| E. globulus subsp. maidenii | Thinning (17) | 15–30 | 300–600 |
| E. exserta | Final felling (10–15) | 15–22.5 | 105–157.5 |
| E. citriodora | Nursery method[b] | 75 | 900 |
|  | Coppice method[c] | 15 | 180 |
|  | Final felling[d] (15) | 7.5 | 90 |

a  Fresh weight basis (leaf + twigs).
b  10,000 trees/ha harvested many times a year.
c  1667 trees/ha harvested twice a year.
d  1667 trees/ha.

Table 8.11 Seasonal variation in oil yields (%)[a]

| Species | Spring | Summer | Autumn | Winter |
|---|---|---|---|---|
| E. globulus | ? | 1.25 | ? | 1.13 |
| E. globulus subsp. maidenii | ? | 1.50 | ? | 1.30 |
| E. exserta | 0.62 | 0.62 | 0.56 | 0.56 |
| E. leizhou No. 1 | 0.43 | 0.61 | 0.57 | 0.62 |
| E. citriodora | 1.34 | 1.84 | 1.75 | 1.96 |

a  Fresh weight basis (distilled from leaf + twigs).

In recent years the potential of *E. smithii* as an important species for oil and fuelwood production has become clear from the results of species and provenance trials in the southern subtropical regions of China (Wang and Wang 1991, Zheng *et al.* 1994, Wang 1997). Its outstanding performance in terms of leaf biomass is already taken advantage of in South Africa, where it has been grown specifically for oil production for many years. Wang and Wang (1991) investigated the growth and above-ground biomass production of *E. bakeri*, *E. dives*, *E. globulus*, *E. macarthurii*, *E. radiata* and *E. smithii* at a site in central Yunnan. After two years under intensive management, growth and biomass yields (both leaf and stem/branch wood) were best for *E. smithii* and *E. globulus*. Oil yields (in the leaf and per hectare) and oil quality (composition) were also determined and are shown in Table 8.12, together with fuelwood yields. *E. smithii* and *E. globulus* again performed best in terms of wood production although the high yield of oil in the leaf of *E. radiata* overcame its slightly inferior leaf biomass to give it the highest yield of oil per hectare. As a dual-purpose crop for oil and fuelwood, *E. smithii* and *E. globulus* have great potential and the former species may come to be much more widely planted in China than it is at present.

The potential of *E. smithii* for oil production has been recognised by CERC who have developed a breeding programme for it. Progeny trials were established in 1995 in Yunnan, Sichuan and Fujian, with 100 families in each trial. Plus trees for leaf oil will be selected from these trials in the next few years. Further tests on the plus trees and the deployment of genetically superior trees will then be carried out.

*Table 8.12* Leaf oil/fuelwood yields and oil composition of six *Eucalyptus* species tested in Yunnan (after Wang and Wang 1991)

| Species | 1,8-Cineole content, relative abundance (%) | Oil yield[a] (%) | Oil yield $(kg\,ha^{-1})$ | Fuelwood yield[a] $(t\,ha^{-1})$ |
|---|---|---|---|---|
| E. smithii | 81.7 | 4.9 | 349.2 | 14.6 |
| E. globulus | 63.5 | <1.6 | 291.6 | 13.4 |
| E. radiata | 44.7 | 9.1 | 396.0 | 5.6 |
| E. bakeri | 90.6 | 8.8 | 295.2 | 3.2 |
| E. dives | 45.1 | 10.1 | 234.9 | 5.4 |
| E. macarthurii | — | 2.0 | 81.6 | 9.0 |

a Moisture-free basis.

*Figure 8.3* Village-scale steam distillation of *Eucalyptus globulus*, near Chu Xiang City, Yunnan Province (photo: S. Midgley).

*Distillation practice*

Distillation of eucalyptus oil from the leaves follows the basic principles of steam distillation which have been described earlier in some detail (Chapter 6). The traditional type of distillation commonly practised in China is briefly described here.

Most of the stills in China are of the above-ground type in which the vessel stands on the ground and water below the charge of leaf is fired directly. The design is simple and suitable for field distillation on both a small and relatively large scale of production. Stills of this type are still used in Guangdong, Guangxi and Yunnan today. An example of village-scale steam distillation of *E. globulus* is shown in Figure 8.3.

The firebox for heating the water consists of a tank resting on a low brick wall on three sides, with one side left open. The wall is about 50 cm high. Wood is the most suitable, and normally the most abundant, fuel available. Spent leaf from previous distillations, once dried, can also be used. The still rests over the firebox and comprises a tank with a cover and a screen in the middle, on which is loaded the leaves. The leaf material is hand cut by the farmers and packed so as to completely fill the available space. The tank is constructed from mild steel sheets, about 2–4 mm thickness for the walls and heavier (5 mm) for the floor. Water is contained in the tank below the screen. When the wood or spent leaf in the firebox has been lit and the water has boiled, steam passes upwards through the fresh charge and distillation commences. The hot mixture of oil vapour and steam is led from the top of the still through a small pipe to the condenser. The latter is a pipe approximately 40–45 m long which runs from the still to a large body of water. The cooled, condensed distillate is then collected in an open drum of 60–200 l capacity, where the oil and water are allowed to separate. The oil is taken off and stored and the water passed back to a constant level tank.

## Medicinal and aromatic uses of *Eucalyptus* in China

### Eucalyptus oil

#### Medicinal oils

The active therapeutic agent and the principal constituent of medicinal-type eucalyptus oils is 1,8-cineole. The medicinal quality of the oil is specified by minimum standards which are defined in different factories. They require a medicinal oil to contain not less than 70 per cent cineole, the same as the Chinese Pharmacopoeia (PPRC 1992), and to be practically free of α- and β-phellandrene. Other requirements of the Chinese Pharmacopoeia are a relative density within the range 0.895–0.920 and a refractive index within the range 1.458–1.468. The limit on heavy metals is 10 ppm.

Although many species of *Eucalyptus* contain 1,8-cineole in their oils, only a limited number are suitable for commercial exploitation; they combine a composition high in 1,8-cineole with consistently high total oil yields. Although most eucalyptus oil is genuinely derived from *Eucalyptus*, the Chinese Pharmacopoeia (PPRC 1992) defines it as 'the volatile oil obtained by steam distillation from the plants of *Eucalyptus globulus* Labill. (Fam. Myrtaceae), *Cinnamomum camphora* (L.) Sieb. (Fam. Lauraceae) or other plants belong[ing] to the same genus of these two families'. Cineole-rich camphor fractions are therefore permitted although, in practice, it is believed that the use of *C. camphora* as a source of 'eucalyptus oil' has declined in recent years.

Examples of Chinese medicines containing eucalyptus oil are shown in Table 8.13. Others include An Ye Tang, Shi Di Shui, Qi Wen You, Qin Liang You and Zhi Ke Tang. People usually buy the ready-made formulations from a traditional medicine shop or general store.

#### Perfumery oils

The lemon-scented gum, *E. citriodora*, is rich in citronellal. Although citronellal itself has limited use as a perfume, it is a valuable intermediate for the production of other fragrances. A few other *Eucalyptus* species contain oils which might be considered for perfumery purposes but *E. citriodora* is the only one ever to have been exploited in China on a large scale.

#### Industrial oils

So-called industrial eucalyptus oils contain piperitone and α-phellandrene as their main constituents. Phellandrene-rich oils are used exclusively for scenting of inexpensive disinfectants, industrial liquid soaps, insecticides and flavouring agents.

*Table 8.13* Examples of Chinese medicines containing eucalyptus oil

| Chinese name | English name | Ingredients | Uses |
|---|---|---|---|
| Bai Hua You | White Flower Embrocation | Winter green oil (40%), menthol crystals (30%), eucalyptus oil (18%), lavender oil (6%), camphor (6%) | Headache, nasal congestion, muscular pain, stomach-ache, abdominal pain, antiseptic, travel sickness, insect bites |
| Feng You Jing | Medicated Oil | Menthol, eucalyptus oil, methyl salicylate, eugenol, camphor | Headache, toothache, inflammation, mosquito bites |
| Fu Biao Qu Feng You | Axe Brand Universal Oil | Menthol crystals (20%), eucalyptus oil (15%), methyl salicylate (15%), other essential oils (12%), camphor (5%) | Headache, rheumatic pain, stomach-ache, colic, giddiness, travel sickness, insect bites |
| Si Ji Run Hou Pian | Even-Cooling Throat Tablets | Menthol crystals (0.001 g), tangerine (0.001 ml), eucalyptus oil (0.0008 ml), peppermint (0.0007 ml), glucose, citric acid | Pharyngitis, tonsillitis |

*Eucalyptus leaves*

Parts of the plant, usually the leaves, are often used in traditional Chinese medicines and *E. robusta* and *E. saligna* have been found to be particularly valuable. Qi (1990) cites three examples:

1. Disinfectant – Boiled water of the leaves of *E. robusta* is used as a disinfecting agent in place of ethyl alcohol after operations. Up to 1970, approximately 100,000 people had been treated in this way at Guangzhou Military Hospital.
2. Treatment of dysentery – In 1973, using Chinese medicines containing *E. saligna* leaves, 100 dysentery patients at a military hospital made a full recovery.
3. Treatment of nephritis – Also in 1973, 115 patients were treated with Chinese medicines containing *E. robusta* and *E. saligna* leaves. Of these, 70 made a full recovery, 25 a partial recovery and 20 were unaffected.

# References

Anon. (1989) China–Australia Afforestation Project at Dongmen State Forest Farm. Technical Communication No. 40, Queensland Department of Forestry, Brisbane.
Bai, J. (1992) *Eucalyptus* tree improvement in southern China – problems and strategies (in Chinese; English summary). *For. Res.*, 5, 574–580.
Bai, J. (1994) Genetic improvement of tropical *Eucalyptus* tree species in China. In A.G. Brown (ed.), *Australian Tree Species Research in China, Proc. Internat. Workshop*, Zhangzhou, Fujian Province, P.R. China, November 1992, ACIAR Proceedings No. 48, pp. 32–49.
Brown, A.G. (ed.) (1994) *Australian Tree Species Research in China, Proc. Internat. Workshop*, Zhangzhou, Fujian Province, P.R. China, November 1992, ACIAR Proceedings No. 48.
Chen, Y., Yang, L., Li, S. and Jiang, Z. (1983) Study on the chemical components of essential oil from the leaves of *Eucalyptus* spp. (in Chinese; English summary). *Chem. Ind. For. Prods.*, 3(2), 16–31.
Cu, J. (1988) Yunnan – the kingdom of essential oil plants. In B.M. Lawrence, B.D. Mookherjee and B.J. Willis (eds), *Flavors and Fragrances: A World Perspective, Proc. 10th Internat. Congr. Essent. Oils Fragr. Flav.*, Washington DC, November 1986, pp. 231–241.

Eldridge, K.G., Davidson, J., Harwood, C. and van Wyk, G. (1993) *Eucalypt Domestication and Breeding*, Clarendon Press, Oxford, UK.

Gong, M. and Ke, J. (1988) *Eucalyptus* diseases in China. In Research Institute of Tropical Forestry (ed.), *The Use of Australian Trees in China, Proc. Workshop*, Guangzhou, Guangdong Province, P.R. China, October 1988, pp. 47–52.

Gu, M. and Chen, P. (1988) *Eucalyptus* pests in China. In Research Institute of Tropical Forestry (ed.), *The Use of Australian Trees in China, Proc. Workshop*, Guangzhou, Guangdong Province, P.R. China, October 1988, pp. 53–63.

Liu, X., Yang, M., Wang, G., Pegg, R.E. and Arnold, R.J. (1996) Development and genetic improvement of China's eucalypt resource. In M.J. Dieters, A.C. Matheson, D.G. Nikles, C.E. Harwood and S.M. Walker (eds), *Tree Improvement for Sustainable Tropical Forestry, Proc. QFRI-IUFRO Conf.*, Caloundra, Australia, October/November 1996, pp. 92–93.

Midgley, S.J. and Pinyopusarerk, K. (1996) The role of eucalypts in local development in the emerging economies of China, Vietnam and Thailand. In K.G. Eldridge, M.P. Crowe and K.M. Old (eds), *Environmental Management: The Role of Eucalypts and Other Fast-Growing Species, Proc. Joint Australian/Japanese Workshop*, CSIRO, Canberra, October 1995, pp. 1–10.

Midgley, S.J. and Yang, M. (1992) The program of technical cooperation with the China Eucalypt Research Centre. Paper prepared for International Workshop, Zhangzhou, Fujian Province, P.R. China, November 1992.

PPRC (1992) Oleum Eucalypti (Anyou), Eucalyptus Oil. In *Pharmacopoeia of the People's Republic of China*, English Edition, Guangdong Science and Technology Press, Guangzhou, p. 129.

Qi, S. (1990) *Eucalypts in China* (in Chinese), Chinese Forest Publishing House, Beijing.

Song, Y. (1991) Production and development prospect of eucalyptus oil in China. In S. Qi (ed.), *Proc. Internat. Eucalypt Symp.*, Zhanjiang, Guangdong Province, P.R. China, November 1990, pp. 163–169.

Wang, H. and Brooker, M.I.H. (1991) *A Key to Eucalypts in China* (in Chinese and English), China Science and Technology Press, Beijing.

Wang, H. and Wang, Z. (1991) Biomass studies in young eucalypt plantations for oil and fuelwood production in southwest China. In *Multipurpose Tree Species Research in Asia, Proc. Workshop*, Los Banos, Philippines, November 1990. Also (in Chinese; English summary) *For. Res.*, 4, 257–263.

Wang, W. (1997) Species and provenance selection for *Eucalyptus* leaf oils production (in Chinese; English summary). *For. Res.*, 10, 104–107.

Wu, B. (1991) Survey for the development of eucalypts in China. In S. Qi (ed.), *Proc. Internat. Eucalypt Symp.*, Zhanjiang, Guangdong Province, P.R. China, November 1990, pp. 19–22.

Wu, J., Wu, K. and Xu, J. (1994) *Eucalyptus* experiments in Guangdong Province. In A.G. Brown (ed.), *Australian Tree Species Research in China, Proc. Internat. Workshop*, Zhangzhou, Fujian Province, P.R. China, November 1992, ACIAR Proceedings No. 48, pp. 101–104.

Wu, K., Wu, J., Xu, J. and Gan, S. (1996) Cross breeding in *Eucalyptus* (in Chinese; English summary). *For. Res.*, 9, 504–509.

Zhang, K. (1997) Development and utilisation of Chinese eucalyptus oil (in Chinese). *J. Chem. Ind. For. Prods.*, 31(2), 27–30.

Zhao, Z.D., Sun, Z., Liang, Z.Q. and Wang, Y. (1997) Gas chromatography of the residue from fractional distillation of *Eucalyptus globulus* leaf oil (in Chinese; English summary). *Chem. Ind. For. Prods.*, 17(2), 37–40.

Zheng, H. (1988) The role of *Eucalyptus* plantations in southern China. In D. Withington, K.G. MacDicken, C.B. Sastry and N.R. Adams (eds), *Multipurpose Tree Species for Small-Farm Use, Proc. Internat. Workshop*, Pattaya, Thailand, November 1987, pp. 79–85.

Zheng, Y., Wang, H., Zhang, R. and Ma, Q. (1994) Trials of *Eucalyptus smithii* and other eucalypt species in Yunnan Province, China. In A.G. Brown (ed.), *Australian Tree Species Research in China, Proc. Internat. Workshop*, Zhangzhou, Fujian Province, P.R. China, November 1992, ACIAR Proceedings No. 48, pp. 116–122.

# 9 Cultivation and production of eucalypts in Africa

## With special reference to the leaf oils

*Paul A. Jacovelli*

## Historical review

### Introduction of eucalypts to Africa

Settlers, missionaries and government officials introduced eucalypts to Africa in the early nineteenth century. The first recorded introduction was in 1828 when the newly appointed Governor of the Cape Colony brought with him from Mauritius nine *Eucalyptus globulus* seedlings. Progeny from these 'blue gums' in the Governor's garden soon found their way to other parts of South Africa. Many other eucalypt species were introduced from Australia to South Africa from around 1860 onwards. The rapid growth, adaptability to a wide range of site conditions and diversity of end uses of the eucalypts encouraged their rapid spread into other parts of Africa. Today, eucalypts are part of the landscape of virtually every African country.

*E. globulus* was for many years the most important eucalypt species in Africa but its use declined due to its susceptibility to premature die-back (often drought related) and the attentions of the eucalyptus snout beetle (*Gonipterus scutellatus* Gyll.), a pest accidentally introduced from Australia. The species is still successful, however, at high elevations in cool, tropical climates such as is found in Ethiopia. The continent's main commercial species are now *E. grandis* (often confused with *E. saligna*) on the more fertile sites, *E. camaldulensis* in drier regions, *E. gomphocephala* in the sandy, arid North African countries (e.g. Libya, Morocco and Tunisia) and *E. robusta* in tropical regions (e.g. Madagascar) (Poynton 1979, FAO 1981).

### Scale of plantings and end-use

Most of these early eucalypt plantings in Africa were around homesteads and farms for fuelwood and pole production, for use as windbreaks and for their ornamental value. The commercial potential of eucalypts in Africa was soon recognised, however, and by the late nineteenth century plantations were being established to meet not only escalating household needs, but also a rapidly expanding industrial requirement (e.g. for the production of pit props for the gold mines of South Africa and to fuel the railways of Angola). By the early 1900s, the planting of eucalypts was rapidly expanding throughout Africa, principally for fuelwood and for the protection of fragile lands. In the latter half of the twentieth century, there was a huge expansion of eucalypt planting for hardwood pulp and sawn timber. Large and small-scale eucalypt plantations (private and state-owned) are now a very important part of many African countries' economies.

Estimates of the eucalypt plantation area in Africa vary markedly due to the difficulties of obtaining reliable (or recent) figures from a number of countries. Pandey (1995) estimated that in 1990 tropical African countries alone had 790,000 ha of eucalypt plantations, the principal countries being Angola (135,000 ha), Madagascar (130,000 ha) and Ethiopia (95,000 ha). Eldridge *et al.* (1993) estimated Ethiopian plantations at over 250,000 ha. Outside the tropical areas there are extensive eucalypt plantations in some Northern and Southern African countries – for example, Morocco (200,000 ha), Tunisia (42,000 ha), Algeria (30,000 ha) and South Africa (538,000 ha) (Davidson 1995). Davidson (1995) estimated a total of 1,636,000 ha of eucalypt plantations in Africa in 1990. A full list of African estimates for 1990, together with more recent data (1995), is given in Appendix 2.

*Eucalyptus oil production in Africa*

There is very little recorded information concerning the use of eucalypts in Africa for their essential oils, though it is likely that local people quickly discovered the therapeutic value of the leaves of certain species. South Africa was probably importing eucalyptus oil from the late nineteenth century (Shiel 1985). Poynton (1979) refers to two publications which indicate that there was an interest in eucalyptus oils in South Africa by the 1930s (De Villiers and Naudé 1932, Van der Riet 1933). From the range of species tested they concluded that there was potential in South Africa for the production of the three main types of eucalyptus oil, namely, medicinal, industrial and perfumery oils.

Obtaining information about the oil industry in Africa is difficult – undoubtedly due to the competitive nature of the business, with the key private growers closely guarding the details of their operations. The first distillation of eucalyptus oil in Africa probably took place using crude stills and the leaves of *E. globulus* trees harvested primarily for timber or fuelwood. The first African plantations dedicated to the production of eucalyptus oil were probably established around the early 1940s in the Democratic Republic of the Congo (formerly Zaire), Rwanda and Burundi. These Central African countries were producing cineole-rich oil from *E. globulus*, *E. smithii*, *E. globulus* subsp. *maidenii* and *E. dives*, as well as other types of oil from *E. citriodora* and *E. macarthurii*. By 1957, these operations were producing 80 t of oil annually (Penfold and Willis 1961, Weiss 1997).

Around the early 1950s, oil plantations were being established on private farms in the highveld region of South Africa and in neighbouring Swaziland. Both cineole and industrial oils were being produced from these plantations: cineole-rich oil mainly from *E. smithii* and *E. radiata*, and industrial oils from the piperitone-rich variant of *E. dives* and the phellandrene-rich variant of *E. radiata*. The total area of plantations dedicated to eucalyptus oil production in South Africa and Swaziland was around 4000 ha by 1990, with over 80 per cent of the area being worked on a short-rotation coppice system. An estimated 470 t of oil per year was being produced in the late 1970s (245 t of cineole-rich oil and 225 t of piperitone-rich oil) (Small 1981). Coppen and Hone (1992) estimated that in 1991, approximately 255 t of cineole-rich oil was produced in Southern Africa (representing over 9 per cent of total world production), with some rectification capability in South Africa. Oil production was often on a stop–start basis, depending on the oil price. The majority of the distilled oil was exported to Australia and Spain for rectification, though there always was (and still is) a small local market for both types of oil.

Plantations of *E. smithii* are believed to have been established in Malawi and Angola for cineole production but little is known of their fate or whether commercial quantities of oil were ever produced. Civil wars in Angola and the Central African countries over the past thirty years or so will undoubtedly have severely affected any eucalyptus oil operations which existed. More recently, private growers in Zimbabwe have grown *E. smithii* and *E. cinerea* on a short-rotation coppice cycle, also for cineole (Poynton 1979, Boland *et al.* 1991). Oil production began there in 1989 but it is not known whether this is continuing today. Also in the late 1980s, a company in the Eastern Highlands of Tanzania conducted trials with oil species and was planning to produce cineole from *E. smithii* on a significant scale.

The economic climate in the late 1980s and 1990s — poor oil prices and rising production costs — meant either that full scale oil production never materialised (in the case of Tanzania) or that some production ceased (as in Swaziland). In South Africa, the area of eucalypt plantations dedicated to oil has now reduced to approximately 1800 ha, most of the reduction occurring in *E. smithii*. One estimate puts the production split at about 1:2 for cineole:piperitone-type oils (V. Davidson pers. comm. 1998).

## Species and provenance selection of oil-bearing eucalypts

### Introduction

Successful oil production operations in Africa have been largely based on plantations cultivated with the primary objective of oil production. These plantations, particularly in Southern Africa, are managed intensively with the aim of maximising oil yields. Other products produced, such as poles and firewood, are seen as useful by-products and utilised (or sold) where possible. These oil plantations have mostly been managed on a short-rotation, coppice system. Dual-purpose plantations — for wood and oil production — have been used to a lesser extent in Africa, though this technique could make the business more appealing to small growers.

The information provided below is based largely on work carried out in the late 1980s and early 1990s at Shiselweni Forestry Company in Swaziland. Working in collaboration with the Natural Resources Institute (Chatham, U.K.), research in Swaziland was aimed at increasing the sustainable yield of cineole-rich oil from the plantations. Matters investigated included the optimisation of source material through species and provenance research, the influence of other factors such as fertiliser application, rotation and seasonal effects on oil yield, field management practices (e.g. spacing, harvesting and transport methods) and distillation conditions (Coppen, Milchard unpubl., Jacovelli 2002).

### Species' characteristics

This section discusses the main eucalypt species used commercially for the production of oil in Africa and those introduced on a trial basis. It must be stressed that whilst research conducted elsewhere provides some indication of what might grow well in a particular area, there is no substitute for performance trials in a region prior to any large-scale plantings for eucalyptus oil production.

Table 9.1 shows the oil characteristics of those species which are, or have been, used for oil production. Yield and compositional data should be regarded as illustrative and not necessarily representative. Dethier *et al*. (1994), for example, in Burundi, only sampled single trees.

E. smithii

On favourable sites where it has been cultivated in Southern and Central Africa, *E. smithii* has often produced higher biomass yields (wood and foliage) than other eucalypts. It appears to be a fairly consistent species throughout its natural range with regard to oil yield and cineole content, though there is some variation between and within provenances in terms of other factors,

Table 9.1 Yields and characteristics of leaf oils obtained from eucalypts growing in Africa[a]

| Species | Country | Data[b] | Main constituent (relative abundance, %) | Oil yield[c] (%) | Reference |
|---|---|---|---|---|---|
| E. camaldulensis | Benin | r | 1,8-Cineole (31–73) | 0.6–1.4 (dry) | Moudachirou et al. (1999) |
| | Burkina Faso | r | α-Phellandrene (24.8) | 1.2 (air-dry) | Samaté et al. (1998) |
| | Burundi | r | 1,8-Cineole (43.3) | 1.5 (mfb) | Dethier et al. (1994) |
| | Egypt | r | 1,8-Cineole (65–75) | 0.7–1.5 | Abou-Dahab (1973) |
| | Morocco | r | 1,8-Cineole (72.4) | 0.7 (mfb) | Zrira et al. (1992) |
| | Nigeria | r | 1,8-Cineole (—) | 0.6 (?) | Osisiogu (1966) |
| | Zimbabwe | r | 1,8-Cineole (ca 75) | ca 2 (mfb) | Doran and Matheson (1994) |
| E. cinerea | South Africa | r | 1,8-Cineole (62.7) | 1.3 | Ndou (1986) |
| | Zimbabwe | c | 1,8-Cineole (62.5) | — | Coppen unpubl. |
| E. citriodora | Benin | r | Citronellal (65.5, 55–90) | 1.5, 2.2–5.9 (air-dry, dry) | Sohounhloue et al. (1996), Moudachirou et al. (1999) |
| | Burundi | r | Citronellal (46.9) | 3.4 (mfb) | Dethier et al. (1994) |
| | Egypt | r | Citronellal (60.5) | 0.7–1.0 | Elkiey et al. (1964) |
| | Ghana | r | Citronellal (60.4) | 2.6 (dry) | Talalaj (1966) |
| | Kenya | r | Citronellal (30–88) | 2.2–8.3 (mfb) | Mwangi et al. (1981) |
| | Madagascar | r | Citronellal (71.2) | — | De Medici et al. (1992) |
| | Morocco | r | Citronellal (56.3) | 1.1 (mfb) | Zrira et al. (1992) |
| | Nigeria | r | Citronellal (—, 50–70) | 4.8 (?), 4.8 | Osisiogu (1966), Weiss (1997) |
| | Rwanda | r | 1,8-Cineole (55.4) | 0.6 | Chalchat et al. (1997) |
| | Zaire, former | c | Citronellal (46–50) | 0.5–0.8 | Weiss (1997) |
| E. dives | Rwanda | r | Piperitone (32.8, 52.2) | 3.5, 3.2 (mfb) | Chalchat et al. (1997), Molangui et al. (1997) |
| | South Africa | c | Piperitone (30–41) | — | Coppen unpubl. |
| | Zaire, former | c | Piperitone (45–55) | 3 | Weiss (1997) |
| | Zimbabwe | r | 1,8-Cineole (67.4) | 9.9 (mfb) | Coppen unpubl. |
| E. elata | South Africa | c | Piperitone (35.2) | — | Coppen unpubl. |
| E. globulus subsp. globulus | Burundi | r | 1,8-Cineole (63.8) | 3.5 (mfb) | Dethier et al. (1994) |
| | Egypt | r | 1,8-Cineole (71–80) | 0.9–1.2 | Yousef et al. (1991) |
| | Madagascar | r | 1,8-Cineole (40.4) | — | De Medici et al. (1992) |
| | Morocco | r | 1,8-Cineole (72.8, 62–82) | 2.1, 1.9–2.7 (mfb) | Zrira et al. (1992), Zrira and Benjilali (1996) |
| | Rwanda | r | 1,8-Cineole (71.2) | 2.6 | Chalchat et al. (1997) |
| | South Africa | r | 1,8-Cineole (48.7) | 1.1 | Ndou (1986) |
| E. globulus subsp. maidenii | Burundi | r | 1,8-Cineole (60.0) | 4.8 (mfb) | Dethier et al. (1994) |
| | Morocco | r | 1,8-Cineole (76.8, 69–80) | 1.4, 2.4–3.9 (mfb) | Zrira et al. (1992), Zrira and Benjilali (1996) |

Table 9.1 (Continued)

| Species | Country | Data[b] | Main constituent (relative abundance, %) | Oil yield[c] (%) | Reference |
|---|---|---|---|---|---|
| | Nigeria | r | 1,8-Cineole (—) | 3.4 (?) | Osisiogu (1966) |
| | Rwanda | r | 1,8-Cineole (70.5) | 2.8 | Chalchat et al. (1997) |
| | South Africa | r | 1,8-Cineole (59.6) | 1.1 | Ndou (1986) |
| E. macarthurii | Angola | r | Geranyl acetate (64.3) | 0.2–1.0 | Cardoso and Proença (1968) |
| | Rwanda | r | Geranyl acetate (50.5) | 0.1 | Chalchat et al. (1997) |
| | Zambia | r | Geranyl acetate (58.0) | 3.2 (air-dry) | Chisowa (1997) |
| E. radiata | South Africa | r | 1,8-Cineole (70.0) | > 8 | Donald (1980) |
| | Swaziland | r | 1,8-Cineole (55–75) | 2.5–4.0 | Jacovelli unpubl. |
| | Tanzania | r | 1,8-Cineole (40–56) | 8.3–10.5 (mfb) | Coppen unpubl. |
| | Zimbabwe | r | 1,8-Cineole (67.3) | 8.5 (mfb) | Coppen unpubl. |
| E. sideroxylon | South Africa | r, c | 1,8-Cineole (51.8, 75) | 0.7, — | Ndou (1986), Weiss (1997) |
| | Zimbabwe | r | 1,8-Cineole (69.1) | 2.7 (mfb) | Coppen unpubl. |
| E. smithii | Rwanda | r | 1,8-Cineole (84.3) | 2.5 | Chalchat et al. (1997) |
| | South Africa | c | 1,8-Cineole (71.1) | — | Coppen unpubl. |
| | Swaziland | c, r | 1,8-Cineole (65–75, 62–71) | —, 5.8–8.4 (mfb) | Jacovelli unpubl., Coppen unpubl. |
| | Tanzania | r | 1,8-Cineole (64–70) | 5.0–6.8 (mfb) | Coppen unpubl. |
| | Zaire, former | c | 1,8-Cineole (—) | 1.0–1.5 | Weiss (1997) |
| | Zimbabwe | c, r | 1,8-Cineole (70.0, 70.7) | —, 5.0 (mfb) | Coppen unpubl. |
| E. staigeriana | Zaire, former | c | Citral (—) | ca 4 | Weiss (1997) |
| | Zimbabwe | r | Citral (27.9) | 5.9 (mfb) | Coppen unpubl. |

a The species listed are those where there has been at some time, or is, commercial production of oil somewhere in Africa or where, in the case of E. camaldulensis, there is the potential for it. Some of the references cited refer to other species, in addition to those indicated here.
b Indicates whether data are research results (r) or relate to commercial production (c).
c Yields are on a fresh weight basis unless otherwise indicated; mfb = moisture free basis. Yields given by Coppen (unpubl.) refer to distillation of leaves only, that is, no twigs or woody material.

particularly biomass production and frost tolerance. Provenance trials in Swaziland found the best overall performers came from Tallaganda State Forest and Narooma (both New South Wales, Australia), whilst South African trials found Wombeyan Road, Larry's Mountain and Mount Dromedary the best provenances on the basis of performance and frost tolerance. E. smithii requires deep soils and a mean annual temperature of >15°C. In Southern Africa, the species appears to be very susceptible to diseases, notably pathogenic fungi causing root rot and stem cankers. Although E. smithii can produce reasonably straight stems, the wood splits badly and it can have spiral grain (ICFR 1997, Jacovelli 2002).

### E. radiata *and* E. dives

*E. radiata* and *E. dives* exhibit significant intraspecific variation, particularly with regard to their oil characteristics. Both species have populations rich in phellandrene or piperitone as well as cineole. Parent trees must be tested first for their oil content and composition before collecting seed of either species for the establishment of oil plantations in order to check that the desired chemotype has been selected. Certain provenances and individual trees of *E. radiata* have shown very high yields of cineole-rich oil in trials in Swaziland: the Nerrigundah region of New South Wales appears very promising. The cineole form of *E. dives* is still harvested in New South Wales (around Tumut, Batlow, Tumbarumba and Rosewood) although most seed imported into Africa has proved to be the piperitone-rich variety (Boland *et al*. 1991, Jacovelli 2002). Oil produced in South Africa is all from the piperitone/phellandrene form of *E. dives*. Both species are small trees with generally poor form (*E. radiata* is often multi-stemmed) and hence they are not good multipurpose trees.

### E. globulus *and its subspecies*

*E. globulus* possesses many desirable traits and these led to its widespread planting in Africa: it is easy to establish (being unpalatable to grazing animals) and is a fast grower, it has good stem form, good fuelwood and pulping properties, and its widespread rooting system controls erosion. However, the species' susceptibility to a range of pests and diseases has limited its planting in Africa. Although *E. globulus* is the main source of cineole-rich oil outside Africa, its oil yield is generally lower than that of many other species cultivated in Africa (Table 9.1). Boland *et al*. (1991), however, report some individuals with up to 4 per cent oil, with an average cineole content of 65–75 per cent. Two subspecies of *E. globulus*, *maidenii* and *bicostata*, appear promising species for certain parts of Africa – the former in warmer and drier situations than suits *E. globulus* and the latter in cooler conditions than *E. globulus*. *E. globulus* subsp. *bicostata* has individuals with higher oil yields than *E. globulus* (Eldridge *et al*. 1993). All species, however, appear susceptible to the eucalyptus snout beetle and in trials in Swaziland *E. globulus* subsp. *bicostata* was very attractive to a number of pests (see below and Jacovelli 2002).

### *Other cineole-rich species*

*E. camaldulensis* is a very adaptable and drought-tolerant species. It has performed well in many of the drier regions of Africa, although on better sites, particularly areas with deeper soils and higher rainfall, its growth falls well behind many other eucalypts. It also produces an excellent fuelwood. The identification of cineole-rich genotypes, particularly from the tropical provenances such as Petford, could lead to this species being used for cineole production in drier regions of Africa (Doran and Brophy 1990).

*E. cinerea* is a promising species for oil production in cooler and moist parts of Africa, and has been used for commercial oil production, albeit on a very limited scale, in Zimbabwe (Figure 9.1). It is a very frost-hardy species and tolerates poor soils (Ndou and von Wandruszka 1986). Weiss (1997) notes that oil from *E. sideroxylon* was once produced commercially in South Africa, where it was used as a flotation agent in the mining industry. *E. badjensis* is another cineole-rich species recently introduced to Africa, where it has performed well in species-provenance trials in South Africa. It has proved to be very frost-tolerant but has also been heavily attacked by the ubiquitous snout beetle (ICFR 1997).

*Figure 9.1  Eucalyptus cinerea* being harvested for oil production in Zimbabwe (photo: J. Coppen).

The mallee eucalypt species tried to date in Africa, particularly *E. polybractea*, grow very slowly. Even though the cineole content of the oil from this and other mallee species is often very high (up to 90 per cent), the overall biomass production, and thus oil yield, is very poor compared to many of the other cineole-rich species cultivated in Africa (Donald 1980).

*Industrial and perfumery oil species*

The use of *E. dives* as a source of industrial (piperitone-rich) oil has been referred to above. *E. elata* is grown commercially in South Africa for its timber and it tolerates poor sites and is frost hardy. Some *E. elata* individuals have a very high leaf oil content (usually high in piperitone and/or phellandrene) but the species is said to be very variable. Considerable screening would be needed before the species could be considered for commercial oil production (V. Davidson pers. comm.).

Poor market prices for the flavouring and perfumery eucalyptus oils have restricted the planting in Africa of *E. citriodora*, *E. macarthurii* and *E. staigeriana* specifically for oil production. *E. citriodora*, however, is cultivated in many African countries and has been grown commercially in Africa for its citronellal-rich oil (Table 9.1). The species is fairly widely planted as an ornamental in Southern Africa, where its fragrant, lemon-scented foliage is also used locally as an insect repellent. *E. citriodora* prefers warmer, humid areas, where it produces a hard, durable timber. There has been only limited and occasional commercial production in Central and Southern Africa of oils from *E. macarthurii* (rich in geranyl acetate) and *E. staigeriana* (rich in monoterpenes, especially neral, geranial, methyl geranate and geranyl acetate; Boland *et al.* 1991) and there is currently little demand for the oil of either species (Robbins 1983, Weiss 1997). *E. macarthurii* is planted on a large scale (for timber production) in the South African highveld on poor soils and in frost-prone areas.

## Silviculture

### Basic information

The site requirements and silvicultural characteristics of the principal oil-bearing eucalypts introduced into Africa are detailed in Table 9.2 (summarised from information contained in Poynton 1979, FAO 1981, Florence 1996, ICFR 1997, Jacovelli 2002).

### Seedling production

In many African countries, small-scale, 'low-tech' nurseries are often appropriate, particularly in remote regions with a poorly developed infrastructure. The shift towards decentralised nursery systems in many developing countries has ensured more effective seedling distribution and greater participation by local people (Shanks and Carter 1994). On the other hand, large, centralised, nurseries have proved successful in some other African countries. These modern nurseries often use either expanded polystyrene (Styrofoam ') or plastic (Unigro®) trays and raise seedlings in a growing medium of composted pine bark or vermiculite (Donald 1986, Jacovelli 1994). Whichever nursery system is chosen to produce seedlings for a eucalyptus oil operation, two factors are crucial to success: the need for plantings to be reasonably concentrated around a central distillation plant (since it is important to minimise the transport cost of the bulky leaf material) and the importance of using the seed from parent trees with the desired oil characteristics.

Three other key points should be borne in mind by potential oil producers:

1. Provenance details must be recorded (and provided by reputable seed merchants) and different seedlots should be kept separate in the nursery and in the field.
2. All unhealthy or under-sized seedlings should be discarded; using poor planting stock will not achieve the objective of establishing uniform, well-stocked plantations.

*Table 9.2* Site requirements and silvicultural characteristics of oil-bearing eucalypts cultivated in Africa

| Species | Min MAR[a] | Frost tolerance[b] | Drought tolerance[b] | Stem form[c] | Wood quality[d] | Pests and diseases[e] | Principle type of oil |
|---|---|---|---|---|---|---|---|
| E. camaldulensis | 400 | m/s | h | 2 | 4 | 3 | Variable[f] |
| E. cinerea | 700 | h | m | 2 | 2 | 3 | 1,8-Cineole |
| E. citriodora | 600 | s | h | 4 | 5 | 4 | Citronellal |
| E. dives | 850 | h | m | 2 | 1 | 4 | Variable[g] |
| E. elata | 1000 | h | m | 3 | 3 | 3 | Piperitone |
| E. globulus | 800 | s | s | 4 | 4 | 1 | 1,8-Cineole |
| E. globulus subsp. maidenii | 800 | s | m | 4 | 4 | 1 | 1,8-Cineole |
| E. macarthurii | 850 | h | m | 3 | 3 | 4 | Geranyl acetate |
| E. radiata | 850 | m | m | 1 | 1 | 4 | Variable[g] |
| E. sideroxylon | 500 | m | h | 2 | 4 | 3 | 1,8-Cineole |
| E. smithii | 850 | h | m | 2 | 2 | 2 | 1,8-Cineole |
| E. staigeriana | 1000 | s | h | 2 | 4 | 3 | Citral |

a Mean annual rainfall (mm).
b Hardy (h), moderately hardy (m) or sensitive (s) to frosts or drought.
c On scale 1 (bushy) to 5 (straight stemmed).
d On scale 1 (poor quality) to 5 (excellent quality).
e On scale 1 (very susceptible) to 5 (rarely affected).
f Only the 1,8-cineole chemotype is likely to be important commercially.
g Numerous chemotypes but in Africa *E. dives* is mainly piperitone and *E. radiata* 1,8-cineole.

3   Through careful planning and knowledge of local conditions the sowing must be timed so that seedlings are ready at the optimum time (for most seasonally dry areas in Africa the best time to plant is at the start of the main rainy season).

*Vegetative propagation*

Vegetative propagation enables the mass multiplication of desirable genotypes to be achieved. Some of the first trials of rooted eucalypt cuttings, on *E. camaldulensis*, took place in Africa (Morocco) in the 1950s. Since then, mass production by rooting cuttings has been carried out in Africa on a commercial scale. From around 1964, French researchers at the Centre Technique Forestier Tropical at Pointe Noire in the Congo successfully mass-produced eucalypt cuttings (mainly the hybrid *E. tereticornis* × *E. grandis*). More recently, in South Africa, pure *E. grandis*, and many *E. grandis* hybrid clones, have been successfully produced on a large scale (Eldridge *et al.* 1993).

For obvious economic reasons, virtually all the vegetative propagation work to date in Africa has concentrated on maximising fibre production. There has been some research, however, with mass propagation of cineole-rich species in Southern Africa. Working with *E. radiata*, Donald (1980) had success rooting cuttings using standard techniques developed for eucalypts (as described by Eldridge *et al.* 1993). However, he reported marked clonal variation in rooting ability. In South Africa, *E. smithii* is classed as a difficult rooting species, along with a number of other 'hard-gums' (cold-tolerant eucalypts).

In Swaziland, in the late 1980s, attempts were made to root cuttings from young coppice shoots of selected high cineole-yielding *E. smithii* and *E. radiata* coppiced trees from plantations. The Oxford Forestry Institute (OFI) successfully propagated shoot-tip cuttings from both species. OFI also raised seedlings from commercial seed of *E. smithii* and *E. radiata* and attempted micropropagation from these. Up to 67 per cent of *E. smithii* and 50 per cent of *E. radiata* clones rooted successfully, although there was a marked clonal variation in rooting ability. These results at least indicate that micropropagation for the rapid multiplication of selected oil-producing eucalypts is possible (Woodward and Thomson 1989).

*Cultivation methods for establishing oil crops*

The topics described below summarise the practices developed mainly in Southern Africa for the establishment of large-scale eucalypt plantations. Much of the work has been carried out with the most important timber species, *E. grandis*, but research has shown that the same principles generally apply to all eucalypt species. What may alter in plantings made specifically for oil production are the spacing, fertiliser regime and rotation.

Establishment costs will vary significantly depending on the local situation. Table 9.3 presents typical inputs involved in establishing eucalyptus oil plantations in Southern Africa where there are significant mechanical inputs. In some African countries it may be more cost-effective (and possibly more socially or environmentally acceptable) to use more labour, particularly for land clearance and planting preparation. The costs will vary considerably depending on site conditions.

*Land preparation*

Thorough land preparation for the successful establishment of eucalypt plantations has long been recommended in Southern Africa. Where there is a rooting impediment – including soils

*Table 9.3* Typical inputs involved in establishing eucalyptus oil plantations in Southern Africa

| Job description | Details | Man-days (per ha) | Machine-hrs (per ha) | Consumables |
|---|---|---|---|---|
| *Land preparation* | | | | |
| Clearing[a] | | 0.5 | 4.0 | |
| Deep ripping[b] | To min. 600 mm | 0.5 | 2.5 | |
| Cultivation | | 0.2 | 1.5 | |
| Pre-plant spray[c] | | 2.0 | 1.5 | Herbicide |
| *Planting/establishment* | | | | |
| Planting[d] | | 6.0 | | Seedlings |
| Infilling/blanking[e] | | 2.0 | | Seedlings |
| Fertilising | See Table 9.4 | 3.5 | | Fertiliser |
| Weeding, year 1 | Manual (line) | 45.0 | | |
| (3 passes) | Chemical | 6.0 | | Herbicide |
| Weeding, year 2 | Mechanical (disc, inter-row) | 0.5 | 1.5 | |

a Varies according to site conditions.
b Necessary on sites with a rooting impediment, including a high clay content.
c Types and dosages depend on current and expected status of weeds.
d Stocking depends on species and objectives.
e Replacement of failures; depends very much on conditions at, and soon after, planting.

with a high clay content – contoured deep ripping is beneficial, not only to increase the effective rooting depth but also to improve the water-holding capacity of the site. On other sites, ploughing and cultivation along the planting line is recommended. The manual preparation of pits is often unsatisfactory for the establishment of eucalypts and is also becoming prohibitively expensive (Schönau 1984).

*Plant espacement*

Most eucalyptus oil plantations in Southern Africa were established initially at a spacing of $9' \times 9'$ (2.74 m $\times$ 2.74 m), equivalent to 1332 stems ha$^{-1}$. In more recent years there has been a tendency to increase the stocking density, with one grower in Swaziland planting at 3.0 m $\times$ 1.5 m (2222 stems ha$^{-1}$). The higher stocking is aimed at increasing the biomass production with short-rotation crops, whilst the rectangular spacing allows mechanical access between the rows for weed control. Foresters have traditionally manipulated stocking to maximise the yield of a particular species on a specific site. The objective with eucalypt plantations established specifically for oil must be to maximise leaf biomass rather than woody material. One possible way of increasing the biomass production might be to establish high density plantations (even greater than 10,000 stems ha$^{-1}$), though more research is needed to assess the sustainability of such high levels of stocking by testing various species on different sites.

*Weed control*

Virtually all eucalypts in cultivation have proved to be intolerant of competition from weeds, particularly grasses, during their establishment. For optimum growth, it is advisable to keep eucalypt plantations completely weed-free until canopy closure. As labour costs escalate in many developing countries, herbicides are increasingly proving cost-effective. The use of grass pre-emergent herbicides (e.g. active ingredient acetochlor) in combination with a non-specific herbicide (especially glyphosate), has proved very cost-effective in Southern African countries, giving up to ten weeks weed control (N.B. advice must always be taken regarding compatibility and labels must always be read for dosages and safety aspects prior to using herbicides).

*Fertilising*

Application of fertiliser at planting has proved to be highly cost-effective for eucalypts in Southern Africa. The full benefit of fertilising, however, is only realised in conjunction with other silvicultural practices, particularly good weed control, the use of healthy planting stock and good site preparation (Schönau 1983, 1984). Table 9.4 gives the fertiliser prescriptions for Southern African conditions (ICFR 1997); under more tropical conditions, boron can also be important. In the intensive eucalyptus oil operations in Southern Africa these prescriptions have generally been followed at planting. No additional fertiliser has been added even though the plantations would be repeatedly harvested on coppice cycles every 12–24 months, sometimes for over twenty years.

An intensive oil operation removes all the leaf and much of the small branch material, and there is clearly a large nutrient demand on the site over successive rotations. The lack of accurate yield data from commercial operations, combined with the many factors that can affect oil yields, makes it difficult to determine whether there has been a decline in oil yield from the intensively managed African plantations as a result of decreasing soil fertility. Studies have shown that whilst eucalypts are very efficient users of available nutrients, the chances of them depleting soil nutrients are increased with shortened rotations and situations where most of the biomass is removed from the site. Grove *et al.* (1996) found that the foliage of eucalypts contains around 20 per cent N and 17 per cent P of the above-ground nutrients. Very short rotations also expose the site to possible losses through erosion and leaching. The nutrient balance and sustainability of eucalyptus oil production under African conditions clearly needs further research (Herbert 1996).

Whilst the addition of fertilisers at establishment is clearly beneficial for most eucalypts' growth (particularly on marginal sites), the effect on overall oil yield and oil composition is not clearly understood. An increase in leaf production does not necessarily produce more oil per unit

*Table 9.4* Fertiliser prescriptions for eucalypts in summer rainfall regions of Southern Africa

| Topsoil organic carbon (%) | N and/or P (per tree) | Standard fertiliser (per tree) |
|---|---|---|
| *Fully cultivated, virgin sites* | | |
| >8 | 15 g P | 140 g single superphosphate (10.5% P) |
| 3–8 | 5 g N + 12 g P | 100 g ammoniated superphosphate or 60 g mono-ammonium phosphate |
| 1–3 | 8 g N + 10 g P | 100 g single superphosphate + 25 g limestone ammonium nitrate or 50 g di-ammonium phosphate |
| *Ripping/pitting (re-establishment)* | | |
| >3 | 8 g N + 12 g P | 60 g di-ammonium phosphate or 125 g NPK (2:3:2, 22%) |
| <3 | 14 g N + 10 g P | 100 g NPK (3:2:0, 25%) or 125 g NPK (3:2:1, 25%) |

area of land. The limited amount of research on this subject indicates that with *E. smithii* in Southern Africa, nitrogen may have a beneficial effect on oil yield (Jacovelli and Coppen 1994). However, recommended dosages and the economics of fertiliser application to oil plantations have not been calculated. A positive response has been obtained to fertiliser applied to *E. grandis* coppice crops in South Africa but the amounts needed have been so large that the operation is considered uneconomic (Schönau 1983). For oil plantations managed intensively on a short-rotation coppice regime, it is likely that nutrients will have to be returned to the system to sustain yields in the long term.

*Rotation*

The optimum rotation for oil production depends on many factors, including:

1 The objective(s) (i.e. whether solely oil production, oil and fuelwood, oil and poles, etc.).
2 The site conditions (particularly the climate and nutrient status).
3 The species or provenance origin of the eucalypt(s).
4 The silvicultural regime (e.g. spacing, weed control, fertilisation, etc.).
5 Utilisation/market factors.

Research in Swaziland has shown that the rotation for the first (seedling) crop of *E. smithii* should be longer (20–30 months, depending on the growth rates) than the subsequent coppice cycle. This presumably gives the root system time to develop well and makes it better able to support repeated coppicing later on. In Swaziland, the *E. smithii* coppice was cut on a cycle of 12–18 months, depending on the growth rates. This cycle was based on the time taken for the crop to reach canopy closure, at which stage the lower leaves begin to be shaded out (and eventually die). How many times a crop can be cut depends on stocking levels, the coppicing ability of the species and the nutrient status of the site. Five-month-old coppiced *E. smithii* on its eighth cycle is shown in Figure 9.2. *E. smithii* plantations in Southern Africa have been repeatedly harvested for oil for over twenty years with little apparent loss of stool vigour. The main cause of yield loss is generally from stool death and the resulting reduction in stocking. As a general rule, plantations with less than 75 per cent stocking should be replanted. The main oil species in Africa are all classed as good coppicing species.

In Swaziland, the stems of *E. smithii* were typically cut about 15 cm above the ground. In South Africa, the practice of one producer of *E. dives* (piperitone variant) has been to cut the stem at knee height (4–5 years after planting, with harvest of the coppice re-growth at 15–18 month intervals thereafter), Figure 9.3.

*Irrigation*

Studies have shown that climatic factors, and particularly rainfall, have a significant effect on oil yield. In South Africa, Donald (1980) predicted yields of over $650\,kg\,ha^{-1}\,yr^{-1}$ by drip irrigating selected high oil-yielding clones of *E. radiata*. However, on a continent where water is frequently scarce, irrigated oil plantations are unlikely to be economically feasible or socially acceptable.

## Pests and diseases

An enormous number of pests and diseases have been identified on eucalypt crops since their introduction to Africa (FAO 1981). Only those that can cause serious losses to the principal oil-producing eucalypts in Africa are discussed here.

*Figure 9.2* Five-month-old coppiced *Eucalyptus smithii* on its eighth cycle, Swaziland (photo: J. Coppen).

*Figure 9.3* Recently harvested *Eucalyptus dives*, showing coppice re-growth from the stump, South Africa (photo: J. Coppen).

The incidence and threat of the many potential pests and diseases can be significantly reduced by practising sound silvicultural techniques so as to reduce the stress on the crop. These techniques have been described elsewhere in this chapter but the most important are:

1   Careful site-species matching, especially in drought and/or frost-prone areas.
2   Ensuring a broad genetic base is maintained by not planting large, contiguous blocks with the same species, provenance or clone.
3   Site amelioration to reduce stress (see section on Silviculture).
4   Planting only healthy, vigorous seedlings at the optimum time.
5   Thorough weed control.

*Insect pests*

*Termites*

Termites, especially in the semi-arid and sub-humid tropics, cause significant yield losses and are often a major constraint on forestry development in these regions. The most serious losses are due to various Macrotermitinae (Termitidae) and exotic trees are most at risk (Cowie *et al*. 1989). In higher elevation, cooler areas of Swaziland and South Africa, where most of the eucalyptus oil plantations are in Africa, termites do not pose a serious threat. Where plantations are established in drier areas, however, control measures have to be taken. In Zimbabwe, mortalities due to termites in *Eucalyptus* are commonly 30–50 per cent but can approach 100 per cent in the absence of any control (Mitchell 1989). The first visible symptoms of termite attack are normally a die back from the branch tips, followed by rapid death of the whole tree. On inspection, the root collar (often from just below the ground) will be ring-barked and many of the roots severed. For many years the organochlorine insecticides (e.g. dieldrin, chlordane and aldrin) were used to protect trees from termites, but their persistence in the environment means that they are not now recommended, even in countries where they are still available. Carbosulfan controlled-release granules (trade name Marshall®/suSCon®; recommended at 1.0 g active ingredient per tree at planting) have been successful in a number of sub-Saharan African countries (Atkinson *et al*. 1991, Canty 1991). Logan *et al*. (1990) discuss other, non-chemical control methods.

*Cutworms and white grubs*

Cutworms (Lepidoptera: Noctuidae) and white grubs (Coleoptera: Scarabaeidae) have caused significant deaths in new plantings of many eucalypts in Southern Africa (including *E. smithii*). Both pests cause damage in the first twelve months after planting and can seriously reduce the stocking of new plantings. Cutworms ring-bark young seedlings at, or just above, ground level whereas white grubs feed on the fine, lateral roots of seedlings. Control can be achieved by spraying deltamethrin (5 per cent suspension concentrate) at a dose of 0.025 g active ingredient per tree (Govender 1993).

*Eucalyptus snout beetle*

The eucalyptus snout beetle (*Gonipterus scutellatus*) is a major pest which influences the choice of which *Eucalyptus* species to plant in many parts of Africa. This defoliating weevil was an unwelcome visitor from Australia around 1916 and, like many pests when away from their natural predators, their population exploded on finding large plantations of susceptible hosts. Most eucalypts growing in Africa were subsequently attacked, with *E. globulus* and *E. globulus* subsp. *maidenii* proving very susceptible. In Southern Africa, the beetle has also heavily attacked *E. smithii*, *E. badjensis* and *E. camaldulensis*. Both the adult and the larvae feed on the leaves and young shoots of susceptible species; severe attacks lead to crown dieback. The severity of attacks by the snout beetle reached epidemic proportions in Africa in the 1920s and lead to the introduction of a parasitic wasp from Australia. This proved to be very successful in controlling the pest except at high elevations or latitudes (Poynton 1979).

### Other insect pests

There are several sap-sucking insects, particularly scales and psyllids, which affect eucalypts to varying degrees. Eucalypt plantations in Malawi, Angola and South Africa have been infested by scale insects. Trial plantings of *E. globulus* subsp. *bicostata* in Swaziland have been heavily infested with a psyllid (unidentified species). Successful control of some species has been achieved by biological means (FAO 1981). Wood-boring, bark beetles (the most important being *Phoracantha semipunctata* Fabr.) have caused serious damage to many eucalypt plantations in Africa. Amongst the susceptible species are *E. globulus* and *E. camaldulensis*. Attacks are more severe where trees are under drought stress.

## Diseases

### Damping-off and root rot disease

Root rot is caused by pathogenic fungi, primarily *Phytophthora*, *Pythium* and *Fusarium* spp., which occur naturally in the soil (some also occur in water sources). They are responsible for serious nursery losses through 'damping-off' and they can also cause significant deaths in young eucalypt plantations. *E. smithii* has proved susceptible in Southern Africa with up to 30 per cent deaths being reported, mostly in the first year after planting. In the nursery, widespread seedling death can occur, with dead patches visible on the stems. Control is mainly by preventative means – for example, by not over-watering, avoiding sowing during very hot periods, treating irrigation water and by using fungicide drenches (such as benomyl and captan). In the field, there is a general wilting of the leaves followed by chlorosis (yellowing) and death of the tree. To reduce root rot losses, cultural practices should be adopted to reduce seedling stress and, ultimately, susceptible species and/or provenances should be avoided (Viljoen *et al*. 1992).

### Botrytis cinerea

This parasitic fungus is welcomed by some producers of sweet wines (being responsible for 'noble rot'), but *Botrytis cinerea* Pers. has caused severe losses in eucalypt nurseries in Africa, as well as during the first year of establishment. Typical signs are a fine web of greyish mould (mycelium) which can be seen on infected plant parts. Control is largely by good nursery management and particularly hygiene: physical damage to seedlings should be minimised (infection is often via wounds), dead plant material should not be left around and humidity should be kept as low as possible (e.g. by ensuring good ventilation and not over-watering). Various fungicides are available to reduce the spread of any outbreaks (e.g. iprodione and benomyl) but chemical resistance has been reported (Nichol 1992).

## Other diseases

*Mycosphaerella muelleriana* (Thüm) Lindau is regarded as a serious leaf pathogen in Southern Africa, causing leaf-spot disease. The fungus has been identified on many eucalypt species and although it rarely causes seedling deaths, a heavy infestation undoubtedly will cause growth losses (Crous and Wingfield 1991). A heavy *Mycosphaerella* infestation occurred on trial plantings of *E. globulus* subsp. *bicostata* in Swaziland, whilst adjacent plots of *E. smithii*, *E. radiata* and *E. dives* were unaffected.

Various other fungal pathogens causing stem cankers have caused serious losses in eucalypt plantations in many countries and are increasingly appearing in Africa. The main culprits in Southern Africa have been *Cryphonectria cubensis* (Bruner) Hodges in hot, humid areas and *Endothia gyrosa* in cooler regions (Swart and Wingfield 1991, Nichol 1992). Areas of dead bark (cankers) can be seen on the stems of infected trees, often stained dark reddish-brown from resin or kino. The only control at present is preventative, by minimising the stress in plantations. *Endothia gyrosa*, in particular, is more severe in drought-stressed trees. Little is known yet concerning oil species' resistance to stem cankers but *E. smithii* oil plantations in Swaziland have been affected by up to 10 per cent of stem cankers (probably *Endothia gyrosa*) (Jacovelli 2002).

## Harvesting and oil distillation techniques

### Harvesting and transport

In contrast to the mechanised eucalyptus oil operations in Australia, the African oil operations have always been very labour intensive. Information provided below relates to a typical oil operation in Swaziland that, until 1994, was producing about $60 \, t \, yr^{-1}$ of cineole-rich oil from around 550 ha of *E. smithii*. The plantations were worked on a short-rotation coppice system to maximise the leaf biomass produced.

Harvesting and transport are important elements in the operation of an oil business because the crop is very bulky. Although the first (seedling) crop is cut by chainsaw (due to the large stem size), the subsequent coppice cuts are made with a hatchet. The harvesting teams comprise one cutter, three de-branchers and two bundlers, with each team producing approximately three tonnes of leaf per day. The de-branchers follow, cutting the branches from the main stem and then collecting the leaf and small branch material into manageable bundles (Figure 9.4). The bundles are laid in rows every fifth tree line, allowing field extraction by tractor-trailer units. The transport teams, comprising a tractor driver and four loaders, then load the bundles with pitchforks onto customised trailers. Four-wheel drive tractors are necessary in order to be able to transport on wet days. The trailers have a flat-deck, low-bed construction with high side-rails to support the bulky material and a capacity of approximately 3 t. The leaf is then off-loaded by hand at the distillation plant. Some of the stem-wood left infield is crosscut (by chainsaw) after 3–6 weeks drying and taken to the distillation plant to fuel the boiler.

Many variations on the above are possible and local conditions and preferences will ultimately determine the optimum harvesting and transport methods used. However, for a successful eucalyptus oil venture the following important factors must be considered:

1. The plantation layout – it should enable easy access to pick up the leaf (and minimise damage to the coppice stools).
2. Infrastructure – road access is needed so that the leaf can be brought out of the plantation quickly.

*Figure 9.4* Harvested seedling trees of *Eucalyptus smithii* (twenty-four months) having their branches trimmed for collection of foliage, Swaziland (photo: J. Coppen).

3 Location – the plantations should be reasonably close to the distillation plant to minimise transport costs.
4 Transport – appropriate vehicles are needed to transport bulky leaf material (e.g. high-sided trailer units).
5 The use of bow-saws or chainsaws for felling.
6 The possibility of extracting whole stems and debranching at the mill.
7 By-products – is there a local market for small poles and/or fuelwood?
8 Compaction problems on sensitive sites caused by vehicles going in-field.

Mechanical harvesting has not been considered appropriate or cost-effective under African conditions to date. The principal oil crops in Africa (*E. smithii*, *E. radiata* and *E. dives*) are not as suited to mechanical harvesting as tough, mallee species such as *E. polybractea*.

*Distillation*

Capital investment in distillation equipment does not have to be great – low-cost stills and ancillary equipment are often adequate (see Chapter 6) and are used in small-scale eucalyptus oil operations in many countries. However, that is not to say that the means by which distillation is achieved should be taken lightly and it is shortsighted, particularly for large-scale operations where a significant investment has been made with plantation establishment, to pay scant attention to it. As much good quality oil as possible should be extracted from the leaf in as efficient and cost-effective a manner as possible. Thus, for example, boilers are best installed to produce steam rather than relying on direct firing of water below the charge of leaf. Part of the distillery in which *E. smithii* leaf is being packed into stills at the (then) Shiselweni Forestry Co. eucalyptus oil operation in Swaziland is shown in Figure 9.5.

The reduced prices being obtained for cineole-rich oil from Africa in the late 1980s highlighted the need for operators to increase the efficiency of their operations. The following factors were found to be important in Swaziland:

1 Fuelwood (for the boiler) is a significant cost of the operation, particularly where there is an alternative market for the wood (e.g. for poles or pulp).
2 Fuelwood should always be air-dried to increase its heating efficiency (preferably for a minimum of six months).
3 Spent leaf can be used to reduce the fuelwood requirement, although this requires cutting and a continuous feed mechanism (a 'Dutch' oven).
4 Tight, uniform packing of the leaf in the stills has proved to be very important so that the steam is forced through, not around, the leaf material.
5 The stills must be well sealed to minimise the loss of oil vapour.
6 It is vital to ensure that safety valves are fitted and functioning to relieve any build-up of pressure in tightly packed stills.
7 The distillation must be closely monitored to ensure that maximum yields of oil are obtained from the charge.
8 An efficient condenser will reduce the amount of water required to cool the distillate (oil vapour and steam) emerging from the still.

Oil operations typically work two shifts per day at the distillery and so it is vital to ensure that there is sufficient leaf brought in to maintain production. Typical charge sizes are 2–3 t of fresh leaf. The distillation takes place under low pressure, around 345 kPa, and usually takes

*Figure 9.5* *Eucalyptus smithii* leaf being packed into stills for distillation, Swaziland (photo: J. Coppen).

around 2 h (longer during cool periods). In some operations the spent leaf is returned to the field and used as a mulch; this is a difficult (and costly) operation, as the material has to be chopped and then transported back to the field, but it does have benefits for the site in terms of reducing soil erosion and the need for weed control.

Most African eucalyptus oil is exported in its crude form (typically around 70–75 per cent cineole in Southern African operations) for re-distillation. Rectification is necessary to remove impurities and increase the cineole content of the oil to meet industry standards. The price differential between the crude oil as it emerges from the still and rectified oil, including high purity oil >98 per cent cineole ('eucalyptol'), will determine whether investment in a rectification plant is justified. At the levels of production of most African growers, and the price of oil in the late 1980s and early 1990s, such an investment could not be justified.

## Oil yields

Table 9.1 lists oil yields obtained in Africa, mainly from research data – there is a paucity of information on yields from commercial operations and published research results are often derived from a very limited number of leaf samples. Moudachirou *et al.* (1999) have demonstrated significant seasonal and site effects, particularly the latter in the case of *E. citriodora* growing in Benin. However, detailed records from the operation in Swaziland, together with research results from elsewhere on the continent, provide a reasonable picture of actual (and achievable) oil yields.

Commercial eucalyptus oil production in Africa has taken place mainly from rain-fed, short-rotation coppice crops. Oil yields range from around 50 kg to over 500 kg of oil $ha^{-1} yr^{-1}$. The lower yields are obtained from poor sites in areas marginal for the particular oil species planted. Conversely, the very high yields have been obtained from high potential sites (e.g. deep, fertile soils and high rainfall) where high stocking levels and very short rotations are possible. In the tropical conditions of the former Zaire, for example, two *E. smithii* crops a year were being harvested, producing up to 600 kg of oil $ha^{-1} yr^{-1}$ (Weiss 1997). From 1979 to 1985, *E. smithii* plantations in Swaziland produced an average of 120 kg of oil $ha^{-1} yr^{-1}$. This is from a region with marginal rainfall (mean annual 845 mm) and an often prolonged dry season. These yields compare favourably with those obtained from natural stands of *E. polybractea* in Australia, where around 30–150 kg of oil per hectare may be produced from an 18-month harvest (Chapter 7).

From commercial oil operations in Southern Africa a number of observations regarding yields have been made, although not all have been proven scientifically:

1 Significant intraspecific variation has been found in *E. radiata*, *E. dives*, *E. citriodora* and *E. globulus* subsp. *bicostata* with regard to both oil yield and oil composition.
2 Maximum oil yields are obtained from short-rotation, coppice crops. Yields are significantly less from operations where oil production is secondary to the production of other products.
3 Site quality and climate, particularly nutrient status and rainfall, appear to be extremely important factors affecting oil yields. In general, the higher the site potential, the higher the oil yields.
4 Higher oil yields have been obtained during the warmer, summer months (many African producers suspend production during the coldest periods).
5 Higher oil yields per hectare are obtained from coppice crops than from seedling crops (due primarily to the increased biomass of multi-stemmed coppice).
6 Younger leaves yield more oil than older leaves.

## Storage, packing and quality control

Crude eucalyptus oil, like some other essential oils, is corrosive and since most African oil is exported, good quality, lined steel drums have to be used for storing and transporting it. Clear labelling to meet national and/or international regulations is also necessary.

Most African-produced cineole-rich eucalyptus oil meets the '70/75' requirement (percentage of cineole) in its crude form. In one commercial oil operation in Swaziland all batches of oil were tested for their 1,8-cineole content prior to being exported using the *o*-cresol crystallisation method (Appendix 5). *E. smithii* oil, however, has a poor reputation with some buyers. Robbins (1983) refers to the low cineole content of some batches (below the industry's minimum requirement of 70 per cent cineole) and its acrid odour, due to the presence of aldehydes (although *E. globulus* crude oil can also contain the undesirable isovaleraldehyde). Distillation trials with *E. smithii* oil in Swaziland showed that its quality could be significantly increased by excluding the initial 5 per cent of the distillate (which contains primarily isovaleraldehyde) from the bulked oil intended for export. The higher price that it fetches offsets any loss in overall yield. Buyers have generally expressed a preference for the cineole-rich oils of *E. radiata* and *E. dives* to that of *E. smithii*, though this is not always reflected in the prices offered.

## Research needs

Despite eucalypts having been cultivated for their essential oils in Africa for over fifty years, comparatively little scientific research has been carried out aimed at improving oil quality and yields. Recent work in Southern Africa, however, has at least identified the priority areas for research, namely:

1. The identification of genotypes with superior oil characteristics, particularly from the principal cineole-rich species *E. smithii*, *E. dives* and *E. radiata*. Also of great interest to Africa is *E. camaldulensis*, due to its ability to grow in drier regions.
2. Trial plantings of other, lesser-known eucalypts with desirable oil characteristics (see Table 9.1).
3. The mass-multiplication (using rooted cuttings or micropropagation techniques) of these desirable genotypes to maximise oil yields.
4. With all selection and breeding work, it is most important to maintain a broad genetic base and to screen for resistance to well-known pests and diseases (particularly root rots and stem cankers).
5. A greater understanding is needed of the nutrient balance and, especially, the longer-term effects on various sites of removing much of the above-ground biomass.
6. The cost-effectiveness of various management options needs assessing, particularly fertilising and spacing regimes, longer rotations and multipurpose plantations.

## Future prospects for eucalyptus oil production in Africa

The recent sharp decline in the production of eucalyptus oil from Africa is due to a number of factors. In the late 1980s and early 1990s, low-priced Chinese oil (mostly from *E. globulus*) dominated international trade in eucalyptus oil and lead to prices being depressed. The competitive advantage of African producers was further eroded around this time by escalating production costs, particularly for labour in what is a labour-intensive industry. Added to this were the problems of increased incidence of disease with the main species, *E. smithii*, and the variation that was found in the quality of seed available from natural stands of possible alternative cineole-rich species (particularly *E. radiata* and *E. dives*).

With such problems, the eucalyptus oil industry in Africa might not appear to have a promising future. There are, however, some very positive signs, not least of which are the excellent oil yields that have been achieved in many parts of Africa with a number of oil-rich eucalypts. More is being learnt, too, about the suitability of *Eucalyptus* species for a wide range of sites and objectives. The impressive biomass production of *E. smithii*, in particular, coupled with the fact that this species yields a consistent quality, cineole-rich oil, makes it a species of exceptional potential.

There are also clear indications from research that selection and multiplication of genotypes with superior oil characteristics could result in very large yield improvements, particularly when combined with good silvicultural practices. The existence of some genotypes of *E. radiata* and *E. dives* possessing high quantities of cineole-rich oil is very encouraging. Studies with *E. kochii* and *E. camaldulensis* have found very high family heritabilities for cineole yield and this bodes well for breeding programmes aimed at increasing the sustainable yields of eucalyptus oils (Barton *et al.* 1991, Doran and Matheson 1994). Breeding programmes in which the emphasis has hitherto been on wood quality and productivity are now beginning to look at the volatile oils and this, too, is encouraging. Thus, selection of individuals for the production of oil from *E. urophylla*, *E. grandis* and *E. urophylla* × *E. grandis* has been reported as part of the CTFT afforestation programme in the Congo (Menut *et al.* 1992).

Another positive sign in recent years has been the resurgence in interest (primarily in developed countries) in natural products – eucalyptus oils are in demand for use in a wide range of therapeutic products and applications, including aromatherapy. With careful planning (and suitable incentives), eucalyptus oil production could become a very appealing small business venture for local farmers in Africa. A rural cooperative approach is likely to suit many African countries but, as Weiss (1997) emphasises, considerable institutional and other support would be needed. Crucial elements would involve the supply of quality seed (or seedlings), silvicultural advice, a centralised distillation plant and marketing assistance. Provided the scale of production could justify the additional investment required, downstream processing, particularly rectification, could become attractive to increase the value of the final product.

## References

Abou-Dahab, A.M. and Abou-Zeid, E.N. (1973) Seasonal changes in the volatile oil, cineole and rutin contents of *Eucalyptus camaldulensis* Dehn. and *E. polyanthemos* Schauer. *Egypt. J. Bot.*, 16, 345–348.

Atkinson, P.R., Tribe, G.D. and Govender, P. (1991) Pests of importance in the recent expansion of *Eucalyptus* plantings in South Africa. In A.P.G. Schönau (ed.), *Intensive Forestry: The Role of Eucalypts, Proc. IUFRO Symp.*, Durban, September 1991, Vol. 2, Southern African Institute of Forestry, Pretoria, pp. 728–738.

Barton, A.F.M., Cotterill, P.P. and Brooker, M.I.H. (1991) Short note: heritability of cineole yield in *Eucalyptus kochii*. *Silvae Genetica*, 40, 37–38.

Boland, D.J., Brophy, J.J. and House, A.P.N. (eds) (1991) *Eucalyptus Leaf Oils. Use, Chemistry, Distillation and Marketing*, ACIAR/CSIRO, Inkata Press, Melbourne.

Canty, C. (1991) Controlled release granules protect eucalyptus trees from termite attack. In A.P.G. Schönau (ed.), *Intensive Forestry: The Role of Eucalypts, Proc. IUFRO Symp.*, Durban, September 1991, Vol. 2, Southern African Institute of Forestry, Pretoria, pp. 739–748.

Cardoso do V., J. and Proença da C., A. (1968) Essential oil of *Eucalyptus macarthuri* Deane & Maiden from Angola (in Portuguese). *Garcia de Orta (Lisboa)*, 16, 423–432.

Chalchat, J.C., Muhayimana, A., Habimana, J.B. and Chabard, J.L. (1997) Aromatic plants of Rwanda. II. Chemical composition of essential oils of ten *Eucalyptus* species growing in Ruhande Arboretum, Butare, Rwanda. *J. Essent. Oil Res.*, 9, 159–165.

Chisowa, E.H. (1997) Chemical composition of the leaf oil of *Eucalyptus macarthurii* Dean & Maiden. *J. Essent. Oil Res.*, 9, 339–340.

Coppen, J.J.W. and Hone, G.A. (1992) *Eucalyptus Oils: A Review of Production and Markets*, NRI Bulletin 56, Natural Resources Institute, Chatham, UK.

Cowie, R.H., Logan, J.W.M. and Wood, T.G (1989) Termite (Isoptera) damage and control in tropical forestry with special reference to Africa and Indo-Malaysia: a review. *Bull. Entomol. Res.*, 79, 173–184.

Crous, P.W.R. and Wingfield, M.J. (1991) *Eucalyptus* leaf pathogens in Southern Africa: a national perspective. In A.P.G. Schönau (ed.), *Intensive Forestry: The Role of Eucalypts*, Proc. IUFRO Symp., Durban, September 1991, Vol. 2, Southern African Institute of Forestry, Pretoria, pp. 749–759.

Davidson, J. (1995) Ecological aspects of eucalypt plantations. In K. White, J. Ball and M. Kashio (eds), *Proc. Regional Expert Consult. on Eucalyptus*, Bangkok, October 1993, Vol. 1, Food and Agriculture Organization of the United Nations, Regional Office for Asia and the Pacific, Bangkok, pp. 35–72.

De Medici, D., Pieretti, S., Salvatore, G., Nicoletti, M. and Rasoanaivo, P. (1992) Chemical analysis of essential oils of Malagasy medicinal plants by gas chromatography and NMR spectroscopy. *Flavour Fragr. J.*, 7, 275–281.

De Villiers, F.J. and Naudé, C.P. (1932) *Oils from South African Eucalypts*, The Government Printers, Pretoria.

Dethier, M., Nduwimana, A., Cordier, Y., Menut, C. and Lamaty, G. (1994) Aromatic plants of tropical Central Africa. XVI. Studies on essential oils of five *Eucalyptus* species grown in Burundi. *J. Essent. Oil Res.*, 6, 469–473.

Donald, D.G.M. (1980) The production of cineole from *Eucalyptus*: a preliminary report. *S. Afr. For. J.*, (114), 64–67.

Donald, D.G.M. (1986) South African nursery practice – the state of the art. *S. Afr. For. J.*, (139), 36–47.

Doran, J.C. and Brophy, J.J. (1990) Tropical red gums – a source of 1,8-cineole-rich eucalyptus oil. *New Forests*, 4, 157–178.

Doran, J.C. and Matheson, A.C. (1994) Genetic parameters and expected gains from selection for monoterpene yields in Petford *Eucalyptus camaldulensis*. *New Forests*, 8, 155–167.

Eldridge, K., Davidson, J., Harwood, C. and van Wyk, G. (1993) *Eucalypt Domestication and Breeding*, Clarendon Press, Oxford, UK.

Elkiey, M.A., Darwish Sayed, M., Hashem, F.M. and Assem, K.A. (1964) Investigation of the volatile oil content of certain *Eucalyptus* species cultivated in Egypt. *Bull. Fac. Pharm., Cairo Univ.*, 3, 97–108.

FAO (1981) *Eucalypts for Planting*, FAO Forestry Series No. 11, Food and Agriculture Organization of the United Nations, Rome.

Florence, R.G. (1996) *Ecology and Silviculture of Eucalypt Forests*, CSIRO, Melbourne, Australia.

Govender, P. (1993) Pests of establishment of eucalypts and wattle. In *ICFR Annual Research Report*, Institute of Commercial Forestry Research, Pietermaritzburg, South Africa, pp. 161–170.

Grove, T.S., Thomson, B.D. and Malajczuk, N. (1996) Nutritional physiology of eucalypts: uptake, distribution and utilization. In P.M. Attiwill and M.A. Adams (eds), *Nutrition of Eucalypts*, CSIRO, Australia, pp. 77–108.

Herbert, M.A. (1996) Fertilizers and eucalypt plantations in South Africa. In P.M. Attiwill and M.A. Adams (eds), *Nutrition of Eucalypts*, CSIRO, Australia, pp. 303–325.

ICFR (1997) *Annual Research Report*, Institute of Commercial Forestry Research, Pietermaritzburg, South Africa.

Jacovelli, P.A. (1994) *Silvicultural Manual for Shiselweni Forestry Company*, SFC, Swaziland.

Jacovelli, P.A. (2002) A review of *Eucalyptus* oil species and provenance research at Shiselweni Forestry Company Ltd., Swaziland. *International For. Rev.* (submitted).

Jacovelli, P.A. and Coppen, J.J.W. (1994) The effect of fertilisers on essential oil production and oil quality in *Eucalyptus smithii* R. Baker coppice. Shiselweni Forestry Company Research Note 12/94 (unpubl.).

Logan, J.W.M., Cowie, R.H. and Wood, T.G. (1990) Termite (Isoptera) control in agriculture and forestry by non-chemical methods: a review. *Bull. Entomol. Res.*, 80, 309–330.

Menut, C., Lamaty, G., Malanda-Kiyabou, G. and Bessière, J.M. (1992) Aromatic plants of tropical Central Africa. VIII. Individual selection of *Eucalyptus* for essential oil production in the Congo. *J. Essent. Oil Res.*, 4, 427–429.

Mitchell, M.R. (1989) Susceptibility to termite attack of various tree species planted in Zimbabwe. In D.J. Boland (ed.), *Trees for the Tropics. Growing Australian Multipurpose Trees and Shrubs in Developing Countries*, ACIAR Monograph No. 10, pp. 215–227.

Molangui, T., Menut, C., Bouchet, P., Bessière, J.M. and Habimana, J.B. (1997) Aromatic plants of tropical Central Africa. Part XXX. Studies on volatile leaf oils of 10 species of *Eucalyptus* naturalized in Rwanda. *Flavour Fragr. J.*, 12, 433–437.

Moudachirou, M., Gbenou, J.D., Chalchat, J.C., Chabard, J.L. and Lartigue, C. (1999) Chemical composition of essential oils of eucalyptus from Benin: *Eucalyptus citriodora* and *E. camaldulensis*. Influence of location, harvest time, storage of plants and time of steam distillation. *J. Essent. Oil Res.*, 11, 109–118.

Mwangi, J.W., Guantai, A.N. and Muriuki, G. (1981) *Eucalyptus citriodora*. Essential oil content and chemical varieties in Kenya. *E. Afr. Agric. For. J.*, 46, 89–96.

Ndou, T.T. and von Wandruszka, R.M.A. (1986) Essential oils of South African *Eucalyptus* species (Myrtaceae). *S. Afr. J. Chem.*, 39, 95–100.

Nichol, N. (1992) *Notes for Identification of Common Pathological Problems in South African Forestry*, ICFR Bulletin 23/92, Institute of Commercial Forestry Research, Pietermaritzburg, South Africa.

Osisiogu, I.U.W. (1966) Essential oils of Nigeria. Part 1: A note on the essential oil content of some *Eucalyptus* species growing in Nigeria. *W. Afr. Pharmacist*, 8(1), 8.

Pandey, D. (1995) *Forest Resources Assessment 1990: Tropical Forest Plantation Resources*, FAO Forestry Paper 128, Food and Agriculture Organization of the United Nations, Rome.

Penfold, A.R. and Willis, J.L. (1961) *The Eucalypts*, World Crop Series, Leonard Hill, London and Interscience, New York.

Poynton, R.J. (1979) *Tree Planting in Southern Africa*, Vol. 2, *The Eucalypts*, South African Forestry Dept., Pretoria.

Robbins, S.R.J. (1983) *Selected Markets for the Essential Oils of Lemongrass, Citronella and Eucalyptus*, Tropical Products [now Natural Resources] Institute Report G171, Natural Resources Institute, Chatham, UK.

Samaté, A.D., Nacro, M., Menut, C., Lamaty, G. and Bessière, J.M. (1998) Aromatic plants of tropical West Africa. VII. Chemical composition of the essential oils of two *Eucalyptus* species (Myrtaceae) from Burkina Faso: *E. alba* Muell. and *E. camaldulensis* Dehnardt. *J. Essent. Oil Res.*, 10, 321–324.

Schönau, A.P.G. (1983) Fertilization in South African forestry. *S. Afr. For. J.*, (125), 1–19.

Schönau, A.P.G. (1984) Silvicultural considerations for the high productivity of *E. grandis*. *For. Ecol. Manage.*, 9, 295–314.

Shanks, E. and Carter, J. (1994) *The Organisation of Small-Scale Tree Nurseries: Studies from Asia, Africa and Latin America*, Rural Development Forestry Study Guide No. 1, Overseas Development Institute, London.

Shiel, D. (1985) *Eucalyptus – The Essence of Australia*, Queensbury Hill Press, Melbourne.

Small, B.E.J. (1981) The Australian eucalyptus oil industry – an overview. *Aust. For.*, 44, 170–177.

Sohounhloue, D.K., Dangou, J., Gnomhossou, B., Garneau, F.X., Gagnon, H. and Jean, F.I. (1996) Leaf oils of three *Eucalyptus* species from Benin: *E. torelliana* F. Muell., *E. citriodora* Hook. and *E. tereticornis* Smith. *J. Essent. Oil Res.*, 8, 111–113.

Swart, W.J. and Wingfield, M.J. (1991) *Cryphonectria* canker of *Eucalyptus* species in South Africa. In A.P.G. Schönau (ed.), *Intensive Forestry: The Role of Eucalypts, Proc. IUFRO Symp.*, Durban, September 1991, Vol. 2, Southern African Institute of Forestry, Pretoria, pp. 806–810.

Talalaj, S. (1966) Essential oil of *Eucalyptus citriodora* grown in Ghana. *W. Afr. Pharmacist*, 8(6), 117–118.

Van der Riet, B. de St. J. (1933) Essential oils of certain South African plants. *J. Chemical, Metallurgical and Mining Society of S. Afr.*, 34(3), 78–87.

Viljoen, A., Wingfield, M.J. and Crous, P.W. (1992) Fungal pathogens in *Pinus* and *Eucalyptus* seedling nurseries in South Africa: a review. *S. Afr. For. J.*, (161), 45–51.

Weiss, E.A. (1997) *Essential Oil Crops*, CAB International, Wallingford, UK.

Woodward, S. and Thomson, R.J. (1989) Micropropagation. In *Oxford Forestry Institute Annual Report, 1988*, Oxford, UK, pp. 20–21.

Yousef, E.M.A., Abou-Dahab, A.M., Badawy, E.S.A. and Imam, M.E. (1991) Non-wood products (essential oils) of *Eucalyptus globulus* Labill. tree. *Bull. Fac. Agric. Cairo Univ.*, 42(4, Suppl. 1), 1419–1436.

Zrira, S.S. and Benjilali, B. (1996) Seasonal changes in the volatile oil and cineole contents of five *Eucalyptus* species growing in Morocco. *J. Essent. Oil Res.*, 8, 19–24.

Zrira, S.S., Benjilali, B.B., Fechtal, M.M. and Richard, H.H. (1992) Essential oils of twenty-seven *Eucalyptus* species grown in Morocco. *J. Essent. Oil Res.*, 4, 259–264.

# 10 Cultivation and production of eucalypts in South America

## With special reference to the leaf oils

*Laércio Couto*

## Introduction

FAO country estimates for the areas of plantation eucalypts in 1990 show that Brazil had the second largest area after India, 3.6 million ha (Pandey 1995). The most recent estimate, also from FAO sources, puts the figure at 3.1 million ha (Appendix 2). Although this massive resource is designed to meet the raw material needs of Brazil's forest-based industries such as timber, pulp and charcoal, it has, nevertheless, indirectly influenced the development of the eucalyptus oil industry in the country, at least in the early days. Apart from China, Brazil has been the only other significant producer and exporter of *Eucalyptus citriodora* oil and this arose from the widespread availability of 'waste' leaf from *E. citriodora* planted primarily for charcoal production. Charcoal is used for fuelling the furnaces in the iron and steel industries and in the manufacture of cement and *E. citriodora* has played an important role in the Brazilian economy (Galanti 1987). In addition to *E. citriodora* oil, oils from *E. globulus* and *E. staigeriana* are produced in Brazil.

Production of eucalyptus oil elsewhere in South America is small compared to that in Brazil and although brief mention is made later in this chapter to Chile, Bolivia, Paraguay and some other countries, the bulk of the discussion concerns methodologies employed in Brazil. Much of what is described will, of course, be applicable elsewhere. Details are given in the form of a case study relating to a company which, until recently, produced *E. citriodora* oil in Minas Gerais state. Supplementary information is provided from other sources concerning this oil and those from *E. globulus* and *E. staigeriana* produced in Brazil.

## Historical review

### Introduction of eucalypts to Brazil and end use

Although eucalypts were planted in the Botanical Gardens of Rio de Janeiro as early as 1824, it was Edmundo Navarro de Andrade who established the first eucalypt plantations in the first decade of the twentieth century (Couto *et al.* 2000). On returning to Brazil from his studies of agronomy in Portugal he brought with him seeds of *E. globulus* for planting. He worked for a railroad company to provide firewood for the steam engines and wood for sleepers and the experimental plots were so successful that they led to over 200 species of *Eucalyptus* being introduced into Brazil.

From 1909 to 1965, about 470,000 ha of eucalypts were planted, 80 per cent of them in São Paulo state and intended mainly as a substitute for native woods for use as fuelwood. Other energy sources such as good quality coal were not available in Brazil. In 1948, the first eucalypt plantations were established in Minas Gerais to provide wood for charcoal for use by the iron and steel industries. Although *Pinus* spp. were also being introduced, partly in response to depletion of the native *Araucaria* as a source of high quality timber, the high yields and quicker returns from short-rotation eucalypts saw them become the dominant plantation species. A federal law passed in 1966, and a further one in 1970, gave considerable financial incentives to companies to plant trees, and accelerated this process of reforestation. The consumption of charcoal by the cement, iron and steel industries has continued to increase although this has been tempered by lower prices of imported coal since 1997 and other economic factors which are beginning to make alternative forms of fuel more attractive.

A further impetus to the planting of eucalypts was the increasing use of short fibre pulp for paper making in the 1970s. The favourable conditions for growth in Brazil, the development of appropriate silvicultural techniques and the advantages gained by using cloned planting stock and hybrids led to the massive plantations that now feed Brazil's pulp mills. Aracruz Celulose S.A. alone has 132,000 ha of eucalypt plantations. In the 1980s, coinciding with growing environmental concerns worldwide on the loss of natural forests and other related issues, laws were passed which removed the financial incentives to the forestry sector provided by the earlier legislation. Today, new plantations are mostly established on previously harvested areas rather than on new land.

Brazil's eucalypt plantations extend from the northern states of Pará and Maranhão to the eastern and southern states of Bahia, Minas Gerais, Espírito Santo, São Paulo, Paraná, Santa Catarina and Rio Grande do Sul. The species grown for pulp are mainly *E. grandis*, *E. urophylla* (and hybrids of these) and *E. saligna*, that is, species containing little or no leaf oil, together with *E. dunnii* and a few oil-yielding species such as *E. globulus*, *E. viminalis* and *E. tereticornis*. Although *E. citriodora* is still the dominant species for charcoal production, there has been some replacement of it by other, faster growing species of eucalypt.

## Eucalyptus oil production in Brazil

Production of essential oils from eucalypts in Brazil started during World War II. It began as a result of a collapse in international trading of citronellal-rich oil produced in Java from citronella. At that time there were, in Brazil, some *E. citriodora* and *E. globulus* plantations and these were used by the first local companies to produce essential oils, mainly citronellal-rich oil from *E. citriodora*. São Paulo state was the most important area for essential oil production and in the 1970s Brazil became the biggest producer of *E. citriodora* oil in the world. Later, Minas Gerais, Espírito Santo, Mato Grosso do Sul and Bahia contributed to the increase in Brazilian production of this type of oil (Romani 1972). *E. citriodora* oil is employed in whole form for fragrance purposes but is also used as a source of citronellal. This in turn is used either as an aroma chemical or for conversion to hydroxycitronellal and other compounds used in perfumery.

*E. globulus*, one of the first species to be introduced in Brazil, is a source of cineole-rich medicinal oil. However, while *E. citriodora* is widely distributed from the north to the south of the country, *E. globulus* is highly dependent on specific climatic and edaphic conditions. Furthermore, *E. globulus* in Brazil is used mainly as a source of leaf oil whereas *E. citriodora* is grown widely for timber and charcoal production and so forms an abundant, ready-made source of 'waste' leaf suitable for distillation. Today, the main eucalyptus oil distilleries in Brazil are located in São Paulo, Minas Gerais, Bahia and Mato Grosso do Sul and use *E. citriodora* leaf as the principal, or sole, raw material.

Of the many hundreds of species of *Eucalyptus* which exist, fewer than twenty have ever been exploited commercially for oil production (Boland *et al.* 1991). For a country as large as Brazil, with continental dimensions and a very large edaphic and climatic diversity, it is perhaps not surprising that of the half dozen most important species for oil production worldwide, two are utilised in this way in Brazil, *E. citriodora* and *E. globulus*. A third species, *E. staigeriana*, is also grown for oil and Brazil is the only such source. *E. citriodora* and *E. globulus* furnish oils in approximately 1–1.5 per cent yield (fresh weight) containing citronellal (65–85 per cent) and 1,8-cineole (around 65 per cent), respectively. *E. staigeriana* oil, in yields of 1.2–1.5 per cent, contains a more complex mixture of terpenes (see later), but with citral as an important fragrance component.

Today, the production of eucalyptus oil in Brazil is carried out by a small number of medium-to-large companies, together with some smaller ones. The technology used in the distillation of the oil is virtually the same for all of them and the main difference is in the way in which they obtain their raw material. Larger companies have their own eucalypt plantations, established specifically for oil production, while the smaller ones rely on 'waste' leaf bought from small landowners or from eucalypt plantations managed for the production of wood (for poles and timber as well as charcoal and firewood). Both large and small companies sometimes enter into joint ventures with those involved in *E. citriodora*-based charcoal production – the leaves collected from harvested areas are sent to the distilleries and a percentage of the profits from the production of oil is paid back for the use of the raw material.

## Oil characteristics

The general characteristics of the oils from the commercially important oil-bearing eucalypts have been described elsewhere in this volume but data relevant to South America are presented in Table 10.1 (see also Baez *et al.* 1992).

*E. staigeriana* oil, which is used in perfumery in whole form, has, as implied by the variable but relatively low figures for citral shown in Table 10.1, a complex composition. Analysis of one such commercial sample found 26.8 per cent limonene, 10.8 per cent terpinolene, 9.6 per cent neral and 12.5 per cent geranial (i.e. 22.1 per cent citral), 4.7 per cent methyl geranate, 4.6 per cent geranyl acetate and 4.7 per cent geraniol (Coppen unpubl.).

Specifications provided by a leading producer of *E. globulus*, *E. citriodora* and *E. staigeriana* oils in Brazil are a minimum 70 per cent 1,8-cineole, minimum 70 per cent citronellal and minimum 20 per cent citral, respectively.

## *Eucalyptus citriodora*: a case study of its cultivation and distillation in Brazil

In some regions of São Paulo state, the lower branches of oil-bearing eucalypts are periodically cropped for oil production, leaving the stems standing for further growth and future use for timber. In parts of Minas Gerais and Bahia, the availability of suitable 'waste' leaf from eucalypt plantations managed for charcoal production is taken advantage of to produce oil when the tree is felled at the end of the rotation. In both cases eucalyptus oil production is a secondary activity. There are also companies, however, who grow eucalypts specifically for oil and, here, the trees are planted and managed more intensively, with much shorter rotations than the normal ones of around seven years.

The company used as a basis for this case study was founded in 1990 and established itself in the western part of Minas Gerais, where it acquired 2500 ha of land in the savanna region. Five million trees of *E. citriodora* were planted specifically for oil production. The company built a

Table 10.1 Yields and characteristics of leaf oils obtained from eucalypts growing in South America[a]

| Species | Country | Data[b] | Main constituent (relative abundance, %) | Oil yield[c] (%) | Reference |
|---|---|---|---|---|---|
| E. camaldulensis | Argentina | r | p-Cymene (31.2) | 0.4 | De Iglesias et al. (1977) |
| | Chile | r | p-Cymene (31.4) | 0.2 | Erazo et al. (1990) |
| E. cinerea | Argentina | r | 1,8-Cineole (69.0) | 0.7 (air-dry) | De Iglesias et al. (1980) |
| | Brazil | r, r | 1,8-Cineole (76.3, 61.0–62.8) | 1.4, 4.1–8.2 (dry) | Silva et al. (1978), Moreira et al. (1980) |
| E. citriodora | Brazil | c | Citronellal (75.7) | — | Coppen unpubl. |
| | Chile | r | Citronellal (85.1) | 3.0 | Erazo et al. (1990) |
| | Uruguay | r | Citronellal (59.2) | 1.3 | Dellacassa et al. (1990) |
| E. globulus subsp. globulus | Brazil | c | 1,8-Cineole (73.6) | — | Coppen unpubl. |
| | Chile | c | 1,8-Cineole (60–75) | 1.2–1.7 | Anon. (1987) |
| | | r | 1,8-Cineole (60.3) | 1.0 | Erazo et al. (1990) |
| | Uruguay | r | 1,8-Cineole (64.5) | 0.8 | Dellacassa et al. (1990) |
| E. piperita | Brazil | r | Piperitone (7.5–40.5) | 0.2–2.1 | Pinto et al. (1982) |
| | | r | 1,8-Cineole (47.4) | — | Garrone (1987) |
| E. staigeriana | Brazil | c, c | Citral (37.2, 22.1) | —, — | Porsch et al. (1965), Coppen unpubl. |
| | Brazil | r | Citral (10.1–63.0) | 0.3–2.5 | Pinto et al. (1976) |
| E. viminalis | Uruguay | r | 1,8-Cineole (43.6) | 0.4 | Dellacassa et al. (1990) |

a The species listed are those where there is commercial production of the oil in South America (not necessarily in the country specified) or where there is, or has been, production elsewhere in the world. Some of the references cited refer to other species, in addition to those indicated here.
b Indicates whether data are research results (r) or relate to commercial production (c).
c Yields are on a fresh weight basis unless otherwise indicated.

distillery (annual production capacity 360 t oil) and infrastructure such as laboratories, offices and a village for its 250 employees.

## Silviculture

### Site selection

*E. citriodora* does not present good leaf biomass yield in areas subject to strong winds since they desiccate the leaves and lead to leaf drop. It is very important, therefore, to select sites for its cultivation which are not exposed to winds. The sites should also be located in flat or gently undulating areas to facilitate mechanised silvicultural and harvesting operations. The stands are best kept below 40–50 ha in size and planted with the contours when necessary.

### Land preparation

*E. citriodora* is very demanding on soil quality and does not grow well where the pH is lower than 5.5 (such as lateritic soils). It is therefore important to analyse soil samples for each plot before planting and to correct soil acidity where necessary. Nutrients such as phosphorus and potassium, and micronutrients such as boron and zinc, should also be added if required. With the annual harvesting of leaf, and the consequent continued removal of plant biomass from the land, occasional foliar analysis is also desirable in order to monitor and rectify any depletion of nutrients and micronutrients.

In the area considered in this case study there were sandy soils and the first operation consisted of locating contour lines in the field to avoid erosion, mainly during the rainy season. To improve the physical and chemical properties of the soil it was common practice to plant legumes one year before planting the eucalypts. These plants are then incorporated into the soil during land preparation. After demarcation of the stands and the building of roads and fire breaks, 1000 kg of lime and 500 kg of natural phosphate per hectare are applied five months and one month, respectively, before planting the eucalypt seedlings. This is followed by arrowing and ploughing the soil to a depth of 30 cm and furrowing to 60 cm. Finally, the soil is levelled with a lighter arrow and the planting lines in contour are marked with furrows 30 cm deep. Before arrowing and ploughing, leaf cutting ants are controlled using sulfluramid-based pesticides.

*Establishment*

Before planting the seedlings a final check is made on the absence of leaf cutting ants. The spacing between rows is 2.8 m and the distance between plants in the same row is 0.75 m. A total of 300 kg of NPK (20:20:20) plus 6 per cent of boron is applied per hectare, distributed along the furrows where the planting holes are located. Planting is carried out during the rainy season as soon as moisture conditions in the soil are satisfactory. At that time, 2 g of pesticide are placed in each planting hole to prevent attack by termites. Replanting of seedlings which do not survive is carried out fifteen days after the initial planting to avoid heterogeneity in the size of the trees later on. New applications of fertiliser are made thirty and sixty days after planting using 50 g of NPK (20:0:20) per seedling. For each 40–50 ha of a new eucalypt plantation it is necessary to have a labourer to tend the site and to prevent attack of the seedlings by leaf cutting ants.

In the region applicable to this case study the planting months are generally November and December, coincident with the rainy season for the western part of Minas Gerais. High temperature and air humidity at this time make it necessary to discontinue harvesting of the leaves, otherwise fermentation of the raw material is promoted and low quality essential oil is produced. Advantage is taken of this 'down time' to undertake maintenance of the distillery and to attend to other necessary work in the eucalypt stands.

*Maintenance*

The most important silvicultural treatment of eucalypt plantations established for essential oil production is weeding. In the study in question this involves dealing mainly with grasses such as *Brachiaria brizantha*. Initially, control was achieved by applying 4 l/ha of Roundup™ twice a year. This resulted in a relatively low tending cost and a good level of soil conservation. Later on, herbicide application was replaced by renting the forest land to local farmers so that the pastures could be used by their cattle. Rental was paid on the basis of 10 per cent of the dollar value of each 15 kg of living animal per month. Besides controlling the grass and other weeds which might compete with the eucalypts, the cattle provide an additional income for the company. Two heads of cattle are enough to control the weeds and avoid the use of 8 l of Roundup per year. The use of cattle to control the weeds, mainly the grasses, was also valuable in reducing the fire risk.

As mentioned earlier, periodic soil and foliar analyses are carried out for individual stands and fertilisers applied when necessary, usually at the beginning of the rainy season. Control of leaf cutting ants is carried out every month of the year.

*Rotation*

The eucalypt stands are grown under an intensive, coppice system of management, solely for the purpose of oil production, and harvesting takes place once a year in different months. There are therefore always stands at different ages and stages of development. There are no special treatments for the coppices or for the stumps but the objective is to obtain the biggest possible production of leaf biomass. The trees never reach more than 5 m in height before they are cut and so irrigation can be used if necessary. The first harvest is usually taken at around 15–16 months (about March) with subsequent harvests at approximately twelve-month intervals. After eleven or twelve years the original stumps are removed and new seedlings planted.

*Silvicultural practice elsewhere in Brazil* [1]

Propagation is from seed which is germinated over a period of 2–3 weeks; the seedlings spend a further three months in the nursery before planting out. Propagation by cuttings has not been successful. *E. globulus* is shy to seed in Brazil and most seed for planting purposes is bought from Australia. Seed of *E. citriodora* and *E. staigeriana* is collected from mature trees within the company's plantations.

As in the case study, contour planting is practised in order to avoid soil erosion. Spacing is 3.5 m × 0.5 m. Fertiliser application is usually too expensive to carry out but 'spent' leaf from the distillery is returned to the fields to provide some replenishment of nutrients (see later, Figure 10.4). Undergrowth around the trees is cleared annually.

*Harvesting*

Initially, the company used to employ manual labour with chain saws to cut the small trees and to remove the branches with the leaves for transportation to the distillery. No stem biomass was utilised in any subsequent stage of the processing and the energy required to produce steam was derived from the residues of distillation. Later, the company adapted a circular saw fitted to a tractor to mechanise the harvesting operation and so reduce labour costs. The potential hazards associated with the use of eight chain saws were also thereby eliminated. With this new equipment the trees are cut leaving a stump 0.8–0.9 m in height. Tests showed that this stump height gave the best results in terms of percentage of subsequent sprouting. About a third of a hectare per hour can be harvested. Each year thereafter the cut is made 2–3 cm above the previous one, resulting in a loss of only 2–3 per cent of the coppicing per year.

Loading of the trucks which take the harvested material to the distillery is also mechanised and approximately 0.6 ha of cut biomass can be loaded per hour. The annual yield of biomass is around 20 t/ha.

---

[1] Supplementary information provided here and later relates to a company which produces eucalyptus oil from *E. globulus*, *E. citriodora* and *E. staigeriana*. All three species are grown specifically for oil on a short-rotation, coppice system of management.

Approximately 30 per cent of the harvesting operation involves manual labour to select the branches and leaves to go for distillation. The calorific value of this spent material after distillation – when used to fuel subsequent distillations – is smaller, and there is less of it, than the residue from the small whole trees which result from the mechanised part of the harvesting. These are chipped at the distillery (see below). In both cases, any surplus residue not fed to the boiler is either sold locally or returned to the fields as organic fertiliser.

Elsewhere in Brazil, the harvesting of coppice-managed eucalypts for oil is sometimes carried out more frequently than every twelve months. One company clips the side branches a year after planting and takes the first full harvest at eighteen months. Thereafter, harvesting of branches and foliage is usually carried out three times in two years. The stems remaining after recent harvesting of *E. citriodora* foliage are shown in Figure 10.1. In the fourth year most of the stems are cut approximately 60 cm from the ground and allowed to sprout; 10 per cent are left uncut and allowed to grow for timber production. The first harvests from the cut stems are taken after twelve months, with subsequent ones as before. Some trees have been harvested for thirty years or more. Most of the harvesting is carried out manually but mechanical grabbers go between the rows loading the heaped foliage onto tractor-trailers for transport to the distillery (Figure 10.2).

*Figure 10.1 Eucalyptus citriodora* being grown specifically for oil, São Paulo state (photo: J. Coppen).

*Figure 10.2* Harvested *Eucalyptus globulus* being loaded for transport back to the distillery (photo: J. Coppen).

## Distillation

Once the trucks containing the harvested biomass arrive back at the distillery each load is weighed and then fed into a chipper to furnish a mix of chipped eucalypt wood and leaves. The chipper has the capacity to process 20 t/h of biomass. This biomass is then loaded into the stills for distillation. Some loss of moisture from the charge before being put into the stills is beneficial since the oil concentration is thereby increased. However, the biomass should not be exposed to the air for too long since there will be some loss of oil through evaporation and/or enzymic oxidation or decomposition.

The distillery in this case study has eight stills, each with a nominal capacity of 1.5 t of mixed, chipped biomass (stem, branches and leaves) and 1.0 t of biomass comprising whole small branches and leaves. When loading the stills with unchipped material it is essential to pack firmly and uniformly so as to avoid air gaps and subsequent channelling of steam when distillation is in progress. This is usually achieved by the workers who load the stills stepping inside them and stamping the biomass down; at the same time a little steam is trickled through the charge.

Once loaded, detachable, insulated lids are attached to the stills, forming a good, steam-tight seal, and distillation proceeds in the manner described in Chapter 6. Steam is generated from a separate boiler at 90–110°C and uses 'spent' leaf from previous distillations as fuel. Although multi-tubular condensers are the most efficient, the company in question uses a simple coiled pipe running through a cold water tank to condense the oil/steam vapours. Each distillation takes about 1.25 h to complete and yields about 1.0–1.25 per cent of oil. This represents a yield of approximately 200 kg of oil per hectare of *E. citriodora*.

For each distillation charge, the oil is analysed to check citronellal content. The oil is allowed to stand to cool for 48 h and the fully separated oil is then transferred to separate containers for storage.

*Figure 10.3* Pair of stills sharing a single lid and condenser (photo: J. Coppen).

*Figure 10.4* Spent eucalyptus leaf from the distillery being loaded onto lorries for return to the fields to serve as slow-acting fertiliser (photo: J. Coppen).

Distillation by another Brazilian company utilises 1- or 2-t stills operating in pairs, with one lid and condenser per pair (Figure 10.3). In this way, one still can be loaded with leaf while the other, already packed, is being distilled. At the end of this distillation the lid is removed and transferred to the second still, where distillation commences, while the first still is unloaded and repacked with fresh material. Distillation usually takes about one hour for *E. citriodora* and a little longer for *E. globulus* and *E. staigeriana*. With year-round harvesting and distillation, average oil yields equate to approximately 140 kg/ha for *E. globulus* and 100 kg/ha for *E. citriodora* and *E. staigeriana*. Spent leaf from the distillation is returned to the fields to act as a slow-acting fertiliser (Figure 10.4).

## Eucalyptus oil production elsewhere in South America

The most recent estimates of eucalypt plantations (see Appendix 2) show that Argentina, Peru, Chile and Uruguay all have substantial areas – 245,000 ha or more. According to Brown (2000), *Eucalyptus* species accounted for 90 per cent of the forest plantation area in Peru in 1995 and 80 per cent of the area in Uruguay. However, none of these countries have ever produced eucalyptus oil on a scale to match Brazil's. The feasibility of exploiting *E. globulus* for oil production in Peru has been investigated (Cano Vela 1980 and Ocana Vidal 1983) but no significant commercial production is known to have developed as a result. Research on oil-bearing eucalypts has been undertaken in Argentina (e.g. Mizrahi *et al*. 1972, Argiro and Retamar 1973, De Iglesias *et al*. 1977, 1980, Mizrahi *et al*. 1997) and there has been some commercial production of oil.

Chile, Bolivia, Paraguay, Uruguay, Argentina and Colombia have all produced oil at one time or other but no reliable data relating to current production are known. Chile and Bolivia produce cineole-rich oil from *E. globulus*, while Paraguay produces *E. globulus* and *E. citriodora* oils. Chile, Paraguay and Argentina all exported oil in 1999 and/or 2000.

### Chile

Although planting of eucalypts on an industrial scale began in Chile in the 1930s it was not until the late 1980s that rates of planting rapidly increased, reflecting the new-found enthusiasm for *Eucalyptus* as a source of wood for pulp and paper making. By 1992, planting had reached almost 41,000 ha per year and estimates at about that time put the total area under eucalypts at 180,000 ha (Davidson 1995, see Appendix 2). More recent estimates put the figure at 245,000 ha (Appendix 2) and it is predicted that there may eventually be 300,000 ha of eucalypt plantations in Chile (Jayawickrama *et al*. 1993). Chilean plantings are almost entirely of *E. globulus*, mostly in the Valparaiso and Bio–Bio Administrative Regions in the centre of the country, although *E. camaldulensis* and non-oil bearing species such as *E. nitens*, *E. delegatensis* and *E. regnans* are starting to be planted, according to climatic preferences.

Such a massive resource of leaf biomass from a species recognised for its oil quality, *E. globulus*, invites exploitation. Chilean production of eucalyptus oil and purified eucalyptol (1,8-cineole) intended for export began in the 1980s (Anon. 1984a, b) and new investment was still taking place in 1991 (Anon. 1991). Oil containing 60–75 per cent 1,8-cineole is distilled in yields of 1.2–1.7 per cent from 'waste' leaf. Yields of such leaf from trees planted for wood production have been estimated at 8–10 t/ha, equivalent to approximately 100–150 kg/ha of oil (Anon. 1987). Other species of eucalyptus have been examined as sources of oil (e.g. Erazo *et al*. 1990) but their limited plantings have meant that none has ever come close to matching *E. globulus* as a commercial source.

## Bolivia

Published accounts of Bolivian oil production are scant but Canadian assistance in the 1980s led to the establishment of farming cooperatives in Cochabamba, whose aims were to produce essential oils from locally grown plants (Eberlee 1991). Three oils were chosen for production – eucalyptus, mint and lemongrass – and by 1991 ten cooperatives had been formed, each using a 5 t capacity still for distillation. In the case of eucalyptus, the nearby university rectified the crude oil and the final product was then sold to various Bolivian end-users. Although carried out on a relatively small scale, it well illustrates the way in which such an operation, if properly organised, can generate much-needed cash income for rural families.

## Research and development

Genetic improvement and the establishment of seed orchards for the production of improved planting stock are vital for the maintenance of a competitive eucalyptus oil industry but more research is needed to quantify the likely gains. Xavier *et al.* (1993) analysed the genetic variability of forty-two-month old Brazilian-grown *E. citriodora* in terms of oil and citronellal content and found that selection within families provided greater gains than selection among progenies. The need for careful provenance selection is illustrated by the analyses of Argentinean and Chilean *E. camaldulensis* (De Iglesias *et al.* 1977, Erazo *et al.* 1990): both sets of plant material yielded oils richer in $p$-cymene than 1,8-cineole (Table 10.1). If this species were to be considered as a potential oil-producing species (rich in cineole) then seed from the well-known Petford provenance would be better planted and tested.

The Department of Forestry of the Federal University of Viçosa, through the Society for Forestry Research, has always been committed to this kind of research and several collaborative studies have been undertaken with the company which has been the subject of the case study above. These have included the genetic improvement of *E. citriodora* for oil production and the evaluation of eighteen other eucalypt species and provenances with potential for essential oil production. In addition to the three species that are currently utilised in Brazil, *E. camaldulensis* has been found to have great potential.

## Acknowledgements

John Coppen is thanked for providing information additional to that of the author's experience in Brazil and for information on eucalyptus oil production elsewhere in South America.

## Reference

Anon. (1984a) New prospects in eucalyptus production. *Chilean For. News* (June), 9–10.
Anon. (1984b) Eucalyptol to be exported by a Chilean company. *Chilean For. News* (October), 6–7.
Anon. (1987) New company appears on the eucalyptus oil scene. *Chilean For. News* (April), 13–14.
Anon. (1991) The fruits of the eucalyptus. *Chilean For. News* (August), 4–5.
Argiro, A.I. and Retamar, J.A. (1973) Essential oils from Tucuman province: *Eucalyptus citriodora* (in Spanish). *Arch. Bioquim. Quim. Farm. Tucuman*, 18, 29–37.
Baez, C.M., Escobar, R.R., Gonzalez, O.C. and Vasquez, V.O. (1992) Mineral nutrition of *Eucalyptus globulus* subsp. *globulus* plants in relation to yield and quality of cineole (in Spanish). *Agricultura Tecnica Santiago*, 52, 475–479.
Boland, D.J., Brophy, J.J. and House, A.P.N. (eds) (1991) *Eucalyptus Leaf Oils: Use, Chemistry, Distillation and Marketing*, ACIAR/CSIRO, Inkata Press, Melbourne.

Brown, C. (2000) *The Global Outlook for Future Wood Supply from Forest Plantations*, FAO Global Forest Products Outlook Study Working Paper Series, GFPOS/WP/03, Food and Agriculture Organization of the United Nations, Rome.

Cano Vela, M.V. (1980) Evaluation and Possibilities for Industry of the Essential Oils of *Citrus Aurantifolia, C. Sinensis* and *Eucalyptus Globulus* in Peru (in Spanish), Thesis, Univ. Nacional Agraria, La Molina, Lima.

Couto, L. and Dube, F. (2001) The status and practice of forestry in Brazil at the beginning of the 21st century: a review. *For. Chronicle*, 77(5), 817–830.

Davidson, J. (1995) Ecological aspects of eucalypt plantations. In K. White, J. Ball and M. Kashio (eds), *Proc. Regional Expert Consult. on Eucalyptus*, Bangkok, October 1993, Vol. 1, Food and Agriculture Organization of the United Nations, Regional Office for Asia and the Pacific, Bangkok, pp. 35–72.

De Iglesias, I.A.D., Catalan, A.N.C., Lascano, L.J. and Retamar, A.J. (1980) The essential oil of *Eucalyptus cinerea*. *Riv. Ital. E.P.P.O.S.*, 62, 113–115.

De Iglesias, D.I.C., de Viana, M.E.L. and Retamar, A.J. (1977) Essential oil of *Eucalyptus rostrata* (*E. camaldulensis*) (in Spanish). *Riv. Ital. E.P.P.O.S.*, 59, 538–540.

Dellacassa, E., Menéndez, P., Moyna, P. and Soler, E. (1990) Chemical composition of *Eucalyptus* essential oils grown in Uruguay. *Flavour Fragr. J.*, 5, 91–95.

Eberlee, J. (1991) A well-oiled industry. *IDRC Reports* (July), 20.

Erazo, S., Bustos, C., Erazo, A.M., Rivas, J., Zollner, O., Cruzat, C. and Gonzalez, J. (1990) Comparative study of twelve species of *Eucalyptus* acclimatised in Quilpué (33° L.S., 5th region, Chile). *Plantes Med. Phyto.*, 24, 248–257.

Galanti, S. (1987) *Produção de Óleo Essencial do Eucalyptus Citriodora no Municipio de Torrinha, Estado de São Paulo*, Monograph, Univ. Federal de Viçosa, Brazil.

Garrone, W. (1987) *Eucalyptus piperita* introduced in Brazil (in Italian). *Essenze Deriv. Agrum.*, 57, 630–633.

Jayawickrama, K.J.S., Schlatter, V.J.E. and Escobar, R.R. (1993) Eucalypt plantation forestry in Chile. *Aust. For.*, 56, 179–192.

Mizrahi, I., Collura, A.M. and Mendonza, L.A. (1972) Primary evaluation of the characteristics of *Eucalyptus fruticetorum* F.v.M. in Argentina (in Spanish). *IDIA Argentina*, 292(April), 68–74.

Mizrahi, I., Rodriguez-Traverso, J., Juarez, M.A., Bandoni, A.L., Muschietti, L. and van Baren, C. (1997) Composition of the essential oil of *Eucalyptus dunnii* Maiden growing in Argentina. *J. Essent. Oil Res.*, 9, 715–717.

Moreira, E.A., Cecy, C., Nakashima, T., Franke, T.A. and Miguel, O.G. (1980) The essential oil of *Eucalyptus cinerea* F.v.M. acclimated in the state of Parana, Brazil (in Portuguese). *Trib. Farm.*, 48, 44–53.

Ocana Vidal, D.J. (1983) Determination of essential oil yield and cineole content of *Eucalyptus globulus* Labill. of the Callejon of Hauylas, Peru (in Spanish), Thesis, Univ. Nacional Agraria, La Molina, Lima.

Pandey, D. (1995) *Forest Resources Assessment 1990: Tropical Forest Plantation Resources*. FAO Forestry Paper 128, Food and Agriculture Organization of the United Nations, Rome.

Pinto, A.J.D'A., Souza, C.J. and Donalisio, M.G.R. (1976) Selection of eucalyptus with emphasis on yield and citral content of the essential oil (in Portuguese). *Bragantia*, 35, 115–118.

Pinto, A.J.D'A., Souza, C.J. and Donalisio, M.G.R. (1982) Cultivation of *Eucalyptus piperita* in Brazil (in French). In *Proc. 8th Internat. Congr. Essential Oils*, Grasse, October 1980, pp. 91–92.

Porsch, F., Farnow, H. and Winkler, H. (1965) The most important constituents of the oil of *Eucalyptus staigeriana*. *Dragoco Report*, 12, 175–177.

Romani, R.A. (1972) *Óleos Essenciais de Eucalipto*, Escola Superior de Agricultura Luiz de Queiroz, Univ. São Paulo, Piracicaba, Brazil.

Silva, G.A., Siqueira, N.C., Bauer, L., Bacha, C.T.M. and Sant'ana, B.M.S. (1978) The essential oil of *Eucalyptus cinerea* F.v.M., Myrtaceae, of Rio Grande do Sul (in Portuguese). *Rev. CCC-UFSM*, 6(3–4), 61–64.

Xavier, A., Borges, R. de C.G., Pires, I.E. and Cruz, C.D. (1993) Genetic variability of essential oil and progeny growth of *Eucalyptus citriodora* Hook. half-sibs (in Portuguese). *Rev. Árvore*, 17, 224–234.

# 11 Cultivation and production of eucalypts in India

## With special reference to the leaf oils

*S.S. Handa, R.K. Thappa and S.G. Agarwal*

## Eucalypts in India

### Historical aspects

Eucalyptus was introduced into India as an ornamental tree in the late eighteenth century by the ruler of Mysore state, Tippu Sultan, who had a great love for gardening. He planted about sixteen species of *Eucalyptus*, given to him by his French friends, on the Nandi Hills (then known as Nandi-Durga) of Karnataka in the period 1782–1802 (Sreenivasa Murthy and Ramakrishnan 1978, Shyam Sunder 1979). Some of these trees have survived and in 1984, one of the *E. tereticornis* trees from Nandi Hills, with a height of 60 m and girth of 4.6 m, was found to have an age of 194 years. Regular planting of *Eucalyptus* in India started in 1856. In 1860, various species were planted on a trial basis in northern India: at Lucknow, Saharanpur, Dehra Dun and Agra in Uttar Pradesh and at Madhopur in Punjab. *E. globulus* was introduced in the Nilgiris of Madras Presidency (now Tamil Nadu state) by Captain Dunn and Captain Cotton (Chaturvedi 1976). Many of these trees, too, still survive, especially along the roadsides.

In the twentieth century, although sporadic attempts were made to grow *Eucalyptus* alongside roads and in gardens it was not until the 1950s that serious efforts were made to plant it, following widespread destruction of *Casuarina* forests on the Nandi Hills, Mysore state, by the fungus *Trichosporium vesiculosum* (Shyam Sunder 1986). Large-scale plantations of *Eucalyptus* in these areas were raised from seed collected from plants grown in the Chickaballapur range near the Nandi Hills. These plants did not clearly resemble any of those grown in the nearby areas and were named *E. chickaballapur*, later changed to '*Eucalyptus* hybrid' or 'Mysore gum'. This form was discovered, and its planting promoted, by M.A. Muthana, Chief Conservator of Forests, Mysore state, between 1948 and 1959 (Rajan 1987). It is believed that the tree is a form of *E. tereticornis* having occasional hybridisation with *E. robusta* (FAO 1981). Originally there was evidence of some admixture of the seed with that of *E. camaldulensis* but it has now stabilised as a separate entity over many generations. Brooker (unpubl.) considers it to be locally naturalised *E. tereticornis*.[1]

The 1960s saw the introduction of *Eucalyptus* hybrid on a massive scale throughout India, from the farthest north to the south. The plantations raised in Karnataka proved it to be the most adaptable of the eucalypts to the varied climatic and edaphic factors which exist (Shyam Sunder 1979, Rajan 1987). It was found to be more drought resistant than the original Australian provenances of *E. tereticornis* and an excellent soil binder, and this has encouraged its introduction into other parts of the world (e.g. Sudan).

---

1 Hereafter in this chapter, because of the very widespread use of the name in India, it is referred to as '*Eucalyptus* hybrid'. However, its near identity with *E. tereticornis* should be kept in mind.

## Present plantations

*Eucalyptus* hybrid has become the most popular and universal eucalypt in all the low rainfall areas of India due to its unique ability to adapt to the many different agroclimatic environments, its ability to withstand drought over long periods, and its excellent coppicing powers. It is grown on the largest scale in most states of the country, principally for meeting the requirements of the pulp and paper industry (Chaturvedi 1976, FAO 1981). Although research aimed at utilising its leaf oil has been carried out (see below) it has not so far been exploited for this purpose on a commercial scale. Oil production presently rests with *E. globulus* and *E. citriodora*.

Since the introduction of *Eucalyptus* into India there have been trials on 170 species, varieties and provenances in many parts of the country (Bhatia 1984), at elevations up to 2200 m and with an annual rainfall range of 400–4000 mm (Chaturvedi 1976). The types of land suitable for growing some of the important oil-bearing eucalypts are shown in Table 11.1 (Sastry and Kavathekar 1990).

In Tamil Nadu *E. globulus* is grown on the high hills, *Eucalyptus* hybrid in the plains and *E. grandis* at medium elevation (Kondas and Venkatesan 1986). The Central Arid Zone Research Institute, Jodhpur, tested 115 species of *Eucalyptus* which were growing in comparatively low rainfall regions of Australia (Muthana *et al.* 1984). Among the most promising species was *E. camaldulensis* and this, too, is now planted. Of late, it is gaining importance and is being grown in arid zones of Rajasthan, Andhra Pradesh and other states on a plantation scale. These five species, namely, *E. tereticornis* (*Eucalyptus* hybrid), *E. globulus*, *E. citriodora*, *E. grandis* and *E. camaldulensis*, have found country-wide acceptance and form the great bulk of eucalypts in India.

Table 11.1 Suitable land types for oil-bearing eucalypts introduced into India[a]

| Species | Land type |
|---|---|
| *E. camaldulensis* | Hills and plains; arid areas; semi-moist non-forest localities; skeletal soils; saline and alkaline soils; kankar pans; waterlogged soils; sand dunes; sandy loam; shallow alluvial soils; ravines; mine sites, especially of lignite, bauxite and tin |
| *E. cinerea* | Dry, cold desert |
| *E. citriodora* | Hills and plains; semi-moist non-forest areas; skeletal soils; ravines; deep, narrow gullies |
| *E. dives* | Hills |
| *E. exserta* | Hills |
| *E. globulus* | Hills; moist to very moist areas; skeletal soils; laterite soils; controls erosion with its dense root system |
| *E. globulus* subsp. *maidenii* | Hills |
| *Eucalyptus* hybrid | Hill slopes; semi-moist to very moist soils; non-forest localities; skeletal soils; laterite soils; coastal sandy soils; deep sandy soils; shallow sandy loam; saline and alkaline soils; sodic soils; ravines; eroded land; dry areas; red soils |
| *E. macarthurii* | Shawalik foothills like Jammu (J&K) and Palampur (HP) areas |
| *E. sideroxylon* | Hills; semi-arid areas; saline and alkaline soils; shallow alluvial soils |
| *E. smithii* | Not successful in India |
| *E. viminalis* | Hills |
| *E. viridis* | Arid zone |

a Only those species which have been, or are, commercial sources of oil in India (*E. citriodora* and *E. globulus*) or elsewhere, or which have the potential for it (*E. camaldulensis* and *Eucalyptus* hybrid), are listed.

*Table 11.2* Main areas of eucalypts in India

| Region | State | Area (ha) | Year |
|---|---|---|---|
| Northeastern | West Bengal[a] | 190,000 | 1987 |
| Northern | Uttar Pradesh | 200,000 | 1990 |
| | Punjab | 95,000 | 1986 |
| | Haryana | 40,000 | 1985 |
| | Bihar | 21,000 | 1983 |
| | Himachal Pradesh | 15,000 | 1969 |
| Central | Madhya Pradesh | 57,000 | 1982 |
| Western | Maharashtra[a] | 155,000 | 1985 |
| | Gujarat[b] | 72,000 | 1983 |
| Southeastern | Orissa | 50,000 | 1984 |
| Southern | Karnataka | 400,000 | 1986 |
| | Tamil Nadu | 90,000 | 1985 |
| | Andhra Pradesh | 60,000 | 1980 |
| | Kerala | 50,000 | 1987 |

a  A further 190,000 ha (West Bengal) and 150,000 ha (Maharashtra) of mixed forest are planted, of which one third is estimated to be *Eucalyptus*.

b  A further 120 million *Eucalyptus* seedlings raised and distributed for planting in rural areas.

Most eucalypts are planted for the production of pulpwood and firewood but, where it is economic to do so, 'waste' leaf from the felled trees of *E. globulus* and *E. citriodora* is distilled for oil. The recovery of cineole-rich oil from *E. globulus* in the Nilgiris marked the beginning of eucalyptus oil production in India and it now forms a valuable cottage industry for the local people. A few high oil-yielding clones of *E. citriodora*, *E. globulus* and *E. camaldulensis* have been successfully cultivated specifically for the production of essential oil, by coppicing younger plants two or three times a year or collecting leaves every alternate year from plantations raised for wood.

Reliable, up-to-date information is not available but estimates of the main areas in India under eucalypts, taken from a variety of published sources, are summarised in Table 11.2.

The problem of acquiring accurate statistics is made more difficult by the diverse nature of the plantings. Apart from large blocks intended for pulp production *Eucalyptus* is widely planted as part of agroforestry and social forestry schemes, as well as in strip plantations alongside roads, railways and canals, and in degraded and otherwise barren areas. In total there are probably around 7.5 million ha of *Eucalyptus* in India. This represents about 8 per cent of the global total (Narwane 1986, Thakekar 1972, Abbasi and Vinithan 1997). Eucalypts are often the dominant plantation tree and in some states, such as Karnataka, Punjab and Haryana, *Eucalyptus* plantations account for nearly 1.5 per cent of the total land under cultivation.

## Principal eucalypts planted in India

### Eucalyptus *hybrid*

*Eucalyptus* hybrid, 'Mysore gum', flourishes from coastal areas to 1000 m altitude, tropical to warm temperate climates, annual rainfall ranges of 400–4000 mm and a wide range of soil types (Champion and Seth 1968, B. Singh 1977). In the north, it has been successfully raised in Himachal Pradesh and Jammu and Kashmir at an altitude of 600–1200 m. In the plains of Punjab, Haryana and Uttar Pradesh, large areas have been brought under cultivation under farm forestry. In Uttar Pradesh, large blocks of plantation are confined to the Tarai and Bhabar tracts, immediately below the submountain region. On the eastern side of India, in Bihar, most of the plantations of *Eucalyptus* hybrid are situated between 100 and 650 m elevation. In West Bengal, it has been planted in areas with annual rainfall varying from 1000 to 1400 mm and summer temperatures up to 48°C. On the western side of India, in Maharashtra region, it covers a wide range of climatic and edaphic conditions, from medium rainfall of 1000 mm to high rainfall of 3700 mm. In Gujarat, and in central India (Madhya Pradesh), *Eucalyptus* hybrid plantations have been raised in areas where annual rainfall varies from 400 to 2000 mm. The southern states of Kerala, Karnataka, Tamil Nadu and Andhra Pradesh have the largest areas of *Eucalyptus* hybrid and rainfall ranges from 500 to 1500 mm in Tamil Nadu and Andhra Pradesh to 1500–3500 mm in Kerala.

E. globulus

*E. globulus* was introduced mainly to create a fuel resource in the Nilgiris plateau (Tamil Nadu) and so protect the evergreen shoals which were being progressively destroyed for firewood. Its reputation as the fastest growing and highest yielding species in India led to its increasing popularity in the region (Kondas and Venkatesan 1986, Samraj and Sharda 1986) and the present area of *E. globulus* in Tamil Nadu is around 20,000 ha, about half the total in India. Although the species is used mainly for rayon and paper pulp, and for fence posts, electricity poles and the construction of farm houses, etc., the distillation of essential oil from its leaves forms a major cottage industry in the Nilgiris. The leaves are collected from plantations every alternate year or from standing trees before felling.

E. citriodora

Of the many eucalypts introduced into India, *E. citriodora* is one of the most valuable. It is particularly suited to the climate of the Nilgiri Hills and large plantations of it are found there, as well as the Annamalai Hills and in Kerala, Punjab, Uttar Pradesh, Andhra Pradesh, Karnataka and Maharashtra. The trees grow well at an elevation up to 200 m, tolerate a rainfall up to 4000 mm, but thrive well even on a poor soil and in a dry climate. About 2000 ha are under *E. citriodora* cultivation in the country, of which more than half are in Tamil Nadu. Although the trees are intended mainly for firewood, timber and pulp, the leaves are used for the extraction of citronellal-rich essential oil. They are also grown as shelterbelts, wind breaks and for ornamental purposes along roads or in gardens.

E. camaldulensis

In the arid zones of Rajasthan, Andhra Pradesh and other parts of the country, *E. camaldulensis* has been planted. The plantations have been established by providing irrigation in the first two years because the sites are sandy and characterised by frequent droughts, high temperatures and desiccating winds. The trees are usually planted alongside canals and roads and in blocks with irrigation facilities. Plantations are managed on a coppice rotation of 7–10 years on good sites and 14–15 years on poor sites.

*Other* Eucalyptus *species*

*E. grandis* has been grown successfully on the high ranges of Kerala. Other species which have been tested include *E. globulus* subsp. *maidenii* and *E. smithii*, both potential sources of oil, and

*E. macrorhyncha*, *E. microtheca*, *E. regnans*, *E. terminalis*, *E. viminalis* and *E. youmanii*. Unfortunately, *E. smithii*, which is grown commercially for oil in South Africa, and has shown good promise in China for oil and fuelwood production, has not been successful in India. The use of *E. macrorhyncha* and *E. youmanii* for the production of rutin in Himachal Pradesh is described later.

## Management practices

In India today, forest management and strategy have become more than forestry in the traditional sense. The focus now is on national development through multipurpose tree cropping, whether it is in the farm, the forests or among the existing different crop patterns. Trees and forests are recognised as one of the most important life supporting systems and, since they provide a much needed protection and production guarantee to the soil, are today the basis of all land management systems. Although there has been much heated debate about the ecological impact of eucalypts (Abbasi and Vinithan 1997), *Eucalyptus*, being a fast growing species, is favoured by foresters and farmers for yielding quick economic returns. In some cases the returns are comparable to, or higher, than those from cash crops such as sugarcane, cotton and grapes (Tewari 1992). Less care and attention is also required compared to other agricultural crops (Kareem *et al*. 1986).

All the principal *Eucalyptus* species cultivated in India have good coppicing ability and most of the plantations are managed under a simple coppice system. Although the use of short rotations – around 7–10 years in most cases, although this is sometimes reduced to six years or less by intensive management, including manuring and irrigation – enables many consecutive coppice crops to be produced, repeated harvesting results in loss of vigour and a consequent decline in wood yields and the number of stumps producing coppice shoots. In practice, therefore, the stumps and roots are usually dug out after four rotations and the area replanted (Neelay *et al*. 1984). In normal plantations farmers also take interim returns in the form of pruned materials.

Several recent studies have been conducted in India, examining ways in which *Eucalyptus* may be integrated with other crops as part of an agroforestry 'package' and so increase and diversify the economic returns from the land on which it is being grown. Various models have been costed in which *E. citriodora* is grown for oil and poles or fuelwood (Shiva *et al*. 1988, Shiva and Jaffer 1990). Scenarios include planting it in blocks within farm forestry programmes, around borders of agricultural fields and interplanted with essential oil-yielding grasses such as lemongrass, citronella and palmarosa. In the latter cases, spacings were 2 m × 2 m and 3 m × 2 m. Earlier, Mathur and Sharma (1984) found that wider spacings for *Eucalyptus* (6 m × 1 m) gave maximum returns on farm lands in Uttar Pradesh, Punjab and Haryana. Unfortunately, no reliable figures are available to estimate the present scale on which agroforestry practices involving eucalypts have been taken up or their profitability. Some farmers are known to intercrop such things as cereals and vegetables, as well as essential oil crops, during the first year and shade-tolerant crops such as ginger, turmeric, pineapple and fodder crops during the later years.

## Pests and diseases

### Insect pests

Besides performing favourably in their introduced habitat when compared to their native habitats (Sehgal 1983, Basu Choudhuri *et al*. 1986), Indian eucalypts have been fortunate in suffering relatively little serious damage from pests and diseases and this has encouraged its planting on a large scale. About sixty species of insects are associated with *Eucalyptus* in India (Sen Sharma and Thakur 1983).

*Celosterna scabrator* Fabr. (Coleoptera: Lamindae), popularly known as Babul stem and root borer, has become a major pest of *Eucalyptus*. Plantations have been reported to be damaged by it in parts of Andhra Pradesh, Madhya Pradesh, Maharashtra and Uttar Pradesh. The adult insect attacks tender shoots of the plant, scraping and feeding in the bark up to the sapwood. Only one egg is laid per plant. On emergence the larvae bore into the stem and root, which are hollowed. The infested plant shows signs of wilting and yellowing and ultimately dies.

Termite attacks are major problems in dry zones. They cause severe injury and mortality to eucalypts in the nursery and to young plants. Mortality has been reported to be about 80 per cent (Nair and Varma 1981). Those termite species which cause damage to eucalypts include *Odontotermes assmuthi* Holmgren, *O. brunneus* (Hagen), *O. distans* Holmgren & Holmgren, *O. feae* Wasmann, *O. indicus* Thakur, *O. microdentatus* Roonwal & Seri Sarma, *O. obesus* Rambur, *O. redemanni* (Wasmann), *O. wallonensis* (Wasmann), *Microcerotermes minor* Holmgren and *Microtermes obesi* Holmgren.

Insecticides such as aldrin, clitordane and heptachlor are the most effective as the mortality to eucalypts is extremely low, ranging from 0 to 4 per cent. They are also effective at the lowest dosage level (0.25 g active ingredient per seedling). Aldrin, however, is no longer manufactured in India and its use is being phased out on environmental grounds. Other pesticides such as those already mentioned and chlordane and chlorpyrifos are equally effective (Rajan 1987, Tewari 1992).

*Diseases*

Two fungal diseases, pink disease (*Corticium salmonicolor*) in plantations (Seth *et al*. 1978, Sharma *et al*. 1984a) and seedling blight (*Cylindrocladium quinqueseptatum*) in nurseries (Sharma *et al*. 1984b), affect eucalypts in high rainfall areas. *Cylindrocladium* blight assumes epidemic proportions in high rainfall areas with the onset of the southwest monsoon and causes devastating damage to seedlings in the nursery, young coppice shoots and one-year-old plants (Anahosur *et al*. 1977, Rattan *et al*. 1982). The disease, which initially appears as isolated leaf spots, causes severe defoliation and stem narcosis and, ultimately, the death of the plant. Although a number of fungicides have been reported to control the disease, the long-term solution probably lies in identifying resistant genotypes (Jayashree *et al*. 1986).

The nursery diseases can be prevented in high rainfall areas by direct sowing in polythene bags instead of seed beds, as they are soil-borne in nature (Sharma and Mohanan 1986). The spread of the disease is reduced considerably by the use of containers since this isolates the fungal propagules which are present (Sharma and Mohanan 1981). BasalineR® (1.0 kg ha$^{-1}$) appears to be a promising herbicide for controlling weeds in *Eucalyptus* nurseries.

Tobacco mosaic disease has been detected in *E. citriodora* in India. The symptoms of the disease are a red coloration of the growing ends on 1–2-year-old plants. The disease adversely affects the quantity and quality of the oil (Sastry *et al*. 1971). Root rot, caused by *Ganoderma lucidum*, spreads from infected woody debris in the soil. Any such debris should be removed immediately or isolated by trenching or interplanting with resistant species. In the submountain region in the north, and on the Central Indian Plateau, *Eucalyptus* is attacked by the sap and heartrot fungus *Trametes cubensis*, which enters through injuries caused by *Celosterna scabrator* (Bagchee 1953).

The threat posed by these diseases in the establishment of eucalypts can be met most effectively by selection of appropriate species and provenances for each geographical area and by adopting proper silvicultural practices (Jayashree *et al*. 1986).

# Eucalyptus oil production

## Introduction

Despite the large number of species of *Eucalyptus* which have been tested in India, and the very large areas of eucalypts which have been planted, only two species, *E. globulus* and *E. citriodora*, have been exploited commercially for oil production (see e.g. Husain *et al.* 1988). Their use for this purpose is described below.

The other, most obvious candidate for exploitation is *Eucalyptus* hybrid, since large quantities of 'waste' leaf are available after felling the trees from industrial plantations established for pulp and timber production. However, although a considerable amount of research has been carried out aimed at determining the composition and properties of the oil and the potential for commercial production (see below), any such potential has not, to date, been realised. Nevertheless, selection of trees with improved oil characteristics could change the situation and enable the resource to be utilised for this purpose.

Silvicultural trials and analyses of *E. macarthurii*, a source of perfumery oil rich in geranyl acetate, have been undertaken (Madan *et al.* 1971, Thappa *et al.* 1979) but despite encouraging results it has never been exploited commercially in India. Essential oils from other Indian eucalypts, none of them produced commercially, are described by Bhalla (1997).

Portable field distillation units employing wood and exhausted leaves as fuel are sometimes used to distil the oil at the site of felling, and this avoids costly transport of bulky foliage to distillation units stationed at one place, as well as reducing the cost of fuel *per se* (Theagarajan *et al.* 1993). Apart from oil production, use of *Eucalyptus* leaves as a more general type of fuel is common and in Karnataka they are preferred by potters for the pottery kilns (Shyam Sunder 1986).

## Eucalyptus *species used for oil production*

### E. globulus

The total production of eucalyptus oil in the country from this source is nearly 400 t annually (S.C. Varshney pers. comm. 1997). In India, *E. globulus* subsp. *globulus* is the only source of medicinal oil. It grows well in Nilgiri, Anamalai and Palani hills in south India, and Simla and Shillong in the north. It is unsuitable for cultivation above 1200 m or at altitudes where the snowfall is heavy. Leaves and twigs are distilled to yield an oil (0.9–1.2 per cent, fresh basis) containing about 60 per cent 1,8-cineole which is very popular in south India. Yields of leaves vary from around 2125 kg to 3375 kg per hectare in the Nilgiris, where the distillation of the oil is a cottage industry. On an average site approximately 10–12 kg ha$^{-1}$ of oil are obtained, Table 11.3 (Samraj and Sharda 1986).

*Table 11.3* Yields of leaf oil and pulpwood from *E. globulus* in Nilgiris, Tamil Nadu

| Site quality | Average height after 10 years (m) | Oil yield (kg ha$^{-1}$) | Pulpwood yield[a] (m$^3$ ha$^{-1}$) |
|---|---|---|---|
| Good | 25–30 | 10–15 | 450–500 |
| Fair | 20–24 | 10–12 | 300–400 |
| Poor | 15–19 | 7–10 | 200–250 |

a Per rotation of ten years.

Manian and Gopalakrishnan (1995, 1997) found that oil production in *E. globulus* is influenced by the ambient temperature and is favoured both seasonally and in altitude terms by warmer temperatures. Foliar analysis indicated a direct relationship between potassium and oil yield.

The leaves are collected and dried in the shade for 3–4 days. Copper stills holding about 360 kg of leaves and 160 l of water per charge are generally used for distillation (Rajan 1987). Each charge takes 8 h for complete distillation. As the oil obtained is somewhat crude, it is purified by redistillation using a small quantity of caustic soda and water. It is colourless, but becomes yellowish-pink on long storage and is rectified by redistillation.

*E. globulus* subsp. *bicostata* has been shown to yield a leaf oil rich in 1,8-cineole (80 per cent) (Suri and Mehra 1991) but there are insufficient areas of it planted to make commercial production possible at the present time.

E. citriodora

The leaf biomass of *E. citriodora* yields a lemon-scented oil which is in good demand in India by the fragrance industry. Despite this, cultivation of *E. citriodora* as an essential oil crop by farmers is declining because of alternative, more remunerative, oil and spice crops. Apart from distillation of leaf from trees grown for wood in places such as the Nilgiris, its present cultivation solely for oil appears to be limited to just a few areas: Calicut (Kerala, southern India), the Ratnagiri Hills (Maharashtra, western India) and the Tarai region of Uttar Pradesh (northern India). Altogether, about 100 t of the oil are produced annually (Singh *et al.* 1983, S.C. Varshney pers. comm. 1997).

Of the three most important oil-bearing eucalypts planted in India, most research (as far as the oil is concerned) has been carried out on *E. citriodora* and a wealth of knowledge exists on both chemical and agronomic aspects. The discussion below concerning agronomic and harvesting aspects relates to *E. citriodora* grown specifically for oil.

*Chemical aspects* In a study in Jammu, northern India, Kapur *et al.* (1982) examined the variability in oil yield and citronellal content amongst twenty-eight individual trees. Trees with a pubescent leaf type were found to give lower yields but higher citronellal content and superior odour (1.9–3.1 per cent yield, fresh basis; 76–86 per cent citronellal) than trees with a non-pubescent leaf (1.4–4.0 per cent yield; 50–75 per cent citronellal). Three to six-month-old mature leaf gave a higher oil yield (with higher citronellal content) than nine-month-old mature leaf, indicating that harvesting is best done at 3–6 month intervals. A consistent pattern of seasonal variation was found with oil yields rising after the summer monsoons and reaching a maximum during the coldest months before declining. Elsewhere, patterns of variation are different (e.g. Nair *et al.* 1983, Rao *et al.* 1984) and, given the wide range of agroclimatic conditions across the country, oil yields and the optimum period for harvesting the leaf will vary from region to region.

*Agronomic aspects* Kapur *et al.* (1982) recommended close planting in spring followed by coppicing the next summer, just before the onset of the monsoons. In this way, a bushy habit was promoted with high leaf biomass. Two or three harvests could then be taken in northwest India (in April and November, with a minor crop in July). If the plants were allowed to grow without any coppicing, then leaf production was poor and, at best, it is two years before the first crop can be taken. The advantages for total oil yield (on a per hectare per year basis) of coppicing with fairly frequent harvesting have also been stated by others (e.g. Singh and Atal 1969, Kavulutlayya 1973).

Higher plant densities give a greater leaf biomass (and therefore oil yield) on a per hectare basis than smaller ones, although the biomass per plant may be less. Thus in Kerala, total leaf and oil yields went down as the plant spacing increased from 2 m × 2 m (equivalent to 2500 plants/ha) to 3 m × 3 m and 4 m × 4 m (Muralidharan and Ramankutty 1982). P. Singh (1977) advocated a spacing of 0.90 m × 0.75 m for raising a commercial crop of *E. citriodora*. Ultimately, the spacing employed is a balance between high density and allowing the seedlings sufficient room to grow and flourish in the time between harvests. Competition from weeds is more of a problem at the higher spacings and sufficient proximity for adjacent plants to form a canopy will help suppress weed growth. Application of nitrogen fertiliser is generally beneficial to leaf production but optimum dose rates appear to be dependent on the nutritional status of the soil (Muralidharan and Nair 1974, Sirsi *et al.* 1984).

*Harvesting*  Depending upon conditions, and whether plantations are irrigated or rainfed, either two or three harvests per year are possible (P. Singh 1977). It is recommended that the first coppicing is done at about 30–45 cm above the ground and subsequent ones at 75–90 cm above ground. Tips of overgrown shoots are pinched once or twice during a season so as to promote growth of side shoots and more leaves. For irrigated plantations there is an increase in fresh weight of 'herb' (leaves + thin branches, approximately 7:3) from 7 t in the first year to 30 t in the second year and 40 t in the third year (per hectare), after which yields stabilise and are 'expected to remain economical for about ten years if properly maintained'. For unirrigated, rainfed areas yields are slightly less. Corresponding yields of oil are claimed to be 50 l, 150 l and 200–250 l/ha (i.e. about 0.5 per cent on a fresh weight basis).

*E. citriodora* oil is also produced by the collection of 'waste' leaf from timber plantations. Yields in this case are reported to be up to 15 kg ha$^{-1}$ (Fernandez and Suri 1981).

Eucalyptus *hybrid*

Early reports of *Eucalyptus* hybrid oil described it as a new source of cineole but also noted its variability in composition (Theagarajan and Rao 1970; see also Theagarajan and Prabhu 1981, 1988). Typically it contains approximately equal parts of α-pinene, β-pinene and 1,8-cineole (about 20 per cent each, Mehra and Shiva 1988) and this and subsequent reports have focused on its potential as a source of pinenes for the manufacture of aroma chemicals rather than its value as a source of cineole. Both α- and β-pinene are important industrial precursors for a wide range of fragrance and other materials. Given the very large areas of *Eucalyptus* hybrid available in India, and its potential for oil production, Verma *et al.* (1978) undertook biomass studies to calculate possible productivity. They stripped the leaves from over 200 trees of a ten-year-old plantation and found an average of 5.9 kg of green leaf per tree. At a stocking of 1200 trees/ha this is equivalent to about 7.1 t of leaf per hectare. The average yield of oil was estimated to be just under 44 kg ha$^{-1}$.

Shiva *et al.* (1984, 1989) sampled and analysed *E. tereticornis* ('*Eucalyptus* hybrid') leaves from seven, eight, nine and ten-year-old plantations at six different sites in Uttar Pradesh. They concluded that distillation of 'waste' leaf from trees felled for pulp and fuelwood at seven years would yield maximum quantities of oil and, thereby, substantial quantities of pinenes. However, pinenes

Table 11.4 Variability in leaf oil yield and composition in E. tereticornis ('Eucalyptus hybrid') according to site and age of trees[a]

| Site[b] | Oil yield | α-Pinene | β-Pinene | α-Terpinene | p-Cymene | 1,8-Cineole |
|---|---|---|---|---|---|---|
| E. Dehra Dun | 1.2–2.1 | 11.1–26.1 | 4.6–18.9 | 20.2–41.0 | 5.3–17.2 | 21.1–31.0 |
| Bijnore | 1.6–1.9 | 14.3–21.6 | 13.6–26.3 | 3.7–13.8 | 10.6–16.6 | 24.1–34.6 |
| Haldwani | 1.6–2.3 | 13.8–29.3 | 13.6–26.7 | 1.5–32.5 | 4.6–11.7 | 11.0–43.7 |
| Pilibhit | 1.0–2.3 | 11.5–24.8 | 13.3–24.0 | 1.1–28.1 | 7.9–44.1 | 8.5–46.6 |
| S. Kheri | 1.4–2.1 | 17.0–37.8 | 4.1–25.8 | 5.0–8.9 | 7.7–24.0 | 18.5–43.2 |
| Bahraich | 1.1–2.6 | 10.7–17.0 | 11.0–31.5 | tr–8.3 | 7.4–24.3 | 27.9–50.7 |

a Figures represent minimum and maximum values at each site for samples taken from plantations aged seven, eight, nine and ten years. Oil yield: , moisture-free basis; composition: , relative abundance.
b All sites are in Uttar Pradesh.

obtained from this source have to compete with cheaper, turpentine-derived pinenes (i.e. ex *Pinus*) and, although there is a shortage of turpentine in India, *Eucalyptus* hybrid oil is not yet being produced commercially. Nor is there evidence that it is likely to be in the near future.

The variability in yield and composition of *E. tereticornis* oil is illustrated in Table 11.4 (Shiva *et al*. 1984, 1989).

## End uses and standards for eucalyptus oil

The primary oils are rectified by redistillation and fractionation to maintain quality. Galvanised iron drums with a capacity of 25 l or 200 l are normally used for storage and marketing of essential oils, including eucalyptus oil. The drums are then packed in wooden crates. For small-scale trading 500 ml and 1.0 l glass and aluminium bottles, packed in corrugated boxes, are commonly used.

### Medicinal oil

*E. globulus* oil is used locally as an antiseptic in the treatment of the respiratory tract, for bronchitis and asthma, and as a rubefacient for rheumatism. It is also used as a component of inhalants, embrocations, gargles, and germicidal and disinfecting preparations. Pure 1,8-cineole is also produced in large quantities by fractional distillation of *E. globulus* oil for use in pharmaceutical preparations.

To comply with the Indian Pharmacopoeia (IP 1996), *E. globulus* oil should be colourless to pale yellow, have an aromatic camphoraceous odour and a camphoraceous pungent taste followed by a feeling of cold, and contain not less than 60 per cent (w/w) cineole. It should also have a weight per ml of 0.897–0.924 (15.5°C), refractive index 1.457–1.469 (20°C) and optical rotation 0° to +10°.

The Bureau of Indian Standards (formerly Indian Standards Institution) also has a requirement of not less than 60 per cent cineole for *E. globulus* oil but the optical rotation range is −5° to +10° (BIS 1992). Industrial sources are pressing for the minimum cineole content to be raised to 70 per cent. Such an oil would then be of international quality and comply with the demands of the British Pharmacopoeia (and some other pharmacopoeias) which specifies a minimum of 70 per cent cineole. Many other new Indian standards have already been introduced as part of a large revision programme and are in tune with ISO and BP standards. The Indian market, too, has come to prefer BIS-marked materials and oil which does not initially conform to BIS standards is often rectified to increase the cineole content.

## Perfumery oil

Citronellal-rich oil from *E. citriodora* is used as a fragrance to perfume soaps, detergents and disinfectants but its main use is as a starting material for the production of citronellol, hydroxycitronellol and hydroxydihydrocitronellal. The latter compound has a fine, flowery odour and is used in large quantities in perfume compositions for creating linden blossom and lily of the valley notes.

To meet the requirements of the Bureau of Indian Standards (BIS 1993) *E. citriodora* oil should be clear and a colourless to pale yellow or greenish yellow liquid with a citronellal content of not less than 70 per cent (w/w). It should also have the following specifications (all at 27°C): relative density 0.850–0.870, refractive index 1.447–1.457, optical rotation $-2°$ to $+6°$ and solubility in two volumes of ethanol (80 per cent v/v). The oil should be packed in clean, dry, air-tight, non-absorbent containers in which it has no action, protected from light and stored in a cool, dry place.

## Use of Eucalyptus *in traditional medicine*

From time immemorial, aromatic plants have been used in Ayurvedic preparations. Either the plant itself is used (in whole or in part, e.g. bark, seeds, leaves or flowers) or an infusion or 'tea'. There is no mention of the use of *Eucalyptus* in ancient Ayurvedic texts since it has only recently been introduced into India but, since then, both Ayurvedic and Unani practitioners have found it to be effective against a wide range of ailments (Singh 1974).

Eucalyptus leaves, kino and oil are used in antiseptic and anti-inflammatory preparations for both external and internal use and for inhalation. Ayurvedic preparations containing eucalyptus oil as a major ingredient are used externally for treating rheumatism, headache, bodyache and abdominal pain (Sharma 1984). It is also used against ailments of the respiratory and gastrointestinal tracts (including indigestion, diarrhoea and dysentery), as a vermicide, in the treatment of skin diseases and as a mouthwash. The cineole-rich oil is used in Ayurvedic preparations for the treatment of typhoid, haemorrhoidal fever, diphtheria, whooping cough, affections of genitourinary organs and for the healing of wounds (Chunekar and Pandey 1988). Tinctures of eucalyptus oil in doses of 10–20 minims are useful in purulent catarrhal affections of the bladder, urethra and vagina. Drop doses of oil in combination with water or tepid milk are protective against cholera.

## Production of extracts other than oil from *Eucalyptus*

### Rutin

*Eucalyptus* leaves are used for composting by farmers (Shyam Sunder 1986) but the only other extractive utilisation of eucalypts is the production of rutin. Rutin, quercetin-3-rhamnoglucoside, is a greenish yellow crystalline powder present in the leaves of a number of *Eucalyptus* species. It is used in both allopathic and Ayurvedic systems of medicine in the treatment of capillary fragility, retinitis, rheumatic fever and haemorrhagic conditions (Thappa *et al*. 1982, 1990). In India, *E. macrorhyncha* and *E. youmanii* have been successfully introduced in Himachal Pradesh as sources of rutin (Tholamani 1969, Bradu *et al*. 1982, Sood *et al*. 1982). In the Palampur area, 12,000 plants have been successfully raised on 4 ha land since 1978 for rutin production. More recently, high density plantations of *E. macrorhyncha* and *E. youmanii* have been established at Mandi (750 m) and Kullu (1700 m) in Himachal Pradesh, and Ooti (Nilgiris) (R. Kapoor pers. comm. 1997).

From the mixed *E. macrorhyncha* and *E. youmanii* plantation about 1500–2250 kg/ha of air dried leaves are harvested annually. The mixed material fetches Rs 13–14 (about US$0.35) per kg and has a rutin content of 6–8 per cent. The Regional Research Laboratory, Jammu, has recently developed a process for the simultaneous extraction of essential oils and rutin from plant materials (Agarwal *et al.* 1998). Oil rich in eudesmol may be obtained as a by-product from rutin production. Total production of rutin in India is presently around 2 t annually – valued at about Rs 1500 (US$38) per kg – and expected to rise considerably in the future.

*Other extractives*

Ursolic acid, which has been isolated from *Eucalyptus* hybrid leaves in 1 per cent yield (Dayal 1987), has a number of potentially useful pharmacological activities. In tests, it has produced a dose-dependent choleretic effect, significant anticholestatic activity against paracetamol-induced cholestasis and marked hepatoprotective activity against paracetamol- and galactosamine-induced hepatotoxicity. Its activity compares favourably with the hepatoprotective drug silymarin. So far, however, ursolic acid production from eucalyptus has not been taken up commercially.

## Current research and future research needs

In India, current industrial and forestry research on *Eucalyptus* is mostly directed at the development of disease-resistant and genetically improved clones or interspecific hybrids for paper, pulp, timber and fuelwood purposes. Improvement to oil-bearing eucalypts receives relatively little attention. Research work conducted at the Forest Research Institute, Dehra Dun, has resulted in the development of two promising interspecific $F_1$ hybrids between *E. tereticornis* and *E. camaldulensis*, FRI-4 and FRI-5. These improved varieties have yielded 3–5 times more wood than the parent species over a ten-year rotation, although the oil composition shows little improvement in commercial terms (Chaudhari and Suri 1991).

While the needs of the country justify the resources which are put into raising the quality and productivity of commercial eucalypt plantations in terms of their primary end use, there is everything to be gained by giving some thought, also, to oil production. While developing improved strains of *Eucalyptus* hybrid or other oil-bearing eucalypts for wood and fibre it requires relatively little extra effort to monitor oil yields and quality (composition). By selecting germplasm for planting which optimises oil properties – but does not adversely affect its silvi-cultural and primary attributes – it may be possible to make the secondary production of oil from 'waste' leaf economically viable where it was only marginal before. Equally, if the production of eucalyptus oil as a primary product from short-rotation, coppiced trees is to withstand the competition from other, more remunerative crops, then annual oil yields of at least 250 kg ha$^{-1}$ need to be realisable. Selected, proven germplasm of *E. globulus* or *E. citriodora* must be made available to farmers to achieve this.

Many research institutes have developed broad-based gene pools of *Eucalyptus*, subject to local environment selection pressures, to produce improved cultivars with respect to growth and disease resistance. Superior provenances and better species for each eco-climatic zone are being selected, germplasm banks are being established and seedling seed orchards and clonal seed orchards are being raised.

Micropropagation by tissue culture has been tried successfully with some species of *Eucalyptus* and thousands of embryoid can be obtained in a single flask. These embryoid can then be developed into plantlets. Plantlet production by somatic embryogenesis is a superior method among various tissue culture techniques. Of the various methods of vegetative propagation tested, side grafting in *E. grandis* and *E. tereticornis*, and cutting from fresh shoots in *E. globulus* have been successful. Progress so far indicates the possibility of increasing biomass production and maintaining it at a level of more than three times the present yields by developing high-yielding and disease-resistant strains through mass selection and recombination breeding.

## References

Abbasi, S.A. and Vinithan, S. (1997) *Eucalyptus: Enduring Myths, Stunning Realities*, Rawat Publications, Jaipur, India.

Agarwal, S.G., Thappa, R.K. and Handa, S.S. (1998) *A Novel Single Step Process for the Simultaneous Extraction of Essential Oils and Rutin from Plant Materials*, Indian Patent 113/Del/1998.

Anahosur, K.H., Padaganoor, G.M. and Hedge, R. (1977) Laboratory evaluation of fungicides against *Cylindrocladium quinquiseptatum*, the causal organism of seedling blight of *Eucalyptus* hybrid. *Pesticides*, 11(6), 44–45.

Bagchee, K. (1953) New and noteworthy diseases of trees in India. The sap and heartrot diseases of *Eucalyptus maculata* Hook. var. *citriodora* Bailey (*E. citriodora*) due to the attack of *Trametes cubensis* (Mint) Sacc. *Indian Forester*, 79, 341–343.

Basu Choudhuri, J. C., Salar Khan, A.M. and Pankajam, S. (1986) Eucalypts – their performance and pest problems. In J.K. Sharma, C.T.S. Nair, S. Kedharnath and S. Kondas (eds), *Eucalypts in India – Past, Present and Future, Proc. Nat. Seminar*, Peechi, Kerala, January 1984, Kerala Forest Research Institute, Peechi, pp. 336–345.

Bhalla, H.K.L. (1997) *Indian Eucalypts and their Essential Oils*, Timber Development Association of India, Dehra Dun.

Bhatia, C.L. (1984) *Eucalyptus* in India – its status and research needs. *Indian Forester*, 110, 91–96.

BIS (1992) *Indian Standard. Oil of Eucalyptus Globulus – Specification* (IS 328:1992), Bureau of Indian Standards, New Delhi.

BIS (1993) *Indian Standard. Oil of Eucalyptus Citriodora – Specification* (IS 9257:1993), Bureau of Indian Standards, New Delhi.

Bradu, B.L., Sobti, S.N. and Atal, C.K. (1982) Prospects of rutin (vitamin P) yielding eucalypts in Himachal Pradesh. In P.K. Khosla (ed.), *Improvement of Forest Biomass*, Indian Society of Tree Scientists, H.P. Agric. Univ., Solan, India, pp. 229–231.

Champion, H.G. and Seth, S.K. (1968) *A Revised Survey of Forest Types of India*, Govt. of India, New Delhi.

Chaturvedi, A.N. (1976) Eucalypts in India. *Indian Forester*, 102, 57–63.

Chaudhari, D.C. and Suri, R.K. (1991) Comparative studies on chemical and antimicrobial activities of fast growing *Eucalyptus* hybrid (FRI-4 & FRI-5) with their parents. *Indian Perfumer*, 35(1), 30–34.

Chunekar, K.C. and Pandey, G.S. (1988) *Bhavprakash Nighantu*, Chaukhambha Bharti Academy, Varanasi, India, p. 827.

Dayal, R. (1987) Occurrence of ursolic acid and related compounds in *Eucalyptus* hybrid leaves. *Curr. Sci.*, 56, 670–671.

FAO (1981) *Eucalypts for Planting*, FAO Forestry Series No. 11, Food and Agriculture Organization of the United Nations, Rome.

Fernandez, R.R. and Suri, R.K. (1981) Studies on the oil of *Eucalyptus citriodora* Hook. grown at Dehra Dun. *Indian Forester*, 107, 243–248.

Husain, A., Virmani, O.P., Sharma, A., Kumar, A. and Misra, L.N. (1988) *Major Essential Oil-Bearing Plants of India*, Central Institute of Medicinal & Aromatic Plants, Lucknow, India.

IP (1996) Eucalyptus oil. In *Indian Pharmacopoeia*, Vol. I, Controller of Publication, New Delhi, p. 310.

Jayashree, M.C., Nair, J.M., Deo, A.D. and Ramaswamy, V. (1986) Relative susceptibility of various eucalypt provenances to *Cylindrocladium* blight. In J.K. Sharma, C.T.S. Nair, S. Kedharnath and S. Kondas (eds), *Eucalypts in India – Past, Present and Future. Proc. Nat. Seminar*, Peechi, Kerala, January 1984, Kerala Forest Research Institute, Peechi, pp. 395–399.

Kapur, K.K., Vashist, V.N. and Atal, C.K. (1982) Variability and utilization studies on *Eucalyptus citriodora* Hook. grown in India. In C.K. Atal and B.M. Kapur (eds), *Cultivation & Utilization of Aromatic Plants*, Regional Research Laboratory, Jammu, India, pp. 446–456.

Kareem, M.K.A., Nair, T.J.C., Deo, A.D., Nair, J.M., Kumar, D. and Ramaswamy, V. (1986) Low cost regeneration technique for *Eucalyptus* in Kerala. In J.K. Sharma, C.T.S. Nair, S. Kedharnath and S. Kondas (eds), *Eucalypts in India – Past, Present and Future, Proc. Nat. Seminar*, Peechi, Kerala, January 1984, Kerala Forest Research Institute, Peechi, pp. 181–187.

Kavulutlayya, M. (1973) Review of research on *Eucalyptus citriodora* in the Govt. Cinchona Department, Tamil Nadu. *Indian Perfumer*, 17(2), 44–47.

Kondas, S. and Venkatesan, K.R. (1986) *Eucalyptus* in Tamil Nadu. In J.K. Sharma, C.T.S. Nair, S. Kedharnath and S. Kondas (eds), *Eucalypts in India – Past, Present and Future, Proc. Nat. Seminar*, Peechi, Kerala, January 1984, Kerala Forest Research Institute, Peechi, pp. 31–35.

Madan, C.L., Vashist, V.N., Gupta, S., Agarwal, S.G. and Atal, C.K. (1971) Introduction of *Eucalyptus macarthuri* in India – cultural practices and chemical variation. *Flavour Ind.*, 2, 246–248.

Manian, K. and Gopalakrishnan, S. (1995) Physiological basis for ecological preference of *Eucalyptus globulus* Labill. (Blue gum). II. Growth and oil production. *Indian Forester*, 121, 300–305.

Manian, K. and Gopalakrishnan, S. (1997) Physiological basis for the ecological preference of *Eucalyptus globulus* Labill. (Blue gum). III. Nutritional status, growth performance and oil production. *Indian Forester*, 123, 1188–1196.

Mathur, N.K. and Sharma, A.K. (1984) *Eucalyptus* in reclamation of saline and alkaline soils in India. *Indian Forester*, 110, 9–15.

Mehra, S.N. and Shiva, M.P. (1988) Fractional distillation of *Eucalyptus* hybrid (*E. tereticornis*) oil of alpha- and beta-pinenes recovered for commercial usage. *Indian Perfumer*, 32(1), 95–108.

Muralidharan, A. and Nair, E.V.G. (1974) Cultural and manurial requirements of *Eucalyptus citriodora* Hook in Wynad, Kerala. *Indian Perfumer*, 18(1), 19–23.

Muralidharan, A. and Ramankutty, N.N. (1982) Effect of plant density and pruning height on *Eucalyptus citriodora* Hook. In C.K. Atal and B.M. Kapur (eds), *Cultivation & Utilization of Aromatic Plants*, Regional Research Laboratory, Jammu, India, pp. 435–437.

Muthana, K.D., Meena, G.L., Bhatia, N.S. and Bhatia, O.P. (1984) Root system of desert tree species. *Myforest* (March), 27–38.

Nair, E.V.G., Nair, K.C., Chinnamma, N.P. and Mariam, K.A. (1983) Optimum time of harvest of *Eucalyptus citriodora* Hooker grown in plains. *Indian Perfumer*, 27(1), 45–55.

Nair, K.S.S. and Varma, R.V. (1981) *Termite Control in Eucalypt Plantations*, Research Report No. 6, Kerala Forest Research Institute, Peechi, India.

Narwane, S.P. (1986) Policy and management of *Eucalyptus* plantings in Maharashtra. In J.K. Sharma, C.T.S. Nair, S. Kedharnath and S. Kondas (eds), *Eucalypts in India – Past, Present and Future, Proc. Nat. Seminar*, Peechi, Kerala, January 1984, Kerala Forest Research Institute, Peechi, pp. 440–443.

Neelay, V.R., Sah, A.K. and Bhandari, A.S. (1984) A study on the growth and coppicing capacity of *Eucalyptus tereticornis* (Mysore gum) in 10 year old plantation. *Indian Forester*, 110, 52–54.

Rajan, B.K.C. (1987) *Versatile Eucalyptus*, Diana Publications, Bangalore.

Rao, B.R.R., Singh, S.P., Rao, E.V.S.P., Chandrasekhara, G. and Ramesh, S. (1984) A note on the influence of location, season and strain on *Eucalyptus citriodora* Hook. oil. *Indian Perfumer*, 28(3–4), 153–155.

Rattan, G.S., Dhanda, R.S. and Randhawa, H.S. (1982) Studies on *Cylindrocladium clavatum* – the cause of seedling disease of *Eucalyptus* hybrid. *Indian Forester*, 109, 562–565.

Samraj, P. and Sharda, V.N. (1986) A review of work done on *Eucalyptus globulus* (blue gum) in Nilgiris. In J.K. Sharma, C.T.S. Nair, S. Kedharnath and S. Kondas (eds), *Eucalypts in India – Past, Present and Future, Proc. Nat. Seminar*, Peechi, Kerala, January 1984, Kerala Forest Research Institute, Peechi, pp. 163–169.

Sastry, K.S.M., Thakur, R.N., Gupta, J.H. and Pandotra, V.R. (1971) Three virus diseases of *Eucalyptus citriodora*. *Indian Phytopathol.*, 24, 123–126.

Sastry, T.C.S. and Kavathekar, K.Y. (eds) (1990) *Plants for Reclamation of Wasteland*, Public and Information Directorate (CSIR), New Delhi.

Sehgal, H.S. (1983) Disease problems of eucalypts in India. *Indian Forester*, 109, 909–916.

Sen Sharma, P.K. and Thakur, M.L. (1983) Insect pests of *Eucalyptus* and their control. *Indian Forester*, 109, 864–881.

Seth, S.K., Bakshi, B.K., Reddy, M.A.R. and Singh, S. (1978) Pink disease of *Eucalyptus* in India. *Eur. J. For. Pathol.*, 8, 200–216.

Sharma, J.K. and Mohanan, C. (1981) A disease complex of *Eucalyptus* in nursery, caused by *Pythium*, *Rhizoctonia* and *Cylindrocladium* and its possible control. In *Proc. Third Internat. Symp. on Plant Pathology*, Indian Agricultural Research Institute, New Delhi, 1981, p. 247 (abstracts of papers).

Sharma, J.K. and Mohanan, C. (1986) Management of seedling diseases of eucalypts in nurseries. In J.K. Sharma, C.T.S. Nair, S. Kedharnath and S. Kondas (eds), *Eucalypts in India – Past, Present and Future, Proc. Nat. Seminar*, Peechi, Kerala, January 1984, Kerala Forest Research Institute, Peechi, pp. 377–383.

Sharma, J.K., Mohanan, C. and Florence, E.J.M. (1984a) Outbreak of pink disease caused by *Corticium salmonicolor* in *Eucalyptus grandis* in Kerala, *Indian Trop. Pest Manage.*, 30, 253–255.

Sharma, J.K., Mohanan, C. and Florence, E.J.M. (1984b) Nursery diseases of *Eucalyptus* in Kerala. *Eur. J. For. Pathol.*, 14, 77–89.

Sharma, P.V. (1984) *Dravyaguna-Vijnana*, Vol. II (Vegetable Drugs), Chaukhambha Bharti Academy, Varanasi, India, p. 311.

Shiva, M.P. and Jaffer, R. (1990) Prospects by models for raising aromatic plants in different forestry programmes for economic uplift. *Indian Forester*, 116, 168–176.

Shiva, M.P., Paliwal, G.S., Chandra, K. and Mathur, M. (1984) Pinene rich essential oil from *Eucalyptus tereticornis* leaves from Tarai and Bhabar areas of Uttar Pradesh. *Indian Forester*, 110, 23–27.

Shiva, M.P., Paliwal, G.S., Jain, P.P., Suri, R.K. and Chaudhari, D.C. (1989) Production potential of pinene rich essential oil from *Eucalyptus tereticornis* leaves. *Indian Perfumer*, 33(2), 84–86.

Shiva, M.P., Prakash, O., Singh, S. and Jaffer, R. (1988) Role of eucalypts in agroforestry and essential oil production potential. *Indian Forester*, 114, 776–783.

Shyam Sunder, S. (1979) *Eucalyptus* in Karnataka. *Myforest*, 15(3), 139.

Shyam Sunder, S. (1986) *Eucalyptus* in Karnataka – past, present and future. In J.K. Sharma, C.T.S. Nair, S. Kedharnath and S. Kondas (eds), *Eucalypts in India – Past, Present and Future, Proc. Nat. Seminar*, Peechi, Kerala, January 1984, Kerala Forest Research Institute, Peechi, pp. 25–30.

Singh, A.K., Bhattacharya, A.K., Singh, K., Dubey, R.K. and Kholai, R.C. (1983) Studies on essential oil content of *Eucalyptus* species grown in Tarai region of Uttar Pradesh, Nainital, for timber. *Indian Forester*, 109, 153–158.

Singh, B. (1977) *Eucalyptus* hybrid (Mysore gum): a review of its performance in some parts of India. In *Proc. Eleventh Silvicultural Conf.*, Dehra Dun, May 1967, pp. 842–852.

Singh, D. (1974) *Unani Dravyagunadarsh*, Vol. II, Ayurvedic and Tibbi Academy, Lucknow, India, p. 600.

Singh, P. (1977) *Cultivation of Eucalyptus Citriodora Hook. for its Essential Oil*, Farm Bulletin No. 9, Central Indian Medicinal Plants Organisation, Lucknow, India.

Singh, P. and Atal, C.K. (1969) Prospects for cultivation of *Eucalyptus citriodora* Hook. in Northwest India. In *Proc. Symp. Raising of Medicinal Herbs*, Regional Research Laboratory, Jammu, India, March 1969, pp. 48–59.

Sirsi, S., Shivashankar, K. and Narayana, M.R. (1984) Effect of levels and methods of nitrogen application on herbage and oil yield of *Eucalyptus citriodora* Hook. *Indian Forester*, 110, 1177–1183.

Sood, R.P., Kalia, N.K., Sobti, S.N. and Atal, C.K. (1982) Cultivation of rutin bearing *Eucalyptus*. In C.K. Atal and B.M. Kapur (eds), *Cultivation & Utilization of Aromatic Plants*, Regional Research Laboratory, Jammu, India, pp. 329–336.

Sreenivasa Murthy, H.V. and Ramakrishnan, R. (1978) *A History of Karnataka*, S. Chand & Co., Ramnagar, New Delhi.

Suri, R.K. and Mehra, S.N. (1991) Chemical examination of oils of some eucalyptus. *Indian Perfumer*, 35(1), 8–12.

Tewari, D.N. (1992) *Monograph on Eucalyptus*, Surya Publications, Dehra Dun, India, pp. 83–93, 100–105.

Thakekar, K.N. (1972) Afforestation for soil conservation in Maharashtra State. In *Man-Made Forests in India. Proc. Symp.*, June 1972, Vol. I, pp. 5–7.

Thappa, R.K., Agarwal, S.G., Dhar, K.L. and Atal, C.K. (1979) *Eucalyptus*: a versatile material for aroma chemicals. Paper presented at *Symfor*-79, Regional Research Laboratory, Jammu, India.

Thappa, R.K., Agarwal, S.G., Dhar, K.L. and Atal, C.K. (1982) Rutin. In C.K. Atal and B.M. Kapur (eds), *Cultivation & Utilization of Aromatic Plants*, Regional Research Laboratory, Jammu, India, pp. 321–328.

Thappa, R.K., Agarwal, S.G., Kalia, N.K. and Kapoor, R.K. (1990) Essential oils from exotic *Eucalyptus*: leaf oils of *E. youmanii*, *E. macrorhyncha*, *E. macarthuri* and *E. cinerea* from Northwest Himalayas (India). *J. Wood Chem. Technol.*, 10, 543–549.

Theagarajan, K.S. and Prabhu, V.V. (1981) Complete tree utilization of a fast growing species – *Eucalyptus* hybrid. *J. Indian Acad. Wood Sci.*, 12, 55–57.

Theagarajan, K.S. and Prabhu, V.V. (1988) Rational utilization of forest resources – *Eucalyptus* biomass. In P.K. Khosla and R.N. Sehgal (eds), *Trends in Tree Sciences*, Indian Society of Tree Scientists, Solan, India, pp. 99–102.

Theagarajan, K.S, Prabhu, V.V. and Kumar, R.V. (1993) Distillation of *Eucalyptus* hybrid oil using portable distillation unit in the field. *Indian Perfumer*, 37(2), 181–187.

Theagarajan, K.S. and Rao, P.S. (1970) *Eucalyptus* hybrid – a new source of cineole. *Indian Forester*, 96, 347–350.

Tholamani, D.A.M. (1969) *Eucalyptus macrorhyncha* in the Nilgiri and the Palani hills of South India – an important source for rutin. *Indian Forester*, 95, 473–474.

Verma, V.P.S., Shiva, M.P., Subrahmanyam, I.V. and Suri, R.K. (1978) Utilisation of *Eucalyptus* hybrid oil from forest plantations. *Indian Forester*, 104, 846–850.

# Part 3
# Biological and end-use aspects

# 12 Chemistry and bioactivity of the non-volatile constituents of eucalyptus

*Takao Konoshima and Midori Takasaki*

## Traditional medicinal uses of eucalyptus

From ancient times, the bark and leaves of various species of *Eucalyptus* have been used as folk medicines for the treatment of such ailments as colds, fever, toothache, diarrhoea and snake bites (Ghisalberti 1996). Bark powder has been used as an insecticide in Africa, and a decoction of leaves has been used for the treatment of colds and bronchitis in Venezuela, Jamaica and Guatemala. In China, the leaves of *Eucalyptus globulus* have been used internally for the treatment of influenza, dysentery, articular pain, tonsillitis and cystitis, and externally for use in the treatment of dermatitis, scabies, erysipelas and burns. The root bark of *E. globulus* has been used as an expectorant. The leaves of *E. robusta*, *E. citriodora*, *E. tereticornis* and *E. exserta* are used for the same purposes as *E. globulus* in China (Jangsu 1977).

Volatile, essential oils have been distilled from the leaves of around 500 species of *Eucalyptus* (Coppen and Dyer 1993) and contain many kinds of terpenic compounds, especially 1,8-cineole (eucalyptol). The oil is used in medicine and perfumery (Coppen and Hone 1992). Eucalyptus oil for medicinal purposes, usually prepared from *E. globulus*, *E. polybractea* or *E. smithii*, contains not less than 70 per cent of 1,8-cineole and is included in the pharmacopoeias of many countries, such as Britain, France, Germany, Belgium, Netherlands, USA, Australia, Japan and China. Eucalyptus oil has antiseptic properties, which have been established unambiguously *in vitro* on many germs, and antimicrobial properties of the oil are described in Chapter 13. Eucalyptol is readily absorbed by the digestive route, as well as by the cutaneous or rectal route, and is eliminated by pulmonary or renal excretion. It is also widely accepted that eucalyptus oil (0.05–0.2 ml/day) has expectorant and mucolytic properties, and stimulates the bronchial epithelium. Despite the absence of clinical trials to demonstrate the undeniable therapeutic action of eucalyptus oil and cineole, both products are ingredients of many proprietary drugs in the form of syrups, lozenges, nasal drops and preparations for inhalation (Bruneton 1995, Chapter 16 this volume). The essential oils of several species of *Eucalyptus* (including *E. globulus*, *E. smithii*, *E. radiata* and *E. citriodora*) are widely used for aromatherapy in Europe and Japan.

## Bioactivity associated with the non-volatile constituents

Although phytochemical studies of *Eucalyptus* have tended to focus on the essential oils, numerous triterpenoids, flavonoids, tannins and other non-volatile compounds have been isolated from the plants (Dayal 1988). The rapid growth and increase in woody biomass of *Eucalyptus* already makes it useful as a resource for the timber and pulp industries and its secondary metabolites are now being recognised as potential renewable natural resources for human health care. Many pharmacological studies have therefore been made and, where bioactivity has been demonstrated, it has

Table 12.1 Biological activities of some *Eucalyptus* species and their non-volatile active constituents

| Biological activity | Species | Active compounds | Reference |
|---|---|---|---|
| Antioxidant | E. cinerea, E. cosmophylla, E. globulus, E. perriniana, E. viminalis | β-Diketone, ellagic acid | Osawa and Namiki (1981, 1985), Osawa et al. (1987) |
| Inhibitory effect on HIV-RTase | E. globulus | Macrocarpal A, B, C, E | Nishizawa et al. (1992) |
| Inhibitory effect on Epstein-Barr virus activation | E. amplifolia, E. blakelyi, E. globulus, E. incrassata, E. grandis, E. tereticornis | Euglobals | Takasaki et al. (1995a), Singh et al. (1998), Umehara et al. (1998) |
| Anti-tumour promotion (cancer chemoprevention) | E. globulus, E. grandis | Euglobal-III, -G1 | Takasaki et al. (1995b) |
| Inhibitory effect on HeLa cell growth | E. sideroxylon | Sideroxylonal A, B | Satoh et al. (1992) |
| Inhibitory effect on aldose reductase | E. macrocarpa | Macrocarpals | Murata et al. (1992), Yamakoshi et al. (1992) |
| | E. sideroxylon | Sideroxylonal A, B | Singh et al. (1996) |
| Inhibitory effect on glucosyltransferase | E. amplifolia, E. globulus | Macrocarpals | Singh and Etoh (1995), Osawa et al. (1995, 1996) |
| Antibacterial | E. macrocarpa | Macrocarpals A–G | Murata et al. (1990, 1992) Yamakoshi et al. (1992) |
| | E. globulus | Macrocarpals ? | Osawa et al. (1996) Navarro et al. (1996) |
| | E. perriniana | Grandinol | Nakayama et al. (1990) |
| | E. sideroxylon | Sideroxylonal A, B | Satoh et al. (1992) |
| | E. citriodora | ? | Muanza et al. (1994) |
| Antifungal | E. citriodora | ? | Muanza et al. (1994) |
| | E. globulus | ? | Giron et al. (1988), Navarro et al. (1996) |
| | E. tereticornis | Flavonoids | Barnabas and Nagarajan (1988) |
| | Various *Eucalyptus* spp. | ? | Egawa et al. (1977) |
| Antimalarial | E. robusta | Robustaol A, robustadial A, B | Xu et al. (1984), Qin and Xu (1986), Cheng and Snyder (1988) |

often been found to be associated with the non-volatile, rather than the volatile, constituents of the plants (Ghisalberti 1996, Singh et al. 1999). Phloroglucinol derivatives in particular (e.g. robustadials, euglobals, macrocarpals and sideroxylonals) have been found to exhibit strong biological activities. These include antibacterial, antioxidant, anti-inflammatory, HIV-RTase inhibitory, antimalarial and anti-tumour promoting activities (Table 12.1).

## β-Diketones and polyphenols

Recently, the correlation between oxidative stress and many diseases and natural processes, including cancer and ageing, has been the focus of investigation, and it has been proposed that taking antioxidants might be effective for the inhibition of such phenomena. At present, only tocopherols have been used with safety as natural antioxidative agents.

In order to isolate new types of antioxidant from natural sources, many kinds of natural products have been screened. It is well known that *Eucalyptus* leaves are covered with a thick wax layer which might provide antioxidative protection. Accordingly, leaf waxes of seventeen *Eucalyptus* species were tested *in vitro* by Osawa and his co-workers and they reported that several species showed a strong antioxidative reaction (Osawa and Namiki 1981). As shown in Table 12.2, leaf waxes of *E. citriodora*, *E. mannifera*, *E. regnans* and *E. robusta* exhibited no antioxidative activity, but the other thirteen *Eucalyptus* species tested did, especially *E. dives*, *E. cosmophylla*, *E. parvifolia*, *E. perriniana*, *E. rubida* and *E. viminalis* (whose induction periods were more than forty days). The concentration of one of the β-diketones in the leaf waxes (*n*-tritriacontane-16,18-dione, which showed strong antioxidative activity) was also determined. A correlation between the concentration of *n*-tritriacontane-16,18-dione and antioxidative activity was observed in all of the species except *E. camaldulensis*, *E. pauciflora* and *E. polybractea*. For another three species (*E. parvifolia*, *E. rubida* and *E. dives*), the concentration of this compound in the leaf waxes was not so high but the activity was very strong. These six species may, therefore, contain antioxidants other than *n*-tritriacontane-16,18-dione. Osawa and Namiki (1985) isolated 4-hydroxy-tritriacontane-16, 18-dione (another strong antioxidant) from the leaf wax of *E. globulus*, in addition to *n*-tritriacontane-16,18-dione (Figure 12.1). In this case, the content of leaf wax was more than 0.3 per cent of the fresh leaves, indicating that *E. globulus* could serve as an abundant source of antioxidants.

*Table 12.2* Antioxidative activity[a] of leaf wax extracted from some species of *Eucalyptus*

| Species | Induction period (days) | β-Diketone in leaf wax[b] | Species | Induction period (days) | β-Diketone in leaf wax[b] |
|---|---|---|---|---|---|
| E. camaldulensis | 20 | ± | E. perriniana | >40 | +++ |
| E. cinerea | 40 | +++ | E. polybractea | 30 | ± |
| E. citriodora | <5 | ± | E. pulverulenta | 30 | ++ |
| E. cosmophylla | >40 | ++++ | E. regnans | <10 | ± |
| E. dives | >40 | + | E. robusta | <10 | ± |
| E. globulus | 40 | ++ | E. rubida | >40 | + |
| E. gunnii | 20 | ++ | E. viminalis | >40 | ++ |
| E. mannifera | <5 | — | α-tocopherol (100 μg) | <20 | |
| E. parvifolia | >40 | + | BHA (100 μg) | >40 | |
| E. pauciflora | 15 | — | Control | 5 | |

a Assayed by thiocyanate method (1 mg of leaf wax extract).
b *n*-Tritriacontane-16,18-dione quantified by TLC chromatoscanner: ++++ >80 μg g, +++ >50 μg g, ++ >30 μg g, + 10–20 μg g, ± <5 μg g of fresh leaf, — not detected.

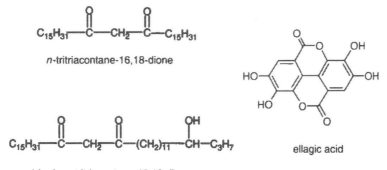

*Figure 12.1* Chemical structures of some antioxidants from *Eucalyptus*.

Table 12.3 Antioxidative activity[a] of ellagic acid (ex *Eucalyptus*)

| | $IC_{50}$ | | |
| --- | --- | --- | --- |
| | $ADP\text{-}Fe^{2+}/NADPH$ | $ADP\text{-}Fe^{3+}/NADPH$ | Adriamycin |
| Ellagic acid | 20 ± 1 | 23 ± 2 | 0.10 ± 0.01 |
| Hexahydroxydiphenic acid | >100 | >100 | 7.0 ± 0.5 |
| Ellagic acid tetraacetate | >100 | >100 | 31 ± 3 |
| α-Tocopherol | 320 ± 3 | 0.80 ± 0.02 | 5.5 ± 0.2 |

a Measured by inhibition of lipid peroxidation in the rat liver microsome system.
$IC_{50}$ = concentration (μM) for 50% inhibition of lipid peroxidation.

Another group of strong antioxidants, polyphenols, especially ellagic acid (Figure 12.1), has been isolated from the bark of *Eucalyptus* as a by-product of the pulp industry (Osawa and Namiki 1983, Osawa *et al*. 1987). As shown in Table 12.3, the antioxidative activity of ellagic acid is considerably stronger than that of α-tocopherol. Ellagic acid was found to be an effective inhibitor of *in vitro* lipid peroxidation by the erythrocyte ghost and microsome test system, and it was the most potent inhibitor of the perferryl-dependent initiation step of NADPH-dependent microsomal lipid peroxidation. Furthermore, ellagic acid strongly inhibited the lipid peroxidation induced by adriamycin, and it was deduced that ellagic acid might be valuable when combined with chemotherapeutic agents such as adriamycin, which normally have the severe side effect of cardiotoxicity caused by lipid peroxidation.

Ellagic acid has also shown inhibitory effects *in vitro* on the mutagenicity and carcinogenicity of benz[*a*]pyrene-7,8-diol-9,10-epoxide, a potent carcinogen (Sayer *et al*. 1982).

Other phenolics shown to exhibit strong antioxidant activity include tannins, flavonol glycosides and acylated flavonol glycosides, all isolated from *E. camaldulensis* (syn. *E. rostrata*) (Okamura *et al*. 1993).

## Macrocarpals

Acquired immunodeficiency syndrome (AIDS) is arguably the most serious disease today, comparable to cancer. Ethanol extracts of leaves and calyces of *E. globulus* have shown significant inhibitory activity against HIV-RTase, and five active compounds, macrocarpals A–E, have been isolated from the active extracts. The macrocarpals have the combined structures of isopentenylphloroglucinol and a sesquiterpene, and macrocarpal B has exhibited the most significant inhibitory effect of HIV-RTase ($IC_{50}$ 5.3 μmol) (Nishizawa *et al*. 1992). The $IC_{50}$ for macrocarpals A, C and E was 10.6, 8.4 and 8.1 μmol, respectively.

The first isolation and structural elucidation of a macrocarpal was that of macrocarpal A, an antibacterial compound, from *E. macrocarpa* (Murata *et al*. 1990). Since then, many other macrocarpals (B–J and am-1) have been isolated from *Eucalyptus* (*E. macrocarpa*, *E. globulus* and *E. amplifolia*), and their antibacterial activities and inhibitory activities on aldose reductase[1] have been reported (Yamakoshi *et al*. 1992, Osawa *et al*. 1995, 1996, Singh and Etoh 1995, Singh *et al*. 1996). Macrocarpals A, B, C and D inhibit aldose reductase in a concentration range of

---

1 Aldose reductase is a target enzyme for the control of diabetic complications such as cataract, retinopathy, neuropathy and nepharcopathy.

2.0–2.8 μmol ($IC_{50}$) (Murata et al. 1992). The inhibitory activities on glucosyltransferase and the cariostatic activities of macrocarpals H, I, J and am-1 (from *E. globulus*) have also been reported (Osawa et al. 1995, 1996). In some cases, detailed configurational and conformational analyses of macrocarpals have been carried out (e.g. Osawa et al. 1997).

The structures of the biologically active macrocarpals are shown in Figure 12.2. There has been confusion in the literature about some macrocarpals. Firstly, two structurally different compounds were given the same name, macrocarpal D, by two groups (Nishizawa et al. 1992, Yamakoshi et al. 1992); the structure containing the seven-membered ring (Nishizawa et al.

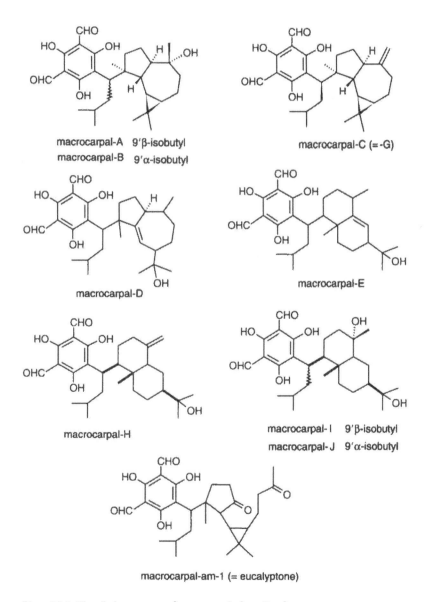

*Figure 12.2* Chemical structures of macrocarpals from *Eucalyptus*.

1992) is now accepted as macrocarpal D. Secondly, two different names (macrocarpal-am-1 and eucalyptone) were given to structurally identical compounds (Osawa et al. 1995, Singh and Etoh 1995). Finally, macrocarpals C and G have now been shown to be identical (Tanaka et al. 1997).

## Sideroxylonals and grandinal

The dimeric phloroglucinol compounds, sideroxylonals A and B (Figure 12.3), isolated from *E. sideroxylon*, have shown strong antibacterial activity against Gram-positive bacteria such as *Staphylococcus aureus* and *Bacillus subtilis*. They have also inhibited the growth of HeLa cells ($IC_{50}$ for A and B = 62.5 μg/ml) and aldose reductase ($IC_{50}$ for A = 1.25 μmol and B = 2.47 μmol) (Satoh et al. 1992). Recently, grandinal (Figure 12.3), which has a similar dimeric phloroglucinol skeleton, was isolated from *E. grandis* (Singh et al. 1997).

## Robustaol and robustadials

The leaves of *E. robusta* have been used for the treatment of malaria as a traditional medicine in China and the benzene-soluble part of the ethanol extract of these leaves showed strong inhibitory effects against the malarial parasite *Plasmodium berghei* (Xu et al. 1984). From this active fraction, robustaol A (a dimeric phloroglucinol) was isolated as an antimalarial constituent, and the structure was confirmed by synthesis (Qin et al. 1981). In addition, robustadials A and B (combined structure of phloroglucinol and monoterpene) were isolated from other fractions exhibiting more potent antimalarial activity, although the reported structures were subsequently revised (Qin and Xu 1986, Cheng and Snyder 1988). The structures of all three compounds are shown in Figure 12.4.[2] Recently, the total synthesis of robustadials A and B and related compounds was carried out (Majewski et al. 1994, Aukrust and Skattebol 1996). Their

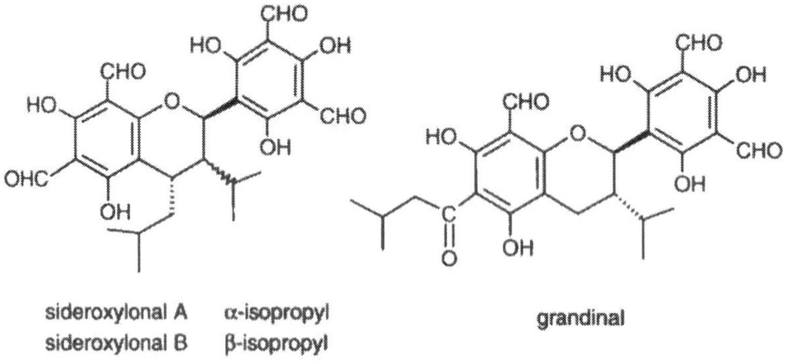

sideroxylonal A   α-isopropyl
sideroxylonal B   β-isopropyl

grandinal

*Figure 12.3* Chemical structures of sideroxylonals and grandinal from *E. sideroxylon* and *E. grandis*.

---

[2] Note that the spectral data of Qin et al. (1981) support the dissimilar aromatic side chains of robustaol A given by them and shown in Figure 12.4; the structure given in the reviews by Ghisalberti (1996) and Singh et al. (1999), which has identical side chains, is incorrect. This has been acknowledged by E. Ghisalberti (pers. comm. 2000 via J. Coppen).

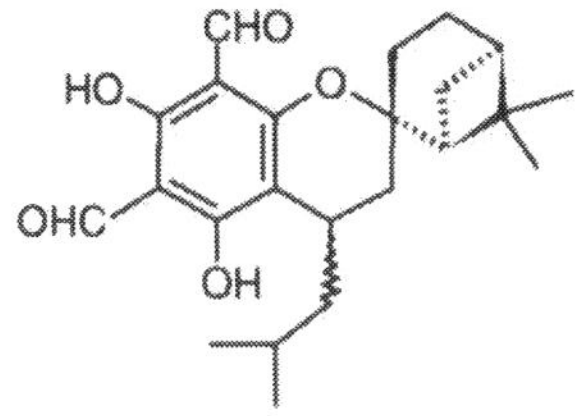


robustadial A β-isobutyl
robustadial B α-isobutyl

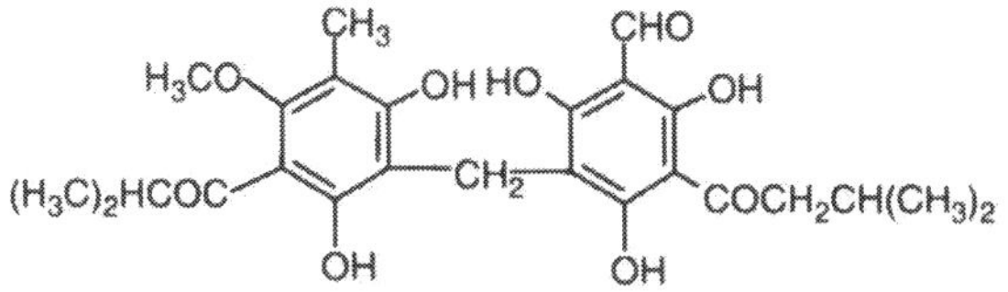


robustaol A

*Figure 12.4* Chemical structures of antimalarial constituents from *E. robusta*.

antimalarial activity is being studied and the development of new antimalarial agents from *Eucalyptus* can be expected.

### *Euglobals*

Euglobals are cycloadducts of formyl phloroglucinol and either mono- or sesquiterpenes, and exhibit a number of pharmacological activities. Like macrocarpals, they are constituents peculiar to certain species of *Eucalyptus*. The phloroglucinol derivatives of *Eucalyptus* have been reviewed elsewhere (Ghisalberti 1996, Singh and Etoh 1997, Singh *et al.* 1999), but the euglobals, which form a large and important group, are described in more detail here. Their distribution in *Eucalyptus* and isolation and structural elucidation using liquid chromatography/mass spectrometry (LC/MS) is discussed first, followed by an elaboration of their biological activities. These latter include anti-inflammatory and anti-tumour promoting activities, and inhibitory effects on Epstein–Barr virus activation.

## Euglobals: analysis and structure elucidation

As indicated above, euglobals are peculiar to *Eucalyptus* and make up a large group of phloroglucinol derivatives. Yields are around 0.03–0.40 per cent (dry weight basis) in the leaves. Euglobal-III was first isolated from the buds of *E. globulus* as a granulation inhibitor and shown to consist of acylphloroglucinol and terpenoid residues (Sawada *et al.* 1980). Other euglobals, with closely related structures, were in the active hexane extract of buds of the plant.

In research on natural products the first essential step is the isolation of pure compounds from an extract. However, this is a difficult and tedious process when the mixture is complex and attempts often fail when the mixture is composed of compounds with closely related structures. In the case of the euglobals from *E. globulus*, eleven euglobals were isolated by Amano *et al.* (1981)

using reversed-phase high performance liquid chromatography (HPLC). The structures of these euglobals were elucidated by means of physicochemical data and X-ray crystallographic analysis and shown to be acylphloroglucinol-monoterpenes (-Ia1, -Ia2, -Ib, -Ic, -IIa, -IIb and -IIc) and acylphloroglucinol-sesquiterpenes (-III, -IVa, -IVb, -V and -VII) (Kozuka et al. 1982a, 1982b). Subsequently, in order to search for new euglobals from other *Eucalyptus* species, a more efficient analytical method using LC/APCI-MS was developed by our group, and about 50 species of *Eucalyptus* were examined (Takasaki 1995). From these analyses it has been concluded that the occurrence of euglobals is restricted to the *Symphyomyrtus* subgenus of *Eucalyptus* (see below). Since the methodologies involved are applicable to other phloroglucinol derivatives they are described below in some detail.

## Analysis of euglobals by LC/MS

Since the 1970s the versatility of HPLC, and its advantages over GC, for the analysis of complex natural organic mixtures, particularly ones containing non-volatile constituents, has been widely recognised. The development of soft-ionisation methods for mass spectrometry has also facilitated the direct measurement of polar organic compounds without the need for prior chemical modification. As a result, complex mixtures of euglobals in *Eucalyptus* have been analysed by LC/MS using atmospheric pressure chemical ionisation (APCI)-MS techniques. A large number of euglobals has been detected by such means and their distribution within the genus *Eucalyptus* has been examined using the following procedures.

## Initial screening

A chloroform extract of the leaves is separated by column chromatography on silica gel into a number of crude euglobal fractions. Each fraction is tested for the presence of euglobals by thin layer chromatography (TLC) and LC/APCI-MS. Since euglobals have a distinctive greenish-yellow fluorescence on the TLC plate when viewed under UV irradiation (365 nm) the primary screening for euglobals is easily carried out (Takasaki et al. 1994c). Using this procedure, euglobals were detected in all but three of forty species belonging to the subgenus *Symphyomyrtus* (the exceptions being *E. aggregata*, *E. mannifera* and *E. polybractea*), while none were detected in fifteen species belonging to the subgenera *Corymbia*[3] and *Monocalyptus*. The results are indicated in the partial classification scheme shown in Figure 12.5.

## Mass spectrometry

After the preliminary separation, each crude fraction is further separated by HPLC and the eluate led directly into the APCI-MS system. In the case of euglobals with an acylphloroglucinol-monoterpene structure, the molecular weight is 386 and the molecular ion peak is at 387 $[M + H^+]$. Acylphloroglucinol-sesquiterpenes have a molecular weight of 454 and a molecular ion at 455 $[M + H^+]$. For *Eucalyptus* species in which no euglobal was detected, no molecular ion peaks were observed at 387 or 455. The fragmentation of euglobals in electron impact mass spectra (EI-MS) is shown in Figure 12.6.

The ion $m/z$ 251, formed by loss of the terpenyl part from the euglobal, is a significant peak and is found in both monoterpene and sesquiterpene types. In the case of the phloroglucinol

---

3 *Corymbia* has been elevated to the rank of a separate genus by Hill and Johnson (1995).

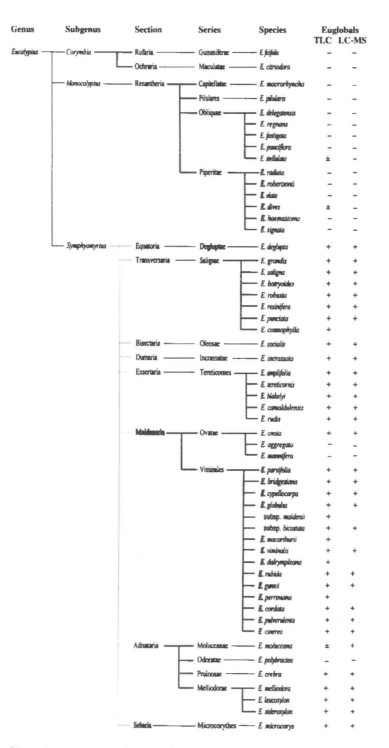

*Figure 12.5* Partial classification of the genus *Eucalyptus* indicating the presence or absence of euglobals.

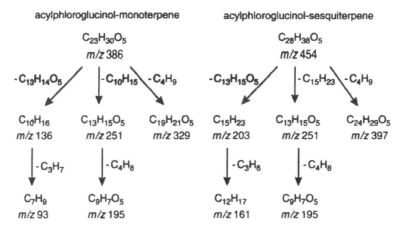

*Figure 12.6* EI-MS fragmentations of euglobals.

dialdehyde type, the fragment ion peak at *m/z* 195 (formed by loss of isobutyl from *m/z* 251) is characteristic. This typical fragmentation of euglobals is also observed in APCI-MS and, together with the total ion chromatogram (TIC) of APCI-MS, provides sufficient information for euglobals to be detected in complex mixtures. The existence of new euglobals may be inferred from a combination of APCI-MS and HPLC retention data. As an illustration of the absence or presence of euglobals in *Eucalyptus* species, the TICs of *E. fastigata* and *E. cordata* are shown in Figure 12.7.

In the case of *E. fastigata*, which belongs to the subgenus *Monocalyptus*, none of the significant TIC peaks showed an [M + H⁺] ion at either *m/z* 387 or *m/z* 455 (Figure 12.7a), confirming the absence of euglobals in the leaves of *E. fastigata*. In the LC/APCI-MS experiments on *E. cordata*, which belongs to the Maidenaria section of the subgenus *Symphyomyrtus*, four TIC peaks (2–5) showed the [M + H⁺] ion at *m/z* 387 and four other peaks (6–9) showed it at *m/z* 455; peaks 2–8 also contained the fragment ion at *m/z* 195 (Figure 12.7b). From these results and the use of HPLC retention data, peaks 2–8 were identified as euglobal-Ib, -Ic, -IIa, -IIb, -III, -IV and -V, respectively, which have a phloroglucinol dialdehyde moiety. Peak 9 was identified as euglobal-VII, having an isovaleroyl group on the phloroglucinol moiety.

In *E. grandis* (subgenus *Symphyomyrtus*), eight TIC peaks exhibited an [M + H] ion at *m/z* 387 but none of them contained ions at *m/z* 455 or *m/z* 195. The euglobals which are present therefore have acylphloroglucinol-monoterpene structures with an isovaleroyl group in the phloroglucinol moiety. From this initial screening preparative HPLC of *E. grandis* enabled five new euglobals to be isolated, euglobal-G1, -G2, -G3, -G4 and -G5 (Takasaki *et al.* 1994c).

### Structures and the structural elucidation of new euglobals

To date, thirty-one euglobals have been isolated from six species of *Eucalyptus* (*E. grandis*, *E. incrassata*, *E. amplifolia*, *E. tereticornis*, *E. blakelyi* and *E. globulus*). Their structures are shown in Figures 12.8a and b.

The structures and stereochemistry of the euglobals, detected and isolated using the screening methods described above, have been elucidated using a variety of spectroscopic techniques,

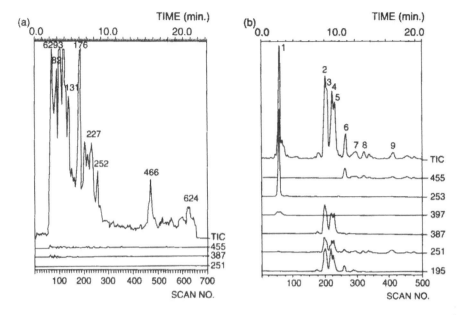

*Figure 12.7* Total ion chromatogram (APCI-MS) of (a) *E. fastigata* and (b) *E. cordata* showing absence and presence, respectively, of euglobals.

among which nuclear magnetic resonance (NMR) has been pre-eminent. In the case of euglobals -G1, -G2, -G3, -G4 and -G5 (isolated from the leaves of *E. grandis*), for example, the results of high resolution MS suggested that they have a common molecular formula ($C_{23}H_{30}O_5$) and phloroglucinol-monoterpene structures. The UV, IR and MS spectra of euglobal-G1 and -G2 were similar to those of euglobal-IIc (from *E. globulus*), indicating the presence of acylphloroglucinol moieties. Extensive $^1$H- and $^{13}$C-NMR experiments finally enabled the stereochemistry and H–H distances of euglobal-G2 to be determined (Takasaki *et al.* 1994c) and these are depicted in Figure 12.9. It was also deduced that euglobal-G1 was a geometrical isomer of -G2 with respect to the formyl and isovaleroyl groups.

Euglobal-G3, -G4 and -G5 were shown to have the same molecular formula as -G1 and -G2. Once again, NMR data were invaluable in deducing their relative stereostructures (Takasaki *et al.* 1990, 1994c). More recently, seven new euglobals (-G6 to -G12) were isolated from *E. grandis* and their structures reported (Singh *et al.* 1998, Umehara *et al.* 1998). Other euglobals which have the acylphloroglucinol-monoterpene structure have been isolated from *E. blakelyi* (euglobal-Bl-1), *E. tereticornis* (euglobal-T1) and *E. amplifolia* (euglobal-Am-1, -2) (Kokumai *et al.* 1991, Takasaki *et al.* 1994b).

Three new euglobals having an acylphloroglucinol-sesquiterpene structure, euglobal-In-1, -In-2 and -In-3, have been isolated from the leaves of *E. incrassata*, along with known euglobals-III and -V (Takasaki *et al.* 1994a, 1997).

---

Including NOE (nuclear Overhauser effect), COSY ($^1$H-$^1$H correlation spectroscopy), COLOC (correlation spectroscopy via long-range coupling), INADEQUATE (incredible natural abundance double quantum transfer) and DEPT (distortionless enhancement by polarisation transfer).

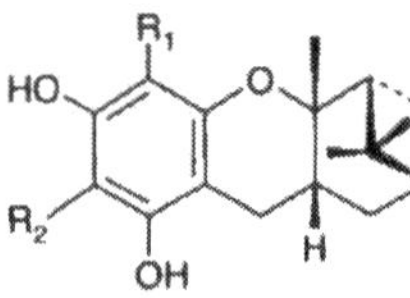

euglobal-G1  $R_1$ = COCH$_2$CH(CH$_3$)$_2$
             $R_2$ = CHO
euglobal-G2  $R_1$ = CHO
             $R_2$ = COCH$_2$CH(CH$_3$)$_2$

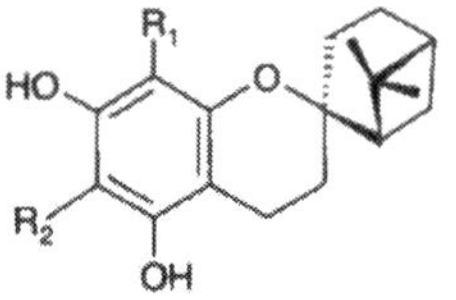

euglobal-G3  $R_1$ = CHO
             $R_2$ = COCH$_2$CH(CH$_3$)$_2$
euglobal-G4  $R_1$ = COCH$_2$CH(CH$_3$)$_2$
             $R_2$ = CHO

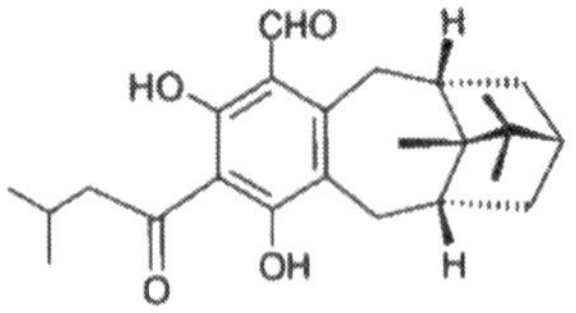

euglobal-G5

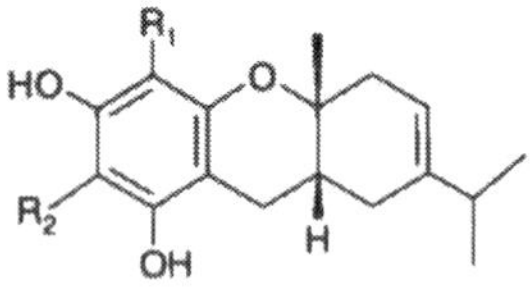

euglobal-G6  $R_1$ = CHO
             $R_2$ = COCH$_2$CH(CH$_3$)$_2$
euglobal-G7  $R_1$ = COCH$_2$CH(CH$_3$)$_2$
             $R_2$ = CHO

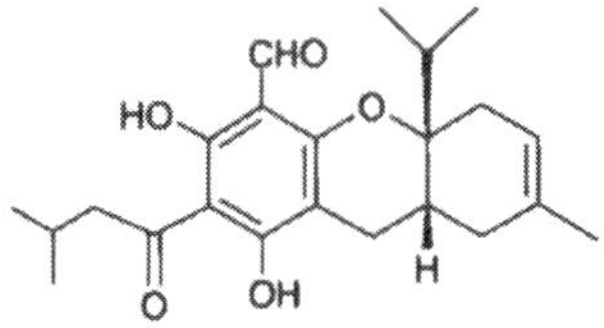

euglobal-G8

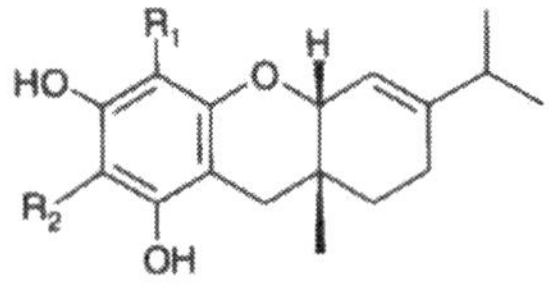

euglobal-G9  $R_1$ = CHO
             $R_2$ = COCH$_2$CH(CH$_3$)$_2$
euglobal-G10 $R_1$ = COCH$_2$CH(CH$_3$)$_2$
             $R_2$ = CHO

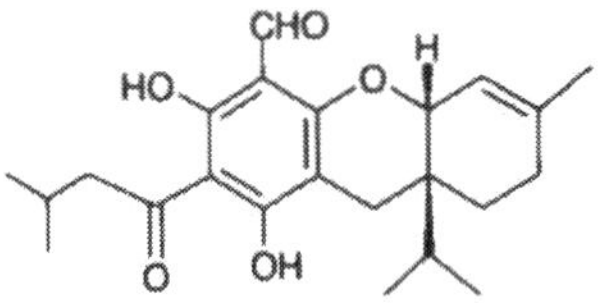

euglobal-G11

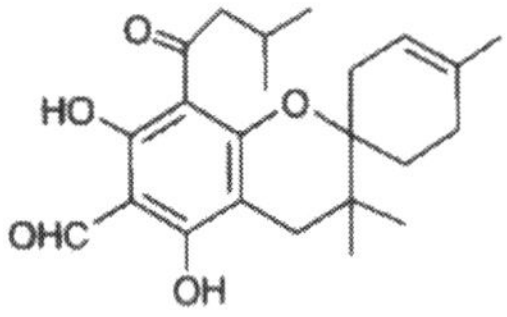

euglobal-G12

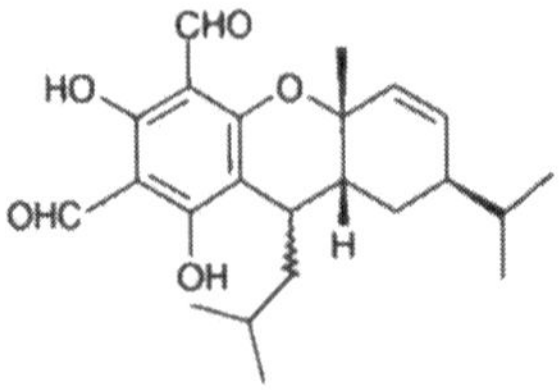

euglobal-Ia$_1$  7β-isobutyl
euglobal-Ia$_2$  7α-isobutyl

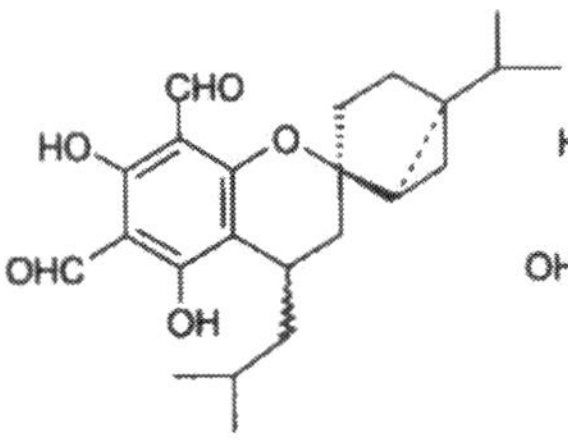

euglobal-Ib     7α-isobutyl
euglobal-B1-1   7β-isobutyl

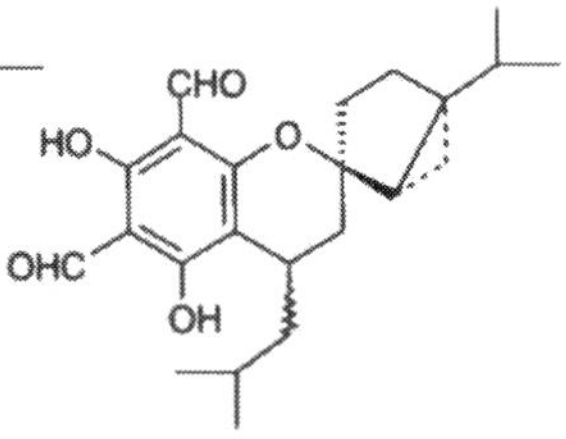

euglobal-Ic    7β-isobutyl
euglobal-IIa   7α-isobutyl

*Figure 12.8a* Chemical structures of euglobals (acylphloroglucinol-monoterpene type) from *Eucalyptus*.

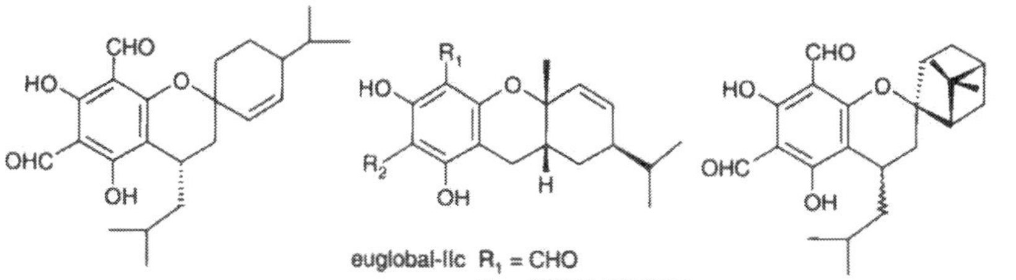

*Figure 12.8a* (Continued).

euglobal-IIb

euglobal-IIc R₁ = CHO
           R₂ = COCH₂CH(CH₃)₂
euglobal-T1 R₁ = COCH₂CH(CH₃)₂
           R₂ = CHO

euglobal-Am-1 7α-isobutyl
euglobal-Am-2 7β-isobutyl

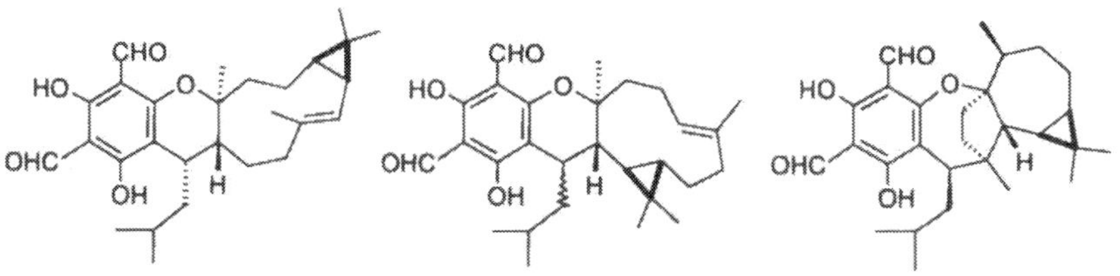

euglobal-III

euglobal-IVa 7α-isobutyl
euglobal-IVb 7β-isobutyl

euglobal-V

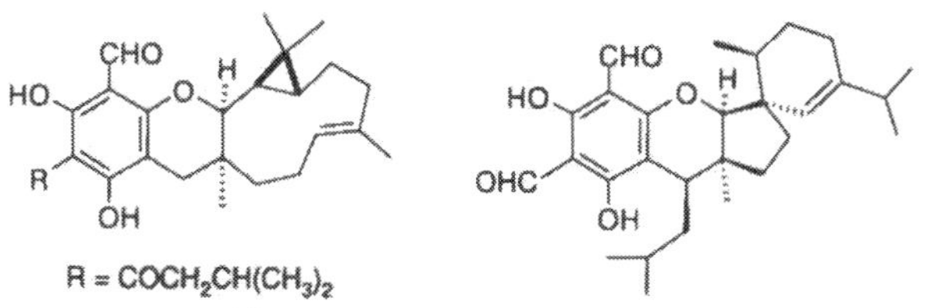

R = COCH₂CH(CH₃)₂
euglobal -VII

euglobal-In-1

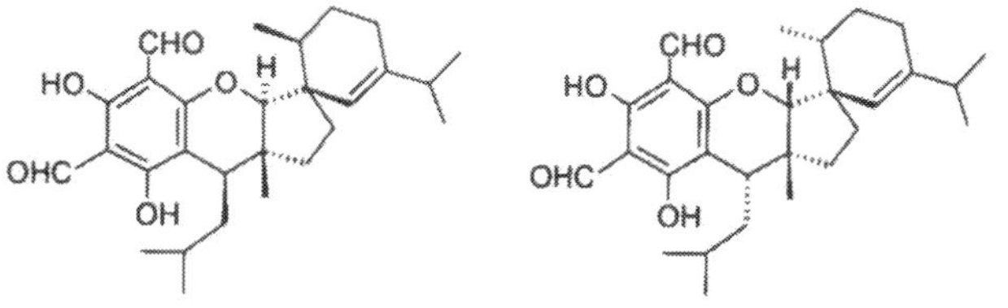

euglobal-In-2

euglobal-In-3

*Figure 12.8b* Chemical structures of euglobals (acylphloroglucinol-sesquiterpene type) from *Eucalyptus*.

## Biogenesis of euglobals

The probable biogenesis of some of the acylphloroglucinol-monoterpene euglobals is indicated in Figure 12.10. They are thought to be derived from cycloaddition of the unstable *o*-quinonemethide A and the appropriate terpene (α-pinene in the case of euglobal-G1, -G2, and -G5; β-pinene in the case of euglobal-G3 and -G4). This generates the chroman or spirochroman skeleton. Euglobal-G6, -G7 and -G8 may be formed from the acylphloroglucinol precursor and γ-terpinene; euglobal-G9, -G10 and -G11 from α-terpinene; and euglobal-G12 from terpinolene. Similar routes may involve euglobal-T1 and -IIc with α-phellandrene and euglobal-B1-1, -Ib, -Ic

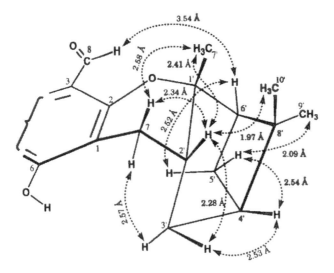

*Figure 12.9* Stereochemistry and H–H distances of euglobal-G2 as deduced from NMR.

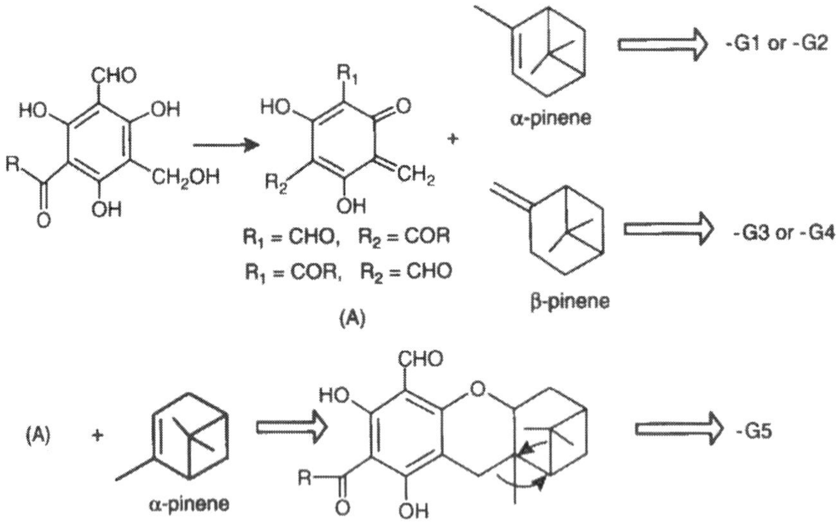

*Figure 12.10* Probable biogenesis of euglobals-G1, -G2, -G3, -G4 and -G5.

and -IIa with sabinene. Biomimetic and other synthetic approaches are gradually shedding light on the pathways to the euglobal skeleton (e.g. Chiba *et al.* 1995, 1996).

## Anti-inflammatory and anti-tumour promoting activities of euglobals

### *Anti-inflammatory effects of euglobals*

The anti-inflammatory effects of eighteen species of *Eucalyptus* have been examined *in vitro* (Takasaki *et al.* unpubl.) by measuring granulation inhibitory activity using the fertile egg

Table 12.4 Anti-inflammatory activity[a] of some *Eucalyptus* species

| Species | Inhibition ratio[b], % (Survival ratio) | | | |
|---|---|---|---|---|
| | Leaves | | Stems | |
| E. viminalis | 66.7 | (11/20) | 58.8 | (15/20) |
| E. cordata | 62.9 | (15/20) | 66.1 | (16/20) |
| E. robusta | 62.7 | (13/20) | 45.1 | (16/20) |
| E. cosmophylla | 57.9 | (18/20) | 57.9 | (15/20) |
| E. radiata | 57.9 | (18/20) | 40.3 | (17/20) |
| E. cinerea | 56.1 | (17/20) | 56.1 | (16/20) |
| E. globulus | 55.6 | (18/20) | 44.4 | (17/20) |
| E. parvifolia | 54.4 | (19/20) | 49.1 | (15/20) |
| E. perriniana | 48.1 | (18/20) | 42.6 | (20/20) |
| E. pauciflora | 46.3 | (16/20) | 20.4 | (19/20) |
| E. gunnii | 44.4 | (19/20) | 18.5 | (16/20) |
| E. regnans | 39.2 | (15/20) | 33.3 | (15/20) |
| E. mannifera | 38.6 | (18/20) | 35.1 | (17/20) |
| E. camaldulensis | 37.5 | (14/20) | 35.7 | (13/20) |
| E. citriodora | 37.0 | (13/20) | 42.6 | (17/20) |
| E. niphophila | 29.8 | (15/20) | 26.3 | (18/20) |
| E. ficifolia | 29.4 | (18/20) | 25.5 | (18/20) |
| E. aggregata | 24.1 | (16/20) | 13.0 | (16/20) |
| Berberine | | 57.1 (17/20) | | |

a Measured as granulation-inhibiting activity.
b Inhibition ratio (%) = (Dry wt in control eggs − Dry wt in test eggs) ÷ (Dry wt in control eggs) × 100, where Dry wt = weight of granulation tissue.
Dose: 100 μg/disc chloroform extract of plant, 25 μg/disc berberine chloride.

method of Otsuka *et al.* (1974). Of the species tested (Table 12.4), eight showed strong inhibitory activity: *E. viminalis*, *E. cordata*, *E. robusta*, *E. cosmophylla*, *E. radiata*, *E. cinerea*, *E. globulus* and *E. parvifolia*. The inhibitory effects of the extracts of leaves were stronger than those of stems. It is notable that apart from *E. radiata* all the active species belong to the *Symphyomyrtus* subgenus of *Eucalyptus* and contain euglobals (Figure 12.5).

Testing of twelve euglobals (-Ia1, -Ia2, -Ib, -Ic, -IIa, -IIb, -IIc, -III, -IVa, -IVb, -V and -VII) isolated from the leaves of *E. globulus* as granulation-inhibiting principles showed that nine of them (Table 12.5) had stronger activities than indomethacin, and similar inhibitory effects to berberine, both well-known anti-inflammatory agents (Kozuka *et al.* 1982a,b). The granulation-inhibiting activity of the euglobals almost certainly derives from the acylphloroglucinol moiety rather than the terpene.

## Inhibitory effects on Epstein–Barr virus activation

Despite substantial progress in our understanding of its development at the molecular level, and in its treatment, cancer remains a tragic disease and one of the major causes of death worldwide. The advancement of cancer chemoprevention, as well as the development of still better methods and drugs for cancer treatment, continues, therefore, to be a much sought-after goal.

The mechanism of chemical carcinogenesis has been explained in terms of either a two-stage or a multi-stage theory; the latter entails initiation, promotion and progression stages (Berenblum 1941). Of these stages, the promotion stage is long-term and consists of a reversible reaction, and the development of inhibitors of promotion has been regarded as the most effective

Table 12.5 Anti-inflammatory activity[a] of euglobals from *Eucalyptus globulus*

| Sample | Dose (μg/disc) | Granulation tissue | | Survival ratio |
|---|---|---|---|---|
| | | Dry weight (mg) | Inhibition ratio[b] (%) | |
| Control | — | 5.6 | — | |
| Euglobal-Ia1 | 12.5 | 2.4 | 57.1 | 18/20 |
| | 6.25 | 3.5 | 37.5 | 18/20 |
| Euglobal-Ib | 12.5 | 2.7 | 51.8 | 15/20 |
| | 6.25 | 3.0 | 46.4 | 16/20 |
| Euglobal-Ic | 12.5 | 2.6 | 53.6 | 16/20 |
| | 6.25 | 3.1 | 44.6 | 17/20 |
| Euglobal-IIb | 12.5 | 2.9 | 48.2 | 16/20 |
| | 6.25 | 3.4 | 39.3 | 16/20 |
| Euglobal-IIc | 12.5 | 2.8 | 50.0 | 18/20 |
| | 6.25 | 3.4 | 39.3 | 17/20 |
| Euglobal-III | 12.5 | 2.7 | 51.8 | 18/20 |
| | 6.25 | 3.4 | 39.3 | 18/20 |
| Euglobal-IVa | 12.5 | 2.4 | 57.1 | 16/20 |
| | 6.25 | 3.5 | 37.5 | 18/20 |
| Euglobal-V | 12.5 | 3.7 | 33.9 | 20/20 |
| | 6.25 | 3.6 | 35.7 | 17/20 |
| Euglobal-VII | 12.5 | 3.3 | 41.1 | 18/20 |
| | 6.25 | 4.1 | 26.8 | 18/20 |
| Indomethacin | 12.5 | 4.1 | 26.8 | 12/20 |
| | 6.25 | 4.2 | 25.0 | 12/20 |
| Berberine HCl | 25.0 | 2.4 | 57.1 | 17/20 |
| | 12.5 | 2.5 | 55.4 | 18/20 |

a, b See footnotes to Table 12.4.

approach to the chemoprevention of cancer. These inhibitors are called anti-tumour promoters. To search for possible anti-tumour promoters (cancer chemopreventive agents) from natural sources, the primary screening of many kinds of natural products has been carried out (Konoshima et al. 1991, 1992, 1996). A suitable method for such screening involves a short-term *in vitro* synergistic assay on Epstein-Barr virus early antigen (EBV-EA) activation induced by a strong tumour promoter, 12-O-tetradecanoylphorbol-13-acetate (TPA). In this assay method, Raji cells carrying the EBV genome are incubated in a medium containing *n*-butyric acid, TPA and various doses of the test compounds. Smears are made from the cell suspension and the EBV-EA inducing cells are stained by means of an indirect immunofluorescence technique. Many compounds that inhibit EBV-EA induction by TPA have been shown to act as inhibitors of tumour promotion in a two-stage carcinogenesis test *in vivo* (Konoshima et al. 1995, Kapadia et al. 1996).

The potential of *Eucalyptus* as a source of anti-tumour promoters has been investigated by Takasaki et al. (1995a,b) and Umehara et al. (1998). In all, twenty-seven euglobals (twenty-one acylphloroglucinol-monoterpenes and six acylphloroglucinol-sesquiterpenes) isolated from the leaves of six species of *Eucalyptus* underwent the primary screening described above. Their inhibitory effects on EBV-EA activation induced by TPA and the viability of Raji cells are shown in Table 12.6. Of the test compounds, euglobal-G1, -G2, -G4, -G5, -Am-2 and -III exhibited significant inhibitory effects (100 per cent inhibition of activation at 1000 mol ratio/TPA and more than 70 per cent inhibition of activation at 500 mol ratio/TPA) and preserved the high

Table 12.6 Screening of euglobals and macrocarpals from *Eucalyptus* as anti-tumour promoters[a]

| Sample | Concentration[b] | | | |
|---|---|---|---|---|
| | 1000 | 500 | 100 | 10 |
| *Acylphloroglucinol-monoterpenes* | | | | |
| Euglobal-G1 | 0.0 (50) | 15.6 (>80) | 70.3 (>80) | 100.0 (>80) |
| -G2 | 0.0 (60) | 25.4 (>80) | 73.3 (>80) | 100.0 (>80) |
| -G3 | 10.5 (>80) | 23.6 (>80) | 65.7 (>80) | 100.0 (>80) |
| -G4 | 0.0 (70) | 26.2 (>80) | 68.4 (>80) | 100.0 (>80) |
| -G5 | 0.0 (70) | 0.0 (>80) | 55.7 (>80) | 93.8 (>80) |
| -G6 | 21.7 (70) | 68.1 (>80) | 80.5 (>80) | 100.0 (>80) |
| -G7 | 22.3 (70) | 64.6 (>80) | 81.9 (>80) | 100.0 (>80) |
| -G8 | 0.0 (70) | 40.8 (>80) | 78.2 (>80) | 91.3 (>80) |
| -G9 | 11.3 (70) | 49.2 (>80) | 84.3 (>80) | 100.0 (>80) |
| -G10 | 0.0 (70) | 44.1 (>80) | 68.5 (>80) | 100.0 (>80) |
| -G12 | 0.0 (70) | 59.9 (>80) | 71.8 (>80) | 100.0 (>80) |
| -T1 | 15.6 (>80) | 66.9 (>80) | 91.3 (>80) | 100.0 (>80) |
| -IIc | 7.8 (>80) | 62.2 (>80) | 89.5 (>80) | 100.0 (>80) |
| -Ia1 | —[c] (0) | 0.0 (20) | 31.5 (>80) | 100.0 (>80) |
| -Ia2 | —[c] (0) | 0.0 (20) | 38.9 (>80) | 89.5 (>80) |
| -B1-1 | 27.3 (70) | 38.1 (>80) | 54.5 (>80) | 100.0 (>80) |
| -Ib | 18.2 (70) | 27.3 (>80) | 51.5 (>80) | 85.1 (>80) |
| -Ic | 12.1 (60) | 30.3 (>80) | 69.7 (>80) | 100.0 (>80) |
| -IIa | 13.6 (60) | 43.7 (>80) | 78.5 (>80) | 100.0 (>80) |
| -IIb | 0.0 (20) | 0.0 (40) | 73.6 (70) | 100.0 (>80) |
| -Am-2 | 0.0 (70) | 15.3 (>80) | 45.2 (>80) | 79.6 (>80) |
| *Acylphloroglucinol-sesquiterpenes* | | | | |
| Euglobal-III | 0.0 (60) | 28.9 (>80) | 80.5 (>80) | 100.0 (>80) |
| -IVa | 0.0 (20) | 0.0 (30) | 59.1 (>80) | 100.0 (>80) |
| -IVb | 0.0 (10) | 0.0 (20) | 68.7 (>80) | 91.3 (>80) |
| -V | 14.7 (70) | 56.4 (>80) | 87.1 (>80) | 100.0 (>80) |
| -VII | 0.0 (20) | 11.4 (30) | 70.2 (>80) | 100.0 (>80) |
| -In-1 | 16.3 (70) | 68.3 (>80) | 90.5 (>80) | 100.0 (>80) |
| *Macrocarpals* | | | | |
| Macrocarpal-A | 20.6 (10) | 48.6 (60) | 66.1 (>80) | 87.4 (>80) |
| -B | 14.2 (20) | 35.7 (60) | 63.2 (>80) | 100.0 (>80) |
| -E | 17.7 (20) | 39.4 (60) | 78.1 (>80) | 100.0 (>80) |
| -am-1 | 33.2 (10) | 67.5 (60) | 88.0 (>80) | 100.0 (>80) |

a Measured by inhibitory effects on Epstein-Barr virus early antigen induction. Values represent relative percentages to the positive control value (100%); values in parentheses are viability percentages of Raji cells.
b Mol ratio TPA (20 ng = 32 pmol).
c Not detected.

viability of Raji cells even at a high concentration. On the other hand, euglobal-Ia1, -Ia2, -IIb, -IVa, -IVb and -VII showed strong cytotoxicity towards Raji cells (exhibiting less than 40 per cent viability of Raji cells at 1000 and 500 mol ratio/TPA). In addition to euglobals, macrocarpals A, B, E and am-1 were tested and showed weak inhibitory effects on EBV-EA activation, similar to those of euglobal-B1-1 and -IIa, with strong cytotoxicity towards Raji cells at high concentration (less than 20 per cent viability at 1000 mol ratio/TPA). The use of indirect immunofluorescence techniques by antigen-antibody reaction requires a high viability of Raji cells (more than 60 per cent viability) to justify a subsequent *in vivo* assay.

The inhibitory effects of euglobal-G1, -G2, -G4, -G5, -Am-2 and -III were stronger than those of glycyrrhetic acid, a well-known potent anti-tumour promoter, and indicate their potential for cancer chemoprevention.

## Anti-tumour promoting activity of euglobal-G1 and -III

In these active euglobals, the yields of euglobal-G1 and -III in the plant are higher than those of the others, and this has enabled the effects of -G1 and -III on the cell cycle of Raji cells treated with TPA to be examined by flow cytometry[5] (Shimomatsuya et al. 1991). The promoter (TPA) increased the percentage of the $G_2$ and M phases of Raji cells and decreased the percentages of both the $G_1$ and S phases in comparison with those of the negative control cultivated without TPA, as shown in Table 12.7. When treated separately with euglobal-G1 and -III, the effects of TPA on the cell phases were opposed and at the highest concentration (32 nmol) the percentages of each phase were near normal. It therefore appears that euglobal-G1 and -III accumulate in Raji cells in the S phase, dependent on the concentration of these compounds, and that, consequently, the percentage of $G_2$ and M phase is restored to a normal value. Therefore, by influencing the cell cycle, euglobal-G1 and -III also strongly inhibit one of the biological activities of the tumour promoter TPA. These in vitro results suggest that euglobal-G1 and -III could be of considerable value as anti-tumour promoters.

To test the potential of euglobal-G1 and -III as anti-tumour promoters, their inhibitory effects in the in vivo two-stage carcinogenesis test in mice have been investigated (Konoshima et al. 1993). One week after initiation of carcinogenic growth with 7,12-dimethylbenz[a]anthracene (DMBA), TPA was applied twice a week to promote the growth. In the test group of mice, each euglobal was applied one hour before the TPA treatment. The inhibitory activities were then evaluated by

Table 12.7 Results of testing euglobal-G1 and -III as anti-tumour promoters by flow cytometry

| Treatment | % of Raji cells | | | Total |
|---|---|---|---|---|
| | Phase $G_1$ | Phase S | Phase $G_2$ + M | |
| Medium only[a] | 61.7 | 27.9 | 10.4 | 100.0 |
| Positive control[b] | 53.6 | 8.4 | 38.0 | 100.0 |
| Euglobal-G1[c] | | | | |
| 32.0 nmol | 60.8 | 28.8 | 10.4 | 100.0 |
| 3.2 nmol | 57.2 | 24.0 | 18.8 | 100.0 |
| 0.32 nmol | 52.7 | 8.2 | 39.1 | 100.0 |
| Euglobal-III[c] | | | | |
| 32.0 nmol | 62.0 | 29.1 | 8.9 | 100.0 |
| 3.2 nmol | 60.5 | 25.6 | 13.9 | 100.0 |
| 0.32 nmol | 54.5 | 8.0 | 37.5 | 100.0 |

a Raji cells cultivated in medium containing 10% fetal calf serum without TPA.
b Treated with TPA (32 pmol) and n-butyric acid.
c Treated with TPA (32 pmol), n-butyric acid and euglobal; 32.0, 3.2 and 0.32 nmol = 1000, 100 and 10 mol ratio/TPA, respectively.

---

[5] Division and proliferation of the cells proceeds via the $G_1$ phase (resting phase after division), S phase (DNA synthetic phase), $G_2$ phase (resting phase before division) and M phase (mitotic phase). The dispersion and distribution of cells in each phase can be measured by flow cytometry.

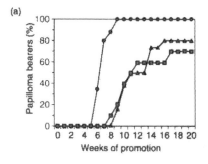 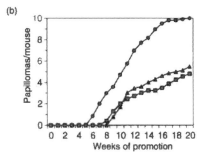

*Figure 12.11* Inhibition of TPA-induced tumour promotion by multiple application of euglobal-G1 and -III. Measured in terms of (a) number of papilloma-bearing mice and (b) average number of papillomas per mouse. Symbols: ● control (TPA alone), ■ TPA + 85 nmol euglobal-G1, ▲ TPA + 85 nmol euglobal-III.

comparing the number (per cent) of papilloma-bearing mice (Figure 12.11a) and the average number of papillomas per mouse (Figure 12.11b) with those of the control group. As Figure 12.11a and b indicate, application of euglobal-G1 and -III before each TPA treatment significantly delayed the formation of papillomas in mouse skin and reduced the number of papillomas per mouse. Euglobal-G1 and -III exhibited about 55 per cent and 44 per cent inhibition, respectively, even at twenty weeks of promotion. The results suggest that euglobal-G1 and -III may be valuable as anti-tumour promoters in two-stage chemical carcinogenesis.

## Conclusion

A diverse range of biological activities has been demonstrated by the many kinds of non-volatile secondary metabolites isolated from *Eucalyptus* species. The types of metabolites and examples of their pharmacological actions have been described above. The activities include antioxidative, antimalarial, antibacterial, antiviral, HIV-RTase inhibitory, aldose reductase inhibitory and cancer chemopreventive activities. Such broad pharmacological profiles provide a stimulus to further research and it is likely that more new compounds will be isolated from an increasing number of *Eucalyptus* species in the future. The ease of cultivation and rapid growth of eucalyptus makes it, potentially, a very valuable natural resource for the commercial production of pharmaceuticals, over and above the present production of eucalyptus oil for medicinal purposes. It is likely that in the years ahead *Eucalyptus* metabolites other than the volatile constituents will form part of the armoury of drugs available to the physician for the treatment or prevention of human diseases.

## References

Amano, T., Komiya, T., Hori, M., Goto, M., Kozuka, M. and Sawada, T. (1981) Isolation and characterization of euglobals from *Eucalyptus globulus* Labill. by preparative reversed-phase liquid chromatography. *J. Chromatogr.*, 208, 347–355.

Aukrust, I.R. and Skattebol, L. (1996) The synthesis of (−)-robustadial A and some analogues. *Acta Chem. Scand.*, 50, 132–140.

Barnabas, C.G.G. and Nagarajan, S. (1988) Antimicrobial activity of flavonoids of some medicinal plants. *Fitoterapia*, 59, 508–510.

Berenblum, I. (1941) The mechanism of carcinogenesis, a study of the significance of cocarcinogenic action and related phenomena. *Cancer Res.*, 1, 807–814.

Bruneton, J. (1995) *Pharmacognosy, Phytochemistry, Medicinal Plants*, Lavoisier, France, pp. 459–461.

Cheng, Q. and Snyder, J.K. (1988) Revised structures of robustadials A and B from *Eucalyptus robusta*. *J. Org. Chem.*, 53, 4562–4567.

Chiba, K., Sonoyama, J. and Tada, M. (1995) Intermolecular cycloaddition reaction of unactivated alkenes and o-quinone methides generated by electrochemical oxidation, a proposed biomimetic approach to the euglobal skeletons. *J. Chem. Soc., Chem. Comm.*, 1381–1382.

Chiba, K., Arakawa, T. and Tada, M. (1996) Synthesis of euglobal-G3 and -G4, *J. Chem. Soc., Chem. Comm.*, 1763–1764.

Coppen, J.J.W and Dyer, L.R. (1993) *Eucalyptus and its Leaf Oils. An Indexed Bibliography*, Natural Resources Institute, Chatham, UK.

Coppen, J.J.W. and Hone, G.A. (1992) *Eucalyptus Oils: A Review of Production and Markets*, NRI Bulletin 56, Natural Resources Institute, Chatham, UK.

Dayal, R. (1988) Phytochemical examination of eucalypts – a review. *J. Sci. Ind. Res.*, 47, 215–220.

Egawa, H., Tsutsui, O., Tatsuyama, K. and Hatta, T. (1977) Antifungal substances found in leaves of *Eucalyptus* species. *Experientia*, 33, 889–890.

Ghisalberti, E.L. (1996) Bioactive acylphloroglucinol derivatives from *Eucalyptus* species. *Phytochemistry*, 41, 7–22.

Giron, L.M., Aguilar, G.A., Caceres, A. and Arroyo, G.L. (1988) Anticandidal activity of plants used for the treatment of vaginitis in Guatemala and clinical trial of a *Solanum nigrescens* preparation. *J. Ethnopharmacol.*, 22, 307–313.

Hill, K.D. and Johnson, L.A.S. (1995) Systematic studies in the eucalypts. 7. A revision of the bloodwoods, genus *Corymbia* (Myrtaceae). *Telopea*, 6, 185–504.

Jangsu College of New Medicine (ed.) (1977) *Dictionary of Chinese Materia Medica*, Shanghai People's Publishing House.

Kapadia, G.J., Tokuda, H., Konoshima, T., Takasaki, M., Takayasu, J. and Mishino, H. (1996) Anti-tumour-promoting activity of *Dryopteris* phlorophenone derivatives. *Cancer Letts.*, 105, 161–165.

Kokumai, M., Konoshima, T., Kozuka, M., Haruna, M. and Ito, K. (1991) Euglobal-T1, a new euglobal from *Eucalyptus tereticornis*. *J. Nat. Prod.*, 54, 1082–1086.

Konoshima, T., Kokumai, M., Kozuka, M., Tokuda, H., Nishino, H. and Iwashima, A. (1992) Anti-tumour-promoting activities of afromosin and soyasaponin I isolated from *Wistaria brachybotrys*. *J. Nat. Prod.*, 55, 1776–1778.

Konoshima, T., Kozuka, M., Tokuda, H., Nishino, H., Iwashima, A., Haruna, M., Ito, K., and Tanabe, M. (1991) Studies on inhibitors of skin tumour promotion. IX. Neolignans from *Magnolia officinalis*. *J. Nat. Prod.*, 54, 816–822.

Konoshima, T., Takasaki, M., Kozuka, M., Nagao, T., Okabe, H., Irino, N., Nakatsu, T., Tokuda, H. and Nishino, H. (1995) Inhibitory effects of cucurbitane triterpenoids on Epstein-Barr virus activation and two-stage carcinogenesis of skin tumour. II. *Biol. Pharm. Bull.*, 18, 284–287.

Konoshima, T., Takasaki, M., Tokuda, H., Masuda, K., Arai, Y., Shiojima, K. and Ageta, H. (1996) Anti-tumour-promoting activities of triterpenoids from fern (1). *Biol. Pharm. Bull.*, 19, 962–965.

Konoshima, T., Terada, H., Kokumai, M., Kozuka, M., Tokuda, H., Estes, J.R., Li, L.P., Wang, H.K. and Lee, K.H. (1993) Studies on inhibitors of skin tumour promotion. XII. Rotenoids from *Amorpha fruticosa*. *J. Nat. Prod.*, 56, 843–848.

Kozuka, M., Sawada, T., Kasahara, F., Mizuta, E., Amano, T., Komiya, T. and Goto, M. (1982a) The granulation-inhibiting principles from *Eucalyptus globulus* Labill. II. The structures of euglobal-Ia1, -Ia2, -Ib, -Ic, -IIa, -IIb and -IIc. *Chem. Pharm. Bull.*, 30, 1952–1963.

Kozuka, M., Sawada, T., Mizuta, E., Kasahara, F., Amano, T., Komiya, T. and Goto, M. (1982b) The granulation-inhibiting principles from *Eucalyptus globulus* Labill. III. The structures of euglobal-III, -IVb and -VII. *Chem. Pharm. Bull.*, 30, 1964–1973.

Majewski, M., Irvine, N.M. and Bantle, G.W. (1994) Stereoselective synthesis of dimethylrobustadials. *J. Org. Chem.*, 59, 6697–6702.

Muanza, D.N., Kim, B.W., Euler, K.L. and Williams, L. (1994) Antibacterial and antifungal activities of nine medicinal plants from Zaire. *Int. J. Pharmacog.*, 32, 337–345.

Murata, M., Yamakoshi, Y., Homma, S., Aida, K., Hori, K. and Ohashi, Y. (1990) Macrocarpal A, a novel antibacterial compound from *Eucalyptus macrocarpa*. *Agric. Biol. Chem.*, 54, 3221–3226.

Murata, M., Yamakoshi, Y., Homma, S., Arai, K. and Nakamura, Y. (1992) Macrocarpals, antibacterial compounds from *Eucalyptus*, inhibit aldose reductase. *Biosci. Biotech. Biochem.*, 56, 2062–2063.

Nakayama, R., Murata, M., Homma, S. and Aida, K. (1990) Antimicrobial compounds from *Eucalyptus perriniana*. *Agric. Biol. Chem.*, 54, 231–232.

Navarro, V., Villarreal, M.L., Rojas, G. and Lozoya, X. (1996) Antimicrobial evaluation of some plants used in Mexican traditional medicine for the treatment of infectious diseases. *J. Ethnopharmacol.*, 53, 143–147.

Nishizawa, M., Emura, M., Kan, Y., Yamada, H., Ogawa, K. and Hamanaka, N. (1992) Macrocarpals: HIV-RTase inhibitors of *Eucalyptus globulus*. *Tetrahedron Letts.*, 2983–2986.

Okamura, H., Mimura, A., Yakou, Y., Niwano, M. and Takahara, Y. (1993) Antioxidant activity of tannins and flavonoids in *Eucalyptus rostrata*. *Phytochemistry*, 33, 557–561.

Osawa, K., Yasuda, H., Morita, H., Takeya, K. and Itokawa, H. (1995) Eucalyptone from *Eucalyptus globulus*. *Phytochemistry*, 40, 183–184.

Osawa, K., Yasuda, H., Morita, H., Takeya, K. and Itokawa, H. (1996) Macrocarpals H, I and J from the leaves of *Eucalyptus globulus*. *J. Nat. Prod.*, 59, 823–827.

Osawa, K., Yasuda, H., Morita, H., Takeya, K. and Itokawa, H. (1997) Configurational and conformational analysis of macrocarpals H, I and J from *Eucalyptus globulus*. *Chem. Pharm. Bull.*, 45, 1216–1217.

Osawa, T., Ide, A., Su, J.D. and Namiki, M. (1987) Inhibition of lipid peroxidation by ellagic acid. *J. Agric. Food Chem.*, 35, 808–811.

Osawa, T. and Namiki, M. (1981) A novel type of antioxidant isolated from leaf wax of *Eucalyptus* leaves. *Agric. Biol. Chem.*, 45, 735–739.

Osawa, T. and Namiki, M. (1983) In J.V. McLoughlin and B.M. Mckenna (eds), *Research in Food Science and Nutrition, Vol. 3*, Boole Press, Dublin, p. 49.

Osawa, T. and Namiki, M. (1985) Natural antioxidants isolated from *Eucalyptus* leaf waxes. *J. Agric. Food Chem.*, 33, 777–779.

Otsuka, H., Tsukui, M., Matsuoka, T., Goto, M., Fujimura, H., Hiramatsu, Y. and Sawada, T. (1974) Studies on anti-inflammatory agents. Anti-inflammatory screening by fertile egg method (in Japanese). *Yakugaku Zasshi*, 94, 796–801.

Qin, G.W., Chen, Z.X., Wang, H.C. and Qian, M.K. (1981) The structure and synthesis of robustaol A (in Chinese). *Acta Chim. Sin.*, 39, 83–89.

Qin, G.W. and Xu, R.S. (1986) Studies on chemical constituents of *Eucalyptus robusta* Sm. (in Chinese). *Acta Chim. Sin.*, 44, 151–156.

Satoh, H., Etoh, H., Watanabe, N., Kawagishi, H., Arai, K. and Ina, K. (1992) Structures of sideroxylonals from *Eucalyptus sideroxylon*. *Chem. Letts.*, 1917–1920.

Sawada, T., Kozuka, M., Komiya, T., Amano, T. and Goto, M. (1980) Euglobal-III, a novel granulation-inhibiting agent from *Eucalyptus globulus* Labill. *Chem. Pharm. Bull.*, 28, 2546–2548.

Sayer, J.M., Yagi, H., Wood, A.W., Conney, A.H. and Jerina, D.M. (1982) Extremely facile reaction between the ultimate carcinogen benzo[a]-pyrene-7,8-diol-9,10-epoxide and ellagic acid. *J. Am. Chem. Soc.*, 104, 5562–5564.

Shimomatsuya, T., Tanigawa, N. and Muraoka, R. (1991) Proliferative activity of human tumours: assessment using bromodeoxyuridine and flow cytometry. *Jpn. J. Cancer Res.*, 82, 357–362.

Singh, A.K., Khare, M. and Kumar, S. (1999) Non-volatile constituents of eucalypts: a review on chemistry and biological activities. *J. Med. Arom. Plant Sci.*, 21, 375–407.

Singh, I.P. and Etoh, H. (1995) New macrocarpal-am-1 from *Eucalyptus amplifolia*. *Biosci. Biotech. Biochem.*, 59, 2330–2332.

Singh, I.P. and Etoh, H. (1997) Biological activities of phloroglucinol derivatives from *Eucalyptus* spp. *Natural Product Sci.*, 3, 1–7.

Singh, I.P., Hayakawa, R., Etoh, H., Takasaki, M. and Konoshima, T. (1997) Grandinal, a new phloroglucinol dimer from *Eucalyptus grandis*. *Biosci. Biotech. Biochem.*, 61, 921–923.

Singh, I.P., Takahashi, K. and Etoh, H. (1996) Potent attachment-inhibiting and -promoting substances for the blue mussel, *Mytilus edulis galloprovincialis*, from two species of *Eucalyptus*. *Biosci. Biotech. Biochem.*, 60, 1522–1523.

Singh, I.P., Umehara, K., Asai, T., Etoh, H., Takasaki, M. and Konoshima, T. (1998) Phloroglucinol-monoterpene adducts from *Eucalyptus grandis*. *Phytochemistry*, 47, 1157–1159.

Takasaki, M. (1995) *Studies of Euglobals from Eucalyptus Plants* (in Japanese), PhD Thesis, pp. 9–21.

Takasaki, M., Konoshima, T., Shingu, T., Tokuda, H., Nishino, H., Iwashima, A. and Kozuka, M. (1990) Structure of euglobal-G1, -G2 and -G3 from *Eucalyptus grandis*, three new inhibitors of Epstein-Barr virus activation. *Chem. Pharm. Bull.*, 38, 1444–1446.

Takasaki, M., Konoshima, T., Kozuka, M., Haruna, M., Ito, K., Crow, W.D. and Paton, D.M. (1994a) Euglobal-In-1, a new euglobal from *Eucalyptus incrassata*. *Chem. Pharm. Bull.*, 42, 2113–2116.

Takasaki, M., Konoshima, T., Kozuka, M., Haruna, M., Ito, K. and Yoshida, S. (1994b) Four euglobals from *Eucalyptus blakelyi*. *Chem. Pharm. Bull.*, 42, 2177–2179.

Takasaki, M., Konoshima, T., Kozuka, M., Haruna, M., Ito, K. and Shingu, T. (1994c) Structures of euglobals-G1, -G2, -G3, -G4 and -G5 from *Eucalyptus grandis*. *Chem. Pharm. Bull.*, 42, 2591–2597.

Takasaki, M., Konoshima, T., Kozuka, M., Yoneyama, K., Yoshida, S., Tokuda, H., Nishino, H. and Iwashima, A. (1995a) Inhibitors of skin-tumour promotion. XIII. Inhibitory effects of euglobals and their related compounds on Epstein-Barr virus activation and on two-stage carcinogenesis of mouse skin tumour. *Biol. Pharm. Bull.*, 18, 288–294.

Takasaki, M., Konoshima, T., Kozuka, M. and Tokuda, H. (1995b) Anti-tumour promoting activities of euglobals from *Eucalyptus* plants. *Biol. Pharm. Bull.*, 18, 435–438.

Takasaki, M., Konoshima, T., Kozuka, M., Haruna, M. and Ito, K. (1997) Euglobals -In-2 and -In-3, new euglobals from *Eucalyptus incrassata*. *Natural Medicines*, 51, 486–490.

Tanaka, T., Mikamiyama, H., Maeda, K., Ishida, T., In, Y. and Iwata, C. (1997) First stereoselective total synthesis of macrocarpal C: structure elucidation of macrocarpal G. *J. Chem. Soc., Chem. Comm.*, 2401–2402.

Umehara, K., Singh, I.P., Etoh, H., Takasaki, M. and Konoshima, T. (1998) Five phloroglucinol-monoterpene adducts from *Eucalyptus grandis*. *Phytochemistry*, 49, 1699–1704.

Xu, R.S., Snyder, J.K. and Nakanishi, K. (1984) Robustadials A and B from *Eucalyptus robusta*. *J. Am. Chem. Soc.*, 106, 734–736.

Yamakoshi, Y., Murata, M., Shimizu, A. and Homma, S. (1992) Isolation and characterization of macrocarpals B-G, antibacterial compounds from *Eucalyptus macrocarpa*. *Biosci. Biotech. Biochem.*, 56, 1570–1576.

# 13 Antimicrobial activity of eucalyptus oils

*Stanley G. Deans*

## Introduction

The preservative properties of the volatile oils and extracts of aromatic and medicinal plants have been recognised since Biblical times, while attempts to characterise these properties in the laboratory date back to the early 1900s (e.g. Hoffman and Evans 1911). Martindale (1910) included '*Eucalyptus amygdalina*' (probably the phellandrene variant of *E. dives*) and *E. globulus* oils, as well as eucalyptol (1,8-cineole), in his study of the antiseptic powers of essential oils and although the 'carbolic coefficients' of eucalyptus oils were not as great as those for oils containing large amounts of phenolics – such as origanum (carvacrol), cinnamon leaf (eugenol) and thyme (thymol) – they did, nevertheless, give some quantitative measure of the antiseptic properties of eucalyptus leaf oils.

Many volatile oils – particularly those of herbs and spices, but including those from *Eucalyptus* – have been used to extend the shelf-life of foods, beverages and pharmaceutical and cosmetic products; their antimicrobial and antioxidant properties have also pointed to a role in plant protection. Such a wide variety of applications, actual or potential, has meant that the antimicrobial properties of volatile oils and their constituents from a large number of plants have been assessed and reviewed (Jain and Kar 1971, Inouye *et al.* 1983, Shelef 1984, Gallardo *et al.* 1987, Janssen *et al.* 1987, Rios *et al.* 1988, Knobloch *et al.* 1989, Péllisier *et al.* 1994, Shapiro *et al.* 1994, Carson *et al.* 1996, Nenoff *et al.* 1996, Pattnaik *et al.* 1996, Baratta *et al.* 1998a,b, Youdim *et al.* 1999). In recent years attempts have been made to identify the component(s) of the oils responsible for such bioactivities (e.g. Deans and Ritchie 1987, Jeanfils *et al.* 1991, Deans *et al.* 1994, Lis-Balchin *et al.* 1998, Daferera *et al.* 2000, Dorman and Deans 2000). Bhalla (1997) has listed the organisms against which oils from Indian eucalypts have been tested.

The level of interest in the antimicrobial properties of volatile oils is just one aspect of the practical potential that such oils have in various protective roles. There also appears to be a revival in the use of traditional approaches to livestock welfare and food preservation in which essential oils play a part (Thomann *et al.* 1997). The use of *Eucalyptus* and its oils in protecting stored food products against insect pests is discussed elsewhere (Chapter 14), as are the non-volatile constituents of *Eucalyptus*, some of which have antibacterial properties (Chapter 12). It is intended here to examine the antimicrobial activity of eucalyptus oils and, where it exists, to assess its potential application to human health care, food preservation and plant protection.

## Antimicrobial activity of eucalyptus leaf oils

Unfortunately, much of the research involving the antimicrobial activity of volatile oils, including that of *Eucalyptus*, has been empirical with many of the oils tested simply because they have been readily at hand. Often, little or nothing has been done to determine the composition of the oils. Occasionally, the botanical source of the oil is not stated or, where it is, the existence of chemical variants (chemotypes) is not acknowledged so that even the compositional type cannot be known with certainty. When commercially available essential oils are tested it is not particularly helpful simply to refer to them by name. As Lis-Balchin *et al.* (1998) have found, oils of the same name can have widely different activities. Testing of 'eucalyptus oil', without any indication of geographic or botanical source, or composition, is therefore of limited value. As far as possible, the examples cited below have been chosen only when the eucalyptus oil tested is from a named species of *Eucalyptus*. Compositional data are also quoted where possible.

A discussion of the antimicrobial activity of individual eucalyptus oil constituents such as cineole and citronellal, and the question of possible structure–activity relationships which might enable one to focus further research on particular types of eucalyptus oil, are deferred till later, after the results of testing whole oils are examined.

## Antifungal activity

### Human pathogens

The volatile oil from *Eucalyptus camaldulensis* (syn. *E. rostrata*) has been the subject of several studies where the target organisms were dermatophytic fungi. Singh *et al.* (1988) tested the oil against four human pathogens, *Trichophyton mentagrophytes*, *Epidermophyton floccosum*, *Microsporum canis* and *M. gypseum*, as well as two storage fungi, *Aspergillus nidulans* and *A. terreus*. At concentrations of 10,000 ppm (1 per cent) the oil showed fungicidal activity towards all the test organisms. In a second study (Ara and Misra 1992), a combination of oils from *E. camaldulensis* and *Juniperus communis* was found to be more effective than either single oil against *Epidermophyton floccosum*, *M. gypseum* and *Paecilomyces variotii*. The minimum inhibitory concentration (MIC) and time taken to inhibit mycelial growth were less with the mixture than with the individual oils, suggesting that there were synergistic interactions between the components present in the two oils.

In a wide-ranging study Pattnaik *et al.* (1996) tested ten essential oils, one of them from *E. citriodora*, against twelve test fungi (mostly human pathogens, with a few plant pathogens): *Alternaria citrii*, *Aspergillus fumigatus*, *A. oryzae*, *Candida albicans*, *Cryptococcus neoformans*, *Fusarium oxysporum*, *F. solani*, *Helminthosporium compactum*, *Macrophomina phaseolina*, *Sclerotium rolfsii*, *Sporothrix schenckii* and *Trichophyton mentagrophytes*. The eucalyptus oil was effective against all the fungi except *M. phaseolina*. Seven of the oils (citronella, lemongrass, patchouli, palmerosa, geranium, orange and aegle) were effective against all twelve fungi, with lemongrass performing the best.

*E. pellita* oil was found to be active against the human dermatophytes *M. gypseum* and *T. mentagrophytes*, and exhibited greater inhibition than the pine oil from *Pinus caribaea* (Duarte *et al.* 1992).

### Organisms involved in food spoilage

The range of fungi that have been screened using eucalyptus oils is very extensive. It has been shown that inhibition of members of the genera *Penicillium* (such as the mycotoxigenic *P. citrinum*) and *Aspergillus* (including aflatoxin-producing *A. flavus*) is readily achievable. Food spoilage fungi are also inhibited, making eucalyptus a potentially useful oil in the preservation of foodstuffs.

*E. globulus* has been tested for its inhibitory effect on the growth of a large number of mould species frequently involved in food spoilage. Benjilali *et al.* (1984) tested it using a micro-atmosphere method, along with five other oils, against thirteen *Penicillium* spp., six *Aspergillus* spp. and sixteen other species. However, while thyme oil was consistently the best performer, the eucalyptus oil was the least effective. Illustrative results for eucalyptus, rosemary and thyme oils, including compositional data for the oils, are shown in Table 13.1. Similar results were obtained using the same oils but an alternative method of testing: *E. globulus* oil was moderately effective against *Byssochlamys nivea*, *Geotrichum candidum*, *Paecilomyces variotii*, *Penicillium purpurogenum* and *Stachybotrys* sp., all spoilage organisms, but was the least effective of the oils overall (Benjilali *et al.* 1986).

The mycotoxigenic *Aspergillus flavus*, responsible for the production of aflatoxins in groundnuts and other crops, has received much attention from researchers. Several groups have included eucalyptus oil in screening volatile oils for possible use as antifungal agents although the results have not been particularly encouraging. Montes-Belmont and Carvajal (1998) investigated the inhibitory effects of eleven volatile oils, including *E. globulus*, against *A. flavus* growing on maize kernels. The antifungal effect of the eucalyptus oil was described as scanty and no further evaluations of it were carried out. Ansari and Shrivastava (1991) examined the inhibitory power of eucalyptus oil at three concentrations: the lower two concentrations caused *A. flavus* growth and toxin production to be inhibited while the highest concentration resulted in complete inhibition of growth. After twelve days incubation, however, toxin production was greater than in the control. It was speculated that the conidia and mycelium were initially under stress, showing poor growth and low toxin production, but that this was followed by high toxin production in the late phase of incubation. Masood and Ranjan (1990) have reported on the ineffectiveness of fungitoxicant after advanced incubation which stimulated synthesis of aflatoxin, and the correlation between stress and aflatoxin biosynthesis clearly requires further investigation.

*Table 13.1* Antifungal activity[a] of eucalyptus, rosemary and thyme oils, with composition in terms of selected constituents (after Benjilali *et al.* 1984)

| Organism | Eucalyptus E. globulus | Rosemary Rosmarinus officinalis | Thyme Thymus capitatus |
|---|---|---|---|
| *Aspergillus flavus* | − | + | +++++ |
| *A. fumigatus* | − | ++ | +++ |
| *A. niger* | + | + | +++++ |
| *A. repens* | ++ | +++ | +++++ |
| *Gliocladium roseum* | + | + | +++++ |
| *Mucor hiemalis* | ++ | ++ | +++ |
| *M. racemosus* | + | + | +++++ |
| *Penicillium clavigerum* | ++ | ++ | +++++ |
| *P. cyclopium* | − | + | + |
| *P. notatum* | + | − | +++++ |
| *P. purpurogenum* | ++ | + | +++++ |
| *Stachybotrys* sp. | ++ | +++ | +++++ |
| Oil composition[b] | | | |
| 1,8-Cineole | 69 | 52 | − |
| Camphene | 29 | 4 | 2 |
| α-Pinene | − | 15 | Trace |
| Camphor | − | 12 | − |
| Carvacrol | − | − | 78 |
| *p*-Cymene | − | − | 15 |

a Measured by volume of oil required to give complete inhibition: >100 μl (−), 100 μl (+), 50 μl (++), 20 μl (+++), 10 μl (++++), 5 μl (+++++).
b Relative abundance.

In tests against the spoilage organisms *Aspergillus niger*, *Penicillium italicum* and *Zygorrhyncus* sp., *E. globulus* oil was found to cause reversible inhibition of spore germination (Tantaoui-Elaraki *et al.* 1993). At the lowest concentration (0.01 per cent), however, the oil actually caused stimulation of germination in *A. niger* conidia and there was evidence to suggest that the fungus responded positively to the stress caused by low concentrations of oil on the mycelium. Both of the other oils tested, origanum and mugwort, were more effective at inhibiting spore germination than the eucalyptus oil and this seems to be in keeping with the observations made by others on the effectiveness of different types of oil, namely, phenolic oils > ketone-rich oils > cineole-rich oils.

*A. niger* and *Zygorrhyncus* sp., as well as the yeasts *Candida albicans* and *Saccharomyces cerevisiae*, and some bacteria, were included by Hajji *et al.* (1993) in a more comprehensive screening process involving twenty-one different *Eucalyptus* oils:

| | | |
|---|---|---|
| *E. amygdalina* | *E. diversicolor* | *E. occidentalis* |
| (syn. *E. salicifolia*) | *E. eximia* | *E. oleosa* |
| *E. astringens* | *E. globulus* subsp. *maidenii* | *E. paniculata* |
| *E. calophylla* | *E. gomphocephala* | *E. piperita* |
| *E. camaldulensis* | *E. longifolia* | *E. sideroxylon* |
| *E. citriodora* | *E. macarthurii* | *E. viminalis* |
| *E. cladocalyx* | *E. macrorhyncha* | |
| *E. dealbata* | *E. melliodora* | |

*E. cladocalyx* oil was the most effective of the oils tested although its yield from the leaves was relatively poor (0.4 per cent) and not indicative of one that could be produced commercially. *E. citriodora* oil was obtained in the highest yield (2.5 per cent) and was moderately effective.

*E. citriodora* oil has been tested, along with palmarosa (*Cymbopogon martinii*), against a variety of organisms, including human and plant pathogens as well as food spoilage organisms: *Alternaria alternata*, three *Aspergillus* spp. (*A. flavus*, *A. fumigatus* and *A. niger*), *Cladosporium cladosporioides*, *Curvularia lunata* and two *Fusarium* spp. (*F. oxysporum* and *F. solani*) (Srivastava *et al.* 1993). Both oils contain geraniol and citronellol. Palmarosa was more effective than the eucalyptus oil at controlling fungal growth, although there was some variability towards different organisms. Both oils caused complete inhibition at 0.01 per cent concentration. In another study involving *E. citriodora* oil and *Cymbopogon martinii* (var. *motia*), Baruah *et al.* (1996) investigated their antifungal activity towards *Fusarium moniliforme*, a post-harvest pathogen of cereal crops. *Mentha piperita* and *Cinnamomum tamala* oils were also tested. Using a disc assay and measuring zones of inhibition on the surface of agar plates, all four oils exhibited activity, with *Cymbopogon martinii* the most effective and *E. citriodora* the next most effective.

*Plant protection*

It would be of great benefit to be able to employ eucalyptus oil as a natural fungicide, one which was biodegradable and able to control some of the important plant pathogens. The potential use of eucalyptus oils in agriculture has been investigated by Singh and Dwivedi (1987) in attempts to control *Sclerotium rolfsii*, the causative organism of foot-rot of barley. Four concentrations, 1000–4000 ppm, were used in the poison food technique, where a number of inhibition parameters were used as a measure of efficacy, including diameter of the colony on agar media, dry weight of hyphal mat and reduction in the viability of sclerotia. Of five different oils tested, *E. globulus* and *Ocimum americanum* (syn. *O. canum*) were the most effective, with MICs of

<4000 ppm. However, further studies showed neem oil (from *Azadirachta indica*) to be more effective against *S. rolfsii* than both *E. globulus* and *O. americanum* oils (Singh *et al.* 1989, Singh and Dwivedi 1990). Nevertheless, *E. globulus* oil showed considerable activity towards ten soil fungi, including the mycotoxigenic *Penicillium citrinum*; it was most active against *Trichoderma viride* (Singh *et al.* 1989).

The leaf oil from *E. camaldulensis* was part of a selection of plant oils assayed for antimycotic activity against the soil-borne fungi *Fusarium moniliforme*, *Phytophthora capsici*, *Rhizoctonia solani* and *Sclerotinia sclerotiorum* (Müller-Riebau *et al.* 1995). The eucalyptus oil, however, was found to have only slight inhibitory effects in the initial bioassay and was not studied further. Oils from *Origanum minitiflorum*, *Thymbra spicata* and *Satureja thymbra* were the most active and fungitoxicity was shown to be due to the presence of carvacrol and/or thymol. *Salvia fruticosa*, which contained twice as much 1,8-cineole as *E. camaldulensis*, had a similar, weakly active profile to the eucalyptus oil.

Eucalyptus oil, along with turpentine and clove oil, completely checked rotting when guava fruits (*Psidium guajava*) inoculated with *Pestalotiopsis versicolor* and *Rhizoctonia solani* were dipped in them. However, they had adverse effects on the keeping quality of the treated fruits, in contrast to mustard oil which checked the rotting and preserved keeping quality (Madhukar and Reddy 1989).

The widespread planting of the so-called '*Eucalyptus* hybrid' in India (essentially *E. tereticornis*), and its leaf oil, have been referred to elsewhere (Chapter 11 this volume). The oil contains variable amounts of pinenes, 1,8-cineole and *p*-cymene. In their work to produce true hybrids with superior growth and wood and oil quality characteristics, Chaudhari and Suri (1991) reported the oil properties of $F_1$ hybrids of reciprocal crosses *E. tereticornis* × *E. camaldulensis* and *E. camaldulensis* × *E. tereticornis*, together with those of the parent species. The data include results from screening the oils for activity against six fungal species: *Aspergillus niger*, *Candida albicans*, *Epidermophyton rubrum*, *Malbranchea pulchella*, *Microsporum gypseum* and *Penicillium notatum*. The four oils were active against all six fungi at a dilution of 1 : 500 and the findings were deemed to bode well for the formulation of control measures against the organisms.

## Antibacterial activity

The antibacterial properties of plant volatile oils have been recognised since antiquity and have been rediscovered in more recent times. Eucalyptus leaf oils have received attention in a number of studies. Deans and Ritchie (1987) examined the antibacterial effects of fifty volatile oils purchased from a commercial supplier, including eucalyptus, on twenty-five different bacterial genera. The culture collection consisted of food spoilage, food poisoning, human, animal and plant disease types, along with indicators of faecal pollution and secondary opportunist pathogens. Eucalyptus oil was most effective against *Flavobacterium suaveolens* and the dairy organism *Leuconostoc cremoris*. However, it was not amongst the ten most inhibitory oils (thyme, cinnamon, bay, clove, bitter almond, lovage, pimento, marjoram, angelica and nutmeg).

Leaf oils from eight Brazilian-grown eucalypts were tested against *Mycobacterium avium* by Leite *et al.* (1998): *E. botryoides*, *E. camaldulensis*, *E. citriodora*, *E. deglupta*, *E. globulus*, *E. grandis*, *E. maculata* and *E. tereticornis*. *M. avium* was sensitive to all the oils at 10 mg/ml but only four of them at 5 mg/ml: *E. citriodora*, *E. maculata*, *E. camaldulensis* and *E. tereticornis*. *E. citriodora* and *E. maculata* oils were particularly rich in citronellal and citronellol. *E. citriodora* was also one of the better oils tested by Hajji *et al.* (1993) against bacteria such as *Bacillus megaterium* and *Staphylococcus aureus* although, like most of the other oils, it was least effective against *Escherichia coli*.

Another screening programme, this time involving seventeen eucalypts growing in Uruguay, had earlier been conducted by Dellacassa *et al.* (1989). Tests were carried out against two Gram-positive bacteria (*Bacillus subtilis* and *S. aureus*) and two Gram-negative ones (*E. coli* and *Pseudomonas aeruginosa*) using volatile oils from the following *Eucalyptus* species:

| | | |
|---|---|---|
| *E. affinis* | *E. diversicolor* | *E. paniculata* |
| *E. amplifolia* | *E. globulus* | *E. pellita* |
| *E. botryoides* | *E. lehmannii?*[1] | *E. punctata* |
| *E. camaldulensis* | *E. longifolia* | *E. sideroxylon* |
| *E. citriodora* | *E. maculata* | *E. tereticornis* |
| *E. cladocalyx* | *E. melliodora* | |

*B. subtilis* and *S. aureus* were most sensitive to the oils and *P. aeruginosa* least sensitive. For the latter, oils from only six of the eucalypts showed any inhibition of its growth. Of the seventeen eucalypts in all, only three (*E. affinis*, *E. cladocalyx* and *E. diversicolor*) inhibited the growth of all four organisms. Overall, oil from *E. globulus* (which had the highest cineole content, 64.5 per cent) was the least effective, only showing some activity towards *S. aureus*. *E. citriodora* oil only inhibited *S. aureus* and *E. coli*. The authors reported no correlation between either 1,8-cineole content or the content of any other constituent and antimicrobial activity, and suggested that the observed activity was due to combinations of more than one oil constituent that are specific to each test bacterium.

In yet another screening study, Kumar *et al.* (1988) evaluated freshly distilled leaf oils from twenty-four species of *Eucalyptus* against eight Gram-positive and seven Gram-negative bacteria, some of which were pathogens. The eucalypts were all growing locally in India but unfortunately no information was provided on the chemical composition of the oils:

| | | |
|---|---|---|
| *E. camaldulensis* (×2) | *E. laevopinea* | *E. rubida* |
| *E. crebra* | *E. leucoxylon* | *E. rudis* |
| *E. dalrympleana* | *E. melanophloia?*[2] | *E. tereticornis* |
| *E. deglupta* | *E. microcorys* | *E. viminalis* |
| *E. globulus* | *E. parvifolia?*[3] | *E. alba* × *E. camaldulensis* |
| *E. goniocalyx* | *E. regnans* | *E. tereticornis* × |
| *E. grandis* | *E. robertsonii* | *E. camaldulensis* |
| '*E. kirtoniana*' | *E. robusta* | '*Eucalyptus* hybrid' |

The Gram-positive bacteria tested included *Bacillus anthracis*, the causative organism of anthrax, *B. subtilis*, *Micrococcus glutamicus*, *Sarcina lutea*, *Staphylococcus aureus* and *Streptococcus pyogenes*. The Gram-negative bacteria included *Enterobacter* sp., *Listeria monocytogenes*, *Proteus vulgaris* and *Pseudomonas* sp. The authors confirmed that in general Gram-negative bacteria are less susceptible than Gram-positive ones. As might be expected, there were differences in inhibitory powers between the oils: *E. tereticornis*, one of the *E. camaldulensis* samples and *E. grandis* were effective against thirteen of the fifteen organisms but *E. melanophloia* and '*Eucalyptus* hybrid' showed no inhibition at all towards any of the Gram-negative bacteria. However, the lack of any compositional data on the oils is a serious weakness in the work – there were marked differences in the effectiveness of the two *E. camaldulensis* oils, for example, for which no explanation is offered.

---

[1] Authors state *E. leshmanii*.
[2] Authors state *E. melanofolia*.
Authors state *E. parviflora* so could also be *E. largiflorens*, for which the former is a synonym.

Of the Gram-positive bacteria, *Micrococcus glutamicus* was the most sensitive while *Streptococcus pyogenes* and *Sarcina lutea* were the most resistant, fewer than half the oils inhibiting them.

In a study of Malagasy medicinal plants De Medici *et al.* (1992) analysed oil from the leaves of *E. citriodora* and *E. globulus*, along with an undefined *Eucalyptus* species. The oils were also tested for activity against *Escherichia coli*. Using undiluted oils, *E. citriodora* oil (71 per cent citronellal) proved to be inactive while *E. globulus* (40 per cent cineole) and *Eucalyptus* sp. (18 per cent cineole, 43 per cent α-pinene) were only weakly active. Comparative results of *E. citriodora* and *E. globulus* oils with those of cineole-rich *Cinnamomum camphora* and *Melaleuca viridiflora*, and eugenol-rich *Ocimum gratissimum*, are shown in Table 13.2.

*E. citriodora* oil had earlier been tested against ten bacteria by Siva Sankara Rao and Nigam (1978). It showed some activity towards eight of them (*Bacillus fumilis*, *Micrococcus* sp., *Pseudomonas solanacearum*, *Sarcina lutea*, *Staphylococcus albus*, *Staphylococcus* sp., *Shigella* sp. and *Xanthomonas campestris*) but not to *Erwinia carotovora* and *Pseudomonas mangifera indica*. It was not as effective as the oil from *Cinnamomum zeylanicum*.

The difficulties in trying to account for the activity of whole oils simply by considering the activity of individual constituents in isolation has been well demonstrated by Low *et al.* (1974). They showed that artificial mixtures of citronellal/citronellol/cineole or citronellal/citronellol in the same concentrations as found in their sample of *E. citriodora* oil (i.e. 90:7.5:2.5 or 90:7.5, respectively) were as effective as the oil itself against *Staphylococcus aureus* and *Salmonella typhi*. This was in contrast to citronellal, citronellol or cineole individually.

The work of Chaudhari and Suri (1991) on *E. tereticornis* and *E. camaldulensis* hybrids and their parents, referred to earlier in connection with antifungal activity, included tests against eight bacteria: *Bacillus mycoides*, *B. pumilus*, *Escherichia coli*, *Proteus vulgaris*, *Salmonella paratyphi*, *Sarcina lutea*, *Shigella nigesta* and *Staphylococcus aureus*. The four oils were active against all the organisms at a dilution of 1:500.

Production of a soap in which the fragrance also serves as a bacteriostatic agent would have several advantages and Morris *et al.* (1979), in the laboratories of International Flavors and Fragrances, investigated over 520 fragrance raw materials with this objective in mind. The test samples included both whole oils and pure aroma chemicals and were initially screened against *Escherichia coli*, *Staphylococcus aureus* and *Candida albicans*. Some 44 per cent of the samples were inhibitory towards a single organism but only 15 per cent were active against all three. A lipophilic diphtheroid, *Corynebacterium* sp., was added to the testing protocol for just over 200 of the original samples, including a commercial cineole-rich eucalyptus oil (70–75 per cent). For eucalyptus oil, as for many of the other materials, the tests gave disappointing results: using the

*Table 13.2* Antibacterial activity of *Eucalyptus*, *Cinammomum*, *Melaleuca* and *Ocimum* oils against *Escherichia coli*, with composition in terms of selected constituents (after De Medici *et al.* 1992)

| Oil | Inhibition[a] | Oil composition[b] | | | | |
| --- | --- | --- | --- | --- | --- | --- |
| | | 1,8-Cineole | α-Pinene | Citronellal | Eugenol | α-Terpineol |
| *Eucalyptus citriodora* | — | 1.0 | 1.4 | 71.2 | 0.3 | 0.4 |
| *E. globulus* | 4.1 ± 0.8 | 40.4 | 32.7 | — | — | 3.0 |
| *Cinnamomum camphora* | 6.8 ± 0.4 | 56.7 | 5.0 | — | 0.3 | 6.6 |
| *Melaleuca viridiflora* | 8.1 ± 0.3 | 72.9 | 7.2 | — | — | 8.2 |
| *Ocimum gratissimum* | 9.3 ± 0.4 | 12.0 | 1.5 | 1.5 | 40.3 | 3.8 |

a Diameter of inhibition on paper disk (mm).
b Relative abundance, .

paper disc diffusion assay there were no zones of inhibition against any of the four test organisms and MIC values were ≥1000 ppm. Even the most promising samples had MICs inferior to that of a commonly used commercial soap bacteriostat and when tested in soaps no reduction of bacterial counts was obtained. The authors concluded that it did not appear to be possible to produce a practical antimicrobial soap fragrance. They also pointed out the need for caution when using experimental zones of inhibition as a measure of antimicrobial activity. Such a method is dependent on the solubility and rate of diffusion of the test sample in the aqueous medium and a sample may be much more bacteriostatic than its zone of inhibition might indicate.

One form of testing which is closer, in practical terms, to the way in which human infections are often spread, namely as bacterial aerosols through coughing and sneezing, has been investigated by Chao et al. (1998). Thieves is a proprietary blend of five volatile oils: E. globulus, Cinnamomum zeylanicum, Citrus limon, Rosmarinus officinalis and Syzygium aromaticum. In well-designed experiments subjected to statistical analysis, the blend's antibacterial activity was tested against three Gram-positive organisms Micrococcus luteus, Pseudomonas aeruginosa and Staphylococcus aureus bioaerosols. Thieves was allowed to diffuse into an enclosed fume hood following spraying of the aerosol-borne bacterial load. P. aeruginosa was the most sensitive to the treatment, with a 96 per cent reduction in bacterial count following a 10-min exposure. Inhibition levels for M. luteus and S. aureus were 82 per cent and 44 per cent, respectively. With further testing the authors suggest that commercial applications could include diffusion of oil aerosols into air-conditioning systems and spraying in enclosed rooms, both at work and at home, to prevent the transmission of illnesses through air-borne bacterial pathogens.

## Composition and structure–activity relationships

The antimicrobial activity of eucalyptus oils and other volatile oils would be expected to reflect their composition, the structural configuration of the constituents and their functional groups, along with potential synergistic interactions between the constituents. Aqueous solubility, and the ability of toxic compounds to penetrate the fungal or bacterial cell wall, is also likely to be an important factor and this, too, will be influenced by the chemical nature of individual compounds within the oil. However, while some general observations can be made about the antimicrobial activity of different classes of terpenes, detailed structure–activity relationships are still not well understood. Carbonylation of terpenoids, for example, is known to increase their bacteriostatic activity but not necessarily their bactericidal activity (Griffin et al. 1999), while alcohols possess bactericidal rather than bacteriostatic activity against vegetative bacterial cells.

In a study by Dorman and Deans (2000), a correlation of the antimicrobial activity of the compounds tested, and their relative percentage composition in the oils, with their chemical structure, functional groups and configuration has confirmed a number of observations concerning structure–activity relationships made by others. Constituents with phenolic structures, for example, such as carvacrol and thymol, were highly active against test bacteria, despite their low capacity to dissolve in water. This may be due to the relative acidity of the hydroxyl group – carvacrol was more active than its methyl ester and, in tests carried out by Knobloch et al. (1989), methyl and acetyl eugenol were less inhibitory than eugenol; alcohols such as geraniol and nerol were also less active than phenolic ones. However, Kurita et al. (1981) found that geraniol was very similar to eugenol (and citronellol) in terms of antifungal activity and only slightly less active than thymol and isoeugenol. The fact that compounds often have markedly different responses towards different organisms sometimes makes generalisations unwise. Citronellol has been found to be relatively inactive towards Escherichia coli and Pseudomonas aeruginosa but it is strongly active against Staphylococcus aureus (Griffin et al. 1999). Some organisms are also more sensitive to relatively small changes in structure than others: terpinen-4-ol and α-terpineol, identical *p*-menthane tertiary alcohols except for the position of the hydroxyl group, have identical activity profiles against *E. coli*, *S. aureus* and *Candida albicans*, but the former retains activity against *P. aeruginosa* while α-terpineol loses it (Griffin *et al.* 1999).

The antimicrobial activity of phenolic compounds is further enhanced by alkyl substitution in the aromatic ring (Kurita *et al.* 1981, Pelczar *et al.* 1988). An allylic side chain appears to enhance the inhibitory effects, chiefly against Gram-negative bacteria. It has been suggested that alkylation alters the distribution ratio between the aqueous and non-aqueous phases, including bacterial phases, by reducing the surface tension or altering the species selectivity. Alkyl-substituted phenolic compounds form phenoxyl radicals which interact with isomeric alkyl substituents (Pauli and Knobloch 1987). This does not occur with etherified or esterified isomers, possibly explaining their relative lack of activity.

Aldehydes generally possess powerful antimicrobial activity. The highly electronegative arrangement of an aldehyde group conjugated to a carbon–carbon double bond appears to enhance activity (Moleyar and Narasimham 1986) and cinnamaldehyde is much more strongly antifungal than benzaldehyde (Kurita *et al.* 1981). Such electronegative compounds may interfere in biological processes involving electron transfer and react with vital nitrogen components such as proteins and nucleic acids, thereby inhibiting growth of the microorganisms. Essential oils rich in cinnamaldehyde or citral have been linked with consistently high antimicrobial activity *in vitro* (Lis-Balchin *et al.* 1998). Kurita *et al.* (1981) and Dorman (1999) also found citral to be moderately active; citronellal, on the other hand, the major constituent of *E. citriodora* oil, was only weakly active. Griffin *et al.* (1999) found citronellal to be active against *S. aureus* and *C. albicans* but relatively ineffective against *E. coli* and *P. aeruginosa*.

Using the contact method, Naigre *et al.* (1996) showed that the bacteriostatic and fungistatic action of terpenoids was increased when there was a keto group present. Griffin *et al.* (1999), however, have found that ketones are variable in their activity – carvone was strongly active against the organisms tested, verbenone was moderately active and menthone was relatively inactive (the latter in some contrast to the work of Dorman and Deans (2000) who found that menthone exhibited modest antibacterial activity, particularly against *Clostridium sporogenes* and *Staphylococcus aureus*). Interestingly, piperitone, a major constituent of *E. dives* oil (piperitone variant), was also quite strongly active (and more so than carvone against *E. coli* and *S. aureus*).

Although Lis-Balchin *et al.* (1998) demonstrated significant antimicrobial activity for several Myrtaceae oils, including eucalyptus, they found no correlation between activity and 1,8-cineole content. *E. globulus* oil (91 per cent cineole) was less active towards bacteria than *E. radiata* (84 per cent cineole) and neither was particularly effective against the fungi tested (*Aspergillus niger*, *A. ochraceus* and *Fusarium culmorum*). *E. citriodora* oil (<1 per cent cineole but high in citronellal) showed much greater antifungal activity.

Terpene hydrocarbons such as α- and β-pinene, limonene and *p*-cymene have been found to be inactive towards a variety of organisms (Kurita *et al.* 1981, Griffin *et al.* 1999), as have terpene acetates such as geranyl, linalyl, neryl and α-terpinyl acetates (Griffin *et al.* 1999). Inactivity has been attributed to low water solubility and low hydrogen bonding capacity.

An increase in activity dependent upon the type of alkyl substituent incorporated into a non-phenolic ring structure was observed by Dorman and Deans (2000). The inclusion of a double bond increased the activity of limonene relative to *p*-cymene, which demonstrated no activity against the test bacteria. In addition, the susceptible organisms were principally Gram-negative, which suggests that alkylation influences Gram reaction sensitivity of the bacteria.

Stereochemistry also has an influence on bioactivity. It has been observed that in some cases α-isomers are inactive relative to β-isomers (e.g. α-pinene cf. β-pinene) and *cis*-isomers are inactive compared to the *trans*-isomers (e.g. geraniol cf. nerol). Compounds with methyl-isopropyl cyclohexane rings are most active and unsaturation in the cyclohexane ring further increases the antibacterial activity, as in terpinolene, α-terpineol and terpinen-4-ol (Hinou *et al.* 1989).

Investigations into the effects of terpenoids upon isolated bacterial membranes suggest that their activity is a function of the lipophilic properties of the constituent terpenes (Knobloch *et al.* 1986), the potency of their functional groups and their aqueous solubility (Knobloch *et al.* 1988, 1989). As noted earlier, the importance of water solubility and hydrogen bonding has been pointed out by others (e.g. Griffin *et al.* 1999). These interacting factors are not easy to unravel and different researchers – using different procedures and different organisms – sometimes obtain results that are difficult to reconcile. In the test conditions used by Knobloch *et al.* (1989), β-pinene showed moderate antibacterial activity towards *Rhodopseudomonas sphaeroides* although it is insoluble in water; cineole was only slightly active, despite having about the same water solubility as citronellal, which was very active; and carvacrol and thymol were also highly active although they were less water soluble than citronellal. The site of action of the terpenoid appears to be at the phospholipid bilayer of the cell and to be caused by biochemical mechanisms catalysed by the bilayers. These processes include the inhibition of electron transport, protein translocation, phosphorylation steps and other enzyme-dependent reactions (Knobloch *et al.* 1986). Terpenoid activity in whole cells appears to be more complex (Knobloch *et al.* 1988).

## Conclusions

Chemotherapeutic agents, used topically or systemically for the treatment of microbial infections of humans and animals, possess varying degrees of selective toxicity. Although the principle of selective toxicity is used in agriculture, pharmacology and diagnostic microbiology, its most dramatic application is the systemic chemotherapy of infectious diseases. Plant products which have been tested appear to be effective against a wide spectrum of microorganisms, both pathogenic and non-pathogenic. Administered orally, these compounds may be able to control a wide range of microbes, but there is also the possibility that they may cause an imbalance in the gut microflora, allowing opportunist pathogenic bacteria, such as coliforms, to become established in the gastrointestinal tract with resultant deleterious effects. Further studies on therapeutic applications of volatile oils, including those from *Eucalyptus*, are needed to investigate these issues, and to complement the substantial number of analytical and *in vitro* bioactivity studies that are being carried out on these natural products.

The potential of eucalyptus oils for use as practical antimicrobial agents remains to be proven. Some results have been encouraging but others have been less so. The variability is as much a reflection of the widely differing conditions, procedures and test organisms used by different workers as it is of the compositional variation in the oils themselves. Mixtures of oils, as used by Chao *et al.* (1998), may be more effective than single oils, although the choice of which oils to combine is no easy matter. Prediction of activity in whole oils (or mixtures) based on that of individual constituents is complicated by the existence of synergistic effects, as noted earlier (Low *et al.* 1974).

*In vitro* studies have shown that oils from some *Eucalyptus* species are effective against a range of pathogens, non-pathogens and spoilage organisms. More comprehensive (and standardised) tests of oils from a greater number of *Eucalyptus* species are needed to determine whether such oils, or formulations containing them, have a major role to play as antimicrobial agents. If they have, then *in vivo* studies are needed to assess their efficacy under clinical conditions. With an increasing public awareness of 'green issues', plant volatile oils, including those from *Eucalyptus*, offer a more eco-friendly alternative to conventional formulations in a number of sectors where antimicrobial action is desirable.

## Acknowledgements

SAC received financial support from the Scottish Executive Rural Affairs Department.

## References

Ansari, A.A. and Shrivastava, A.K. (1991) The effect of eucalyptus oil on growth and aflatoxin production by *Aspergillus flavus. Letts. Appl. Microbiol.*, 13, 75–77.

Ara, R. and Misra, N. (1992) Antifungal activity of mixture of essential oils against some dermatophytes. *Indian Perfumer*, 36, 38–41.

Baratta, M.T., Dorman, H.J.D., Deans, S.G., Biondi, D.M. and Ruberto, G. (1998a) Chemical composition, antibacterial and antioxidative activity of laurel, sage, rosemary, oregano and coriander essential oils. *J. Essent. Oil Res.*, 10, 618–627.

Baratta, M.T., Dorman, H.J.D., Deans, S.G., Figueiredo, A.C., Barroso, J.G. and Ruberto, G. (1998b) Antimicrobial and antioxidant properties of some commercial essential oils. *Flavour Fragr. J.*, 13, 235–244.

Baruah, P., Sharma, R.K., Singh, R.S. and Ghosh, A.C. (1996) Fungicidal activity of some naturally occurring essential oils against *Fusarium moniliforme. J. Essent. Oil Res.*, 8, 411–412.

Benjilali, B., Tantaoui-Elaraki, A., Ayadi, A. and Ihlal, M. (1984) Method to study antimicrobial effects of essential oils: application to the antifungal activity of six Moroccan essences. *J. Food Prot.*, 47, 748–752.

Benjilali, B., Tantaoui-Elaraki, A., Ismaili-Alaoui, M. and Ayadi, A. (1986) A method of studying the antimicrobial properties of essential oils by direct contact in agar medium (in French). *Plantes Med. Phyto.*, 20, 155–167.

Bhalla, H.K.L. (1997) *Indian Eucalypts and their Essential Oils*, Timber Development Association of India, Dehra Dun.

Carson, C.F., Hammer, K.A. and Riley, T.V. (1996) *In vitro* activity of the essential oil of *Melaleuca alternifolia* against *Streptococcus* spp. *J. Antimicrob. Chemother.*, 37, 1177–1181.

Chao, S.C., Young, D.G. and Oberg, C.J. (1998) Effect of a diffused essential oil blend on bacterial bioaerosols. *J. Essent. Oil Res.*, 10, 517–523.

Chaudhari, D.C. and Suri, R.K. (1991) Comparative studies on chemical and antimicrobial activities of fast growing Eucalyptus hybrid (FRI-4 & FRI-5) with their parents. *Indian Perfumer*, 35, 30–34.

Daferera, D.J., Ziogas, B.N. and Polissiou, M.G. (2000) GC-MS analysis of essential oils from some Greek aromatic plants and their fungitoxicity on *Penicillium digitatum. J. Agric. Food Chem.*, 48, 2576–2581.

De Medici, D., Pieretti, S., Salvatore, G., Nicoletti, M. and Rasoanaivo, P. (1992) Chemical analysis of essential oils of Malagasay medicinal plants by gas chromatography and NMR spectroscopy. *Flavour Fragr. J.*, 7, 275–281.

Deans, S.G., Kennedy, A.I., Gundidza, M.G., Mavi, S., Waterman, P.G. and Gray, A.I. (1994) Antimicrobial activities of the volatile oil of *Heteromorpha trifoliata* (Wendl.) Eckl. & Zeyh. (Apiaceae). *Flavour Fragr. J.*, 9, 245–248.

Deans, S.G. and Ritchie, G. (1987) Antibacterial properties of plant essential oils. *Int. J. Food Microbiol.*, 5, 165–180.

Dellacassa, E., Menendez, P., Moyna, P. and Cerdeiras, P. (1989) Antimicrobial activity of *Eucalyptus* essential oils. *Fitoterapia*, 60, 544–546.

Dorman, H.J.D. (1999) Phytochemistry and bioactive properties of plant volatile oils: antibacterial, antifungal and antioxidant activities, PhD Thesis, Univ. Strathclyde, Glasgow.

Dorman, H.J.D. and Deans, S.G. (2000) Antimicrobial agents from plants: antibacterial activity of plant volatile oils. *J. Appl. Microbiol.*, 88, 308–316.

Duarte, A., Rodriguez, A.U. and Quert, Y.R. (1992) Relevant results of the antimycotic action of essential oils from *Eucalyptus pellita* and *Pinus caribaea* against strains of pathogenic fungi (in Spanish). *Rev. Baracoa*, 22, 91–94.

Gallardo, P.P.R., Salinas, R.J. and Villar, L.M.P. (1987) The antimicrobial activity of some spices on microorganisms of great interest to health. IV: seeds, leaves and others. *Microbiol. Aliments Nutrition*, 5, 77–82.

Griffin, S.G., Wyllie, S.G., Markham, J.L. and Leach, D.N. (1999) The role of structure and molecular properties of terpenoids in determining their antimicrobial activity. *Flavour Fragr. J.*, 14, 322–332.

Hajji, F., Fkih-Tetouani, S. and Tantaoui-Elaraki, A. (1993) Antimicrobial activity of twenty-one *Eucalyptus* essential oils. *Fitoterapia*, 64, 71–77.

Hinou, J.B., Harvala, C.E. and Hinou, E.B. (1989) Antimicrobial activity screening of 32 common constituents of essential oils. *Pharmazie*, 44, 302–303.

Hoffman, C. and Evans, A.C. (1911) The use of spices as preservatives. *J. Indian Eng. Chem.*, 3, 835–838.

Inouye, S., Goi, H., Miyouchi, K., Muraki, S., Ogihara, M. and Iwanami, I. (1983) Inhibitory effect of volatile components of plants on the proliferation of bacteria. *Bokin Bobai*, 11, 609–615.

Jain, S.R. and Kar, A. (1971) The antibacterial activity of some essential oils and their combinations. *Planta Medica*, 20, 118–123.

Janssen, M.A., Scheffer, J.J.C. and Baerheim-Svendsen, A. (1987) Antimicrobial activities of essential oils: a 1976–1986 literature review on possible applications. *Pharmaceut. Weekblad (Sci. Ed.)*, 9, 193–197.

Jeanfils, J., Burlion, N. and Andrien, F. (1991) Antimicrobial activities of essential oils from different plant species. *Landbouwtijdschrift-Rev. Agric.*, 44, 1013–1019.

Knobloch, K., Pauli, A., Iberl, B., Weigand, H. and Weis, N. (1989) Antibacterial and antifungal properties of essential oil components. *J. Essent. Oil Res.*, 1, 119–128.

Knobloch, K., Pauli, A., Iberl, B., Weis, N. and Weigand, H. (1988) Mode of action of essential oil components on whole cells of bacteria and fungi in plate tests. In P. Schreier (ed.), *Bioflavour '87*, Walter de Gruyter, Berlin, pp. 287–299.

Knobloch, K., Weigand, H., Weis, N., Scharm, H.-M. and Vigenschow, H. (1986) Action of terpenoids on energy metabolism. In E.J. Brunke (ed.), *Progress in Essential Oils Research*, Walter de Gruyter, Berlin, pp. 429–445.

Kumar, A., Sharma, V.D., Sing, A.K. and Singh, K. (1988) Antibacterial properties of different *Eucalyptus* oils. *Fitoterapia*, 59, 141–144.

Kurita, N., Miyaji, M., Kurane, R. and Takahara, Y. (1981) Antifungal activity of components of essential oils. *Agric. Biol. Chem.*, 45, 945–952.

Leite, C.Q.F., Moreira, R.R.D and Neto, J.J. (1998) Action of *Eucalyptus* oils against *Mycobacterium avium*. *Fitoterapia*, 69, 282–283.

Lis-Balchin, M., Deans, S.G. and Eaglesham, E. (1998) Relationship between bioactivity and chemical composition of commercial essential oils. *Flavour Fragr. J.*, 13, 98–104.

Low, D., Rawal, B.D. and Griffin, W.J. (1974) Antibacterial action of the essential oils of some Australian Myrtaceae with special references to the activity of chromatographic fractions of oil of *Eucalyptus citriodora*. *Planta Medica*, 26, 184–189.

Madhukar, J. and Reddy, S.M. (1989) Efficacy of certain oils in the control of fruit-rot of guava. *Indian J. Mycol. Plant Pathol.*, 19, 131–132.

Martindale, W.H. (1910) Essential oils in relation to their antiseptic powers as determined by their carbolic coefficients. *Perfum. Essent. Oil Res.*, 1, 266–274.

Masood, A. and Ranjan, K.S. (1990) The influence of fungitoxicants on growth and aflatoxin production by *Aspergillus flavus*. *Letts. Appl. Microbiol.*, 11, 197–201.

Moleyar, V. and Narasimham, P. (1986) Antifungal activity of some essential oil components. *Food Microbiol.*, 3, 331–336.

Montes-Belmont, R. and Carvajal, M. (1998) Control of *Aspergillus flavus* in maize with plant essential oils and their components. *J. Food Prot.*, 61, 616–619.

Morris, J.A., Khettry, A. and Seitz, E.W. (1979) Antimicrobial activity of aroma chemicals and essential oils. *J. Am. Oil Chem. Soc.*, 56, 595–603.

Müller-Riebau, F., Berger, B. and Yegen, O. (1995) Chemical composition and fungitoxic properties to phytopathogenic fungi of essential oils of selected aromatic plants growing wild in Turkey. *J. Agric. Food Chem.*, 43, 2262–2266.

Naigre, R., Kalck, P., Roques, C., Roux, I. and Michel, G. (1996) Comparison of antimicrobial properties of monoterpenes and their carbonylated products. *Planta Medica*, 62, 275–277.

Nenoff, P., Haustein, U.F. and Brandt, W. (1996) Antifungal activity of the essential oil of *Melaleuca alternifolia* (tea tree oil) against pathogenic fungi *in vitro*. *Skin Pharmacol.*, 9, 388–394.

Pattnaik, S., Subramanyam, V.R. and Kole, C. (1996) Antibacterial and antifungal activity of ten essential oils *in vitro*. *Microbios*, 86, 237–246.

Pauli, A. and Knobloch, K. (1987) Inhibitory effects of essential oil components on growth of food-contaminating fungi. *Z. Lebensm. Unters. Forsch.*, 185, 10–13.

Pelczar, M.J., Chan, E.C.S. and Krieg, N.R. (1988) *Microbiology*, McGraw-Hill International, New York, pp. 469–509.

Pélissier, Y., Marion, C., Casadebaig, J., Milhau, M., Djenéba, K., Loukou, N.Y. and Bessière, J.M. (1994) A chemical, bacteriological, toxicological and clinical study of the essential oil of *Lippia multiflora* (Verbenaceae). *J. Essent. Oil Res.*, 6, 623–630.

Rios, J.L., Recio, M.C. and Villar, A. (1988) Screening methods for natural products with antibacterial activity: a review of the literature. *J. Ethnopharmacol.*, 23, 127–149.

Shapiro, S., Meier, A. and Guggenheim, B. (1994) The antimicrobial activity of essential oils and essential oil components towards oral bacteria. *Oral Microbiol. Immunol.*, 9, 202–204.

Shelef, L.A. (1984) Antimicrobial effects of spices. *J. Food Safety*, 6, 29–44.

Singh, R.K. and Dwivedi, R.S. (1987) Effect of oils on *Sclerotium rolfsii* causing foot-rot of barley. *Indian Phytopathol.*, 40, 531–533.

Singh, R.K. and Dwivedi, R.S. (1990) Fungicidal properties of neem and blue gum against *Sclerotium rolfsii* Sacc., a foot-rot pathogen of barley. *Acta Bot. Indica*, 18, 260–262.

Singh, R.K., Shukla, R.P. and Dwivedi, R.S. (1989) Studies on fungitoxicity of oils against *Sclerotium rolfsii* Sacc. and soil mycoflora. *Nat. Acad. Sci. Letts. (India)*, 12, 183–185.

Singh, S., Singh, S.K. and Tripathi, S.C. (1988) Fungitoxic properties of essential oil of *Eucalyptus rostrata*. *Indian Perfumer*, 32, 190–193.

Siva Sankara Rao, T. and Nigam, S.S. (1978) Antibacterial study of some Indian essential oils. *Indian Perfumer*, 22, 118–119.

Srivastava, S., Naik, S.N. and Maheshwari, R.C. (1993) *In vitro* studies on antifungal activities of palmarosa and eucalyptus oils. *Indian Perfumer*, 37, 277–279.

Tantaoui-Elaraki, A., Ferhout, H. and Errifi, A. (1993) Inhibition of the fungal asexual reproduction stages by three Moroccan essential oils. *J. Essent. Oil Res.*, 5, 535–545.

Thomann, R., Bauermann, U. and Hagemann, L. (1997) Essential oils and plant substances: an alternative to synthetic growth enhancers in animal feeding. In K.H.C. Baser (ed.), *Proc. 28th Internat. Symp. Ess. Oils*, Eskisehir, Turkey, September 1997, p. 68.

Youdim, K.A., Dorman, H.J.D. and Deans, S.G. (1999) The antioxidant effectiveness of thyme oil, $\alpha$-tocopherol and ascorbyl palmitate on evening primrose oil oxidation. *J. Essent. Oil Res.*, 11, 643–648.

# 14 Eucalyptus in insect and plant pest control

## Use as a mosquito repellent and protectant of stored food products; allelopathy

*Peter Golob, Hiroyuki Nishimura and Atsushi Satoh*[1]

## Introduction

The ease with which essential oils are obtained from aromatic plants and their diverse chemical compositions makes them potential sources of natural pesticides – either through direct toxicity or by repellency – and they have attracted increasing attention among researchers (as reviewed, for example, by Singh and Upadhyay 1993). As natural repellents they have seen a resurgence of interest since the use of synthetic compounds began to displace the large number of essential oils which were formerly used in the 1930s and 1940s (Curtis *et al*. 1990). Their volatility has potential benefits in terms of bringing the pesticide vapour into close contact with the pest while at the same time not leaving residues which might adversely affect the object being protected, be this a crop or food product, or, in the case of, for example, a mosquito repellent, the human body. With the ever-increasing level of air travel the danger of catching malaria and other mosquito-borne diseases is now a worldwide one and not confined to people living in the tropics.

Citronella oil has been used in commercial repellent preparations and is still popular in India. In China, essential oils from *Mentha haplocalyx* and *Clausena kwangsiensis* have been shown to repel mosquitoes (Curtis *et al*. 1990). And essential oils, especially those derived from citrus peels, have been tested as grain protectants for many years (Golob and Webley 1980). However, sufficient research has now been conducted to suggest that eucalyptus oil, its constituents or the leaves from which it is obtained hold, perhaps, most promise of all for use in the battle against insect pests. As well as exhibiting repellency towards mosquitoes, eucalyptus oil has been found to be toxic to mosquito larvae and Corbet *et al*. (1995) suggest that plant essential oils merit further attention as widely available, environmentally benign mosquito larvicides.

In a few cases, eucalyptus oil-based products have already reached the market place. Quwenling, made in China from the waste distillate after extraction of oil from the 'lemon eucalyptus', is used for protection against mosquitoes, and the active ingredient has now found a place in Western human health care as Mosi-guard™. This is referred to in Chapter 16 but is discussed in more detail below.

In many other cases, more research is required, particularly large-scale field trials, to confirm efficacy and determine the economics and viability of production of the most suitable formulation. Nevertheless, there is hope that in the years ahead more eucalyptus-based products and procedures will find practical application in the fight against insect pests. Some of the efforts in

---

[1] The section on the use of eucalyptus as a protectant of stored food products is by Golob. The discussion of eucalyptus as a mosquito repellent and as a source of allelochemicals is by Nishimura and Satoh, with supplementary information relating to quwenling and the development of Mosi-guard™ provided by N. Hill, London School of Hygiene and Tropical Medicine, and described by J. Coppen.

working towards this objective are reviewed below. In the case of insects which damage crops the number of pests is large (see e.g. Kranz et al. 1977) and many insects have been targeted. The discussion, here, is limited to those which damage stored food products, a major problem in the tropics and sub-tropics. This follows a description of eucalyptus-based mosquito repellents and their active ingredients.

Allelopathy, the effect that one plant has on another through the mediation of naturally produced chemicals, has also attracted researchers. Aromatic plants such as *Eucalyptus* release volatile oils from their leaves which appear to have an inhibitory effect on the growth of understorey vegetation and this could lead to the development of natural herbicides. Research aimed at identifying potentially valuable allelochemicals in eucalyptus is therefore also briefly reviewed.

## Eucalyptus as a mosquito repellent

### Introduction

N,N-diethyl-m-toluamide (DEET) has long been used as a repellent against blood-sucking insects such as mosquitoes. However, DEET has several drawbacks, such as an unpleasant odour and its skin penetration (Moody et al. 1986). It also behaves as a solvent towards certain plastics and synthetic rubber and so care has to be taken to avoid contact with glasses and watch straps. There would be advantages, therefore, in finding natural repellents without these undesirable properties. The use of citronella oil was referred to earlier but in an assessment of twelve repellent formulations on the European market, those with citronella as the active ingredient rated considerably worse than those with synthetic active ingredients (Curtis et al. 1990). Neem oil has also been tested (Sharma and Dhiman 1993).

### Eucalyptus citriodora

It has long been known in China that oil from the lemon eucalyptus, '*E. maculata citriodon*', has some repellent effect on mosquitoes.[2] When the active fraction was found to be concentrated in the waste distillate after extraction of the oil the active principle was identified as $p$-menthane-3,8-diol. The product, and its effectiveness against *Aedes aegypti*, *A. albopictus* and *A. vexans*, was first described by Li et al. (1974) (cited, and with data reproduced, in Curtis et al. 1990) and is sold in China as 'quwenling' ('effective repeller of mosquitoes'). Although subsequent studies in the West have found it to be less persistent than DEET (e.g. Schreck and Leonhardt 1991) it has been so successful that it has largely displaced the synthetic dimethyl phthalate from the Chinese market. As well as mosquitoes it gives protection against midges, tabanids and land leeches. In addition to not damaging plastic materials its mammalian toxicity is lower than that of DEET, with mouse oral and dermal $LD_{50}$ values of 3.2 and 12 g/kg, respectively (Curtis et al. 1990).

Using a combination of bioassay and chemical and analytical methodologies, Nishimura et al. (1986) have also searched eucalyptus for compounds with repellent activity against mosquitoes. Using *Aedes albopictus*, the repellent activities of the essential oils of several species of *Eucalyptus*, including *E. citriodora*, were tested. The oils were obtained by steam distilling acetone extracts of the leaves. The results are shown in Table 14.1.

[2] Published accounts all use the name '*Eucalyptus maculata citriodon*' but this is not a recognised species name. However, *E. citriodora* oil is produced in large quantities in China and its constituents are the same as those described for *E. maculata citriodon*. This, together with the fact that Nishimura et al. (1986) have isolated the same active principle from *E. citriodora*, makes it likely that the latter species – or something very close to it chemotaxonomically – is the eucalypt in question.

The essential oils of *E. citriodora* and *E. camaldulensis* exhibited higher repellent activity than the oils from the other species. Column chromatography of the *E. citriodora* oil gave two crystalline compounds with high activity, (±)-*p*-menthane-3,8-*cis*-diol and (±)-*p*-menthane-3,8-*trans*-diol, Figure 14.1 (Nishimura *et al.* 1986, Nishimura and Mizutani 1989). These compounds had earlier been identified by Nishimura *et al.* (1982, 1984) as potent allelochemicals (see below). All of the enantiomers of the *cis* and *trans* diols exhibited approximately the same activity as the natural racemates.

Table 14.1 Repellency of the volatile oils of some *Eucalyptus* species against mosquitoes, *Aedes albopictus*

| Species | Repellency (%)[a] | |
|---|---|---|
| | 1% conc. | 0.1% conc. |
| *E. camaldulensis* | 93 | 60 |
| *E. citriodora* | 90 | 43 |
| *E. radiata* | 54 | 24 |
| *E. globulus* | — | 30 |
| *E. viminalis* | 27 | — |
| *E. pulverulenta* | 16 | — |
| *Cinnamomum camphora*[b] | 0 | 0 |

a Repellency = {(total no. mosquitoes − attracted mosquitoes) ÷ total no. mosquitoes} × 100.
b Oil from this species is known to repel insects such as Coleoptera and was used to compare eucalyptus.

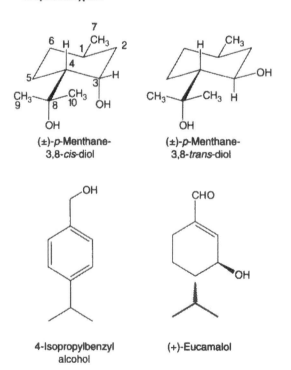

Figure 14.1 Chemical structures of the mosquito repellents from leaves of *E. citriodora* (*p*-menthane-3,8-*cis*- and *trans*-diol) and *E. camaldulensis* (4-isopropylbenzyl alcohol and (+)-eucamalol).

Table 14.2 $ED_{90}$ values for five mosquito repellent formulations, three of them derived from *Eucalyptus citriodora*, against *Anopheles gambiae*

| | $ED_{90}$ | |
| --- | --- | --- |
| | ($\mu$l or $\mu$g product/cm$^2$) | ($\mu$l or $\mu$g a.i./cm$^2$) |
| PMD$^a$ liquid (50% a.i.$^b$) | 0.65 | 0.33 |
| PMD gel (50% a.i.) | 0.67 | 0.34 |
| PMD stick (50% a.i.) | 0.72 | 0.36 |
| Autan stick (20% DEET) | 0.48 | 0.10 |
| Citronella oil (50% a.i.) | 1.37 | 0.68 |

a *p*-Menthane-3,8-diol formulation.
b Active ingredient.

Further developments in the West, building upon the Chinese work on quwenling, have recently led to the introduction of a new repellent preparation Mosi-guard™ Natural. Although still produced from the lemon eucalyptus and containing *p*-menthane-3,8-diol as the active ingredient, it is obtained, this time, from the eucalyptus oil itself, rather than the waste distillate. The extraction process was developed at University College, London, and the final formulation contains a patented mixture of isopulegol and citronellol as well as *p*-menthane-3,8-diol (Trigg and Hill 1996). Laboratory and field testing of the formulation have confirmed its efficacy as a repellent against the malaria vectors *Anopheles gambiae* and *A. funestus*, as well as the biting midge *Culicoides variipennis*, the deer tick *Ixodes ricinis* and the stable fly *Stomoxys calcitrans* (Trigg 1996, Trigg and Hill 1996). In tests against *A. gambiae*, $ED_{90}$ values (the doses calculated to give 90 per cent protection against bites) for three forms of the eucalyptus preparation compared favourably with Autan® stick and were superior to citronella oil (Table 14.2). It was concluded that the level and duration of protection was comparable to that afforded by DEET (Trigg 1996). Acute toxicological studies have demonstrated minimal toxicity in rats (oral and dermal $LD_{50}$ values of 2.4 and $>2$ g/kg, respectively).

### *Eucalyptus camaldulensis*

Guided by bioassay using *Aedes aegypti*, research by Watanabe *et al.* (1993) on the essential oil from *E. camaldulensis* has yielded two mosquito-repellent compounds, 4-isopropylbenzyl alcohol and a new compound, (+)-eucamalol (3-formyl-6-isopropyl-2-cyclohexen-1-ol), Figure 14.1.

(+)-Eucamalol and its 1-epimer were synthesised from (*S*)-(−)-perillaldehyde to determine the absolute configuration and to compare the repellent activities of the two compounds. The absolute configuration of (+)-eucamalol was determined to be (1*R*, 6*R*)-(+)-3-formyl-6-isopropyl-2-cyclohexen-1-ol (Satoh *et al.* 1995).

The repellent and feeding inhibitory activities of synthetic eucamalol and its epimer were evaluated using *Aedes albopictus* as the test mosquito strain (Table 14.3). The activities of the two epimers were approximately the same and both were as effective as DEET (Satoh *et al.* 1995).

## Eucalyptus as a protectant of stored food products

### *Introduction*

In the developed world food crops are quickly removed from the farm and processed by drying, canning or freezing for storage before consumption. In the developing world, harvested produce

Table 14.3 Repellency and feeding inhibition of (+)-eucamalol and its (−)-3-epimer against *Aedes albopictus*

| | Repellency/Feeding inhibition (%)[a] | | |
|---|---|---|---|
| | 500 mg/m$^2$ | 250 mg/m$^2$ | 50 mg/m$^2$ |
| (+)-Eucamalol | 100/100 | 100/100 | 84.2/74.5 |
| (−)-*epi*-Eucamalol | 100/100 | 100/100 | 75.0/65.0 |
| DEET | 100/100 | 100/100 | 80.0/85.0 |

a Repellency calculated as indicated in Table 14.1. Feeding inhibition = {(total no. mosquitoes − blood-sucking mosquitoes) ÷ total no. mosquitoes} × 100.

is infrequently processed. Instead, it is usually stored for long periods as raw commodities or as flour on the farm or in large central storage facilities such as bag stores or silos. Farmers in the tropics and sub-tropics may store grain for a year or more in small granaries at the homestead in order to provide staple food for their family until the next harvest. During storage, durable food commodities, such as cereals and pulses, are susceptible to biodeterioration, particularly as a result of insect infestation, and to avoid severe quality and quantity loss action must be taken to protect commodities against such an insect attack.

Even in some processed foods mites can be a potential problem in storage and may cause gastrointestinal disorders and allergic dermatitis in humans or animals consuming them or coming into contact with them (Perrucci 1995).

During the last fifty years contact insecticides have been widely used to protect stored food produce against insect infestation. Contact insecticides, such as malathion and pirimiphos-methyl, have been used to provide both prophylactic and long-term protection of the produce for large-scale storage and for individual farmers who may only want to protect the contents of a few sacks of grain. Farmers are able to use various types of contact insecticide formulation to protect their stored crop. The dilute dust is specially formulated for use by the individual who has only small quantities of grain to treat and who does not possess any sophisticated application equipment. Such a dust can be applied with a shovel and it requires no prior dilution with water. However, stable dust formulations are difficult to produce and their shelf life is quite short. They are also difficult to package and transport because they are bulky and farmers often complain that they are not available when needed. Other insecticide formulations, which are more appropriate for treating larger quantities of commodity, are used as liquid sprays after being diluted with water or light oil and require spraying equipment for their application.

A major constraint on the use of synthetic protectants, and one which has become an increasing problem in recent years, is that of insect resistance. Resistance occurs when dosages that would have been expected to completely kill a population are no longer effective. In many cases, resistance can be overcome by increasing the dosage but this usually makes the treatment both uneconomic and impractical. The widespread occurrence of resistance has lead to the exclusion of malathion as a storage protectant in many countries. In recent years, resistance has been detected in the use of other storage chemicals, including pirimiphos-methyl and fenitrothion.

Concerns regarding environmental contamination by pesticides, together with the more specific problems associated with the use of insecticides, have led to a search for more natural and environmentally friendly protectants. For on-farm use any alternative must be cheap, locally available and safe to use. Plants fill these criteria in general terms and they have traditionally been used by farmers for insect control, both to protect the growing crop and as storage protectants. Plants are often used after drying, as fine powders, and are applied in a similar manner to dilute dust insecticide formulations. Furthermore, because the parts of plants which are used as protectants contain a mixture of chemical constituents, more than one of which may have an insecticidal effect, they are either less likely to induce resistance or resistance will occur much more slowly in the pest population. However, even though a plant may have been used for generations as a storage protectant (Golob and Webley 1980) it may not necessarily be particularly effective or safe. Exhaustive testing is required to identify any potential toxicant effect which regular use of the plant might cause, and to define optimum methods for application.

Although more than 100 plants have been assessed as grain protectants (Dales 1996) only two, *Chrysanthemum cinerariaefolium* (pyrethrum) and *Azadirachta indica* (neem), have so far been exploited commercially for this purpose. The attractions of essential oils as natural pesticides have already been referred to and some of the research aimed at determining the potential of *Eucalyptus* and its leaf oils for protecting stored food products is reviewed below.

## Eucalyptus as a storage insecticide

There is anecdotal evidence that *Eucalyptus* leaves have been used traditionally in Africa to protect insect infestation in stored grain. For example, farmers in the Teso district of Uganda mix dried powdered leaves with sorghum and grain legumes to prevent insect damage (M. Verstaag pers. comm.). However, there is a dearth of documented evidence describing the practical use of this plant by farmers or its efficacy under these conditions. Most studies on the insecticidal effects of eucalyptus on storage insect pests have been confined to the laboratory – to petri dishes and small glass jars – and so, despite promising experimental work, it is not possible to confirm their real potential as grain protectants.

Researchers have investigated the effects of eucalyptus on pests which attack cereals – including primary feeders which attack whole grain, such as *Sitophilus oryzae* (L.) (Curculionidae: Coleoptera), and secondary feeders which infest flour, such as *Tribolium castaneum* (Herbst.) (Tenebrionidae: Coleoptera) – and insects which damage grain legumes, including *Callosobruchus maculatus* (Fab.) (Bruchidae: Coleoptera) and *Acanthoscelides obtectus* (Say) (Bruchidae: Coleoptera). Several other investigations have been conducted to assess the effects of eucalypts on mites, which occur on both processed and unprocessed food, and on Lepidopteran pests. Research has tended to focus on the effects of leaf oils or leaf extracts, including pure chemicals such as 1,8-cineole, an important constituent of many eucalyptus leaf oils. *Eucalyptus* species which have been tested, together with the pests they have been tested against, are shown in Table 14.4.

## Effects on insect pests of cereals

A recent study (Prates *et al*. 1998) assessed the vapour effects of 1,8-cineole (from *Eucalyptus*) and limonene (derived from *Citrus*) on adult *T. castaneum* and *Rhyzopertha dominica* (Fab.) (Bostrichidae: Coleoptera) in a gas-tight chamber. Cineole led to 100 per cent mortality of *R. dominica* after 24 h but only 58 per cent control of *T. castaneum*. Both insect species were also exposed to contact with treated surfaces, filter papers and small quantities of treated wheat. Contact tests using grain were more sensitive than those using filter paper. Good control was achieved with both compounds. In the vapour phase cineole was more effective against *R. dominica* than against *T. castaneum* and was slightly more effective than limonene against the former. Contact toxicity towards both insects was greater for cineole than limonene. The same research group has studied the effects of cineole and limonene on *S. oryzae* and *S. zeamais* Motsch. (Curculionidae: Coleoptera), another pest of the tropics and sub-tropics (Santos *et al*. 1997).

Table 14.4 Species of *Eucalyptus* (as leaf, oil or extract) tested for efficacy against stored-product insect and arachnid pests

| Insect species | Insect name/type | Commodity type | E. citriodora | E. globulus | E. camaldulensis | E. tereticornis | Unknown[a] |
|---|---|---|---|---|---|---|---|
| *Sitophilus granarius* | Granary weevil | Cereals | | Sharaby (1988) | | | |
| *Sitophilus oryzae* | Rice weevil | Cereals | Gakuru and Foua-Bi (1995, 1996) | Sharaby (1988) | Sarac and Tunc (1995a,b) | Gakuru and Foua-Bi (1995) | Ahmed and Eapen (1986) Sasaki and Calafiori (1987) |
| *Sitophilus zeamais* | Maize weevil | Cereals | | | | | Ahmed and Eapen (1986) |
| *Stegobium paniceum* | Drug store beetle | Flour | | | | | |
| *Rhyzopertha dominica* | Lesser grain borer | Cereals | Thakur and Sankhyan (1992) | | | Dakshinamurthy (1988), Singh *et al.* (1996a,b) | Santos *et al.* (1997) Prates *et al.* (1998) |
| *Tribolium castaneum* | Red flour beetle | Flour | | | | | Santos *et al.* (1997) Prates *et al.* (1998) |
| *Tribolium confusum* | Confused flour beetle | Flour | Thakur and Sankhyan (1992) | | Sarac and Tunc (1995a,b) | | |
| *Acanthoscelides obtectus* | Pulse beetle | Beans | Faroni *et al.* (1995) | Regnault-Roger and Hamraoui (1993), Regnault-Roger *et al.* (1993) | | | Stamopoulos (1991) |
| *Callosobruchus maculatus* | Cowpea weevil | Cowpea, gram | Pajni and Gill (1991) Gakuru and Foua-Bi (1995, 1996) | Pajni and Gill (1991) | | | |
| *Callosobruchus chinensis* | Adzuki bean weevil | Cowpea, gram | | | | | Ahmed and Eapen (1986) |
| *Corcyra cephalonica* | Rice moth | Wheat | | | | | Pathak and Krishna (1991) |
| *Ephestia kuhniella* | Mediterranean flour moth | Wheat | | | Sarac and Tunc (1995a,b) | | |
| *Sitotroga cerealella* | Angoumois grain moth | Paddy | | | | Dakshinamurthy (1988) | |
| *Phthorimaea operculella* | Potato tuber moth | European potato | | | | | Kroschel and Koch (1996) |
| *Tyrophagus longior* | Mite | Processed products, flour | | Perrucci (1995) | | | |
| *Tyrophagus putrescentiae* | Copra mite | Processed products, flour | Miyazaki (1996) | | | | |
| *Dermataphagoides farinae* | Mite | Processed products, flour | Miyazaki (1996) | | | | Gulati and Mathur (1995) |

a Species not declared or tests conducted using extracts, oils or isolates, for example, 1,8-cineole.

Although 1,8-cineole is an obvious candidate for being the active constituent of cineole-rich eucalyptus oils which prove positive in bioassay, possible synergistic effects make it desirable to test whole oils as well as focusing on individual chemicals. Furthermore, although the work of Prates *et al.* provides an indication of the effects that cineole may have under real conditions it does ignore a major consequence of using *R. dominica* as a test insect. This species completes its life cycle inside grain and the larval instars cause a great deal of damage. Any assessment of the effect of eucalyptus on storage pests must include tests against those stages developing internally. Exposing infested wheat grains to cineole vapour, for example, would have added significantly to the conclusions to be drawn from this paper. However, it is an omission that occurs in much of the published work reviewed here.

Ahmed and Eapen (1986), for example, examined the effect of essential oil vapours against adults of three insect pests: *S. oryzae*, another species whose larvae are internal feeders of cereals; *Callosobruchus chinensis* (L.) (Bruchidae: Coleoptera), a pest of small grain legumes whose larvae also develop internally; and *Stegobium paniceum* (L.) (Anobiidae: Coleoptera), a pest whose larvae browse loosely in flour. Fifty young unsexed adults were exposed in desiccators to strips of filter paper impregnated with eucalyptus oil, cineole or one of a selection of ten other essential oils. Both eucalyptus oil and cineole were deemed to give good control of all three species of insect, though they were more effective against the bruchid than against the cereal feeders.

A laboratory study of the persistent toxicity of four plant oils towards storage pests of wheat (Thakur and Sankhyan 1992) showed that at the highest dose tested neem oil was longer lasting in its toxicity towards *R. dominica* than eucalyptus oil (distilled from *E. citriodora*) but that at lower doses eucalyptus oil performed best. However, against *Oryzaephilus surinamensis* (L.) (Silvanidae: Coleoptera) and *Tribolium confusum* J. du Val (Tenebrionidae: Coleoptera) eucalyptus oil gave the best protection at all doses. Treatment involved exposure of 0–7 day old insects to three replicates of grain coated with oil at three concentrations (0.5, 1.0 and 1.5 per cent v/w). Mortality was monitored over a period of sixteen weeks. Partial results for the two oils are shown in Table 14.5.

The effect of oil vapours on the moth pest of cereals, *Corcyra cephalonica* (Stainton) (Pyralidae: Lepidoptera), was determined by Pathak and Krishna (1991). Pairs of either newly emerged larvae or those which were fifteen days old were exposed in small glass containers on a food medium of sorghum and yeast to eucalyptus oil placed in small glass vials in the same enclosure.

*Table 14.5* Comparative effectiveness of *Eucalyptus citriodora* and neem oils against *Rhyzopertha dominica*, *Oryzaephilus surinamensis* and *Tribolium confusum* (after Thakur and Sankhyan 1992)

| | Dose (ml/kg) | Mortality ( ) after time (weeks) | | | | | | | |
| --- | --- | --- | --- | --- | --- | --- | --- | --- | --- |
| | | E. citriodora | | | | Neem | | | |
| | | 2 | 4 | 6 | 8 | 2 | 4 | 6 | 8 |
| R. dominica | 5 | 96.67 | 86.67 | 31.67 | 11.67 | 13.33 | 4.59 | 0 | 0 |
| | 10 | 100.00 | 100.00 | 44.33 | 25.00 | 76.67 | 70.00 | 61.67 | 50.00 |
| | 15 | 100.00 | 100.00 | 58.33 | 36.67 | 100.00 | 100.00 | 100.00 | 100.00 |
| O. surinamensis | 5 | 85.00 | 65.00 | 28.33 | 3.33 | 0 | 0 | 0 | 0 |
| | 10 | 100.00 | 100.00 | 38.33 | 11.67 | 11.67 | 3.33 | 0 | 0 |
| | 15 | 100.00 | 100.00 | 85.00 | 21.67 | 16.67 | 13.33 | 5.00 | 0 |
| T. confusum | 5 | 1.67 | 0 | 0 | 0 | 0 | 0 | 0 | 0 |
| | 10 | 16.67 | 11.67 | 0 | 0 | 1.67 | 0 | 0 | 0 |
| | 15 | 73.33 | 58.33 | 10.00 | 0 | 6.67 | 1.67 | 0 | 0 |

Adult moths were similarly exposed. The exposure of newly emerged larvae led to a reduction in the numbers developing through to adults and to an increase in the development time – for males from thirty days for controls to fifty-six days and for females from thirty-three to sixty-two days. Exposure of older larvae prevented all adult emergence. It was postulated that the effects were more marked on the older larvae because these were much more exposed, moving across or close to the surface of the food substrate, whereas younger larvae were buried more deeply and did not come into contact with the volatiles as readily. Furthermore, the vapour had a marked effect on the reproductive potential because oviposition by the emergent adults of exposed larvae was significantly reduced. Exposure of adults directly to sub-lethal doses did not produce the same effects on oviposition. Neem oil, which was also included in the experiments, did not produce any detrimental effects on the pest. Further work has elaborated on these findings (Pathak et al. 1993).

Sarac and Tunc (1995a) examined the effects of eucalyptus oil vapour against another moth pest, *Ephestia kuhniella* Zeller (Pyralidae: Lepidoptera). Adults and last instar larvae, as well as 2–3 week old adults of *T. confusum* and *S. oryzae*, were exposed to one of four essential oils, including that from *Eucalyptus camaldulensis*. Individuals in glass containers were exposed to the vapours from filter papers treated with the oils. *E. camaldulensis* was very effective against *Ephestia kuhniella* larvae and *S. oryzae* adults, causing 90 and 95 per cent mortality, respectively, at 135 µg/l after five days exposure. It was less effective against *T. confusum*, the same dosage only causing 85 per cent mortality after seven days. *E. camaldulensis* oil was generally less effective than that from anis, *Pimpinella anisum*, but more effective than those from thyme species, *Thymbra spicata* var. *spicata* and *Satureja thymbra*.

When the same oils were impregnated into filter papers with which the insects had direct contact the effect on mortality of *T. confusum* and *S. oryzae* was not as marked (Sarac and Tunc 1995b): *E. camaldulensis* only caused 30 per cent mortality of *T. confusum* after five days exposure (*Ephestia kuhniella* was not tested). However, *E. camaldulensis* oil was a very effective repellent when the adults were presented with a choice of either treated or untreated grain. After 24 h, 74 per cent of the adults settled in the control wheat and only 17 per cent in grain treated with a dosage of 0.25 µl/g oil. As the concentration increased to 1.0 µl/g still fewer adults were attracted (5 per cent). This repellent effect declined as the oil volatilised but, even after four weeks, 50 per cent of the activity was retained. In this respect *E. camaldulensis* oil was better than the others. It is clear from this work that *E. camaldulensis* oil does have potential for use as an insect repellent in stores, though whether this use would be cost-effective remains to be investigated.

The contact effect of eucalyptus leaves mixed with grain has also been observed with other *Sitophilus* species. Sharaby (1988) admixed shade-dried, powdered leaves of *E. globulus* with rice grains at 1–25 per cent concentrations in small petri dishes and exposed the samples to 1–3 day old adults of *S. oryzae* and the temperate species *S. granarius* (L.) (Curculionidae: Coleoptera). He found that the leaves of another Myrtaceae species, *Psidium guajava* (guava), were more toxic to the insects than *E. globulus* but the latter did have an effect. Both plants were more toxic to *S. oryzae* than *S. granarius*. However, in a repellency choice test *E. globulus* was more effective than *P. guajava*: during the course of a week there were never more than 15 per cent of individuals located on the treated grain compared with 60–80 per cent on untreated grain. Furthermore, a sub-lethal dosage of leaf powder of either plant of 15 g/100 g of rice prevented emergence of progeny of both species. Sharaby concluded that *E. globulus* does have potential as a rice protectant.

Sasaki and Calafiori (1987) assessed the effects of dried eucalyptus leaves against *S. zeamais*. They compared application of half a leaf of eucalyptus mixed in 150 g maize with treatments with garlic. Unfortunately, the species of *Eucalyptus* used was not stated but it was found to provide some degree of protection.

Leaf oils from *E. citriodora* and *E. tereticornis* have been tested in the laboratory against *S. oryzae* and found to be relatively ineffective (see later, Table 14.6) (Gakuru and Foua-Bi 1995). On the other hand, *E. citriodora* appeared to show rather more promise when extracts of the leaves were tested. Gakuru and Foua-Bi (1996) coated samples of maize seed with petroleum ether extracts at concentrations of 0.2–1.0 per cent (w/w). The mortality of twenty adults was recorded daily for a week. Although *E. citriodora* was not as effective as four other plant species (*Ocimum basilicum*, *Capsicum fructescens*, *Piper guineense* and *Tetrapleura tetraptera*) it nevertheless gave considerable protection, especially at 1 per cent concentrations (100 per cent mortality was attained after one week) when compared with untreated controls.

A similar experiment was conducted by Singh *et al.* (1996a,b) who exposed 1–5 day old adult pairs of *R. dominica* to wheat mixed with powdered '*E. hybrida*' (=*Eucalyptus* hybrid, i.e. *E. tereticornis*) leaves at 1.0–3.0 per cent (w/w) concentrations. Grain was left in open test tubes for intervals of twenty days and samples were then examined for damage and adult mortality. There were indications that the eucalyptus treatment did provide some degree of protection though the insect mortality data are difficult to interpret and the damage levels found in controls were relatively low (11 per cent after sixty days). Other plants were tested in the study and extracts of asafoetida (*Ferula asafoetida*), neem and *Lantana camara* were all more effective than eucalyptus.

The control of *R. dominica* and the moth *Sitotroga cerealella* (Olivier) (Gelechiidae: Lepidoptera) infesting paddy was investigated by Dakshinamurthy (1988) who treated grain with powdered leaves of *E. tereticornis*. Control of this moth is of particular interest to farmers in the developing world as it inflicts its damage on small-scale farm stores and is not found in large, central stores. However, samples in the study were still only of laboratory size, 100 g for *S. cerealella* experiments and 2 kg for those with *R. dominica*. Treatments were left for four months, after which time damage due to the beetles was assessed and the total number of adult moths counted. *E. tereticornis* provided significant control of both species. In the case of *R. dominica*, only 3 per cent damage was found in treated samples compared with 13 per cent in controls (and 9 per cent in samples treated with neem leaf), while for *S. cerealella* 369 adults emerged from untreated paddy but only seventy-seven in treated samples. In addition to eucalyptus and neem, three other plants were tested – *Vitex negundo*, *Pongamia glabra* and *Ipomea tuberosa* – but *E. tereticornis* was the most effective of all of them. Although the experiments were replicated four times none of the statistical data were presented so it is not possible to take into account replicate variability.

## Effects on insect pests of grain legumes

In store, pulses are attacked by a number of different pests of the Bruchidae family. Small grain pulses, such as cowpea, grams, pigeon pea, chickpea and bambara groundnuts are attacked by species of the genus *Callosobruchus*. The major pest of larger pulses, such as *Phaseolus vulgaris* (common or kidney beans), is *Acanthoscelides obtectus*. Experiments to assess the effects of eucalyptus on these pests have been conducted by several groups of workers.

Pajni and Gill (1991) subjected *C. maculatus* adults to volatiles of eucalyptus oils distilled from *E. citriodora* (83 per cent citronellal) and *E. globulus* (91 per cent 1,8-cineole). Asarone, a constituent of *Acorus calamus* rhizomes, known for its insecticidal properties, and some asarone derivatives, were also tested. Both oils were found to be twice as effective fumigants as asarone. Ahmed and Eapen (1986) examined the vapour effects of cineole and eucalyptus oils against a related species, *C. chinensis*. They found that the vapours gave very effective control of this pest, which was more susceptible than the cereal pests tested (noted earlier).

Although the results of testing *E. citriodora* and *E. tereticornis* oils against *S. oryzae* (Gakuru and Foua-Bi 1995), referred to earlier, were disappointing, tests against *C. maculatus* were more encouraging (Table 14.6). Experiments were conducted using petri dishes containing filter paper impregnated with different quantities of essential oil and cowpea as nutritive support. Of four essential oils tested, *E. citriodora* oil was the most effective, followed closely by *Ocimum basilicum*. Oils from *E. tereticornis* and *Citrus sinensis* were relatively ineffective.

The same researchers (Gakuru and Foua-Bi 1996) confirmed the promise of *E. citriodora* in tests involving extracts of nine plants. They obtained good control of *C. maculatus* when two-day-old adults were exposed to cowpeas treated with 0.2–1.0 per cent extracts of *E. citriodora*. The low dosages did not provide very effective control during the seven-day exposure period but the 1 per cent treatment resulted in 80 per cent control within three days and 90 per cent after four days.

The effects of the volatiles of commercially available eucalyptus oil on *A. obtectus* were assessed in well-designed and replicated experiments by Stamopoulos (1991). In a choice test with individual females, using kidney beans which were either placed in the same arena by themselves or with a vapour diffuser, replicated twenty-seven times, only 3 per cent of the eggs were deposited on beans where the volatiles were present and 97 per cent on the control group. Females moved towards the control seeds within a few minutes of exploration and remained there for six days. In a no-choice test, egg production was significantly reduced by the presence of eucalyptus oil vapour, as was egg hatchability. Three other essential oils were tested (geranium, cypress and bitter almond) but eucalyptus oil was the most effective. Clearly, the repellent effect of eucalyptus oil volatiles against *A. obtectus* can be exploited as a space treatment to protect beans during storage.

Faroni *et al.* (1995) compared the efficacy of mixing leaves of *E. citriodora* and black beans with admixture of ash, sand, black pepper, termite mound soil and synthetic chemicals (including deltamethrin) as protectants against *A. obtectus* infestation. Termite soil and black pepper controlled damage for up to eight months of storage whereas the leaves of *E. citriodora* were only effective in restricting the build-up of insect numbers for twelve weeks.

In southern France, *A. obtectus* is a significant pest of stored *P. vulgaris*. Regnault-Roger and Hamraoui (1993) compared the mortality and fecundity of pairs of adults placed in petri dishes with a small quantity of one of eleven different dried plant materials, generally leaves, and six kidney beans. After twelve days *E. globulus* caused approximately 20 per cent mortality but was not as effective as the most potent plant, *Origanum serpyllum*, which caused 50 per cent mortality. However, *E. globulus* did have a significant effect on oviposition (19.1 ± 8 eggs compared with

*Table 14.6* Effectiveness of *Eucalyptus citriodora* and *E. tereticornis* leaf oils against *Callosobruchus maculatus* and *Sitophilus oryzae* (after Gakuru and Foua-Bi 1995)

| | Dose (ml) | Cumulative mortality (%) after time (days) | | | | | |
| --- | --- | --- | --- | --- | --- | --- | --- |
| | | E. citriodora | | | E. tereticornis | | |
| | | 1 | 2 | 7 | 1 | 2 | 7 |
| C. maculatus | 0.5 | 2.50 | 6.49 | 10.65 | 3.75 | 3.89 | 13.49 |
| | 1.0 | 15.00 | 20.78 | 31.50 | 1.25 | 2.60 | 31.50 |
| | 2.0 | 91.25 | 91.25 | 100.00 | 3.75 | 8.75 | 18.15 |
| S. oryzae | 0.5 | 2.50 | 3.95 | 3.95 | 0 | 2.63 | 2.63 |
| | 1.0 | 0 | 2.60 | 5.63 | 0 | 2.60 | 5.63 |
| | 2.0 | 6.25 | 9.61 | 9.61 | 2.50 | 6.49 | 8.57 |

42 ± 12 for the control). The medium potency of *E. globulus* was confirmed by experiments in which adult *A. obtectus* were exposed to essential oils freshly distilled from twenty-two different plants (Regnault-Roger *et al.* 1993). Fifteen of the oils caused greater mortality than *E. globulus* oil. The most potent group included *Thymus serpyllum* and *Origanum majorana*.

Unlike some other published reports, Regnault-Roger and her co-workers give proper attention to the fact that the effects of essential oils on insects depend upon several things, including, most importantly, their chemical composition. They go on to point out that the composition can vary depending on whether or not there exist chemotypes of the plant in question, the season in which the plant is harvested, the part of the plant which is studied and the conditions used for distillation or extraction. Although bioassay work is necessarily the main element of research, the composition of the oil or extract being tested should also be determined and not relegated to incidental importance as has often been the case in published reports to date. All the oils used in the authors' work were analysed and the results presented. *E. globulus* oil had a cineole (+β-phellandrene) content of 86 per cent.

It would appear that although dried leaves of eucalyptus have an effect on *A. obtectus* mortality and development the volatile components are much more effective.

## Effects on mites

Mites are found associated with many foodstuffs and can be responsible for deterioration both of raw flour and processed material such as dairy and meat products. Several workers have attempted to develop methods using non-chemical techniques, including the use of eucalyptus leaves, to control these pests.

*Tyrophagus putrescentiae* (Schrank) (Acarina: Acaridae) is a cosmopolitan pest of several stored products and is encountered infesting stored grain and flour. Gulati and Mathur (1995) added five pairs of adults to samples of wheat flour to which had been added 5–75 per cent (w/w) powdered eucalyptus leaves (species not specified). Progeny production was observed at the end of the fifteen-day life cycle. There was an approximately linear relationship between dosage and the number of living progeny but even the lowest concentration of 5 per cent gave about 50 per cent control in comparison with untreated flour (Table 14.7). However, when leaf powder was applied at 5 per cent to wheat grain or wheat flour there did not appear to be any effect on either egg hatch or adult emergence.

Perrucci (1995) investigated the control of a related species of mite, *T. longior* (Gervais) (Acarina: Acaridae), a common pest of cured ham, cheese and similar products. The effects of a commercial sample of *E. globulus* oil and its main terpene constituent, 1,8-cineole, were compared with those of lavender and peppermint oils and their principal constituents. Acaricidal

*Table 14.7* Effect of powdered *Eucalyptus* leaves on egg production and other life stages of *Tyrophagus putrescentiae* in wheat flour (from Gulati and Mathur 1995)

| Treatment (% Eucalyptus, w/w) | Living progeny | | | |
|---|---|---|---|---|
| | Egg | Larvae | Nymph | Adult |
| 75 | 2.33 | 0.86 | 1.02 | 5.83 |
| 50 | 4.66 | 1.66 | 1.83 | 14.49 |
| 25 | 8.16 | 1.99 | 2.16 | 25.83 |
| 10 | 42.99 | 6.16 | 11.33 | 61.99 |
| 5 | 51.66 | 9.99 | 15.99 | 69.83 |
| 0 | 98.16 | 14.66 | 20.66 | 78.16 |

activity was measured in petri dishes by (a) direct contact with the uniformly dispersed oil at 0.25–6.0 µl doses and (b) inhalation of vapour over a 24 h period. *E. globulus* oil (analysed and shown to contain 76 per cent 1,8-cineole) was not as effective as the other plant oils and at the highest dosage it resulted in 92 per cent mortality when applied by direct contact (cf. 100 per cent mortality for the other oils at 1.0 µl) and 77 per cent mortality by inhalation. Similarly, 1,8-cineole was much less effective (82 per cent mortality at 6.0 µl by direct contact and 50 per cent mortality by inhalation) than linalool, linalyl acetate, fenchone, menthol and menthone at the same dosage. The author concludes that the absence of harmful residues from essential oils, and the lack of any adverse effects on the stored food itself, warrant further research on the possible use of plant oils for mite control in the food industry.

Miyazaki (1996) also observed the effects of various oil vapours, including that of *E. citriodora*, on *T. putrescentiae*, as well as on *Dermataphagoides farinae* Hughes (Pyroglyphidae: Acaridae), a mite commonly found in house dust. Single replicates of twelve adults were exposed to 5 µl of one of a series of oils in a volatilisation chamber and numbers moving were recorded up to 72 h. *E. citriodora* oil caused 100 per cent immobility of *D. farinae* within 4 h but was much less effective against *T. putrescentiae*, causing a maximum of only 40 per cent immobility (obtained after 24 h).

*Effects on insect pests of other stored products*

The potato tuber moth, *Phthorimaea operculella* Zeller (Gelechiidae: Lepidoptera) is a widely distributed pest of solanaceous crops in the tropics and sub-tropics. Kroschel and Koch (1996) conducted trials to control this pest using a variety of treatments including synthetic insecticides, *Bacillus thuringiensis*, and several plants, including leaves of unspecified eucalypts. Potatoes were laid on, or completely wrapped in, leaves and the degree of tuber infestation was noted. Although the treatments were only replicated twice and were not subjected to statistical analysis, there were clear indications that the leaves had a deterrent effect and that wrapping the tubers in leaves created a barrier to infestation. Despite very much better protection being obtained by applying *B. thuringiensis* or the pyrethroid insecticide, fenvalerate, wrapping tubers in eucalyptus leaves could provide a simple, cheap method of preserving tubers during storage.

*Conclusion*

The work of Kroschel and Koch (1996) represents the only testing of *Eucalyptus* under simulated practical conditions. Even this work lacks sufficient replication to allow firm conclusions to be drawn. However, it is apparent from laboratory experimentation that this genus does have potential for use in the protection of stored grain and other food products against insect pest attack. In particular, the volatile constituents of the leaves have demonstrated their potential for use in the vapour phase as short-term repellents. Repellency appears to be more strongly associated as a function of this plant than the ability to cause mortality directly. However, the effects may well cause mating and developmental disruption and so reduce reproductive potential. Although repellency is generally a short-term effect, it may last for many weeks when extracts are applied directly to grain and so may be of benefit to farmers storing grain for several weeks or months. However, in order to determine whether this is the case, and to assess the cost-effectiveness of using eucalyptus treatments under other storage conditions, more testing under practical storage conditions needs to be undertaken.

Given the diversity in composition of eucalyptus leaf oils it would also be advantageous to examine a greater number of *Eucalyptus* species for insecticidal effects. So far only a handful of species have been investigated.

# Allelopathy in eucalyptus

## Introduction

Since Molish (1937) defined the term allelopathy, many scientists have been concerned with the exploration and exploitation of allelochemicals (Whittaker 1972, Muller and Chou 1972, Rice 1984, Putnam and Tang 1986, Harborne 1988, Inderjit *et al*. 1995). In the widest sense the term allelopathy refers to the biochemical interactions which occur between plants and includes both deleterious and advantageous interactions. Rice (1984) regards it as an all-embracing term to cover most types of biochemical interactions, including those between higher plants and microorganisms. Typically, allelopathic inhibition results from a combination of allelochemicals which interfere with several physiological processes in the receiving plant or microorganism.

Allelopathy has the potential to enhance crop production and lead to the development of a more sustainable agriculture, including weed and pest control, through crop rotation, residue management and a variety of approaches in biocontrol (Einhellig 1995). Trees of the genus *Eucalyptus* are frequently surrounded by a grass-free zone and this has led to a search for possible allelochemicals in *Eucalyptus* species. The results to date indicate that eucalypts may well be a practical, commercial source of such chemicals in the future. In its simplest form this might entail use of the powdered leaves as a natural herbicide. Alternatively, and with a greater understanding of their mode of action, the allelochemicals themselves or suitable derivatives could be used as selective herbicides.

On the other hand, allelopathic effects may incline towards less use of certain eucalypts in agroforestry schemes. On the basis of reduced germination of some Ethiopian crops in experiments with aqueous leaf extracts, Lisanework and Michelsen (1993) suggested that planting of *E. camaldulensis* in integrated land use systems should be minimised while *E. globulus* had less of a potentially detrimental effect.

## Allelochemicals in E. citriodora

*Eucalyptus citriodora* (lemon-scented gum) is often surrounded by bare, grass-free ground and it seems likely that secondary metabolites are exuded from both fresh and fallen leaves which inhibit the growth and seed germination of surrounding plants. In the course of research on *Eucalyptus* metabolites which exhibit biological activities, the authors have studied the allelochemicals from *E. citriodora* leaves.

Extraction of the leaves with 70 per cent aqueous acetone, followed by column chromatography in which the inhibitory activity of fractions was monitored against germinating seeds and seedlings of lettuce (*Lactuca sativa* cv. Wayahead), garden cress (*Lepidium sativum* L.), green foxtail (*Setaria viridis* L.) and barnyard grass (*Panicum crus-galli* L.) furnished two crystalline inhibitors. These were subsequently identified as the ($\pm$)-*p*-menthane-3,8-*cis*- and *trans*-diols shown in Figure 14.1 (Nishimura *et al*. 1982, 1984).

Both diols, as well as citronellal and citronellol, were determined quantitatively in the leaves of growing *E. citriodora* by gas chromatography and gas chromatography combined with mass spectrometry. The variation in concentration of the constituents in the seedlings with ontogenetic age is shown in Figure 14.2. The diols were absent from juvenile tissue until thirteen months (approximately fifty nodes) when they then increased to 4.5 mg/g (*cis*-diols) and 2.2 mg/g (*trans*-diols) in fresh adult leaves (twenty-one months old). Interestingly, both diols occur as racemates. Levels of ($\pm$)-citronellal, on the other hand, the major constituent of *E. citriodora* essential oils, gradually increased at first but after thirteen months decreased

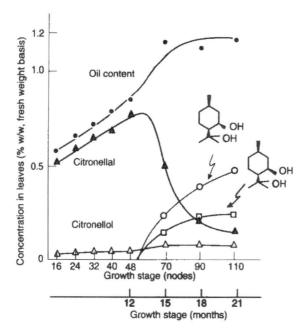

*Figure 14.2* Variation in oil content and concentration of the allelochemicals *p*-menthane-3,8-*cis*- and *trans*-diol and other constituents in *E. citriodora* with ontogenetic age.

dramatically. This suggests that the *cis*- and *trans*-diols are formed biogenetically by cyclisation of citronellal. However, it is not clear whether they are produced enzymatically. The diols were not artifacts since the pH of the homogenised tissues was neutral (Nishimura *et al.* 1984).

In order to evaluate the allelopathic effect, samples of soil were collected at an *E. citriodora* grove. Extracts of the soil were analysed and shown to contain *cis* and *trans* *p*-menthane-3, 8-diols, but in much lower amounts than in *E. citriodora* leaves: approximately 15 ppm compared to about 4600 ppm in the leaves. The low concentration of the diols in the soil suggests either that their extraction was inefficient and/or that they were partially transformed by microorganisms in the soil.

The chiral compounds, (+)-*cis*, (−)-*cis*, (+)-*trans* and (−)-*trans*, were prepared to compare their germination and growth inhibitory activities against lettuce seeds and seedlings, respectively. The biological activity of racemic (natural product) and chiral compounds of the *cis* isomers was found to be much higher than that of the *trans* isomers. Furthermore, the synthetic (+)-*cis* isomer had a higher inhibitory activity than the optical antipode, suggesting that a receptor site involved in the germination and growth of lettuce is also chiral (Nishimura *et al.* 1982).

The inhibitory activities of *p*-menthane-3,8-*cis*-diol against seed germination of several plants are shown in Figure 14.3 (Nishimura *et al.* 1984). Concentrations of 100–300 ppm ($5.8 \times 10^{-4} - 1.7 \times 10^{-3}$ M) inhibited seed germination in lettuce, garden cress, green foxtail and barnyard grass. However, even at high concentrations (300 ppm) the diol had very little inhibitory effect on the germination of *E. citriodora* itself or of rice. The patterns of inhibition of hypocotyl growth of seedlings of the same plants were similar to those of seed germination: very little inhibitory effect on *E. citriodora* and rice but significant inhibition of the other plants at 100–300 ppm (green foxtail experiencing the greatest inhibition and garden cress the least). In

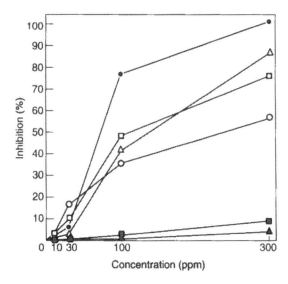

*Figure 14.3* Germination inhibition of *p*-menthane-3,8-*cis*-diol against seeds of some higher plants. Experimental error is within 7 per cent. Symbols: ● lettuce, △ garden cress, □ green foxtail, O barnyard grass, ■ rice, ▲ *E. citriodora*.

terms of its use as a potential herbicide, it is interesting that the biological activity of the *cis* diol is very selective against higher plants.

Studies by other workers have further demonstrated the inhibitory effects of *E. citriodora* (e.g. Kohli *et al*. 1988, 1998, Singh *et al*. 1991). Particularly encouraging results have been obtained in India in tests to combat the noxious weed *Parthenium hysterophorus* (Kohli *et al*. 1998). The oil from *E. citriodora* was more effective in causing injury to the weed than *E. globulus* oil.

### Allelochemicals in other Eucalyptus species

Of the other eucalypts studied, *E. camaldulensis* has attracted much attention. Del Moral and Muller (1970) investigated allelochemicals in *E. camaldulensis* which is surrounded by bare ground and identified 1,8-cineole, α- and β-pinene, and α-phellandrene as volatile inhibitors. This role for the terpenes was established from observations of their presence in all the various phases of the interaction. Thus:

1. Terpenes are present in significant quantities in the leaves of *E. camaldulensis*.
2. They are constantly turned over by the eucalypt and a vapour cloud of volatile essence hangs around the plants.
3. They occur in the soil surrounding the plants.
4. They remain in the dry soil until rains activate soil microorganisms which degrade them.
5. They can be transported into plant cells through the waxy coatings of seeds or roots.
6. They have a significant effect on the germination of seeds of those annuals which grow in the adjacent grassland area.

The presence of other possible allelochemicals in *E. camaldulensis* has been investigated by Nishimura (1989). Fractionation of acetone extracts of leaves, with monitoring of growth inhibitory activity, resulted in the isolation of spathulenol (Figure 14.4).

The leaves of *E. viminalis*, *E. delegatensis* and *E. pauciflora*, which form colonies in nature, have also been found to contain potent allelochemicals, mostly terpenoids and phenolics (Figure 14.4) (Nishimura unpubl.).

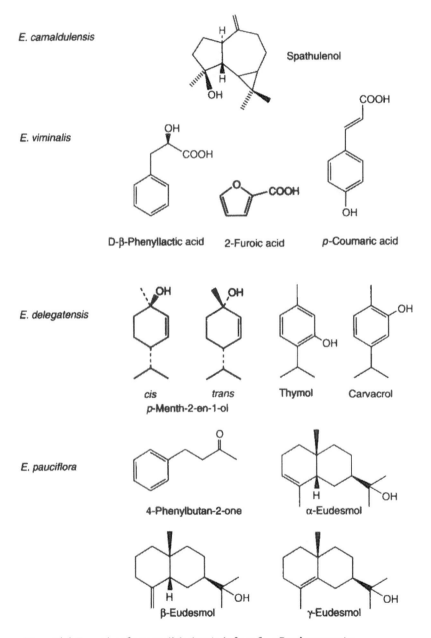

Figure 14.4 Examples of potent allelochemicals from four *Eucalyptus* species.

# References

Ahmed, S.M. and Eapen, M. (1986) Vapour toxicity and repellency of some essential oils to insect pests. *Indian Perfumer*, 30, 273–278.

Corbet, S.A., Danahar, G.W., King, V., Chalmers, C.L. and Tiley, C.F. (1995) Surfactant-enhanced essential oils as mosquito larvicides. *Entomol. Experiment. Applic.*, 75, 229–236.

Curtis, C.F., Lines, J.D., Lu Baolin and Renz, A. (1990) Natural and synthetic repellents. In C.F. Curtis (ed.), *Appropriate Technology in Vector Control*, CRC Press, Boca Raton, USA, pp. 75–92.

Dakshinamurthy, A. (1988) Effect of certain plant products on storage pests of paddy. *Trop. Sci.*, 28, 119–122.

Dales, M.J. (1996) *A Review of Plant Materials Used for Controlling Insect Pests of Stored Products*, NRI Bulletin 65, Natural Resources Institute, Chatham, UK.

Del Moral, R. and Muller, C.H. (1970) The allelopathic effects of *Eucalyptus camaldulensis*. *Am. Midl. Nat.*, 83, 254–282.

Einhellig, F.A. (1995) Allelopathy: current status and future goals. In Inderjit, K.M.M. Dakshini and F.A. Einhellig (eds), *Allelopathy, Organisms, Processes and Applications*, ACS Symposium Series 582, American Chemical Society, Washington DC, pp. 1–24.

Faroni, L.R. D'A., Molin, L., de Andrade, E.T. and Cardoso, E.G. (1995) Utilisation of natural products in the control of *Acanthoscelides obtectus* in bean storage (in Portuguese). *Rev. Brasil. Armazenamento*, 20, 44–48.

Gakuru, S. and Foua-Bi, K. (1995) Compared effect of four plant essential oils against cowpea weevil (*Callosobruchus maculatus* Fab.) and rice weevil (*Sitophilus oryzae* L.) (in French). *Tropicultura*, 13, 143–146.

Gakuru, S. and Foua-Bi, K. (1996) Effects of plant extracts on the cowpea weevil and the rice weevil (in French). *Cahiers Agricultures*, 5, 39–42.

Golob, P. and Webley, D.J. (1980) *The Use of Plants and Minerals as Traditional Protectants of Stored Products*, TPI Report G138, Tropical Products Institute, London (now Natural Resources Institute, Chatham, UK).

Gulati, R. and Mathur, S. (1995) Effect of *Eucalyptus* and *Mentha* leaves and *Curcuma* rhizomes on *Tyrophagus putrescentiae* (Schrank) (Acarina: Acaridae) in wheat. *Exp. Appl. Acarology*, 19, 511–518.

Harborne, J.B. (1988) Biochemical interaction between higher plants. In *Introduction to Ecological Biochemistry*, Academic Press, London, pp. 277–301.

Inderjit, Dakshini, K.M.M. and Einhellig, F.A. (eds) (1995) *Allelopathy, Organisms, Processes and Applications*, ACS Symposium Series 582, American Chemical Society, Washington DC.

Kohli, R.K., Batish, D.R. and Singh, H.P. (1998) Eucalypt oils for the control of parthenium (*Parthenium hysterophorus* L.). *Crop Protection*, 17, 119–122.

Kohli, R.K., Chaudhry, P. and Kumari, A. (1988) Impact of *Eucalyptus* on parthenium, a weed. *Indian J. Range Manage.*, 9, 63–67.

Kranz, J., Schmutterer, H. and Koch, W. (eds) (1977), *Diseases, Pests and Weeds in Tropical Crops*, Verlag Paul Parey, Berlin.

Kroschel, J. and Koch, W. (1996) Studies on the use of chemicals, botanicals and *Bacillus thuringiensis* in the management of the potato tuber moth in potato stores. *Crop Protection*, 15, 197–203.

Li, Z., Yang, J., Zhuang, X. and Zhang, Z. (1974) Studies on the repellent quwenling (in Chinese). *Malar. Res.*, 6.

Lisanework, N. and Michelsen, A. (1993) Allelopathy in agroforestry systems: the effects of leaf extracts of *Cupressus lusitanica* and three *Eucalyptus* spp. on four Ethiopian crops. *Agrofor. Syst.*, 21, 63–74.

Miyazaki, Y. (1996) Differences in susceptibilities of mites (*Dermataphagoides farinae* and *Tyrophagus putrescentiae*), found in house dust, to exposure to several leaf oils. *Mokuzai Gakkaishi*, 42, 532–533.

Molish, H. (1937) *Der Einfluss einer Pflanze auf die andere-Allelopathie*, Fischer, Jena.

Moody, R.P., Sidon, E. and Franklin, C.A. (1986) Skin penetration of the insect repellent DEET (N,N'-diethyl-*m*-toluamide) in rats, rhesus monkeys and humans: effect of anatomic site and multiple exposure. Paper presented at 6th Internat. Congr. of Pesticide Chemistry, Ottawa, Paper 8A/7E-06.

Muller, C.H. and Chou, C.H. (1972) Phytotoxins: an ecological phase of phytochemistry. In J.B. Harborne (ed.), *Phytochemical Ecology*, Academic Press, London, pp. 201–216.

Nishimura, H. (1989) Petroleum plants. In O. Kitani and C.W. Hall (eds), *Biomass Handbook*, Gordon and Breach Science Publishers, Amsterdam, pp. 114–132.

Nishimura, H., Kaku, K., Nakamura, T., Fukuzawa, Y. and Mizutani, J. (1982) Allelopathic substances, (±)-*p*-menthane-3,8-diols isolated from *Eucalyptus citriodora* Hook. *Agric. Biol. Chem.*, 46, 319–320.

Nishimura, H. and Mizutani, J. (1989) Economically useful ingredients and clonal propagation of *Eucalyptus citriodora* plant. *The Proc. Hokaido Tokai Univ., Sci. Engineer.*, (2), 57–65.

Nishimura, H., Mizutani, J., Umino, T. and Kurihara, T. (1986) New repellents against mosquitoes, *p*-menthane-3,8-diols in *Eucalyptus citriodora* and related compounds. Paper presented at 6th Internat. Congr. of Pesticide Chemistry, Ottawa, Paper 2D/E-07.

Nishimura, H., Nakamura, T. and Mizutani, J. (1984) Allelopathic effects of *p*-menthane-3,8-diols in *Eucalyptus citriodora*. *Phytochemistry*, 23, 2777–2779.

Pajni, H.R. and Gill, M. (1991) Use of new pesticides of plant origin for the control of bruchids. In F. Fleurat-Lessard and P. Ducom (eds), *Proc. 5th Internat. Working Conf. on Stored Product Protection*, Bordeaux, France, September 1990, Vol. 3, pp. 1671–1677.

Pathak, P.H., Gurusubramanian, G. and Krishna, S.S. (1993) Changes in postembryonic development and reproduction in *Corcyra cephalonica* Stainton (Lepidoptera: Pyralidae) as a function of eucalyptus and neem oil vapour action during rearing or on adults. *Z. Angew. Zool.*, 80, 345–352.

Pathak, P.H. and Krishna, S.S. (1991) Post-embryonic development and reproduction in *Corcyra cephalonica* (Stainton) (Lepidoptera: Pyralidae) on exposure to eucalyptus and neem oil volatiles. *J. Chem. Ecol.*, 17, 2553–2558.

Perrucci, S. (1995) Acaricidal activity of some essential oils and their constituents against *Tyrophagus longior*, a mite of stored food. *J. Food Prot.*, 58, 560–563.

Prates, H.T., Santos, J.P., Waquil, J.M., Fabris, J.D., Oliveira, A.B. and Foster, J.E. (1998) Insecticidal activity of monoterpenes against *Rhyzopertha dominica* (F.) and *Tribolium castaneum* (Herbst). *J. Stored Prod. Res.*, 34, 243–249.

Putnam, A.R. and Tang, C.S. (1986) *The Science of Allelopathy*, John Wiley, Chichester, UK.

Regnault-Roger, C. and Hamraoui, A. (1993) Efficiency of plants from the south of France used as traditional protectants of *Phaseolus vulgaris* L. against its bruchid *Acanthoscelides obtectus* (Say). *J. Stored Prod. Res.*, 29, 259–264.

Regnault-Roger, C., Hamraoui, A., Holeman, M., Theron, E and Pinel, R. (1993) Insecticidal effect of essential oils from Mediterranean plants upon *Acanthoscelides obtectus* Say (Coleoptera: Bruchidae), a pest of kidney bean (*Phaseolus vulgaris* L.). *J. Chem. Ecol.*, 19, 1233–1244.

Rice, E.L. (1984) *Allelopathy*, Academic Press, New York.

Santos, J.P., Prates, H.T., Waquil, J.M. and Oliveira, A.B. (1997) Evaluation of plant-origin substances on the control of stored product pests (in Portuguese). *Pesquisa em Andamento Centro Nacional de Pesquisa de Milho e Sorgo*, No. 19.

Sarac, A. and Tunc, I. (1995a) Toxicity of essential oil vapours to stored-product insects. *Z. Pflanzenkr. Pflanzenschutz*, 102, 69–74.

Sarac, A. and Tunc, I. (1995b) Residual toxicity and repellency of essential oils to stored-product insects. *Z. Pflanzenkr. Pflanzenschutz*, 102, 429–434.

Sasaki, E.T. and Calafiori, M.H. (1987) Repellent effect of garlic (*Allium sativum* L.) on *Sitophilus* species in stored maize. *Ecossistema*, 12, 30–33.

Satoh, A., Utamura, H., Nakade, T. and Nishimura, H. (1995) Absolute configuration of a new mosquito repellent, (+)-eucamalol and the repellent activity of its epimer. *Biosci. Biotech. Biochem.*, 59, 1139–1141.

Schreck, C.E. and Leonhardt, B.A. (1991) Efficacy assessment of quwenling, a mosquito repellent from China. *J. Am. Mosq. Control Assoc.*, 7, 433–436.

Sharaby, A. (1988) Evaluation of some Myrtaceae plant leaves as protectants against the infestation by *Sitophilus oryzae* L. and *Sitophilus granarius* L. *Insect Sci. Applic.*, 9, 465–468.

Sharma, V.P. and Dhiman, R.C. (1993) Neem oil as a sandfly (Diptera: Psychodidae) repellent. *J. Am. Mosq. Control Assoc.*, 9, 364–366.

Singh, D., Kohli, R.K. and Saxena, D.B. (1991) Effect of eucalyptus oil on germination and growth of *Phaseolus aureus* Roxb. *Plant and Soil*, 137, 223–227.

Singh, G. and Upadhyay, R.K. (1993) Essential oils: a potent source of natural pesticides. *J. Sci. Ind. Res.*, 52, 676–683.

Singh, H., Mrig, K.K. and Mahla, J.C. (1996a) Effect of different plant products on the fecundity and emergence of lesser grain borer *Rhyzopertha dominica* in wheat grains. *Ann. Biol. (Ludhiana)*, 12, 96–98.

Singh, H., Mrig, K.K. and Mahla, J.C. (1996b) Efficacy and persistence of plant products against lesser grain borer *Rhyzopertha dominica* in wheat grains. *Ann. Biol. (Ludhiana)*, 12, 99–103.

Stamopoulos, D.C. (1991) Effect of four essential oil vapours on the oviposition and fecundity of *Acanthoscelides obtectus* (Coleoptera: Bruchidae): laboratory evaluation. *J. Stored Prod. Res.*, 27, 199–203.

Thakur, A.K. and Sankhyan, S.D. (1992) Studies on the persistent toxicity of some plant oils to storage pests of wheat. *Indian Perfumer*, 36, 6–16.

Trigg, J.K. (1996) Evaluation of a eucalyptus-based repellent against *Anopheles* spp. in Tanzania. *J. Am. Mosq. Control Assoc.*, 12, 243–246.

Trigg, J.K. and Hill, N. (1996) Laboratory evaluation of a eucalyptus-based repellent against four biting arthropods. *Phytotherapy Res.*, 10, 313–316.

Watanabe, K., Shono, Y., Kakimizu, A., Okada, A., Matsuo, N., Satoh, A. and Nishimura, H. (1993) New mosquito repellent from *Eucalyptus camaldulensis*. *J. Agric. Food Chem.*, 41, 2164–2166.

Whittaker, R.H. (1972) The biochemical ecology of higher plants. In E. Sondheimer and J.B. Simeone (eds), *Chemical Ecology*, Academic Press, New York, pp. 43–70.

# 15 Chemical ecology of herbivory in eucalyptus

## Interactions between insect and mammalian herbivores and plant essential oils

*Ivan R. Lawler and William J. Foley*

## Introduction

*Eucalyptus* dominates more than 90 per cent of Australian forests and woodlands and supports a wide range of endemic vertebrates and invertebrates. Nonetheless, few insects, and even fewer mammals, eat the foliage to any appreciable extent (Landsberg and Cork 1997). *Eucalyptus* foliage contains low amounts of dietary nitrogen, which is essential for the maintenance and reproduction of all animals (Cork and Foley 1991), but also appreciable amounts of essential oils and phenolic compounds. Both these groups of compounds are believed to have a significant influence on the acceptability of foliage as food and the nutritional quality of that foliage (Hume 1982).

There has long been speculation about the degree and nature of the influence of *Eucalyptus* oils on the food choice of herbivores. This interest has been driven by a number of considerations. Firstly, the restricted diet of koalas suggests that a knowledge of its nutrition and food choices will contribute substantially to its conservation and the conservation of other sympatric species (Cork and Sanson 1990).

Secondly, efforts to establish plantations of *Eucalyptus* for wood fibre and sawn timber are plagued by the damage inflicted on plants by pest insects and mammals (Montague 1994, Patterson *et al*. 1996). Although this is countered by extensive poisoning campaigns, current research is seeking to incorporate naturally resistant genotypes into plantation management systems (Farrow 1993). Determining the role of essential oils and related compounds in conferring resistance to herbivores has substantial economic consequences.

Finally, the dominance of *Eucalyptus* in the Australian environment, a variable level of herbivory and the substantial quantities of plant secondary metabolites contained in the foliage, has spurred a number of studies seeking a more fundamental understanding of the evolutionary interactions between plants and herbivores (Morrow and Fox 1980, Morrow *et al*. 1976, Cork and Foley 1991). For example, Morrow and Fox (1980) used *Eucalyptus* and its volatile oils to test whether herbivory was controlled by the metabolic costs of detoxification. These types of studies have made important contributions to broader ecological theory.

In this chapter, we review studies which have examined the role of *Eucalyptus* oils in determining the feeding of Australian insects and mammals or which have measured the consequences of ingested oils. We conclude by evaluating the role of essential oils in conferring natural herbivore resistance on *Eucalyptus* and the prospects of selecting high yielding genotypes for use in breeding programmes.

## Components of eucalyptus oils relevant to herbivores

Previous chapters have reviewed the nature of eucalyptus oils and discussed the diversity of mono- and sesquiterpenes present in different species. Several different techniques have been used to extract eucalyptus oils from plants eaten by herbivores, but most studies have used steam distillation. However, steam distillation can volatilise compounds other than essential oils and may also lead to some rearrangements or hydrolysis of naturally occurring compounds. Since some of these non-terpene compounds are involved in significant ecological interactions, we review their nature and effects below.

### Benzaldehyde

Benzaldehyde (Figure 15.1a) almost certainly arises from the hydrolysis of cyanogenic glycosides in leaf tissue and in some species (e.g. *E. yarraensis*, Boland *et al.* 1991), can dominate the steam-volatile extractives. The only cyanogenic glycoside that has been isolated from *Eucalyptus* is prunasin (Figure 15.1b) (Finnemore *et al.* 1935, Pass *et al.* 1998, E.E. Conn pers. comm.). Concentrations in the young expanding tips of *E. cladocalyx* can exceed 10 mg/g dry mass (Gleadow *et al.* 1998). Sheep have been reported to be poisoned after eating *E. cladocalyx* foliage (Everist 1981) and koalas after eating *E. viminalis* foliage, but in neither case has the link with cyanogens been proven (Southwell 1978).

Of potentially more importance ecologically is the observation that many eucalypts are polymorphic for prunasin. These include *E. viminalis*, *E. orgadophila*, *E. ovata* and *E. polyanthemos*. This provides a powerful tool for investigating fundamental questions of the cost of maintaining plant secondary metabolites and may provide explanations of variable herbivore behaviour (R. Gleadow and I. Woodrow pers. comm.).

### Acylphloroglucinols

*Eucalyptus* and several other genera of the Myrtaceae contain fully substituted acylphloroglucinols, some of which are steam volatile and frequently isolated during phytochemical investigations aimed principally at the volatile essential oils (Boland *et al.* 1991, Ghisalberti 1996). The most widespread compound is torquatone (Figure 15.1c) (Bowyer and Jeffries 1959) which can comprise up to 40 per cent of the steam-volatile constituents of *E. torquata* (Bignell *et al.* 1994). A range of other compounds varying in the level of oxidation of the nuclear carbons has been described including jensenone (Figure 15.1d), which dominates the steam-volatile extract of *E. jensenii* leaf (Boland *et al.* 1991, 1992). These compounds, together with the acylphloroglucinol-terpene adducts described below, can have substantial biological activity in a number of different systems.

### Acylphloroglucinol–terpene adducts

In the past 10–15 years a number of compounds have been identified which result from terpenes bonded to fully substituted acylphloroglucinol derivatives (Chapter 12 this volume, Ghisalberti 1996, Pass *et al.* 1998). These are known as euglobals (e.g. euglobal III, Figure 15.1e) (Kozuka *et al.* 1982a,b) and macrocarpals (e.g. macrocarpal G, Figure 15.1f) (Yamakoshi *et al.* 1992). None are steam volatile but, given that common essential oils form part of the structure, and given their apparent importance in some ecological interactions (Pass *et al.* 1998, Lawler *et al.* 1998), some weight will be given to them in this chapter.

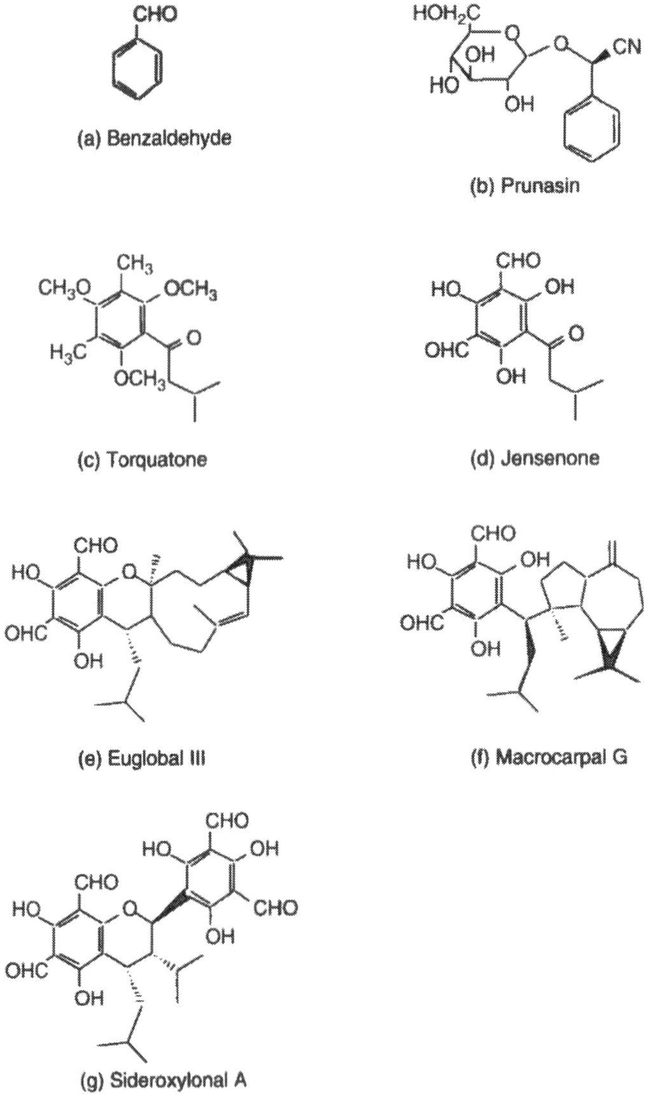

(a) Benzaldehyde

(b) Prunasin

(c) Torquatone

(d) Jensenone

(e) Euglobal III

(f) Macrocarpal G

(g) Sideroxylonal A

*Figure 15.1* Structures of some non-terpenoid compounds extracted from *Eucalyptus* leaves by conventional steam distillation for essential oils and of possible antifeedant compounds with similar structures.

Although there are undoubtedly more compounds to be discovered, those already known are formed with only a restricted number of terpenes. The euglobals contain a monoterpene moiety derived most commonly from β-pinene, although compounds with sabinene and β-phellandrene have also been identified. A second group of euglobals are formed with sesquiterpenes, most commonly bicyclogermacrene. Macrocarpals differ from euglobals in that they lack the ether linkage between the aromatic and terpenoid parts of the structure. All terpene adducts of known macrocarpals are derived from sesquiterpenes.

*Dimeric acylphloroglucinols*

A limited number of compounds have been described which can generically be termed dimers of the simple acylphloroglucinols. These include sideroxylonal A (Figure 15.1g) and B, robustaol A and grandinal. Of these, only the effect of sideroxylonals on insect and mammal feeding has been investigated, with strong indications of high deterrency (see below).

*Biosynthesis of acylphloroglucinol compounds*

Little is known of the biosynthesis of the simple acylphloroglucinol structure itself. The euglobals and dimeric compounds are presumed to be formed by Diels–Alder condensations but there have been no formal synthetic studies to date (Ghisalberti 1996). The generation of a macrocarpal is thought to involve a carbocationic species which acts as a cationic initiator in the cyclisation of the sesquiterpene precursors (Ghisalberti 1996). Acylphloroglucinols occur in all subgenera of *Eucalyptus* (Eschler, Pass, Willis and Foley unpubl.).

## Ecological factors affecting the concentration of oils in *Eucalyptus* foliage

The concentration of volatile essential oils in foliage may be influenced by a number of extrinsic factors including soil type, light and atmospheric $CO_2$. Knowing how essential oils and other plant secondary metabolites respond to these factors is important because (a) we need to be able to predict how climate change will affect animal–plant interactions, and (b) several key theories of animal–plant interactions are couched in terms of the relationship between the level of herbivory and the resources available to the plant.

Most influential of these theories are those of resource availability or carbon-nutrient balance (Bryant *et al.* 1983, Coley *et al.* 1985). These propose that the concentration of carbon-based plant secondary metabolites is governed by the availability of carbon to other nutrients. That is, under conditions where carbon is plentiful relative to other nutrients (e.g. high light, low soil nutrients) growth will be limited by these other nutrients, leaving the excess carbon to be diverted to activities other than growth, such as carbon-based defences (Coley *et al.* 1985).

The few available data suggest that *Eucalyptus* phenolics behave in a manner consistent with these theories but that the same is not true for essential oils. Lawler *et al.* (1997) grew seedlings of *E. tereticornis* under conditions of varying carbon and nutrient supply and measured the concentration of steam-volatile essential oils in the foliage (Figure 15.2). The concentration of leaf essential oils in *E. tereticornis*, and also in *E. citriodora* (Lawler, Foley and Woodrow unpubl.), actually increased with high soil nutrients. Increased light also increased foliar essential oil concentrations but only under high soil nutrient conditions (Lawler *et al.* 1997). Elevated atmospheric $CO_2$ had no effects on foliar essential oil concentration in *E. tereticornis* or *E. citriodora* but Adamson and Woodrow (pers. comm.) found elevated foliar essential oils in *E. nitens* seedlings grown under elevated $CO_2$.

The oil content of seedlings differs both quantitatively and qualitatively from that of the foliage of the mature plant (Chapter 4 this volume, Boland *et al.* 1982). For example, seedling leaves of *E. delegatensis* contained only about 0.1 per cent (dry weight basis) of steam-volatile essential oils, whereas mature foliage contains 10–20 fold more oil (Boland *et al.* 1982). These differences are significant because seedlings may often be more susceptible to herbivory than adult plants. Attempts to select for, or breed, plants resistant to herbivores on the basis of their foliar terpene concentrations may founder unless the relationship between the concentration of essential oils and herbivory in seedling plants is considered. The differences between seedling

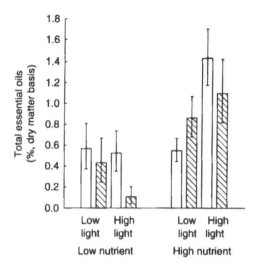

*Figure 15.2* Effect of variation in soil nutrients, light and atmospheric $CO_2$ on essential oil content of *Eucalyptus tereticornis* leaves. Unshaded bars represent plants grown at ambient (350 ppm) $CO_2$ levels while shaded bars represent plants grown at elevated (800 ppm) $CO_2$ levels (after Lawler *et al.* 1997).

and adult plants are most probably a consequence of the low energy reserves contained in seedlings, together with the relatively high metabolic cost of synthesis and maintenance of essential oils (Gershenzon 1994), but this is an area that needs further research.

## Diet choice in vertebrate herbivores

The vertebrate herbivores that eat *Eucalyptus* foliage extensively are all marsupials. These include arboreal folivores such as koalas (*Phascolarctos cinereus*), greater gliders (*Petauroides volans*), common ringtail possums (*Pseudocheirus peregrinus*) and common brushtail possums (*Trichosurus vulpecula*), and some macropodines (kangaroo family) such as the swamp wallaby (*Wallabia bicolor*), the Tasmanian pademelon (*Thylogale billardieri*) and the red-necked wallaby (*Macropus rufogriseus*). Although kangaroos can inflict severe damage on *Eucalyptus* plantations (Montague 1994), there are no data examining their interactions with *Eucalyptus*. In contrast, the reliance of koalas and the other arboreal folivores on *Eucalyptus* foliage, and in particular that of a restricted subset of species, has led to many studies of the relationship between the volatile oil content of the foliage and food choice. The rationale for these studies has ranged from the distinctive odours of the oils, to their alleged thermogenic properties and their potential germicidal effects on the microflora of the gut (Pratt 1937, Fleay 1937).

However, to date there has been no convincing or conclusive evidence that simple essential oils play any role in diet selection by koalas or any other marsupial herbivore. For example, Southwell (1978) could find no association between the level of defoliation of individual *Eucalyptus* trees and their concentration of the total steam-volatile oils or the proportion of 1,8-cineole in these oils.

Betts (1978) claimed that the ratio of cineole to sesquiterpenes in *E. rudis* was an important determinant of food choice in koalas eating that species but his data explained only 18 per cent of the observed variation. Furthermore, there was no explanation as to why this fraction or ratio should be more or less important than any other component. Pratt (1937) argued that cineole had a thermogenic effect, whereas α-phellandrene exerted a cooling effect on the animals which ate the compound, and that these effects correlated with food choice in koalas in different locations. Neither the supposed thermogenic effects nor the alleged pattern of food choice has stood up to closer examination (Southwell 1978).

Hume and Esson (1993) argued that koalas required a minimum threshold of oil in the diet but there was no clear correlation between the total steam-volatile concentrations and the relative feeding effort by koalas. In this case, feeding effort was allocated to one of four categories but animals always had some favoured foliage available. Zoidis and Markowitz (1992) found no correlation between the concentration of cineole in foliage and food intake by koalas in the San Diego Zoo. In contrast to the study of Hume and Esson (1993), Lawler *et al.* (1998) compared koala feeding on different individual trees of *E. ovata* and *E. viminalis* and found that intakes of foliage were higher when the essential oil content was significantly lower than Hume and Esson's suggested threshold. For individual trees within each *Eucalyptus* species, intakes decreased in a manner strongly consistent with the total essential oil concentration (Figure 15.3). A similar pattern was also observed for common ringtail possums feeding on the same trees (Figure 15.4a).

Although these correlative approaches appear to be a reasonable way to test the basis of food choice in mammals, they are unlikely to succeed where the essential oil profile of the foliage is complex, as in *Eucalyptus*. This is because we have to decide *a priori* what dietary components are important, and simply because a foliage smells strongly to man does not mean that it smells the same way to an animal. In addition, what is toxic to man may not be toxic to a particular herbivore. For example, black colobus monkeys (*Colobus satanus*) can eat large quantities of toxic alkaloid-rich leaves which would be fatal to non-adapted species (McKey *et al.* 1981). Similarly, while we have shown above that ingestion of *Eucalyptus* leaves correlates strongly with the essential oil content of those leaves, for their body sizes, the marsupial folivores of *Eucalyptus* can eat over ten times the amount of essential oils known to cause fatality in humans (McLean and Foley 1997), and can do this for sustained periods (Lawler unpubl.). Correlative evidence does not imply any causal link; a correlation may arise simply because the concentration of oils is correlated with the true limiting factor.

Most studies of the role of essential oils in regulating feeding in other vertebrate herbivores have been similarly inconclusive (e.g. Duncan *et al.* 1994) or do little to demonstrate causal

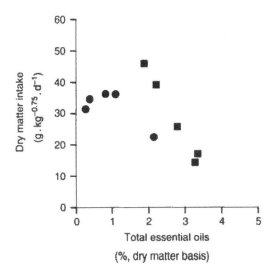

*Figure 15.3* Relationship between food intake of koalas and essential oil content of leaves. Circles represent *Eucalyptus ovata* and squares represent *E. viminalis* (after Lawler *et al.* 1998a).

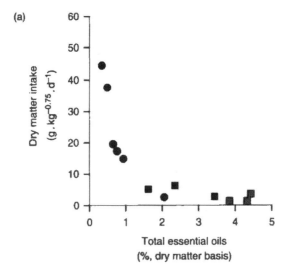
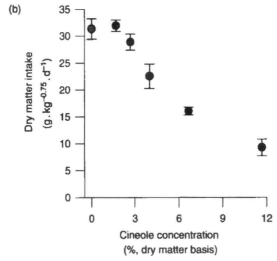

*Figure 15.4* Relationship between food intake of common ringtail possums and essential oil content of: (a) leaves, (b) an artificial diet. In (a), circles represent *Eucalyptus ovata* and squares represent *E. viminalis*. Note that in leaf diets data are for total essential oils, while in the artificial diet cineole was used since in a range of species, including *E. viminalis*, it is the predominant component of the oil (after Lawler *et al.* 1998a).

relationships. Clearly a better option is a bioassay involving the use of pure compounds added to an artificial diet in isolation. In this way, other unknown differences between diets can be excluded and cause and effect can be identified conclusively.

### Bioassays with marsupials fed isolated essential oils

The few studies that have fed isolated eucalyptus oils to marsupial herbivores do not support the notion that essential oils, in themselves, are important influences on diet selection. Pass *et al.*

(1998) isolated oils from both palatable and unpalatable forms of *E. ovata*. The main difference was quantitative: unpalatable trees had higher total oil contents though the compositions of the oils were similar, with cineole comprising about 80 per cent of the total extract. Although animals preferred not to feed on either the whole extract or cineole-treated food when given a choice, there was no reduction in food intake when only cineole-treated food was offered. These results are consistent with studies of Krockenberger (1988) in which he fed cineole to both common ringtail and common brushtail possums. No effect on food intake, total diet digestibility or nitrogen metabolism was detected. M. Harvey and I.D. Hume (unpubl.) fed common brushtail possums volatile monoterpenes and sesquiterpenes extracted from *E. haemostoma* leaves at 3.5–4.0 per cent (dry matter) levels in the diet but they, too, could detect no effect of these fractions on food intake except when the animals were given a choice between oil-rich diets and the basal diet alone.

The study by Lawler *et al.* (1998) mentioned earlier has presented perhaps the strongest evidence of a negative correlation between food intake by marsupial folivores and the essential oil content of the leaves. This was achieved by looking at the differences in intake of individual trees within each *Eucalyptus* species. This reduced the interference of qualitative differences (i.e. differences in the constituent compounds of the oil) with the quantitative differences (i.e. actual amounts) in feeding between subject trees which may occur when several species are used. However, the same study also refuted the role of essential oils as true toxins. Using no-choice experiments they showed that the amounts of essential oil required to be added to an artificial diet to reproduce the decrease in food intakes seen in leaves were much higher than those found in leaves (Figure 15.4). Hence the bioassay showed that the essential oils were not the cause of the effect, as may have been implied by the correlation. This is further emphasised by data which show that both common ringtail possums and common brushtail possums can rapidly be conditioned to eat diets containing greater than 12 per cent cineole (dry weight basis) (Lawler *et al.* 1999 – see below, Boyle unpubl.).

Although the difference between two-choice and no-choice experiments suggests that there is some metabolic cost involved in ingesting essential oils, this cost is either so small or dissociated sufficiently from the ingestion process that it has no effect on food intake. Integrating the above observations, a possible role for essential oils in diet choice of herbivores was suggested by Lawler *et al.* (1998). There is a strong smell and taste associated with essential oils and food intakes are strongly correlated with leaf essential oil levels. However, since they apparently do not present an insurmountable challenge to the animals after ingestion, perhaps they are used by the animals as a cue to the toxin levels of the leaves.

Lawler *et al.* (1999) tested whether the avoidance by animals of high essential oil diets could be the result of a conditioned food aversion, that is, does the aversion result from the animals learning to associate the sensory effects of the essential oils with the negative effects of another compound also found in *Eucalyptus* leaves? An initial test was made on twelve ringtail possums of reduction in food intake when cineole was added to the diet (again in a no-choice situation): all possums were found to reduce food intake significantly (Figure 15.5a). Half the animals (treatments) were then 'taught' to eat cineole by increasing the amount of cineole in the diet daily until it reached over 10 per cent of the dry weight of the diet after twelve days. The remaining animals (controls) were fed only the basal diet over the same period. At the end of this time they were again tested: those exposed to increasing cineole showed little reduction in intakes, while the control group matched their previous levels (Figure 15.5b). The treatment animals were then taught the aversion to cineole again by adding both cineole and jensenone (an acylphloroglucinol derivative) in corresponding amounts to the diet so that the effects of the jensenone would be associated with the smell and taste of the cineole. Finally, both groups were

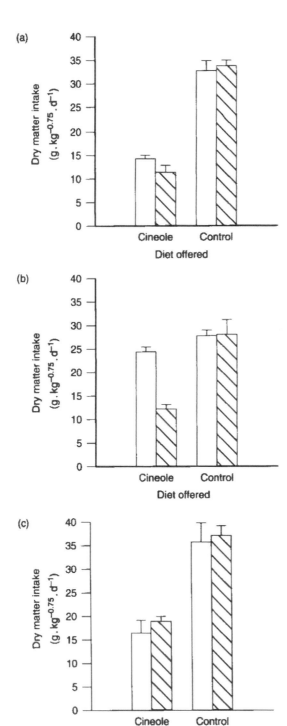

*Figure 15.5* Evidence for a conditioned food aversion to cineole in common ringtail possums. Data shown are food intakes by the same animals of an untreated diet and one to which cineole was added. Data are for two groups of animals, one of which (treatment) was acclimated to a high-cineole diet while the other (control) was fed only the untreated diet throughout. Three stages of the experiment are represented: (a) initial test, (b) post acclimation, (c) after reconditioning (see text for explanation). Unshaded bars represent treatment animals and shaded bars represent control animals (after Lawler *et al.* 1998b).

tested again with only cineole added to the diet: the treatment group again behaved in a manner not significantly different from the controls (Figure 15.5c). Lawler *et al.* (1999) also present matching data for common brushtail possums fed cineole in artificial diets. It is clear from these experiments that the avoidance of *Eucalyptus* essential oils by marsupial folivores can be the result of a learned association with negative effects not caused by the essential oils themselves. The value of essential oils as deterrents of these animals appears to lie in their strong, overriding smell and taste, which is easily associated with negative effects.

## Effects of ingested essential oils on vertebrate herbivores

Even if essential oils do not actually deter mammals from eating *Eucalyptus* foliage, they may still impose a metabolic cost on the animal and lead to some disruption of metabolic and digestive processes. The germicidal effects of many constituents of *Eucalyptus* foliage (Boland *et al.* 1991) have suggested to some workers the potential for deleterious interactions with intestinal bacteria (Fleay 1937).

The marsupial herbivores that eat *Eucalyptus* foliage rely to a small extent on the digestion of foliar cell walls in the hindgut for part of their energy intake, but the great majority of digestible energy is derived from the constituents of the cells. For example, fermentation measured *in vitro* in the caecum and/or proximal colon of koalas (Cork and Hume 1983) and greater gliders (Foley *et al.* 1987) contributed only about 10 per cent of both species' digestible energy intake. Nonetheless, if ingested essential oils disrupt this process by killing the caecal bacteria, then the animal might find itself unable to digest an essential part of its diet. Early *in vitro* studies in deer by Nagy *et al.* (1964), Oh *et al.* (1968) and Connolly *et al.* (1980) showed that many components of the oils of Douglas Fir significantly inhibited the growth and activity of rumen microbes derived from sheep and deer *in vitro*.

It is difficult to sustain an analogous argument for marsupials, namely, that the microbial populations of koalas, greater gliders, and ringtail and brushtail possums are all found in the caecum and colon. Indeed, Foley *et al.* (1987) showed that most oils were absorbed in the simple stomach and small intestine, thus avoiding for the most part interactions with microbes. *Eucalyptus* essential oils could still have a deleterious effect in those species which house their microbial population in the stomach (e.g. foregut fermenters such as goats and wallabies) but the evidence from studies of similar essential oils in sheep and deer (Cluff *et al.* 1982, White *et al.* 1982) suggests that the oils are again rapidly absorbed, or else volatilised and eructated from the rumen very soon after they are ingested. There is little reason to suspect that the situation would be different with wallabies and *Eucalyptus*. Overall, the evidence suggests that *Eucalyptus* essential oils have little direct effect on the digestive processes of animals that normally eat the foliage.

## Metabolic transformations of Eucalyptus essential oils by mammals

Absorbing ingested oils from the gut avoids interactions with microbes but these materials must still be detoxified and excreted, and doing so can be energetically expensive. Several authors (Cork *et al.* 1983, Foley 1987) have argued that the poor conversion of digested energy from a diet of *Eucalyptus* foliage into metabolisable energy is due, in part, to the excretion of oils and their metabolites in the urine.

There have been a number of studies of the metabolic fate of ingested *Eucalyptus* oils in folivorous marsupials. Part of the rationale for these studies has been a curiosity about the pathways of degradation (e.g. Bull *et al.* 1993) but amongst ecologists the interest has been driven by the notion that understanding the metabolites might help in measuring the cost of ingesting oil components.

In koalas, brushtail possums and greater gliders 95–98 per cent of volatile oils are absorbed from the digestive tract. These oils are believed to be modified in the liver by oxidation or hydrolysis, possibly conjugated with a small molecule, and then excreted via the urine or bile (Hume 1982). Several recent studies have been reviewed by McLean and Foley (1997). The outstanding trend in studies by McLean *et al.* (1993, 1995, 1997) is that mammals which are capable of including *Eucalyptus* foliage in their diets appear to have developed a significantly greater capacity to produce highly oxidised metabolites of ingested monoterpenes than have other animals. These highly polar products can then be excreted in the urine without the need for conjugation with glucuronic acid or glycine.

For example, McLean *et al.* (1995) studied the metabolic fate of dietary citronellal in greater gliders and common ringtail possums. They found that this monoterpene was cyclised and oxidised to *trans*-3,8-dihydroxy-*p*-menthane-7-carboxylic acid, in contrast to rabbits which do not oxidise the cyclised product. McLean *et al.* (1997) examined the metabolism of *p*-cymene in rats, common brushtail and ringtail possums, and greater gliders. They observed eleven urinary metabolites and grouped these according to the number of oxygens added per molecule of *p*-cymene. Rats excreted metabolites containing from one to four oxygens, whereas brushtails excreted metabolites with two to four oxygens, and common ringtails and greater gliders only excreted metabolites with three or four oxygens. This trend is also apparent when considering metabolites of α- and β-pinene formed by common brushtails and koalas (Southwell *et al.* 1980). One advantage of excreting unconjugated products is that the loss of carbohydrates and amino acids which are used as conjugates is avoided. This may be a significant advantage in the case of nutrient-poor diets such as *Eucalyptus* foliage.

Recently, Carman and co-workers (Carman and Klika 1992, Bull *et al.* 1993, Carman and Rayner 1996) made a detailed study of the fate of cineole in common brushtail possums. Common brushtails produce a range of metabolites, principally 9-hydroxycineoles and cineole-9-oic acids, as well as other alcohols and diols. Of particular interest is the chirality of the major metabolites because females produce a higher ratio of S/R enantiomers than males. Carman and Klika (1992) argue that these could be involved in pheromonal signalling. Clearly there remain a number of exciting avenues to explore in our understanding of the metabolism of *Eucalyptus* essential oils by folivorous marsupials.

## Diet choice in insect grazers of *Eucalyptus*

Feeding by insect herbivores on *Eucalyptus* foliage is just as variable as feeding by vertebrate herbivores. Several studies have identified marked interspecific (Morrow and Fox 1980), intraspecific (e.g. Journet 1980, Edwards *et al.* 1993, Stone and Bacon 1994, Patterson *et al.* 1996) and even intra-individual (Edwards *et al.* 1990) differences in the amount of defoliation. A review by Ohmart and Edwards (1991) details the nature of the plant–animal interaction. They concluded that plant secondary metabolites of *Eucalyptus* have very little effect on patterns and amounts of insect herbivory and on the survival and performance of insect herbivores. However, since the time of Ohmart and Edwards' review, continuing studies have gathered strong correlative evidence to suggest that insect herbivory is affected by foliar essential oils, in particular cineole (Edwards *et al.* 1993, Stone and Bacon 1994). We briefly review the accumulation of this evidence below.

Morrow and Fox (1980) were the first to suggest that there was an association between the foliar essential oil content and the level of damage in two alpine eucalypts, *E. stellulata* and *E. pauciflora*. However, their study of five *Eucalyptus* species showed that the distribution of

Table 15.1 Relationship between essential oil content of leaves and grazing damage by insect herbivores of *Eucalyptus* (from Morrow and Fox 1980)

| Species | Essential oil content[a] | | p value of difference |
|---|---|---|---|
| | Heavy grazing | Light grazing | |
| E. pauciflora | 1.149 ± 0.150 | 1.253 ± 0.253 | ns[b] |
| E. bridgesiana | 3.866 ± 0.423 | 3.214 ± 0.363 | ns |
| E. melliodora | 4.223 ± 0.700 | 2.973 ± 0.546 | ns |
| E. viminalis | 3.740 ± 0.658 | 6.864 ± 1.466 | $0.025 < p < 0.05$ |
| E. dives | 9.909 ± 0.870 | 13.389 ± 1.512 | $0.013 < p < 0.025$ |

a  w/w, dry matter basis.
b  Not significant.

damage, eggs and insects, as well as the feeding rates, growth and survival of insects, demonstrated few consistent effects of total essential oil content (Table 15.1). The average oil yield of the species used in the experiments ranged from 1.2 per cent to 11.6 per cent (dry matter basis) with yields in some individual plants as high as 20 per cent. In *E. dives* and *E. viminalis* there was an association between yield and damage. The plants that suffered low levels of damage had oil yields significantly greater than those suffering high levels of damage. However, there was no significant difference in total oil yields between the two levels of damage for the other three species, *E. pauciflora*, *E. bridgesiana* and *E. melliodora*. In fact, there was a tendency toward lower oil yields for lightly damaged trees in the latter two species. Morrow and Fox (1980) argued that antiherbivore effects may only become apparent if a threshold in total oil concentration is exceeded. However, it can be seen from Table 15.1 that the heavily grazed *E. dives* had more than twice the total oil yield of the lightly grazed trees of *E. pauciflora*, *E. bridgesiana* and *E. melliodora*.

Recent studies by Edwards *et al.* (1993) and Stone and Bacon (1994) have shown a strong negative association between insect herbivory and leaf essential oils, while another (Patterson *et al.* 1996) has shown no relationship. Recent work has taken into account the composition of the oils, whereas Morrow and Fox only considered the total oil yield. For example, Edwards *et al.* (1993) found that the proportion of cineole in the leaf essential oil best explained differences in levels of herbivory by Christmas beetles (Figure 15.6). Although Patterson *et al.* (1996) found no effect of essential oils in *E. regnans* on herbivory by *Chrysophtharta bimaculata*, the concentration of cineole in that species is very low. The authors also suggested that perhaps paropsine beetles are less affected by *Eucalyptus* essential oils than are other insect herbivores. It should be noted also that a high proportion of the *E. camaldulensis* trees examined by Edwards *et al.* (1993) had both a high proportion of cineole in the leaves and a high level of defoliation by *Anoplognathus*, a non-paropsine beetle (Figure 15.6c).

The above studies suffer from the same shortcomings as do those relating mammal feeding preferences to leaf essential oils, namely, the studies are all correlative and do not show cause and effect. If essential oils, including cineole, are to be conclusively shown to be deterrents to insect feeding, then bioassays with a range of insect species need to be performed. To date we are not aware of any studies in which either whole oil extract or isolated cineole have been fed to any foliage-feeding insect. Previously it might have been argued that direct addition of cineole would lead to an unnatural situation because the headspace vapour pressure would be excessive. It has been shown that high vapour pressures of *Eucalyptus* essential oils are highly toxic to insects (Sarac and Tunc 1995, J. Seymour pers. comm.). However, there are now a number of methods available, such as microencapsulation (Clancy *et al.* 1992), that may overcome these difficulties and make these sorts of experiments more feasible.

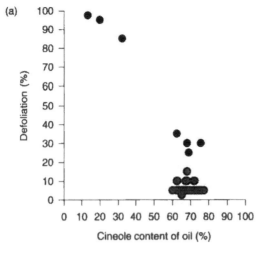
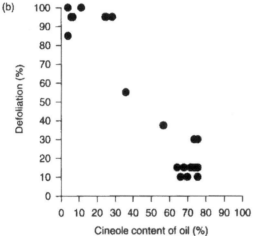
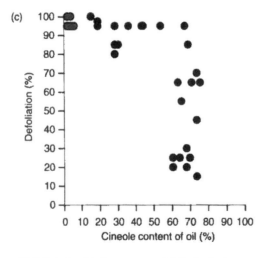

*Figure 15.6* Relationship between tree defoliation by insects and proportion of cineole in the essential oil of the leaves of three *Eucalyptus* species: (a) *E. melliodora*, (b) *E. sideroxylon*, (c) *E. camaldulensis* (after Edwards *et al.* 1993).

It may be that cineole is only effective in synergy with other oil components or it may serve as a cue to the presence of some other toxic molecules. Cineole is clearly metabolised by some of the major insect herbivores of *Eucalyptus* (see below). However, it remains to be seen whether this need to metabolise cineole is limiting the ability of insects to ingest cineole-rich diets, as appears now not to be the case for marsupial folivores of *Eucalyptus* (see above).

## Metabolic transformations of Eucalyptus essential oils by insects

Insects use a diverse range of strategies for coping with the essential oils they encounter in their food. Some insects cope with the high essential oil content of *Eucalyptus* leaves by feeding around oil glands (Ohmart and Edwards 1991) or sequestering them within the body (Morrow *et al*. 1976). Others may simply tolerate the oils, while still others possess mechanisms for metabolising and detoxifying them.

Morrow and Fox (1980) showed that the gas chromatographic profile of faeces from both *Paropsis atomaria* and *Anoplognathus montanus* fed on a range of *Eucalyptus* species was not significantly different from that of the ingested oils. They concluded that the oils are tolerated but not detoxified, even though many of the oil components induce polysubstrate membrane oxidases in other species. Similarly, Southwell *et al*. (1995) found that *Paropsisterna tigrina* produced frass with an essential oil profile almost identical to that of the ingested leaf when fed *Melaleuca* leaf, unless cineole was a significant component of the diet. For high-cineole diets, the beetles (adults and larvae) oxidised the bulk of the cineole to (+)-2β-hydroxycineole while other oil components were unchanged. Ohmart and Larsson (1989) found that larvae of *P. atomaria* absorbed or converted the majority of essential oils of *E. blakelyi* leaves, of which most (>75 per cent) was again cineole. They found no evidence of sequestration of the oils and concluded that the oils are metabolised by the larvae. However, proportions of the two other main components of the oil (limonene and α-pinene) were similar in the frass and the leaves, which appears to be consistent with the findings of Southwell *et al*. (1995) that cineole is metabolised while the other components pass through the gut unchanged. Oil budgets were provided for total oils and for cineole alone, and it was not clear whether the absorption of cineole entirely explained the effect seen in the total oils. Ohmart and Larsson (1989) also drew attention to the fact that when Morrow and Fox (1980) fed leaf of *E. elata* to *P. atomaria* the major component of the oil, piperitone, was removed and presumably metabolised.

Just how the essential oils are metabolised by insects is still unknown. Monoterpene compounds are known to induce mixed function oxidase (MFO) activity (Moldenke *et al*. 1983) which accords with studies by Rose (1985) who found that insects with high MFO activity invariably fed on hosts (including *Eucalyptus*) containing monoterpene compounds. The study of Southwell *et al*. (1995) is the most recent study in this area and concludes with the statement that the oxidase source (microbial or gut enzymes) is unknown. Clearly then, there is much to be learnt about insect metabolism of *Eucalyptus* essential oils.

## The role of terpene–acylphloroglucinol adducts in food choice of mammals and insects

Data on the role of essential oils in mediating feeding by folivorous mammals and insects on *Eucalyptus* have not been clear and we are now beginning to realise that perhaps their role has been overstated. Nevertheless, as described above, they may play a very important role due to their very close relationships with another group of compounds, the acylphloroglucinols, which have only relatively recently been discovered, and for which evidence of an antiherbivore role is accumulating rapidly, for both mammalian and insect herbivores.

Earlier, we described the structures of these compounds and their molecular relationship with the essential oils. It is unfortunate that our current state of knowledge is such that we cannot discuss in greater detail the biosynthesis of these compounds or provide a mechanistic explanation of the relationship. Nevertheless, we do know that all of the currently identified active deterrent acylphloroglucinol compounds contain either a terpene side chain or, at the very least, an isoprene unit (the basic building block of terpenes). There is a strong correlation between the concentrations of terpenes and of acylphloroglucinol compounds in four *Eucalyptus* species that

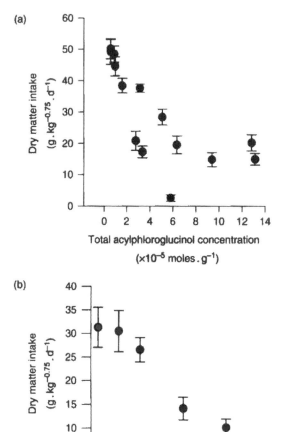

Figure 15.7 Relationship between food intake of common ringtail possums and total acylphloroglucinol/macrocarpal G content of: (a) *Eucalyptus ovata* leaves, (b) an artificial diet. Note that data for leaves are total acylphloroglucinols as there is no existing assay specifically for macrocarpals (although mass spectrometry shows that acylphloroglucinols in *E. ovata* are predominantly macrocarpals) (after Lawler *et al.* 1998a).

have been examined closely (Foley, Matsuki and Floyd unpubl.). In mammalian herbivores, at least, the secondary correlation between food intake and essential oil concentration is due to the primary relationship between the oils and the acylphloroglucinols. The development of the evidence for this is discussed in this section.

Pass *et al.* (1998) investigated the difference between palatable and unpalatable forms of *E. ovata* by a process of bioassay-guidedND fractionation. They found that the differences between the two forms in their ability to deter feeding by common ringtail possums were principally due to the higher concentration of a compound called macrocarpal G (Figure 15.1f), an acylphloroglucinol with an adduct of bicyclogermacrene – a relatively uncommon sesquiterpene found in *Eucalyptus*. This led Lawler *et al.* (1998) to conduct more comprehensive experiments with ringtail possums, examining a range of palatabilities of individual *E. ovata* trees to the animals. As discussed above, the resultant range of intakes correlated strongly with the essential oil content of the leaves (Figure 15.4a) but bioassay experiments showed that essential oils could not cause the effects seen (Figure 15.4b). In contrast, bioassays with purified macrocarpal G reproduced the range of reduced intakes in concentrations that corresponded closely with those found in the leaves (Figure 15.7) and it was concluded that macrocarpal G was the primary cause of feeding deterrence in *E. ovata*.

Difficulties in the extraction and purification of macrocarpals led us to investigate the deterrent properties of related compounds. The initial focus was on jensenone (Figure 15.1d) as it was listed by Boland *et al.* (1991) as a major component of the steam distillate of *E. jensenii*, and was thus likely to be relatively easy to extract and purify in suitable quantities. This was achieved and a number of experiments have been carried out with jensenone added to the diets of common ringtail and common brushtail possums. These have shown jensenone to be a potent antifeedant in molar quantities very similar to those found for macrocarpal G. Further work has indicated that their mechanism of action is due to stimulation of the animals' emetic systems (Lawler, Pass and Foley unpubl.).

More recent work has focused on sideroxylonals (dimers of jensenone) as they are found in variable concentrations in a number of *Eucalyptus* species, are often the predominant

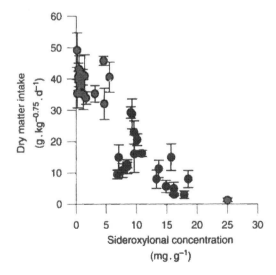

*Figure 15.8* Relationship between food intake of common ringtail possums and sideroxylonal content of *Eucalyptus polyanthemos* leaves (Lawler unpubl.).

acylphloroglucinol in those species (Eschler, Pass and Foley unpubl.), and methods have been developed to quantify precisely the amounts found in the leaves. These compounds, too, show strong correlations with the range of food intakes by common ringtail possums (Figure 15.8) and preliminary evidence suggests that this applies also to koalas (Moore and Foley unpubl.). Bioassay experiments have again supported the conclusion that sideroxylonals are responsible for feeding deterrency in those plants where concentrations are sufficiently high (Lawler, Eschler and Foley unpubl.). Once again we see the correlation between essential oils and both food intakes and sideroxylonal concentrations, suggesting that the oils do not have a causal antifeedant effect.

There are, so far, few data on the effects of acylphloroglucinols on insect feeding. Correlations between defoliation and sideroxylonal concentrations of leaves have been shown (Foley and Floyd unpubl.), but bioassays have not yet been conducted. Hence the same problems as described previously with correlative studies are encountered, and we cannot yet say that acylphloroglucinols are the cause of deterrency in insect-resistant trees. Work in this area should advance rapidly since the problems of headspace concentrations associated with essential oils do not occur with acylphloroglucinols.

## Efficacy of selecting high essential oil genotypes in conferring natural herbivore resistance on *Eucalyptus* plantations

The selection of herbivore-resistant genotypes for *Eucalyptus* plantations may be a practical approach to protecting them from damage by both vertebrate and invertebrate herbivores (Floyd and Farrow 1994). However, we conclude with a word of caution on selecting for genotypes containing high concentrations of essential oils to confer this resistance as suggested by Farrow (1993) and Stone and Bacon (1994).

There is now some evidence that for mammalian folivores essential oils are not the actual deterrents, but appear so because their concentrations are correlated with acylphloroglucinol compounds. Whether this is also the case for insects cannot currently be assessed, due to the difficulties associated with mixing free essential oils into basal diets. There is no doubt that essential oils are absorbed and metabolised, in some cases, by both insect and mammalian folivores, though there is little compelling evidence that the rate of metabolism is limiting to food intakes.

In any case, the selection of resistant genotypes is a sound approach. Difficulties may arise, however, if this resistance is then attributed to the wrong compound and selection is based on that criterion rather than resistance *per se*. It appears that the concentrations of essential oils and of acylphloroglucinol compounds are closely related, so selection of high essential oil genotypes may be successful, as long as that selection also selects for correspondingly high acylphloroglucinol contents. However, our knowledge of the biosynthesis of acylphloroglucinols, and its relationship with the biosynthesis of essential oils, is sorely lacking at present. There is a distinct danger that at some point in the artificial selection process the relationship between concentrations of the two groups of compounds may break down, and that high essential oil genotypes may no longer contain high acylphloroglucinol concentrations. This danger is especially significant where genetic engineering may be used to manipulate a single factor, such as cineole production, unless flow-on effects on related compounds are also assessed.

We believe that in this instance mammalian folivores at least, and perhaps insects, would rapidly learn, within a local area, that high essential oil contents are not associated with negative effects and will commence to feed on those plants at a high rate. It now appears that mammalian folivores of *Eucalyptus* avoid high essential oil diets as the result of a conditioned food aversion, and there is strong evidence to suggest that these aversions are lost rapidly where there is repeated exposure to the taste stimuli in the absence of the negative feedback. This would be the case for a high essential oil/low acylphloroglucinol *Eucalyptus* plantation.

Whether this effect would be seen for insect herbivores of *Eucalyptus* is less certain. For the reasons discussed above, the actual role of essential oils in determining resistance of foliage against any insect herbivore remains uncertain. Furthermore, if, indeed, essential oils prove not to be the causal agents then there is less information on the insect's ability to learn to eat these leaves over either a single generation or evolutionary time. That is, there may be an immediate behavioural response to encountering edible high essential oil plants, or there may be a reduction in selection for avoidance due to reduced mortality or increased performance of those insects feeding on high essential oil plants. It was shown by De Little and Madden (1975) that *Chrysoptharta bimaculata*, a significant pest of Tasmanian *Eucalyptus* plantations, may choose to oviposit on species on which larval growth and survival rates were subsequently shown to be lower (Baker 1995) than those on other species present. Thus there may be the opportunity for at least the larval stages to encounter plants rich in essential oils and to learn that these plants are a suitable food source, again undermining the resistance conferred by the essential oils.

It is important then, that in the attempts to develop these herbivore-resistant plantations the questions and gaps in our knowledge outlined in this chapter are addressed. At the very least, caution is advised in this endeavour, and the resistance of genotypes intended for use should be checked regularly and directly on the animals concerned, until we have confidence in the attributes to which we attribute herbivore resistance.

## Acknowledgements

We thank Dr Bart Eschler, Dr Darren Schliebs and Mr Bruce Clarke for their assistance with laboratory work and discussion of the chemistry and role of acylphloroglucinols in animal feeding.

## References

Baker, S. (1995) A comparison of feeding, development and survival of larvae of the leaf beetle *Chrysoptharta bimaculata* (Olivier) (Coleoptera: Chrysomelidae) on *Eucalyptus nitens* and *Eucalyptus regnans*. In Open Forum and Symp. Conf., Ecological Society of Australia, Hobart, September 1995, p. 26.

Betts, T.J. (1978) Koala acceptance of *Eucalyptus globulus* Labill. as food in relation to the proportion of sesquiterpenoids in the leaves. In T.J. Bergin (ed.), *The Koala*, Zoological Parks Board of NSW, Sydney, pp. 75–85.

Bignell, C.M., Dunlop, P.J., Brophy, J.J. and Jackson, J.F. (1994) Volatile leaf oils of some south-western and southern Australian species of the genus *Eucalyptus*. Part II – Subgenus Symphomyrtus, Section Dumaria, Series Torquatae. *Flavour Fragr. J.*, 9, 167–171.

Boland, D.J., Brophy, J.J., Flynn, T.M. and Lassak, E.V. (1982) Volatile leaf oils of *Eucalyptus delegatensis* seedlings. *Phytochemistry*, 21, 2467–2469.

Boland, D.J., Brophy, J.J. and Fookes, C.J.R. (1992) Jensenone, a ketone from *Eucalyptus jensenii*. *Phytochemistry*, 31, 2178–2179.

Boland, D.J., Brophy, J.J. and House, A.P.N. (eds) (1991) *Eucalyptus Leaf Oils. Use, Chemistry, Distillation and Marketing*, ACIAR/CSIRO, Inkata Press, Melbourne.

Bowyer, R.C. and Jeffries, P.R. (1959) Studies in plant chemistry. I. The essential oils of *Eucalyptus caesia* Benth. and *E. torquata* Luehm. and the structure of torquatone. *Aust. J. Chem.*, 12, 442–446.

Bryant, J.P., Chapin, F.S. III and Kleinig, D.A. (1983) Carbon/nutrient balance of boreal plants in relation to vertebrate herbivory. *Oikos*, 40, 357–368.

Bull, S.D., Carman, R.M., Carrick, F.N. and Klika, K.D. (1993) 7-hydroxy-1,8-cineole and 7-cineolic acid – 2 new possum urinary metabolites. *Aust. J. Chem.*, 46, 441–447.

Carman, R.M. and Klika, K.D. (1992) Partially racemic compounds as brushtail possum urinary metabolites. *Aust. J. Chem.*, 45, 651–657.

Carman, R.M. and Rayner, A.C. (1996) Chiral 2α,4-dihydroxy-1,8-cineole as a possum urinary metabolite. *Aust. J. Chem.*, 49, 1–6.

Clancy, K.M., Foust, R.D., Huntsberger, T.G., Whitaker, J.G. and Whitaker, D.M. (1992) Technique for using microencapsulated terpenes in lepidopteran artificial diets. *J. Chem. Ecol.*, 18, 543–560.

Cluff, L.K., Welch, B.L., Pederson, J.C. and Brotherson, J.D. (1982) Concentration of monoterpenoids in rumen ingesta of wild mule deer. *J. Range Manage.*, 35, 192–194.

Coley, P.D., Bryant, J.P. and Chapin, F.S. III (1985) Resource availability and plant antiherbivore defense. *Science*, 230, 895–899.

Connolly, G.E., Ellison, B.O., Fleming, J.W., Geng, S., Kepner, R.E., Longhurst, W.M., Oh, J.H. and Russell, G.F. (1980) Deer browsing of Douglas Fir trees in relation to volatile terpene composition and *in vitro* fermentability. *Forest Sci.*, 26, 179–193.

Cork, S.J. and Foley, W.J. (1991) Digestive and metabolic strategies of arboreal mammalian folivores in relation to chemical defenses in temperate and tropical forests. In R.T. Palo and C.T. Robbins (eds), *Plant Chemical Defenses and Mammalian Herbivory*, CRC Press, Boca Raton, Louisiana, USA.

Cork, S.J. and Hume, I.D. (1983) Microbial digestion in the koala (*Phascolarctos cinereus*, Marsupialia), an arboreal folivore. *J. Comparative Physiol.*, 152, 131–135.

Cork, S.J., Hume, I.D. and Dawson, T.J. (1983) Digestion and metabolism of a natural foliar diet (*Eucalyptus punctata*) by an arboreal marsupial, the koala (*Phascolarctos cinereus*). *J. Comparative Physiol.*, 153, 181–190.

Cork, S.J. and Sanson, G.D. (1990) Digestion and nutrition in the koala. In A.K. Lee, K.A. Handasyde and G.D. Sanson (eds), *The Biology of the Koala*, Surrey Beatty and Sons, Sydney, pp. 129–144.

De Little, D.W. and Madden, J.L. (1975) Host preference in the Tasmanian eucalypt defoliating Paropsini (Coleoptera: Chrysomelidae) with particular reference to *Chrysophtharta bimaculata* (Olivier) and *C. agricola* (Chapuis). *J. Aust. Entomol. Soc.*, 14, 387–394.

Duncan, A.J., Hartley, S.E. and Iason, G.R. (1994) The effect of monoterpene concentrations in sitka spruce (*Picea sitchensis*) on the browsing behaviour of red deer (*Cervus elaphus*). *Can. J. Zool.*, 72, 1715–1720.

Edwards, P.B., Wanjura, W.J. and Brown, W.V. (1993) Selective herbivory by Christmas beetles in response to intraspecific variation in *Eucalyptus* terpenoids. *Oecologia*, 95, 551–557.

Edwards, P.B., Wanjura, W.J., Brown, W.V. and Dearn, J.M. (1990) Mosaic resistance in plants. *Nature*, 347, 434.

Everist, S.L. (1981) *Poisonous Plants of Australia*, Angus and Robertson, Sydney.

Farrow, R.A. (1993) Chemical defences in eucalypts and how they can be exploited to reduce insect attack. In S.A. Corey, D.J. Dall and W.M. Milne (eds), *Pest Control and Sustainable Agriculture*, CSIRO, Melbourne, pp. 313–316.

Finnemore, H., Reichard, S.K. and Large, D.K. (1935) Cyanogenetic glucosides in Australian Plants. III. *Eucalyptus cladocalyx. J. Proc. Roy. Soc. N.S.W.*, 69, 209–214.

Fleay, D. (1937) Observations on the koala in captivity. Successful breeding in Melbourne Zoo. *Aust. Zool.*, 9, 68–80.

Floyd, R.B. and Farrow, R.A. (1994) The potential role of natural insect resistance in the integrated pest management of eucalypt plantations in Australia. In S.C. Halos, F.F. Natividad, L.J. Escote-Carlson, G.L. Enriquez and I. Ubmoh (eds), *Forest Pest and Disease Management*, Seameo Biotrop (Biotrop Special Publication No. 53), Bogor, Indonesia, pp. 55–76.

Foley, W.J. (1987) Digestion and energy metabolism in a small arboreal marsupial, the greater glider (*Petauroides volans*), fed high-terpene *Eucalyptus* foliage. *J. Comparative Physiol.*, 157B, 355–362.

Foley, W.J., Lassak, E.V. and Brophy, J. (1987) Digestion and absorption of *Eucalyptus* essential oils in greater glider (*Petauroides volans*) and brushtail possum (*Trichosurus vulpecula*). *J. Chem. Ecol.*, 13, 2115–2130.

Gershenzon, J. (1994) Metabolic costs of terpenoid accumulation in higher plants. *J. Chem. Ecol.*, 20, 1281–1328.

Ghisalberti, E.L. (1996) Bioactive acylphloroglucinol derivatives from *Eucalyptus* species. *Phytochemistry*, 41, 7–22.

Gleadow, R., Foley, W.J. and Woodrow, I.A. (1998) Enhanced $CO_2$ alters the relationship between photosynthesis and defence in cyanogenic *Eucalyptus cladocalyx* F. Muell. *Plant Cell Environ.*, 21, 12–22.

Hume, I.D. (1982) *Digestive Physiology and Nutrition of Marsupials*, Cambridge University Press, Cambridge, UK.

Hume, I.D. and Esson, C. (1993) Nutrients, antinutrients and leaf selection by captive koalas (*Phascolarctos cinereus*). *Aust. J. Zool.*, 41, 379–392.

Journet, A.R.P. (1980) Intraspecific variation in food plant favourability to phytophagous insects: psyllids on *Eucalyptus blakelyi* M. *Ecol. Entomol.*, 5, 249–261.

Kozuka, M., Sawada, T., Kasahara, F., Mizuta, E., Amano, T., Komiya, T. and Goto, M. (1982a) The granulation-inhibiting principles from *Eucalyptus globulus* Labill. II. The structures of euglobal-Ia1, Ia2, -Ib, -Ic, -IIa, -IIb and -IIc. *Chem. Pharm. Bull.*, 30, 1952–1963.

Kozuka, M., Sawada, T., Mizuta, E., Kasahara, F., Amano, T., Komiya, T. and Goto, M. (1982b) The granulation-inhibiting principles from *Eucalyptus globulus* Labill. III. The structures of euglobal-III, -IVb and -VII. *Chem. Pharm. Bull.*, 30, 1964–1973.

Krockenberger, A. (1988) *Metabolic Cost of Detoxification of 1,8-Cineole (Eucalyptol) in Two Arboreal Folivores, the Ringtail Possum (Pseudocheirus peregrinus) and the Brushtail Possum* (Trichosurus vulpecula), BSc Hons. Thesis, Univ. of Sydney.

Landsberg, J. and Cork, S.J. (1997) Herbivory: interactions between eucalypts and the vertebrates and invertebrates that feed on them. In J.E. Williams and J.C.Z. Woinarski (eds), *Eucalypt Ecology: Individuals to Ecosystems*, Cambridge University Press, Cambridge, UK, pp. 342–372.

Lawler, I.R., Foley, W.J., Eschler, B., Pass, D.M. and Handasyde, K. (1998) Intraspecific variation in secondary metabolites determines food intake by folivorous marsupials. *Oecologia*, 116, 160–169.

Lawler, I.R., Foley, W.J., Woodrow, I.E. and Cork, S.J. (1997) The effects of elevated $CO_2$ atmospheres on the nutritional quality of *Eucalyptus* foliage and its interaction with soil nutrient and light availability. *Oecologia*, 109, 59–68.

Lawler, I.R., Stapley, J., Foley, W.J. and Eschler, B. (1999) Ecological example of conditioned food aversion in plant-herbivore interactions: effect of terpenes of *Eucalyptus* leaves on feeding by common ringtail and brushtail possums. *J. Chem. Ecol.*, 25, 401–415.

McKey, D.B., Gartlan, J.S., Waterman, P.G. and Choo, G.M. (1981) Food selection by black colobus monkeys (*Colobus satanus*) in relation to food chemistry. *Biol. J. Linn. Soc.*, 16, 115–146.

McLean, S., Boyle, R., Foley, W.J. and Davies, N.W. (1997) Comparative metabolism of *p*-cymene in three arboreal marsupials and the rat. In *Proc. ASCEPT. 31st Annual Scientific Meeting*, Canberra, December 1997, oral 3–4.

McLean, S. and Foley, W.J. (1997) Metabolism of *Eucalyptus* terpenes by herbivorous marsupials. *Drug Metabolism Rev.*, 29, 213–218.

McLean, S., Foley, W.J., Davies, N.W., Brandon, S., Duo, L. and Blackman, A.J. (1993) The metabolic fate of dietary terpenes from *Eucalyptus radiata* in the common ringtail possum (*Pseudocheirus peregrinus*). *J. Chem. Ecol.*, 19, 1625–1643.

McLean, S., Foley, W.J., Davies, N.W., Brandon, S. and Peacock, E.J. (1995) A novel metabolite of citronellal formed in arboreal marsupials fed *Eucalyptus citriodora*. Meeting of Australasian Pharmaceutical Sciences Association (APSA), Auckland, December 1994. Abstr. publ. *New Zealand Pharmacy*, 15(3), Supplement, p. 8.

Moldenke, A.F., Berry, R.E. and Terriere, L.C. (1983) Cytochrome P-450 in insects. V. Monoterpene induction of cytochrome P-450 and associated monooxygenase activities in the larva of the variegated cutworm *Perdroma saucia* (Hübner). *Comparative Biochem. Physiol.*, 74C, 365–371.

Montague, T.L. (1994) Wallaby browsing and seedling palatability. *Aust. For.*, 57, 171–175.

Morrow, P.A., Bellas, T.E. and Eisner, T. (1976) *Eucalyptus* oils in the defensive oral discharge of Australian sawfly larvae (Hymenoptera: Pergidae). *Oecologia*, 24, 193–206.

Morrow, P.A. and Fox, L.R. (1980) Effects of variation in *Eucalyptus* essential oil yield on insect growth and grazing damage. *Oecologia*, 45, 209–219.

Nagy, J.G., Steinhoff, H.W. and Ward, G.M. (1964) Effects of essential oils of sagebrush on deer rumen microbial function. *J. Wildlife Manage.*, 28, 785–790.

Oh, H.K., Jones, M.B. and Longhurst, W.M. (1968) Comparison of rumen microbial inhibition resulting from various essential oils isolated from relatively unpalatable plant species. *Appl. Microbiol.*, 16, 39–44.

Ohmart, C.P. and Edwards, P.B. (1991) Insect herbivory in *Eucalyptus*. *Ann. Rev. Entomol.*, 36, 637–657.

Ohmart, C.P. and Larsson, S. (1989) Evidence for absorption of eucalypt essential oils by *Paropsis atomaria* Olivier (Coleoptera: Chrysomelidae). *J. Aust. Entomol. Soc.*, 28, 201–205.

Pass, D.M., Foley, W.J. and Bowden, B. (1998) Vertebrate herbivory on *Eucalyptus* – identification of specific feeding deterrents for common ringtail possums (*Pseudocheirus peregrinus*) by bioassay-guided fractionation of *Eucalyptus ovata* foliage. *J. Chem. Ecol.*, 24, 1513–1527.

Patterson, K.C., Clarke, A.R., Raymon, C.A. and Zalucki, M.P. (1996) Performance of first instar *Chrysophtharta bimaculata* larvae (Coleoptera: Chrysomelidae) on nine families of *Eucalyptus regnans* (Myrtaceae). *Chemoecology*, 7, 94–106.

Pratt, A. (1937) *The Call of the Koala*, Robertson and Mullens, Melbourne.

Rose, H.A. (1985) The relationship between feeding specialisation and host plants to aldrin epoxidase activities of midgut homogenates in larval Lepidoptera. *Ecol. Entomol.*, 10, 455–467.

Sarac, A. and Tunc, I. (1995) Toxicity of essential oil vapours to stored-product insects. *J. Plant Diseases Protect.*, 102, 69–74.

Southwell, I.A. (1978) Essential oil content of koala food trees. In T.J. Bergin (ed.), *The Koala*, Zoological Parks Board of NSW, Sydney, pp. 62–74.

Southwell, I.A., Flynn, T.M. and Degabriele, R. (1980) Metabolism of $\alpha$- and $\beta$-pinene, $p$-cymene and 1,8-cineole in the brushtail possum, *Trichosurus vulpecula*. *Xenobiotica*, 10, 17–23.

Southwell, I.A., Maddox, C.D.A. and Zalucki, M.P. (1995) Metabolism of 1,8-cineole in tea tree (*Melaleuca alternifolia* and *M. linariifolia*) by pyrgo beetle (*Paropsisterna tigrina*). *J. Chem. Ecol.*, 21, 439–453.

Stone, C. and Bacon, P.E. (1994) Relationships among moisture stress, insect herbivory, foliar cineole content and the growth of river red gum *Eucalyptus camaldulensis*. *J. Appl. Ecol.*, 31, 604–612.

White, S.M., Welch, B.L. and Flinders, J.T. (1982) Monoterpenoid content of pygmy rabbit stomach ingesta. *J. Range Manage.*, 35, 107–109.

Yamakoshi, Y., Murata, M., Shimizu, A. and Homma, S. (1992) Isolation and characterization of macrocarpals B-G, antibacterial compounds from *Eucalyptus macrocarpa*. *Biosci. Biotech. Biochem.*, 56, 1570–1576.

Zoidis, A.M. and Markowitz, H. (1992) Findings from a feeding study of the koala (*Phascolarctos cinereus adjustus*) at the San Francisco Zoo. *Zoo Biol.*, 11, 417–431.

# 16 Eucalyptus oil products
## Formulations and legislation

*Judi Beerling, Steve Meakins and Les Small*

Introduction

Eucalyptus oils are being used with increasing frequency in a variety of products found in the supermarket or pharmacy. 'With extract of Eucalyptus' or 'With Eucalyptus essential oil' claims are becoming more common on the labels of modern consumer products such as cosmetics, toiletries and household products due to the ever-increasing interest in natural or botanical ingredients. Eucalyptus oil may be used as an active ingredient to provide scientifically provable benefits – such as nasal decongestion or antibacterial effects – or at much lower dosages to impart more esoteric or folkloric connotations to the product concerned. Eucalyptus oils are also used as components of perfumes to provide a medicinal-type note to the fragrance.

*Eucalyptus globulus*, or Blue Gum, oil was a traditional Australian aboriginal remedy for infections and fevers. It is now used all over the world for relieving coughs and colds, sore throats and other infections. Its main constituent, 1,8-cineole, is mucolytic (i.e. it thins out and relaxes the flow of mucus) and is excreted through the lung surface. *E. radiata* oil is sometimes preferred by aromatherapists for its more pleasant smell while *E. smithii* oil is sometimes preferred due to a perception that it is better tolerated by the skin. *E. radiata* and *E. smithii* oils have also been shown to be useful for treating disorders of the respiratory system, although with some differences in their uses. A steam inhalation with Eucalyptus[1] is not only an effective cold treatment because it relieves nasal and bronchial congestion, but because it is also claimed to inhibit proliferation of the cold virus (Davis 1990).

When applied as a diluted oil on the skin Eucalyptus has a warming and slightly anaesthetic effect. Massaging with such an oil, therefore, will help to relieve respiratory infections, pain caused by rheumatic joints, neuralgia, fibrositis and muscular aches. Burns, blisters, insect bites and skin infections such as abscesses are also claimed to respond positively to the topical application of the essential oil or extract. It is also said to be valuable for easing the symptoms of shingles, chickenpox and cold sores.

Another common name for eucalyptus is Fever Tree. Baron Ferdinand von Müller, the German botanist and explorer who was Director of the Botanical Gardens in Melbourne from 1857 to 1873, is credited with making the properties of eucalyptus known outside Australia. Due to the medicinal smell of the leaves, nineteenth century colonists planted the trees in fever-ridden areas in various parts of the world in an attempt to drive away insects and contagious

---

[1] The term 'Eucalyptus' (i.e. spelt with an initial capital letter) is used in this chapter to denote a formulation or product containing eucalyptus oil, or the oil itself. Likewise, other oil or resinoid ingredients – such as Geranium, Lavender, Benzoin, etc. – are generally spelt with a capital letter.

diseases. Since the high water-holding capacity of the trees' root systems coincidentally tended to dry out marshy soils, disease-carrying insects such as mosquitoes, which need standing water to breed, were effectively deterred.

Topically applied oils from certain eucalypts also repel mosquitoes from the skin. A preparation from *E. citriodora*, Lemon-scented Gum, has been used in China for many years under the name Quwenling. In Europe, a formulation based on Quwenling is now being used in 'natural' insect repellent products such as Mosi-guard™ Natural, marketed by MASTA in the UK (see below; see also Chapter 14 for a more detailed discussion of Quwenling and Mosi-guard).[2]

## The use of eucalyptus oils in perfumery

### *Cineole-rich oils*

Although more generally associated with medicinal usage, *E. globulus* oil is the most common eucalyptus oil used in perfumery. It consists mainly (approximately 70–75 per cent) of 1,8-cineole (sometimes referred to as eucalyptol and, occasionally, as cajeputol). Cineole has a molecular weight of 154 and boils at 176°C. The remainder of the oil consists of a complex mixture of monoterpenes, sesquiterpenes, sesquiterpenols, aliphatic alcohols, monoterpenones and aldehydes (see Chapter 5).

Apart from contributing to the fragrance, 1,8-cineole has two important pharmacological properties: it stimulates the mucous-secreting cells in the nose, throat and lungs and it has an antiseptic effect. *E. globulus* oil is a low-cost essential oil which is comparatively stable in soap (and has long been known for its antiseptic properties, e.g. Martindale 1910) and Poucher (1991) reports an effective antiseptic natural soap perfume. It consists of:

| | |
|---|---|
| Geranium | 200 |
| Lavender | 200 |
| Lemon | 200 |
| Cassia | 100 |
| Clove | 100 |
| Rosemary | 80 |
| Eucalyptus | 50 |
| Thyme | 50 |
| Vetivert | 20 |
| Total | 1000 |

It is often the case that small amounts of Eucalyptus are used to claim aromatherapy effects, even though the contribution to the overall fragrance theme is minimal. However, in a floral fragrance for soap the inclusion of only 0.5 per cent of Eucalyptus can have a lifting effect on the fragrance.

Eucalyptus oil is used in many Australian products, such as fabric softeners and shampoos, in order to support a product-marketing theme. Australian wool wash detergents also often include eucalyptus oil because of its renowned stain-removing properties. However, in most countries an overt eucalyptus note would not be acceptable.

---

[2] Mention of any product name in this chapter is for illustrative purposes only and is not intended as a recommendation. There may be other products which are just as effective for the intended use.

1,8-Cineole is stable in hypochlorite bleach and actually stabilises the chlorine loss. Eucalyptus oil, or cineole itself, is therefore often used in fragrances designed to be incorporated in bleach products at levels between 5 and 35 per cent.

Thus, it can be seen that eucalyptus oil, and its constituent 1,8-cineole, have a valuable place in the perfumer's palette because of the powerful message delivered by such a characteristic odour.

As an alternative to using the oil itself, a solvent extract (termed an 'absolute') can be produced from the leaves and branches of *E. globulus*. This is usually solid at room temperature and has an aromatic-fruity note rather than the typical eucalyptus odour. It has found use in alcoholic fragrances but is more expensive to produce than the oil and is not so widely used.

Eucalyptus citriodora *and other oils*

*E. citriodora* oil, like citronella oil, has a lemon-like odour and contains citronellal as the principal constituent. However, when a 'clean' citrus effect is required, one that can be seen as very functional smelling, then *E. citriodora* oil is the product of choice. Hence its use in laundry soaps and powders; it is especially popular in Africa for this purpose. It is also used in other low-cost soaps, perfumes, air fresheners and disinfectants.

*E. staigeriana* oil from Brazil also has a lemon-like character – this time due to the presence of citral – and although it is used in perfumery its limited availability and rather higher price means that it is not used on anything like the scale of *E. citriodora* oil.

## *The safety of eucalyptus oils in fragrances*

Although the fragrance industry is covered by normal consumer protection laws, it has also established a self-regulatory mechanism to ensure that the ingredients used in fragrances do not pose a risk to the consumer. This self-regulatory system involves the close cooperation of two major international fragrance organisations: the Research Institute for Fragrance Materials (RIFM) and the International Fragrance Association (IFRA). The addresses and other contact details for both organisations are given in Appendix 7.

### *Research Institute for Fragrance Materials*

RIFM was established in 1966 by the American Fragrance Manufacturing Association as a non-profit making, independent body whose task is to check the safety of fragrance ingredients. To date, RIFM has tested over 1300 fragrance materials, including all of the commonly used ingredients. The test results for each material examined are reviewed by an independent international panel of toxicologists, pharmacologists and dermatologists and the results are published as monographs in the journal *Food and Chemical Toxicology*. RIFM also collates all the information available for an ingredient from the scientific literature and from the aroma chemical manufacturers for inclusion in these monographs. Should there be any cause for concern about the use of an ingredient this is immediately signalled to the industry by RIFM through the publication of an advisory letter, which is then acted upon by IFRA.

The type of basic tests carried out by RIFM include acute oral toxicity; acute dermal toxicity if the oral toxicity is significant; skin irritation and sensitisation; and phototoxicity if the material absorbs in the UV range. Where there is a need, more detailed studies are undertaken involving sub-chronic feeding, dermal absorption and metabolic fate. Through IFRA, RIFM also collects from the industry consumer exposure data on fragrance ingredients. This ensures that the test data it has are relevant to the market situation and also provides guidance on the requirements for future research. Thus, RIFM will undertake a review of its safety data, or instigate further research, if the results of these surveys indicate that a particular ingredient is occurring in a wider range of products and/or at higher concentrations than was the case when it was first examined.

*International Fragrance Association*

IFRA was established in 1973 by trade associations representing over 100 fragrance manufacturers in fifteen countries. It represents the scientific and technical expertise of the industry and is responsible for issuing and up-dating the 'Code of Practice' upon which the whole self-regulation policy is based. IFRA is funded by these fragrance manufacturing companies, who all agree to abide by the code of practice whilst they remain members of the association. This code of practice has many functions, including setting standards for good manufacturing practice within the industry, as well as standards for quality control and for labelling and advertising. It also sets limits on, or prohibits the use of, certain ingredients.

Although IFRA and RIFM are independent of each other they work closely together and it is only after considerable discussion between them that restrictions or prohibitions are imposed. It is always the IFRA board in conjunction with the technical advisory committee who make the final decision in such matters, as they are ultimately responsible for the implementation of any restrictions, by way of the code of practice. The most common cause for a use restriction is the ability of some materials to be skin sensitisers. Unlike skin irritation, which usually disappears soon after the irritant has been removed, skin sensitisation involves the activation of the immune system and reactions can persist for much longer after the initial exposure and become more severe on subsequent contact. The strict code of practice applied by IFRA not only protects the consumer but also protects the health and well-being of those employed within the industry.

E. globulus *and* E. citriodora *oils*

Eucalyptus oil of the type from *E. globulus*, and that from *E. citriodora*, have both been the subjects of RIFM monographs (Opdyke 1975, Ford *et al.* 1988). Opdyke (1975) also gives a separate monograph on eucalyptol. Neither oil has been found to be harmful, irritating or sensitising. Both oils were tested on human volunteers in 48-h closed patch tests at a concentration of 10 per cent in petrolatum and neither was irritating. *E. globulus* oil was also tested undiluted and still gave a negative result. When subjected to a human maximisation test, again at a level of 10 per cent in petrolatum, neither oil was found to be sensitising (Kligman 1966, Kligman and Epstein 1975). Use of these oils in fragrance compositions is therefore not restricted by IFRA.

## The use of eucalyptus oils in consumer products

*Insect repellents*

As noted in the introduction, *E. citriodora* oil has been used as a 'natural' insect repellent. Depending on the product formulation it is used in, Lemon Eucalyptus (known as Quwenling in China) is up to four or five times more effective and longer-lasting than citronella oil (from *Cymbopogon nardus*), one of the best known natural insect repellents. *p*-Menthane-3,8-diol is the main active component of Quwenling and this can be isolated and used as a highly effective insect repellent. *E. citriodora* oil contains up to 80–90 per cent citronellal, along with geraniol, both of which are known to have insect repellent activity but tend to dilute the much higher activity of the *p*-menthane-3,8-diol.

The Mosi-guard Natural insect repellent spray produced by MASTA in the UK contains 'Extract of Lemon Eucalyptus' and claims on the label:

> Approved and recommended by the London School of Hygiene and Tropical Medicine. Field trials have shown effective protection for 6 h after a single application in mosquito infected areas. Also protects against many other biting insects. Mosi-guard Natural is made from a natural and renewable resource. It is kind to your skin and has no adverse effects on fabrics and surfaces. Apply Mosi-guard Natural to exposed areas as often as required.

*E. citriodora* was traditionally used for perfuming the linen cupboard by putting the dried leaves in a small cloth sachet. During the last century, its use as an insect repellent was recognised, especially against silverfish and cockroaches.

The use of eucalyptus oil as a natural flea repellent, sprayed around the home three times a week as a dispersion in water, has also been advocated. Pets and their bedding can be sprayed with the same solution. An equine shampoo, containing eucalyptus oil as a natural insect repellent, is also available for sale on the Internet.

The insect repellency properties of eucalyptus have also been reportedly used in mosquito mats in India.

## Medicinal uses

Today, the number and diversity of medicinal products containing eucalyptus oil which are marketed is very large indeed. This is illustrated in Table 16.1 which lists some, but by no means all, of the products which are available in Australia, Europe, Thailand and the United States. Note that there are many more products which contain eucalyptol (cineole) and although these are derived from eucalyptus oil they are not included in Table 16.1.

Eucalyptus is widely used for the treatment of asthma, bronchitis, catarrh, colds, influenza and sinusitis. Vaporisers and aromatherapy oil burners can be used in the home to distribute the essential oil odour around the room, helping to purify the air and aid breathing. There are also numerous inhalation/chest rub products on the market, one of the most well-known being Vicks VapoRub™ (Proctor & Gamble), which has been marketed for over eighty years. The oils used in this, and similar products, which have demonstrated a clinically significant contribution to the overall nasal decongestant effects of the product, are Eucalyptus, Camphor and Peppermint (menthol). Vicks VapoRub ointment contains as active ingredients: Camphor B.P. 5.46% w/w, Menthol B.P. 2.82% w/w, Turpentine oil B.P. 4.71% w/w and Eucalyptus oil B.P. 1.35% w/w in a petrolatum base. Non-medicinal ingredients are typically Cedar leaf, Nutmeg and Thymol.

A typical formula for a decongestant oil is:

| | |
|---|---|
| Wintergreen | 27 |
| Camphor | 23 |
| Peppermint | 20 |
| Turpentine | 18 |
| Eucalyptus | 8 |
| Cedar leaf | 2 |
| Nutmeg | 2 |
| Total | 100 |

This is then mixed with pure petroleum jelly for use as a chest rub or encapsulated for use with hot water as a decongestant.

Table 16.1 Examples of pharmaceutical products containing eucalyptus oil which are marketed in Australia, Europe, Thailand and the United States[a, b]

| Product name | Manufacturer | Treatment |
| --- | --- | --- |
| *Australia* | | |
| Alcusal Sport | Alcusal | Sports injuries |
| Analgesic Rub | J. McGloin | Muscle and joint pain |
| Biosal | Yauyip | Arthritic pain |
| Bosisto's Eucalyptus Inhalant | Felton Grimwade & Bickford | Cold symptoms |
| Bosisto's Eucalyptus Rub | Felton Grimwade & Bickford | Muscular aches and pains |
| Dencorub | Carter Wallace | Musculoskeletal pain |
| Dentese | J. McGloin | Toothache |
| Euky Bear Eu-Clear Inhalant | Felton Grimwade & Bickford | Cold symptoms, respiratory tract congestion |
| Euky Bearub | Felton Grimwade & Bickford | Cold symptoms, nasal congestion, muscular aches and pains, insect bites |
| Logicin Chest Rub | Sigma Pharmaceuticals | Respiratory tract congestion |
| Methyl Salicylate Compound Liniment | J. McGloin | Muscle pain |
| Metsal | 3M Pharmaceuticals | Rheumatic pain, arthritis, sprains and strains |
| Vicks Vapodrops | Procter & Gamble | Nasal congestion, sore throat |
| Vicks VapoRub | Procter & Gamble | Cold symptoms |
| Zam-Buk | Key Pharmaceuticals | Minor skin disorders |
| *France* | | |
| Balsofletol | Laboratoires Pharmascience | Nose and throat infections |
| Édulcor eucalyptus et menthol | Laboratoires Pierre Fabre | Sore throat |
| Eucalyptine Le Brun | Laboratoires Janssen | Coughs |
| Inongan | Laboratoires Fumouze | Muscle pain |
| Kamol | Whitehall | Localised pain and injury |
| Nazophyl | Laboratoires Médecine Végétale | Rhinitis, rhinopharyngitis, sinusitis |
| Pulmax | Laboratoires Zyma | Respiratory tract congestion |
| Pulmoll au menthol et à l'eucalyptus | Sterling Midy | Throat disorders |
| Sirop Pectoral | Laboratoires Oberlin | Coughs |
| Thiopon Balsamique | Laboratoires Amido | Upper respiratory tract infections |
| Trophirès Composé | Millot-Solac | Respiratory tract disorders, coughs, fever |
| Tussipax à l'Euquinine | Laboratoires Thérica | Coughs |
| Vicks Pastilles | Laboratoires Lachartre | Sore throats |
| Vicks Vaporub | Laboratoires Lachartre | Respiratory congestion |
| *Germany* | | |
| Aerosol Spitzner N | W. Spitzner | Respiratory tract disorders |
| Angocin percutan | Repha | Respiratory tract disorders, muscle and joint pain |
| Aspecton-Balsam | Krewel-Werke | Coughs, cold symptoms |
| Babiforton | Plantorgan | Respiratory tract catarrh |
| Babix-Inhalat N | Mickan Arzneimittel | Respiratory tract disorders |
| Baby-Transpulmin | Asta Medica | Cold symptoms |
| Bronchicum Sekret-Löser | A. Nattermann | Upper respiratory tract disorders |
| Bronchodurat | G. Pohl-Boskamp | Bronchitis, cold symptoms |
| Bronchoforton | Plantorgan | Bronchitis, sinusitis |
| Divinal-Broncho-Balsam | Divinal Arzneimittel | Respiratory tract disorders |

Table 16.1 (Continued)

| Product name | Manufacturer | Treatment |
| --- | --- | --- |
| Dolo-cyl | Pharma Liebermann | Musculoskeletal and joint disorders, sports injuries |
| Emser Balsam echt | Siemens | Respiratory tract disorders |
| Emser Nasensalbe N | Siemens | Colds, hay fever |
| Ephepect | Bolder Arzneimittel | Bronchitis, catarrh |
| Eucabal-Balsam S | Esparma Pharmazeutische Fabrik | Respiratory tract disorders |
| Eucafluid N | Steigerwald Arzneimittelwerk | Musculoskeletal and joint disorders |
| Eufimenth N mild | Lichtenstein Pharmazeutica | Respiratory tract disorders |
| Hevertopect | Hevert-Arzneimittel | Respiratory tract disorders |
| Hustagil | Dentinox | Coughs, respiratory tract disorders |
| Inspirol N Solution | Lyssia | Mouth and throat disorders |
| Kneipp Erkältungs-Balsam N | Kneipp-Werke | Coughs, respiratory tract disorders |
| Leukona-Eukalpin-Bad | Dr Atzinger Pharmazeutische | Bath additive, respiratory tract disorders |
| Liniplant | W. Spitzner | Upper respiratory tract disorders |
| Logomed Erkältungs-Balsam Creme | Logomed Pharma | Upper respiratory tract disorders |
| Makatussin Balsam Menthol | Roland Arzneimittel | Catarrh, bronchitis |
| Melrosum Inhalationstropfen | A. Nattermann | Cold symptoms |
| Mentholon Original N | Richard Schöning | Catarrh |
| Nasivin Intensiv-Bad | E. Merck | Respiratory tract congestion |
| Nervfluid S | Fides Vertrieb Pharm. Präparate | Nerve, muscle and joint pain |
| Palatol | Pascoe Pharm. Präparate | Catarrh |
| Piniol N | W. Spitzner | Respiratory tract disorders |
| Pulmotin-N | Serum-Werk Bernburg | Respiratory tract disorders |
| Pumilen-N | E. Tosse | Colds, nasal congestion |
| Repha-Os | Repha | Inflammatory disorders of oropharynx |
| Retterspitz Aerosol | Retterspitz | Respiratory tract infections |
| Salviathymol N | Galenika Dr Hetterich | Inflammatory disorders of oropharynx |
| Sinuforton (oral drops) | Plantorgan | Sinusitis |
| Thymipin N (cream) | Zyma | Coughs, respiratory tract disorders |
| Tumarol-N | Robugen | Colds |
| Tussipect | Beiersdorf | Coughs, respiratory tract disorders |
| *Italy* | | |
| Antipulmina | Lisapharma | Respiratory tract disorders |
| Capsolin | Parke Davis | Joint and muscle pain |
| Eucalipto Composto | Dynacren | Nasal congestion |
| Fomentil | SIT | Respiratory congestion |
| Rinostil | Laboratorio Chimico Deca | Inflammation of nose and throat |
| Sloan | Parke Davis | Musculoskeletal and joint pain |
| Vicks Gola | Proctor & Gamble | Oropharyngeal disorders |
| Vicks Vaporub | Proctor & Gamble | Upper respiratory tract disorders |
| *Spain* | | |
| Bellacanfor | Bicther | Rheumatic and muscular pain |
| Gartricin | Cantabria | Mouth and throat inflammation |
| Lapiz Termo Compositum | Domenech Garcia | Rheumatic and muscular pain |
| Porosan | Reig Jofre | Upper respiratory tract disorders, soft tissue and muscular disorders |
| Termosan | Domenech Garcia | Upper respiratory tract disorders, soft tissue disorders, muscular pain |
| Vicks Vaporub | Procter & Gamble | Upper respiratory tract disorders |

Table 16.1 (Continued)

| Product name | Manufacturer | Treatment |
|---|---|---|
| *Switzerland* | | |
| Antiphlogistine | Carter-Wallace | Pain, inflammation |
| Baby Libérol | Galactina | Colds, fever |
| Baume Kytta | Whitehall-Robins | Musculoskeletal and joint disorders |
| Bismorectal | Vifor | Oropharyngeal disorders |
| Bradoral | Zyma | Oropharyngeal disorders |
| Bronchoforton N | Sterling Health | Respiratory tract disorders |
| Capsolin | Parke, Davis | Musculoskeletal pain |
| Demo pâtes pectorales | Demopharm | Coughs, bronchitis |
| Embropax | Taphlan | Peri-articular and soft tissue disorders |
| GEM | Piraud | Coughs, throat disorders |
| Huile analgésique 'Polar-Bär' | Panax Import, F. Ruckstuhl | Headache |
| Huile Po-Ho A. Vogel | Bioforce | Homeopathic preparation |
| Kemeol | Interdelta | Nasal congestion |
| Libérol | Galactina | Muscle and joint pain, chilblains, cold symptoms |
| Makatussin (oral drops) | Makapharm | Coughs, catarrh |
| Massorax | Laboratoire RTA | Joint and soft tissue disorders |
| Nasobol | Sodip | Respiratory tract disorders |
| Pasta boli | Spirig | Musculoskeletal and soft tissue pain |
| Phlogantine | G. Streuli | Musculoskeletal inflammation, bronchitis, skin disorders |
| Pinimenthol | Piniol | Influenza, muscle and joint pain |
| Pirom | Solmer | Muscle and joint pain, headache, sports injuries, insect bites |
| Pulmex | Zyma | Respiratory tract disorders |
| Rhinothricinol | Laboratoires Plan | Rhinopharyngeal infections |
| Roliwol | ECR Pharma | Muscle and joint pain |
| Sloan Baum | Warner-Lambert | Musculoskeletal and joint disorders |
| Tumarol | Renapharm | Cold symptoms |
| Vicks VapoRub | Procter & Gamble | Upper respiratory tract disorders |
| *Thailand* | | |
| Golden Cup Balm | Golden Cup Pharmaceutical | Muscular rheumatism, strains, insect bites, eczema |
| Golden Lion Balm | Jack Chia Industries | Minor muscular aches and pains |
| Golden Snake Balm | L.P. Standard Laboratories | Minor muscular aches and pains, sprain, headache, insect bites |
| Neotica Analgesic Balm | Thai Nakorn Patana | Muscular aches and pains, sprains, sports injuries, insect bites |
| White Monkey Holding Peach Balm | Monkey Holding Peach Brand | Minor muscular aches and pains, sprain, insect bites |
| *United Kingdom* | | |
| 9 Rubbing Oils | Potter's Herbal Supplies | Rheumatic and muscular pain |
| Aleevex | Pure Plant Products | Cold symptoms |
| Cabdrivers | Opal Products | Coughs |
| Catarrh Cream | Weleda | Nasal congestion |
| Catarrh Pastilles | Healthcrafts | Cough and cold symptoms |
| Chymol Emollient Balm | Rosmarine Manufacturing | Chapped skin, chilblains, bruises, sprains |
| Cupal Baby Chest Rub | Cupal | Bronchial congestion |
| Deep Heat Rub | Mentholatum | Rheumatic and muscular pain |
| Dragon Balm | Gerard House | Muscular aches |
| Fisherman's Friend | Lofthouse of Fleetwood | [Confectionery] |
| Gonne Balm | G.R. Lane Health Products | Muscular pain |

*Table 16.1* (Continued)

| Product name | Manufacturer | Treatment |
|---|---|---|
| Mackenzies Smelling Salts | Cox Pharmaceuticals | Catarrh, head colds |
| Medicinal Gargle | Weleda | |
| Mentho-lyptus | Warner-Lambert Confectionery | Cold symptoms, sore throats |
| Mentholease | Warner-Lambert Health Care | Hay fever |
| Merothol | Marion Merrell Dow | Mouth and throat infections, nasal congestion |
| Nicobrevin | Intercare Products | Aid to smoking withdrawal |
| Olbas Oil | G.R. Lane Health Products | Bronchial and nasal congestion, hay fever, muscular pain |
| Olbas Pastilles | BCM for G.R. Lane Health Products | Colds, coughs, catarrh, sore throats, flu, catarrhal headache, nasal congestion |
| Oleum Rhinale | Weleda | Catarrh, sinus congestion |
| Penetrol | Seton Healthcare | Nasal congestion |
| Potter's Decongestion Pastilles | Ernest Jackson | Blocked noses, coughs, catarrhal colds |
| Proctor's Pinelyptus Pastilles | Ernest Jackson | Throat irritation, coughs |
| Sanderson's Throat Specific | Sandersons | Sore throat, catarrh |
| Snufflebabe | J. Pickles & Sons | Congestion |
| Soothene | G.R. Lane Health Products | Minor skin disorders |
| Strepsils Menthol & Eucalyptus | Crookes Healthcare | Sore throat, mouth infections, nasal congestion |
| Tixylix Inhalant | Novartis Consumer Health | Nasal congestion |
| Vicks Vaporub | Proctor & Gamble (H&B Care) | Congestion |
| Zam-Buk | Roche Consumer | Wounds, skin abrasions |
| *United States* | | |
| Capastat | SmithKline Beecham | Sore throat |
| Eucalyptamint | Ciba Pharmaceutical | |
| Hall's Sugar Free Mentho-Lyptus | Warner-Lambert | Sore throat |
| Massengill | Beecham Products | Vaginal disorders |
| Maximum Strength Flexall 454 | Chattem Consumer Products | Pain, musculoskeletal disorders |
| Mexsana | Schering-Plough | Nappy rash |
| Robitussin Cough Drops | Whitehall-Robins | Coughs |
| Unguentine Ointment | Mentholatum | Pain in minor burns |
| Vicks Menthol Cough Drops | Procter & Gamble | Sore throat |
| Vicks VapoRub | Procter & Gamble | |
| Vicks Victors Dual Action Cough Drops | Richardson-Vicks | Coughs |

*Source*: RPS (1996), Ody (1996) and retail outlets in the UK and Thailand.

a Products containing ingredients listed as eucalyptol/cineole are not included.
b The list is intended to be illustrative rather than exhaustive. Some products may no longer be marketed; or they may be used for the treatment of conditions different, or additional, to those described; or they may no longer be labelled as containing explicitly 'eucalyptus oil'. Company names may also have changed.

Hall's Mentho-Lyptus™ cough suppressants, Listerine™ mouthwash, Olbas™ oil and various nasal inhalers are further examples of products utilising the refreshing and expectorant properties of Eucalyptus. A recent eucalyptus product, Eucalyptamint, has been promoted in the United States as a treatment for muscle soreness. Researchers at the University of California are reported to have tested the ointment and discovered that it increases the blood flow to muscle tissue (Internet 2000).

Tigerbalm™ is one of the most popular liniments, traditionally used in Asia but now widely available in the West. A Chinese merchant is said to have composed the first Tigerbalm, based

Table 16.2 Active ingredients and essential oils used in some proprietary preparations containing eucalyptus oil (%)[a]

| | Tiger® Liniment | Golden Cup Balm | Kwan Loong® Oil | Neotica Analgesic Balm | Gome Balm | Golden Snake Balm | Golden Lion Balm (White) | Golden Lion Balm (Red) | Olbas Oil® | Olbas® Pastilles |
|---|---|---|---|---|---|---|---|---|---|---|
| Methyl salicylate | 38 | 15 | 15 | 15 | 9 | 1 | — | — | — | — |
| Menthol | 8 | 12 | 25 | 10 | 4 | 10 | 11.6 | 15 | 4.1 | 0.1 |
| Camphor | 15 | 15 | 10 | 2 | 2 | 22 | 8.0 | 8 | — | — |
| Eucalyptus | 6 | 3 | 10 | 2 | 2 | 3 | 13.7 | 6 | 35.45 | 1.16 |
| Peppermint | — | 4 | — | — | — | 16 | 16.7 | 10 | 35.45[b] | 1.12 |
| Clove | — | 0.5 | — | — | — | 2 | 6.0 | 4 | 0.1 | 0.0025 |
| Spike lavender | 5 | — | 7 | — | — | — | — | — | — | — |
| Turpentine | — | — | — | — | 7 | — | — | — | — | — |
| Cajuput | — | 1 | — | — | — | — | — | 10 | 18.5 | — |
| Cinnamon | — | 1 | — | — | — | — | — | — | — | — |
| Wintergreen | — | — | — | — | — | — | — | — | 3.7 | 0.047 |
| Juniper berry | — | — | — | — | — | — | — | — | 2.7 | 0.067 |

[a] Tiger Liniment and Kwan Loong Oil are manufactured by Drug Houses of Australia (Asia) Pte Ltd, Singapore. Manufacturers of other preparations are given in Table 16.1.
[b] Dementholised mint oil.

on the ointment that tradition says was used by Genghis Khan's Mongolian horsemen and which was very effective against saddle and backache.
A typical formula is:

| | |
|---|---|
| Peppermint | 25 |
| Wintergreen | 20 |
| Camphor | 15 |
| Eucalyptus | 15 |
| Lavender | 15 |
| Vegetable oil | 10 |
| Total | 100 |

This is then mixed with pure petroleum jelly (adding the above oil at about 15% w/w). A small spot is applied to the forehead to alleviate headaches, or it can be rubbed on aching muscles or insect bites. However, in Tigerbalm products currently marketed in the UK eucalyptus oil appears to have been replaced by cajeput oil (from *Melaleuca cajeputi*) which also contains a high level of cineole and has similar warming, expectorant and analgesic properties.

Further examples of active ingredients and essential oils used in some preparations containing eucalyptus oil, and the proportions used, are given in Table 16.2.

Other applications of eucalyptus oil include its use in a footbath for sore feet (one teaspoon in a bowl of hot water) and addition to the water used in saunas or steam rooms to invigorate and relieve congestion. It can also be used to prevent blistering caused by burns or over-exposure to the sun, to soothe itching caused by insect bites or rashes and to aid healing of sores, cuts and abrasions.

*Aromatherapy uses*

The reported medicinal and aromatherapeutic properties of Eucalyptus are legendary. Amongst the properties attributed to *E. globulus* and other cineole-rich oils by various aromatherapy texts are the following:

| | |
|---|---|
| Analgesic | Deodorant |
| Antidiarrhoeal | Depurative (cleanses the blood) |
| Anti-inflammatory | Digestive |
| Antineuralgic (relieves or reduces nerve pain) | Diuretic (aids production of urine) |
| | Expectorant |
| Antiphlogistic (checks or counteracts inflammation) | Febrifuge (reduces fever) |
| | Hypoglycemiant (reduces blood sugar levels) |
| Antirheumatic | Insecticidal |
| Antiseptic | Parasiticidal |
| Antispasmodic (prevents and eases spasms or convulsions) | Prophylactic (prevents disease or infection) |
| Antiviral | |
| Bactericidal | Rubefacient (causes redness of the skin) |
| Balsamic (soothing, having qualities of a balsam) | Stimulant |
| | Vermifuge (expels intestinal worms) |
| Cicatrisant (promotes healing by formation of scar tissue) | Vulnerary (helps heal wounds and sores by external application) |
| Decongestant | |

Aromatherapy texts also recommend Eucalyptus in cases of boils, pimples, dandruff, athlete's foot, measles, nervous exhaustion or fatigue and various other complaints. In addition to its physical benefits, Eucalyptus is claimed to act on the emotional centre of the brain. With the properties of centring, balancing and stimulation, it can be used for treating exhaustion, congestive headaches, an inability to concentrate, mood swings and temper tantrums.

In aromatherapy, Eucalyptus is said to blend well with Angelica, Benzoin, Black pepper, Cedar wood, Coriander, Geranium, Juniper, Lavender, Lemon, Lemongrass, Marjoram, Melissa, Peppermint, Pine, Rosemary, Sandalwood, Spearmint, Tea tree, Thyme and Wintergreen. It is said to combine well with Lemon for mental clarity and with Myrtle for building up the immune system. The heating effect of Eucalyptus can often be balanced by mixing it with a 'cooling' oil such as Lavender or Peppermint (to reduce fever).

*Personal care products*

Table 16.3 lists some examples of personal care products which contain eucalyptus oil, together with the benefits claimed for them. They are found mainly in the UK market. The list includes some products (mainly bath additives) which may also be described as having medicinal or aromatherapeutic actions. Two other types of product, a general purpose cleaner and a disinfectant, are shown at the bottom of Table 16.3.

*Typical formulations for personal care products*

Eucalyptus oils can easily be incorporated into detergent or soap-based products or those based on alcohol or vegetable/mineral oils. However, their strong odour may conflict with other fragrances which are used in the products. Furthermore, in common with other essential oils, eucalyptus is not water-soluble and so cannot easily be incorporated in purely aqueous products without adding a solubiliser (usually a type of non-ionic detergent).

One way of circumventing any odour and water solubility problems is to use one of the many eucalyptus herbal extracts, widely available around the world, where solvents such as propylene/butylene glycol, glycerol or ethanol are used to extract the fresh plant material. In these cases, and depending on the extraction methods used, the components of the extract will differ significantly from those of the essential oils. However, under the sixth Amendment to the EU Cosmetics Directive the INCI (International Nomenclature for Cosmetic Ingredients) name used on European product packaging is the same for both oil and extract, namely the Linnaean botanical name of the plant species used, for example, *Eucalyptus globulus*. However, the solvent used, such as propylene glycol, will also appear somewhere in the ingredient listing. In the USA, the CTFA[3] INCI name 'Eucalyptus Globulus Extract' or 'Eucalyptus Globulus Oil' must be used until such time as the two sets of nomenclature are fully harmonised.

Formulations and methods of preparation for some typical personal care products containing Eucalyptus are given below, where European INCI names are used throughout.

---

3 The Cosmetic, Toiletry and Fragrance Association, the leading US trade association for the personal care products industry, with more than 500 member companies.

Table 16.3 Examples of personal care and other products which contain eucalyptus oil

| Product name | Manufacturer[a] | Product description | Benefit claimed |
|---|---|---|---|
| Baby Breatheasy Bath and Breatheasy Cream | Johnson & Johnson | Mild baby bath foam and skin cream with decongestant action. Contains Eucalyptus, Rosemary and Menthol | Helps to relieve blocked nasal passages and aids restful sleep |
| Breathe Clear with Extract of Eucalyptus | [For] Sainsbury's Supermarkets Ltd | Foam bath | Soothing, aids relaxation and makes breathing easier. Eucalyptus oil is renowned for its head-clearing properties |
| Carex Bath & Body Wash; Liquid Handwash; Antibacterial Moisturising Hand Gel | Cussons (International) Ltd | Green, soothing Aloe vera and Eucalyptus variant of the range of shower and bath body cleanser, liquid hand soap and waterless antibacterial hand gel | Dermaclens™ is a unique new moisturising system with antibacterial action, derived from natural oils |
| Radox Solutions 'Spring Unwind' | Sara Lee H & BC | Bath and massage oil containing Juniper berry and Eucalyptus | Relieves muscle tension |
| Radox Herbal Bath Breathe Easy | Sara Lee H & BC | Bath foam with Menthol and Eucalyptus | Renowned for helping to clear the head |
| Original Radox Herbal Bath Muscle Soak | Sara Lee H & BC | Bath foam with Eucalyptus | Contains Eucalyptus, well known to relieve congestion. Helps to relieve stiff, aching muscles and clear the head. |
| Radox Bath Salts Vapour Therapy | Sara Lee H & BC | Water-softening bath salts with Eucalyptus | Contains Eucalyptus which acts as a decongestant |
| Fresh! Green Devil Bath Bonanza | The Boots Co. | Bath 'bomb' – large effervescing bath tablet | Recharge yourself with this refreshing Eucalyptus soak |
| Eucalyptus Spa Bath | Aubrey Organics, USA (via Internet) | Bath oil containing blend of Eucalyptus and Menthol | Soothes minor muscle strain and tension. Helps open up blocked sinuses and clear the head |
| Relax-R-Bath | Aubrey Organics, USA (via Internet) | Herbal bath emulsion. Contains soothing blend of roots and herbs, including ground Eucalyptus leaf | Soothes sore muscles and eases tension. Ideal for athletes, dancers and for anyone after a hard day's work |

Table 16.3 (Continued)

| Product name | Manufacturer[a] | Product description | Benefit claimed |
| --- | --- | --- | --- |
| G'Day Eucalyptus Bath Bar | Aubrey Organics, USA (via Internet) | Vegetable soap base containing Eucalyptus, Menthol and other essential oils | Wake up your senses with this invigorating deodorant soap bar. Menthol and Eucalyptus tone and deodorise |
| Breathe Easy Headache Mask | Mother Earth Herbals (via Internet) | Mask that can be chilled or heated. Ingredients: Flax seed, Eucalyptus and Peppermint | Designed to provide relief for the pain and discomfort of sinus/cold congestion and allergies |
| Blue Gum Shampoos and Aqua Conditioners | CanCan Products | Range of shampoos and conditioners | Based on Blue Gum – Eucalyptus essential oil – added to promote healthy hair and scalp |
| Original Source: Eucalyptus & Mint Conditioner Shampoo Bath Foam Facial Wash | Health & Beauty Solutions Ltd. | Intensive hair conditioning treatment and various detergent-based products. Fragranced by natural Eucalyptus and Peppermint oils | Inspired by herbal remedies based on the healing power of plants of the Australian outback. Stimulating vapours give an exhilarating and refreshing sensation |
| Fresh! Bar Belle Massage Bar | The Boots Co. | Massage bar that melts on the skin | With refreshing Mint and Eucalyptus to work out and freshen up |
| NSR (Natural Sports Rub) | Aubrey Organics, USA (via Internet) | Warm-up/rub-down massage lotion. Compounded with seven essential oils, including Eucalyptus | Has a warming/cooling action. Eases minor muscle tightness and tension to help relaxation after a workout. Helps warm up and loosen muscles before a workout |
| Hand Salve – a farmer's friend | Burt's Bees, USA | Moisturising and healing hand salve with Comfrey, Rosemary, Lavender and Eucalyptus | Soothes the skin and promotes healing. For dry, chapped and weather-beaten skin |

| Product | Supplier | Description | Notes |
|---|---|---|---|
| Buzz Off Natural Insect Repellent | Mother Earth Herbals (via Internet) | Insect repellent oil. Contains Eucalyptus and other essential oils | An all natural alternative for fighting off 'pesky insects' |
| *Hairwars* – Healthy head and hair promoter | Tisserand Aromatherapy Products Ltd | Leave-on, vegetable-based hair oil treatment for head lice | An original aromatherapy formulation. Contains Tea tree, Eucalyptus and Kanuka oils as a 'natural solution to one of life's little problems' |
| Head Lice Treatment | Lincs Aromatics (via Internet) | Complete treatment for head lice. Oil, rinse and fine comb pack | Contains Lavender, Geranium and Eucalyptus |
| Soothing Anti-Itch Herbal Remedy | Aubrey Organics, USA (via Internet) | Skin-soothing balm containing Aloe vera, Canadian Willowherb extract and essential oils including Eucalyptus, Lavender, Tea tree and Oregano | Relieves itching and inflammation of rashes and insect bites; soothes symptoms of poison oak and ivy. Helps clear skin fungi and yeast overgrowth and ease symptoms of eczema, seborrhoea, psoriasis and other skin conditions |
| King of Shaves Shaving Oil. (Shaving gels are also produced containing Eucalyptus) | Knowledge and Merchandising (KMI) Inc. Ltd | Shaving oil containing Camphor, *Eucalyptus globulus*, *Mentha viridis* (Spearmint) and Menthol | Total sensitive skin protection |
| Ajax with Eucalyptus Earth Aware Household Cleaner | Colgate, France Aubrey Organics, USA (via Internet) | General-purpose cleaner General-purpose cleaner (incl. laundry detergent). Contains Rosemary, Sage and Eucalyptus oils | With essences of Eucalyptus Cleans most surfaces in the home without harming health or the environment |
| Myo Powerfresh Ultra Concentrate, Eucalyptus | Wolseley Castle, Australia | Disinfectant | With real Eucalyptus oil. Hospital grade |

a All products are manufactured in the UK except where indicated otherwise.

*Formulation 1 – Aching muscle relief bath foam* This formulation is designed to relieve congestion and help soak away aches and pains due to everyday stresses and strains. It should be poured into the bath under warm running water.

|  | % w/w |
|---|---|
| Aqua (*purified water*) | to 100.00 |
| Sodium laureth sulfate, 28% aq. solution (*foaming agent*) | 45.00 |
| Cocamidopropyl betaine, 40% aq. solution (*mild cleanser*) | 10.00 |
| Cocamide DEA (*foam booster/stabiliser*) | 2.00 |
| Sodium chloride (*thickener*) | 0.50 |
| *Eucalyptus globulus* (as essential oil) | 0.50 |
| Tetrasodium EDTA (*sequestering agent*) | 0.10 |
| Preservative | as required |
| Parfum (*fragrance*)/other essential oils | as required |
| Dyes | as required |
| Citric acid (*pH adjuster*) | to pH 6.00 |

Preparation: Mix the ingredients in the order given, stirring well after each addition. Adjust the pH as shown.

*Formulation 2 – Clear hand cleanser gel* This clear antimicrobial gel has been designed to hygienically cleanse the hands without the need to use soap and water.

|  | % w/w |
|---|---|
| *Phase A* | |
| Aqua (*purified water*) | to 100.00 |
| Alcohol (Ethanol, DEB 100) (*solvent/antimicrobial*) | 65.00 |
| Acrylates/C10-30 alkyl acrylate crosspolymer (*thickener*) | 0.50 |
| Glycerin (*moisturiser*) | 1.50 |
| Triclosan (*bacteriostat*) | 0.25 |
| Triethanolamine (*pH neutraliser*) | 0.50 |
| *Phase B* | |
| Parfum (*fragrance*) | 0.30 |
| *Eucalyptus globulus* (as essential oil) | 0.10 |

Preparation: Disperse the Acrylates/C10-30 alkyl acrylate crosspolymer in the water and add the remaining ingredients of phase A. Pre-mix phase B, add to phase A and stir gently until completely mixed.

*Formulation 3 – Hair conditioning treatment with Eucalyptus and Mint* This thick, creamy hair conditioner is designed to remoisturise the hair and prevent static build-up. The eucalyptus and mint essential oils are designed to stimulate the scalp.

|  | % w/w |
|---|---|
| *Phase A* | |
| Aqua (*purified water*) | to 100.00 |
| Hydroxyethyl cellulose (high viscosity grade) (*thickener*) | 1.00 |
| Panthenol (*Pro-vitamin B5 hair conditioning agent*) | 0.50 |
| *Phase B* | |
| Cetyl alcohol (*emulsion stabiliser/bodying agent*) | 3.00 |
| Dialkyl ester ethyl hydroxyethyl methylammonium methosulfate (*biodegradable hair conditioning agent*) | 2.00 |
| Isopropyl myristate (*emollient oil*) | 2.00 |
| *Phase C* | |
| Preservative | as required |
| Citric acid or Triethanolamine (*pH adjuster*) | to pH 6–6.5 |
| *Eucalyptus globulus* (as essential oil) | 0.40 |
| *Mentha piperita* (peppermint oil) | 0.20 |
| Parfum (*fragrance*) | as required |

Preparation: Heat the water of phase A to 75°C, stir rapidly and slowly add the hydroxyethyl cellulose; stir until fully hydrated. Heat phase B to 75°C, add phase A to phase B with stirring and continue stirring until cool. Finally add phase C.

*Oral health care*

The therapeutic properties of eucalyptus oils and extracts have meant that their use in oral health care products such as toothpastes, mouthwashes and chewing gums is becoming more common. Euthymol™ by Warner Lambert, a traditional style toothpaste, is based on the renowned natural antibacterial properties of Eucalyptus and thymol. Eucalyptus is believed to kill bacteria that cause halitosis and is said to be particularly effective against breath problems caused by strong-smelling food and drink. Microbiological testing by Quest has demonstrated that *E. globulus* oil (with an 80–85 per cent 1,8-cineole content) has moderate antimicrobial activity against both Gram-positive and Gram-negative bacteria.

The Eucalyptus may thus be present as an active agent, for example, as a decongestant (along with menthol) in the Wrigley's Airwaves 'vapour release' chewing gum, or as part of a more complex flavour compound. Eucalyptus is not common as a predominant flavour ingredient in oral care products as it has a medicated note that is not liked by some people, particularly in North America. Clark (2000) cites a 1992 Wall Street Journal article which reported that the major brands of mouthwash which contain eucalyptol were all losing market share. However, Eucalyptus is used to give lift and some freshness to oral care flavours.

One interesting application of Eucalyptus is in a mouthwash to combat or reduce snoring. A product called 'Good Night Stop Snore' is available by mail order in the UK. This is claimed to be an effective natural remedy for snoring if used every night. The mouthwash contains a blend of natural essential oils including Peppermint, Thyme, Fennel and Eucalyptus. These are said to have the combined effect of clearing bronchial congestion, whilst toning and increasing blood supply to the soft palate.

Another interesting oral care use is in an 'aromatherapeutic' chewing gum, designed to impart a stress-free mind as well as fresh breath. Peace of Mind™ Gumballs are marketed by Origins (Estée Lauder) in the USA as part of their Sensory Therapy line. They contain a blend of relaxing essential oils, including Peppermint, Basil and Eucalyptus. The latter is added for its 'tranquil effect'.

Research by Osawa and Yasuda at the Lotte Central Laboratory in Japan found that extracts of the dried leaves of *E. globulus* showed appreciable antibacterial activity against the bacteria that cause dental caries and periodontal disorders (Anon 1996). The researchers identified eight compounds that exhibited antibacterial behaviour, three of which (macrocarpals) had never been isolated before (see also Chapter 12). These macrocarpals have greater antibacterial activity than thymol, commonly used as an oral antibacterial agent. The extracts also act as inhibitors of glucosyltransferases, used by cariogenic bacteria such as *Streptococcus mutans* to synthesise insoluble glucans. These adhere to the tooth surface, forming dental plaque and leading to acid attack of the enamel. Further work was planned to involve *in vivo* testing as a possible new natural cariostatic drug for the maintenance of oral health.

In a draft review by the Food and Drug Administration Dental Plaque Subcommittee in the USA, a fixed combination of menthol, Eucalyptus, thymol and methyl salicylate is classified as a Category I, safe and effective oral care ingredient. The final report was due to have been published during 1999.

*Cleaning uses*

Eucalyptus oil can be used in a wide variety of ways as a cleaning agent. It is perhaps best known in Australia as a wool wash (Chapter 7) but other applications in the home include use in carpet shampoos, floor and general hard surface cleaning, spot and stain removal from clothes and furniture, as a solvent for the removal of chewing gum and for general air freshening. Elsewhere, it can be used as a leather or vinyl cleaner and for the removal of tar marks from motor vehicle paintwork, residues from stickers, labels or sticky tape, and grease or paint from the hands (when it will also eliminate obnoxious odours). These properties utilise the ability of cineole and other monoterpenes to act as a solvent for other oily or greasy molecules.

A general purpose cleaner which contains Eucalyptus is included in Table 16.3.

## Legislation in Europe and the United States

In Europe, *E. globulus* oil has to be labelled for bulk handling purposes under the Dangerous Substances Directive (67/548/EEC) as its flashpoint is 48°C and it is classified as Flammable (R10). It is not classified for either health or environmental effects. *E. citriodora* oil is not classified for health, environmental or physico-chemical (flammability) effects. More detailed information on packaging and labelling requirements for the handling and transportation of eucalyptus oils, particularly those arising from European legislation, is provided in Appendix 6.

The situation in the United States is slightly different to that in Europe. Drums of either of the two oils are required to be labelled for worker protection with the phrase 'May irritate skin and eyes'. For transportation purposes *E. globulus* oil is classified as a flammable liquid while, again, *E. citriodora* is not classified.

At the present time, under the sixth Amendment to the European Cosmetics Directive (96/335/EC), if either *E. globulus* or *E. citriodora* oils are used in a cosmetic product they need to be listed as an active ingredient on the label of that product under their Linnaean name.

The European Biocidal Products Directive (98/8/EC) which is being implemented at the present time may also have some labelling consequences for eucalyptus oils. Eucalyptus oil, in particular that from *E. citriodora*, is known to have insect repellent properties and repellents are included in this legislation. The exact consequences for consumer product labelling are as yet unclear, but it would appear that if biocidal properties are claimed for a product then the active ingredient must be named on the label, and listed on the annexes to the directive. However, legal opinion in Europe (Commission DGIII) is of the view that this will not affect cosmetic products since a cosmetic product cannot be a biocidal product at the same time. In this case, neither cosmetic products nor their ingredients will have to comply with any of the requirements of this directive.

The use of these oils is not regulated or restricted in consumer products in the United States.

## Acknowledgements

The authors would like to acknowledge the assistance of Anthony C. Dweck of Dweck Data for allowing us access to his database on natural plant materials and John Coppen for his compilation of Tables 16.1 and 16.2.

## References

Anon (1996) Taking a eucalyptus leaf out of the koala's book. *Chem. Brit.*, 32(12), 17.
Clark, G. (2000) Eucalyptol. *Perfum. Flavor.*, 25(May/June), 6, 8–10, 12, 14–16.
Davis, P. (1990) Eucalyptus. In *Aromatherapy, An A-Z*, C.W. Daniel, Saffron Walden, UK, pp. 122–124.
Ford, R.A., Letizia, C. and Api, A.M. (1988) *Eucalyptus citriodora* oil. *Food Cosmet. Toxicol.*, 26, 323.
Internet (2000) Herbal Remedies. Eucalyptus: The Australian Antiseptic. www.healthyideas.com/healing/herb/rem/971216.herb.html.
Kligman, A.M. (1966) The identification of contact allergens by human assay. III. The maximization test. A procedure for screening and rating contact sensitizers. *J. Invest. Derm.*, 47, 393.
Kligman, A.M. and Epstein, W. (1975) Updating the maximization test for identifying contact allergens. *Contact Dermatitis*, 1, 231.
Martindale, W.H. (1910) Essential oils in relation to their antiseptic powers as determined by their carbolic coefficients. *Perfum. Essent. Oil Rec.*, 1, 266–274.
Ody, P. (1996) *Handbook of Over-the-Counter Herbal Medicines*, Kyle Cathie, London.
Opdyke, D.L.J. (1975) Eucalyptol; Eucalyptus oil. *Food Cosmet. Toxicol.*, 13, 105–106; 107–108.
Poucher, W.A. (1991) *Poucher's Perfumes, Cosmetics and Soaps, Vol. 1, The Raw Materials of Perfumery*, Chapman & Hall, London, p. 159.
RPS (1996) *Martindale, The Extra Pharmacopoeia*, 31st edn, Royal Pharmaceutical Society, London.

## Selected bibliography

Arctander, S. (1960) *Perfume and Flavor Materials of Natural Origin*, Elizabeth, USA (available from Allured Publishing, Carol Stream, USA).
Chevallier, A. (1996) *The Encyclopedia of Medicinal Plants*, Dorling Kindersley, London.
CTFA (1997) *International Cosmetic Ingredient Dictionary & Handbook*, 7th edn, The Cosmetic, Toiletry & Fragrance Association, Washington DC, USA.
Curtis, S. (1996) *Neal's Yard Remedies. Essential Oils*, Aurum Press, London.
Grieve, M., *A Modern Herbal*, Eucalyptus, Internet at www.botanical.com.
Hill, C. (1997) *The Ancient and Healing Art of Aromatherapy*, Hamlyn, London.
Lawless, J. (1995) *The Illustrated Encyclopedia of Essential Oils*, Element Books, Shaftesbury, UK.

Leung, A.Y. and Foster, S. (1996) *Encyclopedia of Common Natural Ingredients Used in Food, Drugs and Cosmetics*, John Wiley & Sons, USA.

Miller, L. and Miller, B. (1995) *Ayurveda and Aromatherapy*, Lotus Press, USA.

Ody, P. (1997) *100 Great Natural Remedies*, Kyle Cathie, London.

Wells, F.V. and Billot, M. (1981) *Perfumery Technology: Art, Science, Industry*, Ellis Horwood, Chichester, UK.

# 17 Production, trade and markets for eucalyptus oils

*John J.W. Coppen*

## Introduction

The botanical, chemical, genetic and other aspects of oil-bearing eucalypts are discussed in detail elsewhere in this volume, as are the uses and types of formulations of eucalyptus oil. It is intended, here, to examine the production, trade and markets for the oil. With the possible exception of rutin, for which little or no information on trade and markets – to the extent that trade exists – is available, the volatile oil is the only aromatic/medicinal product of eucalyptus around which has been built an industry of any size. Some of the non-volatile constituents have considerable potential in the pharmaceutical and related fields but their utilisation in this way is, as yet, some way off from commercial realisation. Eucalyptus oil, on the other hand, is one of the largest volume and most widely used essential oils, produced in many parts of the world and on widely differing scales. It is a well established article of commerce and for many countries a valuable source of foreign exchange.

The different types of eucalyptus oil, and the species from which they are obtained commercially, are first summarised. For each of the major producing countries or regions, production is then described briefly in terms of locations and species utilised – other chapters have dealt with the subject in greater depth but some description is necessary to facilitate the rest of the discussion. Where possible, the levels of production are quantified and trade statistics are used to indicate the scale and directions of international trade. Price trends are then examined in some detail and, finally, some remarks are made on the outlook for the industry.

## Principal sources and uses of eucalyptus oil

### Plant sources

Eucalyptus oils are clear liquids with aromas characteristic of the particular species from which they are obtained. The oils are colourless when refined but usually slightly yellow when first distilled from the leaf. The composition of the oil, and therefore its olfactory and other properties, is dependent, mainly, on genetic rather than environmental factors. The species of *Eucalyptus* from which the oil is obtained is, therefore, the most important factor determining its composition and quality and the use, if any, to which it is put.

Of the hundreds of species of *Eucalyptus* that have been shown to contain volatile oil in their leaves, probably fewer than twenty of these have ever been exploited commercially for eucalyptus oil production. About a dozen species are presently utilised in different parts of the world, of which six account for the greater part of world oil production: *E. globulus*, *E. exserta*, *E. polybractea*, *E. smithii*, *E. citriodora* and *E. dives* (piperitone variant). The oils are conveniently classified

Table 17.1 Commercial sources of eucalyptus oil: main species and countries of production[a]

| Species | Producing country |
|---|---|
| Medicinal | |
| E. globulus[b] | China, Portugal, Spain, India, Brazil, Chile |
| E. exserta[c] | China |
| E. polybractea | Australia |
| E. smithii | South Africa |
| E. radiata[d] | South Africa, Australia |
| Perfumery | |
| E. citriodora | China, Brazil, India |
| E. staigeriana | Brazil |
| Industrial[e] | |
| E. dives (piperitone variant) | South Africa |
| E. olida | Australia |

a  Oils from a few species listed here (*E. radiata*, *E. staigeriana* and *E. olida*) are produced in much smaller quantities than others. Some other species, such as *E. dives* (cineole variant), *E. viridis* and *E. camaldulensis* are exploited commercially but only intermittently and/or on a small scale and are not listed. Small producers such as Bolivia, Paraguay and Uruguay are also not listed.
b  Mainly subsp. *globulus* but including subsp. *maidenii* in China.
c  Includes *E. leizhou No. 1*, a local land-race in China considered to be a natural hybrid of *E. exserta*.
d  Still sometimes referred to in commerce as *E. australiana* or *E. radiata* var. *australiana*.
e  Both *E. dives* and *E. olida* oils (or their constituents) are used for fragrance purposes so could alternatively be regarded as perfumery oils.

according to their composition and/or main end-use: medicinal (cineole-rich), perfumery and industrial. Of these, the most important in volume terms is the 'medicinal' type.

The main species currently exploited for oil, and the countries where this occurs, are listed in Table 17.1. Note that there are some other species and countries which are not listed because oil production is only intermittent and/or on a small scale. *E. globulus* is the principal species and is often the one against which others are judged in terms of quality (cineole content) and productivity. *E. cneorifolia* (medicinal), *E. macarthurii* (perfumery) and *E. radiata* (phellandrene variant; industrial) have been used in the past. *E. cinerea* oil (medicinal) was produced on a small scale in Zimbabwe in the early 1990s but its present status is not known. Western Australian species such as *E. kochii* and others may come to be exploited as a source of eucalyptus oil on a massive scale in Australia in the future (see below) but at the present time this remains speculative.

## Medicinal oils

### Present uses

The medicinal properties of eucalyptus oil were known to the aborigines of Australia thousands of years ago, but were first exploited commercially by Bosisto. He began production in Australia in 1852, extolling the oil's virtues for the treatment of a wide range of ailments. Since then, medicinal eucalyptus oil has remained the most important of the three types of oil, although Australia is no longer the dominant producer that it was.

The value of eucalyptus oil for medicinal purposes lies in its cineole content (strictly, 1,8-cineole, to distinguish it from the alternative 1,4-cineole, but the term 'cineole' is commonly applied to the former and is used here). This largely determines, also, the price that it fetches. While many other essential oils are referred to simply by name in the trade or in manufacturers' or dealers' product and price lists, medicinal eucalyptus oil is invariably specified in terms of cineole content. 'Eucalyptus oil China 80 per cent' and 'eucalyptus oil 70/75 per cent (or 80/85 per cent) Spain/Portugal' are typical descriptions. The higher grades are usually rectified forms of the crude oil. Essentially pure cineole (about 98 per cent+), which commands the highest price, is known as 'eucalyptol'. The general norm for cineole content, typified by the international standard (ISO) for 'oil of *Eucalyptus globulus*' and the British Pharmacopoeia specification for 'eucalyptus oil' is a minimum of 70 per cent (see Appendix 5). Any species of *Eucalyptus* which yields an oil with less than about 60–65 per cent cineole in the crude oil is unlikely to be of interest to a buyer or rectifier of such oils.

Providing a medicinal oil satisfies the appropriate pharmacopoeial requirements it may be sold as such, neat, in pharmacies and other retail outlets. In the home, eucalyptus oil is used in liquid form in a variety of ways, either medicinally or as a disinfectant, cleaner or deodoriser about the house. More commonly the oil is formulated before sale and used in the form of sprays, lozenges, ointments or other, compound oils or solutions (see Chapter 16). Although termed 'medicinal' oils, the fragrant properties of cineole-rich oils also make them attractive to use in certain soaps and bath oils.

*Possible future uses*

Research has also been carried out on other non-medicinal uses of cineole-containing eucalyptus oils. In Australia and Japan the use of such oils as fuel additives for petrol–alcohol mixtures has been explored but to date there has been no widespread application of this nature.

More recently, research has been undertaken in Western Australia which has far-reaching implications for eucalyptus oil supply and consumption if the potential indicated by the results to date is realised. Trials in the early 1990s (see e.g. Eastham *et al*. 1993) aimed at utilising *Eucalyptus* to combat the problem of soil salinity in the wheatbelt of Western Australia were sufficiently promising that a long-term programme of research and development was formulated. Any perennial planted on a sufficiently large scale to have any prospect of combating salinity would need to pay its way in terms of cash return. Work coordinated by the Western Australian Department of Conservation and Land Management (CALM) has shown that if native eucalypts of the 'mallee' type (multi-stem eucalypts which are amenable to coppicing, such as *E. kochii*) were planted, they could be repeatedly harvested and processed in an integrated manner to produce activated carbon, renewable energy and eucalyptus oil (RIRDC 1999). By mid-2000 some 17 million seedlings had been planted and a demonstration-scale plant was about to be constructed to undertake processing trials (J. Bartle pers. comm.). The extent of revegetation required to have an impact on salinity can be measured in millions of hectares and this inevitably means large-volume markets for the products. For eucalyptus oil the existing 'medicinal' markets would be insufficient to absorb the large volumes produced and a new market became the focus of attention: industrial solvents (Bartle *et al*. 1996, Bartle 1999). In particular, it is aimed to replace, at least partially, the environmentally damaging 1,1,1-trichloroethane with environmentally friendly cineole-rich eucalyptus oil. Before its production was discontinued in 1996 under international convention, around 700,000 t of trichloroethane were produced annually. CALM believes that eucalyptus oil has the opportunity to enter the market as a suitable, 'natural' replacement product. Even just 1 per cent of the market would require 7000 t of eucalyptus oil, around twice the size of the present worldwide consumption of cineole-type oil.

## Perfumery oils

Of the two perfumery oils listed in Table 17.1, that from *E. citriodora* is produced in the greatest volume. It contains citronellal as the major component – usually about 65–85 per cent – and is employed in whole form for fragrance purposes, usually in the lower-cost soaps, perfumes and disinfectants. However, it is also used as a source of citronellal for the chemical industry. The citronellal obtained by fractionation of the crude oil is then employed as such as an aroma chemical or converted to hydroxycitronellal. The latter compound finds major usage as a perfumery material. Other, minor, constituents of the oil, such as citronellol, are recovered during fractionation for subsequent use by the fragrance industry.

Both citronellal and the derived hydroxycitronellal are produced from two other sources, citronella oil and turpentine. To some extent, therefore, there is competition between these sources and *E. citriodora* oil. Price and slight odour differences caused by accompanying minor constituents in the chemical isolates determine customer preference. 'Natural' citronellal derived from eucalyptus or citronella oils may have some marketing advantages over 'synthetic', turpentine-derived citronellal in flavour applications.

*E. staigeriana* oil, produced only in Brazil, is utilised solely in whole form for fragrance purposes. It has a lemon-type character. No single chemical predominates as it does in most other commercial eucalyptus oils, although limonene and the higher-boiling citral (neral+geranial) together account for about 50–60 per cent of the total terpenes (see Appendix 4).

Oil from *E. macarthurii* is a rich source of geranyl acetate but has only ever been available in very small quantities and appears no longer to be produced.

## Industrial oils

The term 'industrial' is usually used to indicate that the oil is employed as a source of chemical isolate(s), that is, one or more constituents of the oil is isolated by a process of fractionation or some other means and then used in its own right for some application or further processed into marketable derivatives. (In this sense, *E. citriodora* oil might be considered an industrial oil if it is used as a source of citronellal). However, if cineole-rich oil were to be produced on a very large scale for use as an industrial solvent, as discussed above, common parlance would probably ensure that this, too, was referred to as an industrial oil.

The piperitone variant of *E. dives* yields an oil rich in piperitone and phellandrene. Since the mid-1960s, most of the world's supply of this oil has originated in South Africa and, until the early 1990s, much of it was shipped to Australia for fractionation. The lower boiling 'phellandrene' fraction (comprising α-phellandrene and other monoterpenes) was recovered and sold, mainly for use as a cheap fragrance source, while the piperitone (which accounts for about 40 per cent of the oil) was recovered for use as a starting material for the production of menthol, a major flavour ingredient. Although this route to menthol was in competition with natural menthol obtained directly from 'Japanese' mint (*Mentha arvensis*) and that obtained synthetically from petroleum-based raw materials, and is no longer used, some fractionation does still occur.

Oil distilled from the leaf of *E. olida* (originally referred to as *Eucalyptus* sp. nov. aff. *campanulata*) contains about 98 per cent *E*-methyl cinnamate, a chemical used as a flavour and fragrance material. Commercial production began in Australia in the late 1980s and to date this remains the sole source. Given the relatively small market for *E*-methyl cinnamate this is likely to remain the case for the foreseeable future.

## Production, consumption and trade by country/region

### Introduction

Any attempt to accurately quantify and analyse production and consumption trends for eucalyptus oil is fraught with difficulties. Unlike some of the larger commodities, or some other essential oils such as the citrus oils, quantitative information is not always available or accessible. Where data are available their accuracy and reliability have to be judged as best one can. There are, fortunately, some published trade statistics and this gives some cause for optimism that one can, at least, be talking within the right orders of magnitude. The finer detail is less certain. Trade statistics may identify importing countries but if these same countries are also processors and/or re-exporters then it is difficult to quantify actual consumption in geographical terms. The same is true of producing countries: since production data are rarely available, any export statistics will give no clues as to domestic consumption. China's exports of eucalyptus oil are discussed below but their own, internal consumption can only be guessed at. And trade statistics for 'eucalyptus oil' say nothing about the trade in formulated products containing eucalyptus oil.

Nevertheless, and in the face of these deficiencies, some useful observations can be made. Before examining published trade statistics some further words of caution on their use and interpretation are appropriate. Firstly, the data are only as good as the customs' returns allow. That is to say, if the exporter chooses not to describe his shipment as eucalyptus oil – for whatever reason – then it clearly will not be recorded as such and the official returns will underestimate exports. Occasionally, items are misnamed by the shipper or misclassified by the customs, which can result in either inflated or deflated figures. Small quantities of 'eucalyptus oil' of Indonesian origin, for example, which appear in some countries' import statistics are probably cajeput oil (from *Melaleuca*) rather than eucalyptus oil.

Secondly, items of trade are not always separately specified in the statistics of the country concerned. In all systems of customs classification, a numbering hierarchy groups commodities according to type and becomes increasingly more specific as the number of digits increases.[1] If the commodity is a major item then it usually specified, but otherwise it is included with similar commodities under a general heading. Unfortunately, this is often the case for eucalyptus oil and although, thankfully, China lists it separately, many other countries or regions simply include it in the all-embracing 'essential oils' or under an 'essential oils, not elsewhere specified' (NES) heading. Amongst the importing countries, Japan groups eucalyptus oil with fifteen other oils under a single essential oil heading and Hong Kong and Singapore, important intermediate destinations for Chinese eucalyptus oil, include it with other essential oils. In these circumstances, if the essential oil(s) coming from any of the countries of origin listed can reasonably be expected to be solely or mainly eucalyptus oil then it *may* be possible to draw some tentative conclusions. Thus if the figures given in Table 17.2 – which shows imports of essential oils into Japan for the period 1992–1999 and the contributions of three eucalyptus oil producers to these imports – are reliable, an upper limit of between 2 and 7 t annually can be put on imports of Australian eucalyptus oil. These figures are close to the levels of exports of eucalyptus oil from

---

[1] Most countries now use the Harmonised Commodity Description and Coding System (usually known simply as the Harmonised System) of the Customs Cooperation Council. A few countries show the SITC number (Standard International Trade Classification, Revision 3) of the United Nations alongside the HS number. For eucalyptus oil the HS number is usually 330129xx (where xx is e.g. 60 for China, 10 for the USA and 14 for India). The SITC number, if it is used, is of the form 5513xxx.

Table 17.2 Volume of imports of essential oils[a] into Japan, showing quantities from China, Spain and Australia, 1992–1999

|  | 1992 | 1993 | 1994 | 1995 | 1996 | 1997 | 1998 | 1999 |
|---|---|---|---|---|---|---|---|---|
| Total (t) | 220 | 237 | 168 | 192 | 184 | 166 | 131 | 149 |
| *Of which from* | | | | | | | | |
| China | 50 | 64 | 61 | 55 | 61 | 55 | 33 | 61 |
| Spain | 37 | 28 | 37 | 28 | 29 | 32 | 31 | 34 |
| Australia | 5 | 2 | 4 | 4 | 5 | 6 | 7 | 4 |

Source: National statistics.

a Defined as 'Essential oils (bay leaf, cananga, citronella, eucalyptus, fennel, star anise, petit-grain, rosemary, rosewood, ylang–ylang, cinnamon leaf, ginger grass, palmerosa, thyme, gyusho and lemongrass oils)'.

Australia to Japan for the earlier period 1982/83–1989/90 (Coppen and Hone 1992) and the more recent period 1997–2000 (Table 17.7, below). In the case of imports into Japan from China and Spain, eucalyptus oil may well fall within the upper limits indicated in Table 17.2 but it would be unwise to speculate as to exactly how much might represent eucalyptus oil.

Since 1993, European Union statistics (*Eurostat*), which previously listed eucalyptus oil separately, have not done so. In this case, fortunately, a reasonable picture of the relative importance of the member states as importers of eucalyptus oil can be gained from the Chinese export statistics and this is discussed in more detail below. Of the other producing and exporting countries, South Africa does not separate eucalyptus oil from other essential oils and neither do the smaller South American producers such as Paraguay.

Finally, with the exception of export data for China up to 1994 – which separates *E. citriodora* oil from other types of eucalyptus oil (see below) – no distinction is made in trade statistics between different botanical sources of the oil.

## People's Republic of China

Eucalyptus has been planted extensively in southern areas of China, almost 1 million ha in all. Around 150,000 ha of *E. globulus* (mainly subsp. *globulus* but also subsp. *maidenii*) occur in the south-western area, mainly in Yunnan province, and this is the chief source of Chinese eucalyptus oil. The main producing regions in Yunnan are Dali, Kunming and Chuxiong in the north of the province. *E. globulus* is also grown in southern Sichuan province and parts of Guangxi and Guangdong. China is the world's largest supplier of *E. citriodora* oil and this, as well as *E. exserta* oil, is produced from plantations growing in Guangxi and Guangdong; additional sources are the Leizhou Peninsula and Hainan Island. The potential of *E. smithii* as a dual-purpose crop – oil and fuelwood – is being recognised (see Chapter 8) and this species may begin to be utilised as a source of oil in the future, as it already is in South Africa (see below).

Opinion is divided on how much so-called Chinese eucalyptus oil is derived genuinely from *Eucalyptus* and how much is 'ex camphor'. The camphor tree, *Cinnamomum camphora*, yields an oil which, on fractionation, furnishes a cineole-rich fraction (Coppen 1995). This material is difficult to distinguish chemically from genuine eucalyptus oil, although it has a different odour and is excluded from most direct pharmaceutical use outside China by virtue of its slightly negative optical rotation.[2] In the recent past it has been suggested that *C. camphora* has been over-logged

---

2 The British and European Pharmacopoeias, for example, stipulate an optical rotation range of 0° to +10°. Eucalyptus oil 'ex camphor', however, is explicitly allowed in the Chinese Pharmacopoeia (see Appendix 5).

and is no longer used as a source of medicinal eucalyptus oil but trade sources continue to assert that it is. In China, eucalyptus oil 'ex camphor' is used primarily as a feedstock for eucalyptol production.

The scale and great geographical spread of oil production in China makes it exceedingly difficult to quantify it with any accuracy. Wang and Green (1990) put annual production of all types of eucalyptus oil at about 3000 t and this is usually the sort of level (sometimes up to 3500 t) that is cited by other sources in the literature or in the trade. The authors also stated that over 1000 t of oil were exported. However, Chen (Chapter 8) puts the figure for production much higher, at 4000 t, of which about 80 per cent (i.e. 3200 t) is exported. Sewell (1998) states that production of cineole oil varied between about 2000 and 4000 t annually in the early and mid-1990s, a very wide variation. He also estimated production of *E. citriodora* oil during the same period to have been in excess of 1000 t/year.

Table 17.3 shows officially reported exports of eucalyptus oil from China in volume, value and unit value terms for the period 1991–2000. The rising trend in exports through the 1980s (Chapter 8) continued into the early 1990s (Table 17.3) and peaked in 1994 (at 4400 t). For the years 1991 and 1993–2000, annual exports averaged just under 3600 t. These figures are in line with those put forward by Chen (Chapter 8). Exports were valued at US$15.2 million in 1995; the higher volume of exports in 1994 was offset in value terms by the lower average price. Care must be taken in using the export data in Table 17.3 to assess levels of imports of Chinese oil into the destination countries shown – considerable quantities are shipped to Hong Kong (not re-integrated with the People's Republic of China until 1997), Singapore and Taiwan but almost all of this is re-exported (often, in the case of Singapore and Taiwan, as eucalyptol). For 1991, where the split between *E. citriodora* oil and other eucalyptus oil is known (see footnote to Table 17.3), France can be seen to have taken a higher proportion of the perfumery oil, while the other destinations took, largely, the cineole oil.

Table 17.4 cites data from another official source which separates *E. citriodora* oil from 'eucalyptus oil'. Where it is possible to compare years in Tables 17.3 and 17.4, it can be seen that total volumes for 1991 are similar, those for 1993 and 1994 in Table 17.3 appear to exclude the quantities of *E. citriodora* oil shown in Table 17.4, and those for 1995 and 1996 are identical for both Tables. It is not clear, therefore, even using official statistics, as to whether the term 'eucalyptus oil' is all-inclusive. Whatever the truth it seems likely that, firstly, exports of *E. citriodora* oil have been between 450 t and about 750 t annually and, secondly, since 1993 exports of eucalyptus oil (all types) have been at least around 3500 t/year and they may have been significantly higher, that is, in accordance with the estimates for production and exports of Chen (Chapter 8).

Outside of China, Vietnam is the only other Southeast Asian (mainland) producer of eucalyptus oil although, of course, on a much smaller scale. Sewell (1998) put production as high as 250 t in 1990.

## United States (consumption)

In the absence of recent data for member states of the European Union, it is not possible to know whether the United States is the biggest single market for eucalyptus oil. France, Germany and Spain all imported more than the USA in the early 1990s (see below) and are undoubtedly still major importers but all four countries process and re-export eucalyptus oil. It is likely that in terms of actual consumption the USA is the largest consumer but that one or more of the EU countries remain(s) the biggest primary destination(s). The volume, value and unit value of imports of eucalyptus oil into the USA, with origins, for the period 1991–2000 are shown in Table 17.5.

372  J.J.W. Coppen

Table 17.3 Volume, value and unit value of exports of eucalyptus oil from China, and destinations, 1991–2000[a]

| | 1991 | 1992 | 1993 | 1994 | 1995 | 1996 | 1997 | 1998 | 1999 | 2000 |
|---|---|---|---|---|---|---|---|---|---|---|
| Volume (t) | 3159[b] | ns[c] | 3414 | 4420 | 3858 | 3659 | 3690 | 2811 | 3784 | 3449 |
| Value ('000 US $) | 14,784 | ns | 9247 | 10,107 | 15,215 | 12,893 | 12,197 | 9916 | 11,468 | 12,986 |
| Unit value (US $/kg) | 4.68 | ns | 2.71 | 2.29 | 3.94 | 3.52 | 3.31 | 3.53 | 3.03 | 3.77 |
| *Of which to (t)* | | | | | | | | | | |
| Hong Kong | 1444[b] | ns | 1031 | 1445 | 1321 | 1432 | 1133 | 381 | 217 | 377 |
| Singapore | 358 | ns | 609 | 692 | 651 | 456 | 77 | 106 | 183 | 119 |
| Taiwan | —[d] | ns | — | 63 | 205 | 223 | 287 | 43 | 218 | 138 |
| Japan | 5[b] | ns | 16 | 11 | 15 | 29 | 32 | 7 | 15 | 11 |
| Thailand | 17 | ns | 26 | 11 | 25 | 27 | 33 | 45 | 26 | 28 |
| Indonesia | — | ns | 3 | 11 | — | — | 63 | 46 | 324 | 387 |
| Malaysia | 5 | ns | 3 | 29 | — | — | 11 | — | 5 | 12 |
| Vietnam | — | ns | 28 | — | — | 7 | 14 | 3 | — | 3 |
| Philippines | — | ns | — | 4 | 10 | 4 | 6 | — | 25 | 11 |
| Australia | 118 | ns | 84 | 140 | 158 | 70 | 170 | 192 | 196 | 266 |
| India | — | ns | 14 | 71 | — | 46 | 160 | 87 | 353 | 74 |
| Pakistan | 10 | ns | 13 | 13 | 4 | 5 | 4 | 12 | 20 | 15 |
| Sri Lanka | — | ns | 1 | — | 1 | — | — | 4 | 3 | 3 |
| Bangladesh | — | ns | — | 1 | 4 | 2 | 5 | 4 | — | — |
| France | 454[b] | ns | 302 | 408 | 227 | 153 | 486 | 399 | 391 | 399 |
| Germany | 261[b] | ns | 394 | 325 | 325 | 186 | 166 | 379 | 540 | 339 |
| UK | 135 | ns | 298 | 332 | 385 | 374 | 324 | 245 | 421 | 399 |
| Spain | 122[b] | ns | 74 | 331 | 139 | 331 | 181 | 221 | 251 | 302 |
| Netherlands | 53[b] | ns | 86 | 134 | 164 | 81 | 74 | 73 | 122 | 154 |
| Switzerland | 5 | ns | 11 | 11 | 19 | 19 | 72 | 38 | — | 8 |
| Italy | 9 | ns | 3 | 8 | — | — | 15 | 14 | 24 | 14 |
| Portugal | — | ns | 14 | — | — | — | 58 | 115 | — | — |
| Denmark | — | ns | 11 | 2 | 6 | 6 | 2 | 2 | 18 | 12 |
| Sweden | 1 | ns | 1 | 1 | 1 | 1 | — | 1 | — | — |
| South Africa | — | ns | — | 3 | — | — | 4 | — | — | 5 |
| USA | 153[b] | ns | 360 | 347 | 166 | 203 | 292 | 358 | 395 | 343 |
| Canada | — | ns | — | — | 18 | 3 | 4 | 7 | 22 | 8 |
| Mexico | — | ns | 7 | 7 | 7 | — | — | — | — | 15 |
| Argentina | 5 | ns | 17 | 14 | — | — | 14 | 29 | 14 | — |
| Others | 4 | ns | 8 | 6 | 7 | 1 | 3 | — | 1 | 7 |

Source: *China Customs Statistics Year Book*.

a  For 1991, source data separate 'Eucalyptus oil' and '*Eucalyptus citriodora* oil' but these are combined here. For 1993–1997, source data are described as 'Essential oils of eucalyptus'.
b  Of which *E. citriodora* oil represents (t) 619 (total), 209 (Hong Kong), 1 (Japan), 303 (France), 16 (Germany), 35 (Spain), 21 (Netherlands) and 34 (USA).
c  Eucalyptus oil not specified separately.
d  Indicates that country not recorded (so assumed to be nil or negligible).

Imports rose year-on-year from 1991 to 1995 and then followed a sequence of peaks and troughs in the second half of the decade; in 1998 and 2000 imports reached just over 700 t/year (valued at US $3.2 and US $2.9 million, respectively). Imports averaged 510 t/year over the whole period (cf. 325 t/year for 1983–1990; Coppen and Hone 1992).

China is clearly seen to be the main source of US imports of eucalyptus oil but Brazil has also been a consistent, albeit much smaller, supplier. Of the other primary producers, Australia and South Africa have been the most important.

Table 17.4 Volume, value and unit value of exports of eucalyptus oil from China, 1989–1996[a]

|  | 1989 | 1990 | 1991 | 1992 | 1993 | 1994 | 1995 | 1996 |
|---|---|---|---|---|---|---|---|---|
| **Eucalyptus oil** | | | | | | | | |
| Volume (t) | 1344 | 2354 | 2521 | 2654 | 3316 | 4482 | 3858 | 3659 |
| Value ('000 US $) | 9560 | 13,026 | 12,763 | 10,781 | 9651 | 11,309 | 15,215 | 12,893 |
| Unit value (US $/kg) | 7.11 | 5.53 | 5.06 | 4.06 | 2.91 | 2.52 | 3.94 | 3.52 |
| **E. citriodora oil** | | | | | | | | |
| Volume (t) | 451 | 454 | 459 | 505 | 780 | 744[b] | ns[c] | ns |
| Value ('000 US $) | 1410 | 1260 | 1792 | 2046 | 2508 | 2432 | ns | ns |
| Unit value (US $/kg) | 3.13 | 2.78 | 3.90 | 4.05 | 3.22 | 3.27 | ns | ns |

Source: *Almanac of China's Foreign Economic Relations and Trade*.

a  For 1989–1994, eucalyptus oil and *Eucalyptus citriodora* oil are listed separately under 'Essential oils'. For 1995 and 1996, eucalyptus oil is described as 'Eucalyptus oil, natural' and *E. citriodora* oil is no longer listed separately.
b  Source data show figure of 9744 which cannot be right; it is assumed here that correct figure should be 744, which is consistent with the value figure.
c  Not specified separately.

Table 17.5 Volume, value and unit value of imports of eucalyptus oil into the USA, and origins, 1991–2000

|  | 1991 | 1992 | 1993 | 1994 | 1995 | 1996 | 1997 | 1998 | 1999 | 2000 |
|---|---|---|---|---|---|---|---|---|---|---|
| Volume (t) | 326 | 347 | 454 | 478 | 504 | 402 | 604 | 704 | 576 | 707 |
| Value ('000 US $) | 1709 | 1583 | 1859 | 1407 | 2582 | 1931 | 2591 | 3161 | 2261 | 2909 |
| Unit value (US $/kg) | 5.24 | 4.56 | 4.09 | 2.94 | 5.12 | 4.80 | 4.29 | 4.49 | 3.93 | 4.11 |
| *Of which from (t)* | | | | | | | | | | |
| China | 222 | 268 | 386 | 389 | 387 | 306 | 452 | 485 | 482 | 466 |
| Hong Kong | 18 | —[a] | — | 14 | 3 | 4 | — | 5 | — | — |
| Singapore | 2 | — | 13 | — | — | — | — | — | — | — |
| Taiwan | 32 | — | — | — | 1 | 5 | 26 | 21 | 7 | 55 |
| Indonesia | 4 | — | — | — | 2 | — | — | — | — | 15 |
| Australia | 3 | 16 | 4 | 7 | 10 | 16 | 2 | 1 | 12 | 74 |
| India | — | — | — | — | — | — | 4 | — | 20 | — |
| France | 1 | 2 | 7 | 7 | 26 | 4 | 7 | 4 | 3 | 4 |
| UK | 5 | 8 | 10 | 8 | 14 | 1 | 2 | 12 | 10 | — |
| Germany | — | 3 | 6 | 6 | 8 | 5 | 1 | 10 | 11 | — |
| Spain | — | — | — | — | 2 | — | 6 | 13 | — | — |
| Switzerland | 15 | 7 | — | — | — | — | 1 | 12 | — | — |
| South Africa | — | 14 | 12 | 19 | 7 | 9 | 9 | — | — | 1 |
| Swaziland | 7 | — | — | — | — | — | — | — | — | — |
| Canada | — | — | 1 | — | — | 2 | — | — | — | 2 |
| Jamaica | 6 | 5 | 1 | — | 3 | 4 | 3 | 2 | 8 | 8 |
| Brazil | 3 | 22 | 14 | 28 | 33 | 38 | 82 | 138 | 18 | 80 |
| Chile | 2 | — | — | — | 2 | 3 | — | — | — | — |
| Paraguay | 5 | — | — | — | — | 5 | 9 | — | — | — |
| Guatemala | — | — | — | — | 4 | — | — | — | — | — |
| Bolivia | 1 | — | — | — | — | — | — | — | — | — |
| Others | — | 2 | — | — | 2 | — | — | 1 | 5 | 2 |

Source: US Department of Commerce, Bureau of Census. Provided by Global Trade Information Services, Inc.

a  Indicates nil or negligible.

### European Union (consumption)

Imports of eucalyptus oil into the European Union from 1990 to 1992, the last year for which eucalyptus oil is reported separately, are shown in Table 17.6. Origins and destinations within the EU are also shown. Total imports averaged just over 2600 t/year (cf. just under 1800 t/year in the period 1983–1990; Coppen and Hone 1992).

As expected, China is by far the biggest supplier but South Africa is notable amongst the other producing countries. France, Germany, Spain and the UK are the biggest importers but all are processors, blenders and re-exporters. This introduces yet another difficulty in properly interpreting the statistics because there will be a measure of double counting – imports into

Table 17.6 Volume of imports of eucalyptus oil into the European Union, and origins and destinations, 1990–1992

|  | 1990 | 1991 | 1992 |
| --- | --- | --- | --- |
| Total (t) | 2643 | 2382 | 2853 |
| *Of which from (t)* | | | |
| China | 1794 | 1784 | 2070 |
| Hong Kong | 47 | 66 | 47 |
| Singapore | 1 | 9 | 5 |
| Australia | 10 | 12 | 8 |
| India | 57 | 5 | — |
| Sri Lanka | —[a] | 4 | 1 |
| South Africa | 164 | 127 | 176 |
| Swaziland | 15 | — | — |
| Namibia | 21 | — | — |
| Zimbabwe | — | 2 | — |
| USA | 20 | 5 | 2 |
| Brazil | 31 | 23 | 34 |
| Paraguay | 6 | — | — |
| Chile | 1 | 9 | — |
| Spain | 175 | 128 | 210 |
| Portugal | 117 | 66 | 106 |
| UK | 78 | 37 | 106 |
| Germany | 40 | 52 | 44 |
| France | 35 | 31 | 17 |
| Netherlands | 29 | 18 | 24 |
| Italy | 1 | — | 2 |
| Belgium/Lux. | — | 2 | — |
| Switzerland | 1 | 1 | 1 |
| Total (t) | 2643 | 2382 | 2853 |
| *Of which to (t)* | | | |
| France | 863 | 750 | 723 |
| Germany | 542 | 627 | 701 |
| Spain | 529 | 499 | 717 |
| UK | 432 | 285 | 440 |
| Netherlands | 108 | 90 | 79 |
| Portugal | 41 | 16 | 61 |
| Denmark | 22 | 22 | 16 |
| Ireland | 1 | 4 | 5 |
| Greece | 1 | — | 2 |

Source: *Eurostat.*

a Indicates nil.

France from, say, China which are processed and then re-exported to other countries within the EU in the same year will be counted twice. Imports into the EU from EU countries accounted for between 14 and 18 per cent of total imports for the years 1990–1992.

The levels of direct imports into the EU countries from China during 1991–2000 can be seen from Table 17.3 but, as noted earlier, it is impossible to know how much more Chinese eucalyptus oil is going into them via Hong Kong and Singapore. The most one can say is that the figures for the main EU countries are under-estimates, especially so for the years prior to 1998.

### Portugal and Spain

In Portugal, very large areas of eucalypts exist, distributed mainly along the Atlantic coast north of Lisbon and the Tagus valley. The vast majority, and sole source of oil, is *E. globulus*, all of which has been planted to meet the needs of the large domestic pulp industry. The main concentrations of eucalypt plantings in Spain are in the south of the country, in the forests of Huelva, where the main species is again *E. globulus*. In both cases, oil production is from the 'waste' leaf remaining after the trees have been felled. The main distillery groups, in addition to producing crude oil themselves, buy oil from a number of small, independent distilleries.

At the beginning of the 1990s, Coppen and Hone (1992) estimated that annual production in Portugal and Spain was around 150–200 and 50 t or less, respectively, but on a declining trend. The fall in output was a result of rising labour costs and severe competition from low-priced Chinese oil. This trend has continued and although both Spain and Portugal remain exporters of eucalyptus oil products, including eucalyptol (1,8-cineole), much of the starting oil is imported from China and elsewhere. For the three years 1990–1992, Spanish imports of Chinese oil averaged 470 t/year (out of total average imports of 580 t, Table 17.6). Present indigenous production is probably less than 100 t/year for Portugal and Spain combined (and could be as little as 50 t).

### Australia

After the development of large-scale commercial operations during the latter part of the nineteenth century, Australian production of eucalyptus oil reached a peak in the 1940s and has since declined. However, in the face of increasing production elsewhere in the world, the introduction of mechanised harvesting enabled the Australian industry to become more efficient and it has consolidated in two main geographical areas: near West Wyalong, New South Wales, and the Inglewood area of Victoria. In both cases, 'cleaned' natural stands of *E. polybractea*, a high quality source of cineole-rich medicinal oil, are utilised, complemented by smaller areas of plantation. Some oil is produced by independent distillers from species such as *E. radiata* and *E. dives* outside of these two regions but it is neither substantial nor regular. In the last decade, small plantations of *E. olida* have been established to meet demand for natural *E*-methyl cinnamate; production is probably no more than 10 t/year.

Domestic production of cineole oil is around 100 t year or less; McKelvie *et al.* (1994) put it at around 80 t year. However, it is supplemented by imports of lower quality oils, particularly from China, which are rectified and then blended with locally produced oils. Much of the subsequent re-exports are in the form of the final, formulated product, rather than the whole oil. If the Chinese trade statistics are accurate, Australian imports of eucalyptus oil averaged 155 t/year for the years 1991 and 1993–2000 (Table 17.3) with a rising trend. Additional quantities of Chinese oil may be imported via Hong Kong and Singapore. Piperitone-containing *E. dives* oil is no longer imported from South Africa, as it used to be, for menthol production.

Table 17.7 Volume of exports of eucalyptus oil from Australia, and destinations, 1997–2000

|  | 1997 | 1998 | 1999 | 2000 |
|---|---|---|---|---|
| Total (t) | 121 | 93 | 91 | 147 |
| *Of which to* | | | | |
| New Zealand | 5 | 8 | 12 | 17 |
| Thailand | 29 | 5 | 13 | 21 |
| Malaysia | 3 | 6 | 10 | 7 |
| Indonesia | —[a] | — | — | 5 |
| Singapore | 6 | 13 | 17 | 11 |
| Hong Kong | 6 | 5 | 8 | 6 |
| Japan | 6 | 7 | 4 | 9 |
| South Korea | — | — | 1 | 1 |
| Sri Lanka | 1 | — | — | — |
| Germany | 36 | 31 | — | 7 |
| UK | 10 | 3 | 1 | 6 |
| Netherlands | 2 | 1 | 3 | 6 |
| Belgium/Lux. | — | — | 2 | 1 |
| France | — | — | 1 | 3 |
| Spain | — | — | — | 1 |
| Poland | 2 | — | — | — |
| Russia | — | 9 | — | — |
| USA | 11 | 4 | 18 | 33 |
| Canada | 2 | 1 | — | 11 |

Source: Global Trade Information Services, Inc.
a Indicates nil or negligible.

Recent Australian exports (1997–2000) are shown in Table 17.7. They averaged 113 t/year and apart from North America and Europe, regional destinations, as one would expect, are seen to be important.

The plans to produce very large quantities of cineole-rich oil in Western Australia have already been referred to. If the plans come to fruition, eventual oil production in Australia would exceed that of China.

*Africa*

Southern Africa remains a significant producing region for eucalyptus oil although it has declined in both size and diversity since the report of Coppen and Hone (1992). Production now rests almost entirely with South Africa. Swaziland, previously a significant producer of oil – mainly the medicinal type from short-rotation, coppiced *E. smithii* – effectively ceased production when the principal company involved decided to concentrate on its more profitable timber operations. Previously, some 70–80 t of oil were produced annually, most of which was exported to Australia for rectification and blending. Zimbabwean production began in 1989 and only ever amounted to around 10 t of oil annually, mostly from coppiced *E. smithii* but a little from *E. cinerea*. Prevailing low prices for the oil made the operation very marginal economically and it is believed, though not certain, that production has ceased.

With the exception of a small quantity of medicinal oil produced from *E. radiata* in Cape Province, South African oil production is centred in the eastern Transvaal, close to the border with Swaziland. Although large areas of *E. smithii* have been planted in the south-eastern

Transvaal and Natal for timber purposes, and could be utilised for oil production if there were a wish to do so, existing production of oil from this species is derived almost entirely from short-rotation, coppiced trees in which oil is the sole product. In the late 1980s/early 1990s, production volume was split approximately equally between *E. smithii* and *E. dives* (piperitone type): 150–180 t annually each species (Coppen and Hone 1992). Since then, the industry has contracted. Although *E. dives* oil is no longer used for menthol production, with consequent loss of the Australian market for this purpose, it continues to be exported for use in flavour and fragrance manufacture, mainly to France and Germany. The greatest decline has been in the area given over to production of cineole-rich oil, under pressure from competing, low-priced oil from China. One South African source has put the production split at about 1 : 2 for cineole : piperitone-type oils, representing approximately 135 t of cineole oil and 200 t of *E. dives* oil annually (V. Davidson pers. comm. 1998). The figure for cineole oil is divided between *E. smithii* (100 t) and *E. radiata* (35 t). However, another estimate has put annual oil production at around 30–50 t (*E. smithii*) and 100–120 t (*E. dives*) (C. Teubes pers. comm. 1998).

For the years 1990–1992, imports of South African eucalyptus oil into the European Union averaged 155 t/year (Table 17.6) with France, Germany, Spain and the UK as the main (but irregular) destinations. During the 1980s, France and Spain were the main EU destinations while imports into Australia (including those from Swaziland) consistently exceeded 100 t/year (Coppen and Hone 1992). The USA has been a small but fairly regular destination for South African eucalyptus oil, importing (with a few exceptions) between 5 and 15 t throughout the 1980s and 1990s (Coppen and Hone 1992 and Table 17.5); imports were, however, nil or small in the period 1998–2000.

Most of the *E. radiata* and *E. dives* oils produced are exported but the *E. smithii* oil – at least, much of it – is rectified locally for domestic consumption. *E. dives* oil is occasionally rectified: the phellandrene fraction is exported and the piperitone used locally. In order to meet increasing demand in domestic markets, South Africa also imports Chinese oil (80 per cent cineole) for compounding a variety of fragrance and flavour materials.

Outside of Southern Africa, the Democratic Republic of the Congo (formerly Zaire) once produced small amounts of oil but no longer does so. Recent research on oil-yielding eucalypts in several of the Francophone countries of West and Central Africa (such as Benin, the Congo, Rwanda and Burundi) has indicated a desire to utilise for oil production those species already planted for afforestation, fuelwood and other purposes. However, at this stage, the commercial production of oil from such sources remains speculative.

### South America

Brazil has about 3 million ha of eucalypts but most of this is not of oil-bearing species. Nevertheless, it produces significant quantities of eucalyptus oil, particularly *E. citriodora* oil. Production takes place in the states of São Paulo, Minas Gerais, Bahia and Mato Grosso do Sul. Leaf is harvested either from trees grown specifically for oil on a coppice system, or by collection as 'waste' leaf from plantations established for other purposes (for the production of charcoal for use by the steel industry, for example). Oil is also produced from *E. globulus* and, to a lesser extent, *E. staigeriana* (for which Brazil is the only source). Brazil is by far the largest producer of eucalyptus oil in South America.

Brazilian production of eucalyptus oil is difficult to quantify. Coppen and Hone (1992) estimated *E. citriodora* oil, the major production, to be 400–600 t/year at the beginning of the 1990s. Much less cineole oil and perfumery oil from *E. staigeriana* is produced, possibly around

50 t/year or less of each. However, a more recent estimate put production of all oils at 750 t in 1985 and 970 t in 1996 (L. Couto pers. comm. 1999).

Brazil has a large domestic market for eucalyptus oil but there is still a surplus for export: annual exports averaged 215 t in the period 1983–1990 (Coppen and Hone 1992) and 195 t for the years 1992–2000 (Table 17.8). Levels of exports during the latter period were characterised by a trough in the middle years (1994–1996) followed by an upward trend to 2000. For the years for which a breakdown of destinations is available (1997–2000), the nearby markets of the USA, Mexico and Colombia have been important (though Mexico and Colombia less so towards the end of the period). Further afield, Spain and Sweden have been consistent importers but the UK, France and the Netherlands imported significant quantities of Brazilian oil in 2000. Shipments to Spain reflect more its position as a fractionator/processor of oils than a consumer, and the need to maintain a viable feedstock for this purpose, from whatever sources are available. Although a net exporter, Brazil also imports eucalyptus oil and these imports averaged 70 t annually for 1992–1999.

Elsewhere in South America, Chile, Bolivia, Paraguay, Uruguay and Colombia have all produced oil at one time or other but on a very much smaller scale than Brazil. Sewell (1998) states that in the mid-1990s annual production of cineole-rich oil was 20 t in each of Bolivia, Uruguay and Colombia and 25 t in Paraguay; the source of this information is not given. Chile emerged as a modest producer of eucalyptus oil in the mid-1980s, using stands of *E. globulus* which is widely planted. The main areas of *E. globulus* are in the Valparaiso and Bio-Bio Administrative Regions in the centre of the country. Annual production was about 80–100 t in the early 1990s, most of which was exported, either as the whole oil or in the form of purified 1,8-cineole. Bolivia began production of cineole-rich oil from *E. globulus* in the 1980s but this has been on a very

*Table 17.8* Volume of exports of eucalyptus oil from Brazil, and destinations, 1992–2000

|  | 1992 | 1993 | 1994 | 1995 | 1996 | 1997 | 1998 | 1999 | 2000 |
|---|---|---|---|---|---|---|---|---|---|
| Total (t) | 206 | 225 | 140 | 146 | 133 | 188 | 215 | 223 | 275 |
| *Of which to* | | | | | | | | | |
| USA | na[a] | na | na | na | na | 90 | 124 | 18 | 78 |
| Mexico | na | na | na | na | na | 30 | 13 | 12 | 6 |
| Colombia | na | na | na | na | na | 24 | 12 | 8 | 9 |
| Paraguay | na | na | na | na | na | 3 | 2 | 2 | — |
| Argentina | na | na | na | na | na | —[b] | 8 | 2 | 2 |
| Bolivia | na | na | na | na | na | — | — | 1 | — |
| Venezuela | na | na | na | na | na | — | 1 | — | — |
| Spain | na | na | na | na | na | 24 | 27 | 148 | 57 |
| Germany | na | na | na | na | na | 1 | 1 | 6 | 7 |
| France | na | na | na | na | na | — | 5 | — | 16 |
| UK | na | na | na | na | na | — | — | — | 45 |
| Netherlands | na | na | na | na | na | — | — | — | 12 |
| Sweden | na | na | na | na | na | 12 | 17 | 15 | 27 |
| Turkey | na | na | na | na | na | 4 | 5 | — | — |
| India | na | na | na | na | na | — | — | 10 | 11 |
| Singapore | na | na | na | na | na | — | — | — | 5 |
| Australia | na | na | na | na | na | — | — | 1 | — |

Source: National statistics (1992–1996) and Global Trade Information Services, Inc. (1997–2000).

a Not available.
b Indicates nil or negligible.

small scale, operating through a series of farming cooperatives; most of the oil has been used by domestic industries. Paraguay produces *E. globulus* and *E. citriodora* oils. Chile and Paraguay appear occasionally as suppliers of oil in US import statistics (Table 17.5) and both were exporting in 2000.

*India*

Eucalypts have been planted on a large scale in India – mainly for fuelwood, pulp, pole and afforestation purposes – but the dominant species is *E. tereticornis* ('*Eucalyptus* hybrid'), which is not ideally suited to oil production. Both *E. globulus* and *E. citriodora*, however, have also been planted and these two species form the basis for Indian eucalyptus oil production.

Much research has been carried out on *E. citriodora* grown under Indian conditions and it has been widely promoted as a crop suitable for small-scale cultivation for oil production. Despite this it remains the lesser of the two species as a source of oil. Reliable estimates of Indian production are not easy to come by and some figures which have appeared in the Indian literature have differed wildly, with no indication of the sources on which they are based. Earlier estimates by Coppen and Hone (1992) put production at around 50 t of *E. citriodora* oil and 150–200 t of cineole-rich *E. globulus* oil annually. Handa *et al.* (Chapter 11) have cited a personal communication which puts annual production at about twice these levels, 100 and 400 t, respectively.

Levels of Indian exports of eucalyptus oil, as recorded in the annual trade statistics, are small and erratic and this is in line with domestic consumption which, although not possible to quantify, is known to be high. Exports varied from nil (in 1991/92) or less than one tonne (in 1996/97 and 1997/98) to 39 t (in 1998/99) during the period 1990/91–1998/99. In the three years where there were any significant exports, the biggest single destinations were Spain (8 t in 1990/91), France (14 t in 1994/95) and Germany (22 t in 1998/99). However, India is a net importer of eucalyptus oil, sometimes significantly so: annual imports averaged almost 100 t during the period 1992/93–1998/99 and reached 226 t in 1994/95 and 205 t in 1997/98, almost all of it from China (or of Chinese origin).

Price trends

It is clear from what has been said earlier that world production and trade in eucalyptus oil is dominated by the People's Republic of China, particularly of cineole-rich oil. It is also the main source of *E. citriodora* oil although Brazil is also a major producer. Chinese sources put total annual production of eucalyptus oil at around 4000 t with exports of about 3200 t. Official trade statistics indicate that even these figures may be under-estimates – exports have averaged 3600 t/year in recent years and exceeded 4000 t in 1994. The question as to how much of this is genuine eucalyptus oil rather than 'ex camphor' remains, although in terms of consumption it is all used as if it *were* eucalyptus oil. In broad terms, Chinese exports during the 1990s have been around 3000 t/year for cineole-type ('medicinal') oil and 500 t/year for *E. citriodora* (perfumery) oil, although there have been deviations above and below these levels.

Chinese eucalyptus oil production, and any events which may affect it, are therefore very influential in terms of prices. With the passage of time, any eucalyptus oil prices that are quoted soon become out of date and of only historical interest. Nevertheless, an examination of price trends is worthwhile and demonstrates that eucalyptus oil is, indeed, a low-priced oil compared to most other essential oils.

## Cineole-rich oils and eucalyptol

Prices for standard grade Chinese 80 per cent medicinal oil and eucalyptol for the period 1981–1993 are illustrated graphically in Figure 17.1.[3] As would be expected, price movements for eucalyptol tend to parallel those for 80 per cent oil. The prices of comparable grades of Spanish/Portuguese eucalyptus oil are invariably higher than that of Chinese 80 per cent oil.

The very large volumes of oil produced, combined with demand in the West and the ability to earn valuable foreign exchange, encouraged China to pursue a policy of aggressive export pricing throughout much of the 1970s and 1980s. After two exceptional high price peaks during 1973 and 1974, standard grade Chinese 80 per cent medicinal oil settled down to around US $4.00–5.50/kg (CIF major world ports) before rising at the end of 1987. During the early 1990s, the centrally controlled economy gave way to economic liberalisation and this resulted in a plethora of new businesses and trading companies, each eager to acquire a share of the eucalyptus oil trade. The disruption of traditional trading practices caused, in turn, some volatility in prices. At the same time (1990–1991), the opportunities for foreign multinational companies to manufacture and market in China became greater and this led to a decrease in domestic demand for traditional flavours such as eucalyptus. Increased supplies of Chinese-origin eucalyptus oil to the world market coincided with a downturn in demand caused by the recession in 1991–1992. All these factors contributed to the continuing fall in price of Chinese 80 per cent medicinal oil from the high US $9.50–10.00/kg mark at the end of 1988 to the US $5.00/kg level at the end of 1991 and historically low levels around US $2.50/kg in early 1994. Prices then recovered and in the recent past were below US $4.00/kg during 1999 but reached US $6.00/kg towards the end of 2000 before beginning to fall. Recent prices are shown in Table 17.9.

Along with the normal ebbs and flows of seasonal changes in the supply of crude eucalyptus oil, and the consequent firming or weakening of prices, come the less predictable changes caused by climatic events or natural disasters. Floods or unseasonably cold weather in Yunnan, for example, can hit the eucalyptus harvest – either directly or by diverting farmers to other, more pressing activities. Shortages of supply of eucalyptus oil 'ex camphor' can also have an

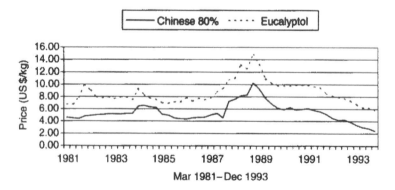

*Figure 17.1* Price trends for Chinese 80 per cent eucalyptus oil and eucalyptol, 1981–1993.

---

3 The graphs (and those of Figure 17.2) are based on average prices per quarter (CIF Main European Port for Figure 17.1). Source: London dealer.

*Table 17.9* Price movements for Chinese 80 per cent cineole eucalyptus oil over two years, 1998–2000 (US $/kg, CIF Main European Port)

| Date | Price | Date | Price | Date | Price |
|---|---|---|---|---|---|
| Sep 1998 | 3.50 | Oct 1999 | 3.70 | May 2000 | 4.50 |
| May 1999 | 3.45 | Nov 1999 | 3.85 | Jul 2000 | 4.75 |
| Jun 1999 | 3.20 | Dec 1999 | 3.60 | Aug 2000 | 5.25 |
| Aug 1999 | 3.15 | Jan 2000 | 3.75 | Sep 2000 | 6.00 |
| Sep 1999 | 3.55 | Mar 2000 | 4.40 | Oct 2000 | 5.75 |

Source: London dealer.

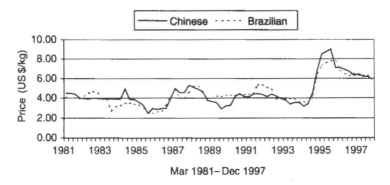

*Figure 17.2* Price trends for Chinese and Brazilian *Eucalyptus citriodora* oil, 1981–1997.

effect: eucalyptol producers may be forced to switch to *E. globulus* oil as a feedstock, with a consequent effect on the availability and price of Chinese 80 per cent oil.

E. citriodora *oil*

Prices for Chinese and Brazilian *E. citriodora* oil for the period 1981–1997 are illustrated graphically in Figure 17.2. Neither source has been consistently lower priced than the other although, as noted, Chinese prices are CIF, compared with Brazilian FOB. For most of the period prices were in the range US $3–5/kg but towards the end of 1994 they rose fairly sharply and had doubled to US $7.80 (Brazilian) and US $9.00/kg (Chinese) by late 1995 before falling and levelling out at just over US $6/kg during 1997. For most of 2000, Chinese *E. citriodora* oil was in the range US $5.50–6.00/kg; in October 2000, one London dealer was quoting US $5.75/kg.

The outlook

Given the uncertainty of the plans for Western Australian eucalyptus oil production, China will, for the foreseeable future, continue to be the dominant force in eucalyptus oil supplies and events in China will be of prime importance in determining prices, particularly for cineole-type oils. However, with the liberalisation in China came a keenness to diversify and develop value-added industries, fuelled in part by the desire to meet the domestic demands of an increasingly consumer-conscious society but also by the desire of foreign fragrance and flavour companies to invest in a part of the world which, as well as being the centre of production of many of the major essential oils, has an indigenous technical expertise and can support relatively low-cost operations. So today, while Chinese eucalyptus oil remains a high-volume, relatively low-priced oil, exported for use in blending and compounding by others, there has been an expansion and consolidation in China of its own refining and marketing activities, particularly in and around Kunming, Yunnan, and often through joint ventures with foreign companies.

As Smith (1998) has observed in the context of the fragrance and flavour industry:

> The world has become, in recent years, an increasingly uncertain place to do business. The economic turmoil seen in many countries, and particularly in Asia Pacific and the former Soviet Republics, has sent echoes reverberating around the financial markets of the world. As information technology develops, these markets move more rapidly than ever before and capital becomes increasingly mobile. ... At the same time there is more opportunity than ever before for those companies with the will and courage to go after it. Market deregulation, the lowering of international tariff barriers, the development of global markets and the emergence of new markets are features of the international business landscape that provide prospects for growth and development in abundance.

The increasing globalisation imposes greater demands on both smaller companies and suppliers of eucalyptus oil (and any other essential oils) in the developing world. Jackets (1998) has indicated that the move towards global purchasing contracts, together with global quality standards, including the implementation of ISO 9000, has led the major companies to install stricter control on the suppliers, who must also be capable of providing the necessary documentation.[4] All this suggests that those in the eucalyptus oil business – whether they are suppliers, dealers, processors, compounders or manufacturers of final products – who can adapt positively to the changes that are taking place will survive and prosper. In the West, increasing public awareness of 'green' issues and attention to product labelling will ensure that natural products such as eucalyptus oil will continue to find favour in the marketplace. In the East, the massive potential of increasing consumer demand in countries such as China, India and those of the former Soviet Union beckons, though it may be some time before it becomes a reality.

## References

Bartle, J.R. (1999) Why oil mallee? In *Oil Mallee Profitable Landcare, Proc. Oil Mallee Association Seminar*, Perth, Western Australia, March 1999, pp. 4–10.

Bartle, J.R., Campbell, C. and White, G. (1996) Can trees reverse land degradation? In *Farm Forestry and Plantations: Investing in Future Wood Supply, Proc. Australian Forest Growers Conf.*, Mount Gambier, Australia, September 1996, pp. 68–75.

Coppen, J.J.W. (1995) *Flavours and Fragrances of Plant Origin*, Non-Wood Forest Products Series, No. 1, Food and Agriculture Organization of the United Nations, Rome.

Coppen, J.J.W. and Hone, G.A. (1992) *Eucalyptus Oils. A Review of Production and Markets*, Bulletin 56, Natural Resources Institute, Chatham, UK.

---

4 The burden on the supplier arising from increasing legislation centred on environmental and safety concerns is discussed in Appendix 6.

Eastham, J., Scott, P.R., Steckis, R.A., Barton, A.F.M., Hunter, L.J. and Sudmeyer, R.J. (1993) Survival, growth and productivity of tree species under evaluation for agroforestry to control salinity in the Western Australian wheatbelt. *Agrofor. Syst.*, 21, 223–237.

Jackets, P.A. (1998) The combined effects of technology and the global shift to market economies. In *Global Markets: Present and Future, Proc. IFEAT Internat. Conf. Essential Oils and Aromas*, London, November 1998, pp. 208–220.

McKelvie, L., Bills, J. and Peat, A. (1994) *Jojoba, Blue Mallee and Broombush: Market Assessment and Outlook*, ABARE Research Report 94.9, Australian Bureau of Agricultural and Resource Economics, Canberra.

RIRDC (1999) *Integrated Tree Processing of Mallee Eucalypts*. Report by Rural Industries Research and Development Corporation, Barton, ACT, Australia.

Sewell, M. (1998) Eucalyptus oil: a review of world production and trade. In *Proc. Internat. Seminar on Yunnan's Trade Development*, Kunming, China, April 1997, China-EU Centre of Agricultural Technology (CECAT), Beijing, pp. 109–122.

Smith, C.M. (1998) World markets and the current business context. In *Global Markets: Present and Future, Proc. IFEAT Internat. Conf. Essential Oils and Aromas*, London, November 1998, pp. 1–5.

Wang, H. and Green, C.L. (1990) A review of eucalyptus oil: production, market and its prospect in the world. *World For. Res.*, 3(2), 71–76.

# 18  Research trends and future prospects

*Erich V. Lassak*

## Introduction

There are at present approaching 1000 or so fully described species of *Eucalyptus* and this number continues to grow. Included amongst these species are the 'bloodwoods', previously of the sub-genus *Corymbia* but which were elevated to a separate genus, *Corymbia*, by Hill and Johnson (1995). For the purpose of this chapter, as throughout this book, and in line with the preference of Brooker (Chapter 1), species of *Corymbia* will be retained here under the genus *Eucalyptus*.[1] Owing to the unusually large size of this almost exclusively Australian genus, and to the fact that many botanical descriptions have been published in relatively little known and less accessible Australian journals, many names, now obsolete, continue to appear in the literature. A case in point is the frequent use of the old name *E. rostrata* instead of *E. camaldulensis*. The compilation and dissemination of a regularly up-dated list of species and synonyms would be of great assistance to researchers working in this field, and lessen the chance of duplication occurring when different workers investigate the same species under two or more different names. The task would be made easier using electronic forms of communication.

The factors which have influenced past research into the chemistry of eucalypts (and in particular of their essential oils) combine, perhaps more than in most other plant genera, pure science with applied science, that is, aspects related to their commercial utilisation. A wide range of topics has been discussed in preceding chapters of this volume and it is evident that *Eucalyptus* will continue to provide new or improved commercial opportunities which, in turn, will be a driving force for further research. This chapter examines certain aspects of past work, highlights some areas of current research and indicates where additional research might be required for a better understanding of the chemistry of *Eucalyptus* and for a continuing expansion of its economic potential. Although earlier research focused on the volatile constituents of eucalypts – their essential oils – and these have largely been the basis upon which the industry exploiting the aromatic and medicinal uses of eucalyptus has been built, research in the last two decades has demonstrated the commercial potential, particularly in the pharmaceutical field, of the non-volatile constituents. Both types of compounds are therefore discussed below.

## Extraction of eucalyptus leaf oils

### *Developments in large-scale commercial extraction*

What developments have there been in the methodologies used for large-scale extraction of eucalyptus oils and are there any pointers for the future? Steam distillation (or in rare cases

---

[1] *Corymbia citriodora*, previously named *Eucalyptus citriodora*, is the source of the commercially important *E. citriodora* oil and will, without any doubt, continue to be traded under its old name.

hydrodistillation) at atmospheric pressure is the traditional and most widely used method for the extraction of eucalyptus oils. More or less freshly harvested foliage, including terminal branchlets, is used without any pre-treatment. Distillation times are of the order of several hours depending on the nature of the oil being extracted: 1,8-cineole or citronellal-rich oils require shorter distillation times than, for example, piperitone-rich oils.

Since eucalyptus oil glands are buried deep within the leaf tissue they require longer times to rupture and to release the oil than oil cells located in the epidermis of leaves or flowers (as found, for example, in most plants of the family Lamiaceae). Consequently, various attempts have been made to reduce the distillation time by crushing or grinding the leaves, and thus at least partly releasing the oil, before steam distilling them. Whilst the time required for complete oil extraction may, indeed, be shortened, production costs are usually increased. The chief reason for this increase in costs is the additional machinery and processing time required. Furthermore, oil recovery tends to be significantly poorer as some of the lower-boiling components of the oil evaporate and are lost during the comminution process (Provatoroff 1972, Abbott 1989).

Continuous steam distillation, as opposed to the normal batch process, has also been attempted by several Australian eucalyptus oil producers. However, the technical difficulties proved sufficiently serious that this approach had to be abandoned.

The possibility of producing eucalyptus oils by expression from some of the higher oil-yielding species was investigated at the Museum of Applied Arts and Sciences in Sydney (where most of the eucalyptus oil research in Australia was carried out up to the late 1970s). Whilst some liquid was obtained by the application of quite high hydraulic pressures, no significant quantities of essential oil separated, even after centrifugation.

Some recent Western Australian research has examined the possibility of using solar energy to drive off essential oil and water vapour from foliage contained in very large plastic film chambers, and to separate the oil after condensation of the vapours. The technique apparently shows some promise (Giles 1998).

It is the author's opinion that for Australian producers, and perhaps others where labour costs are high, the Australian method of steam distilling foliage which has been mechanically harvested directly into a mobile still is well-established and proven in terms of simplicity and cost effectiveness: labour costs are kept to a minimum, fuel costs are almost nil as the air-dried spent foliage provides all the energy required to raise steam, and quantities of cold water are usually readily available. Low production costs are essential since virtually all commercial eucalyptus oils have much lower unit values than essential oils from other plant families. Outside Australia where, for various reasons, mechanised harvesting may not be possible or appropriate, there is still scope for greater efficiency in terms of improved oil yields, lower production costs, etc., using conventional methods of distillation. Significant improvements can be gained by adopting many of the 'good manufacturing practices' advocated by Denny (Chapter 6). Attempts to develop novel methods of extraction, even if technically feasible, may ultimately prove not to be viable commercially and thus wasteful as far as effort expended and financial cost are concerned.

*Small-scale laboratory extraction methods*

Long before any commercial operation can be contemplated, and in order to provide necessary data on oil yields and composition, plant material must be analysed in the laboratory. But the analytical procedure itself (as well as the sampling) must be sound if the conclusions drawn are to be valid. What methods are available and do alternatives to traditional ones which have been used in recent years offer any advantages?

Perusal of the recent literature shows that the traditional method of essential oil extraction by steam entrainment, that is, hydrodistillation with or without cohobation and, less frequently, true steam distillation, is no longer the sole method used. Other methods, often claimed to be milder owing to the use of lower extraction temperatures and the absence of acidic still waters (and therefore less likely to cause heat or pH-induced artifact formation such as the hydrolysis of esters, skeletal rearrangements, etc.), include solvent extraction and supercritical fluid extraction, vacuum distillation, and even the extraction of the oil directly from the leaf by means of capillaries inserted into the oil glands. The choice of extraction method will depend on the number of plant samples to be extracted, the resources available and the intended use of the results thus obtained.

## Traditional methods

Hydrodistillation and steam distillation, being technologically the simplest, and also the cheapest, are quite appropriate for experiments aimed at the eventual commercial utilisation of the oil. The primary constituents of the major commercial eucalyptus oils are 1,8-cineole, citronellal, piperitone and α-phellandrene; methyl cinnamate is the main component of a minor oil. None of these compounds are significantly affected by hydrodistillation and, since all commercial oils containing these substances are very cheap, large-scale solvent extraction or vacuum distillation techniques must be ruled out on grounds of cost. However, distillation times utilised by different groups of workers have varied widely, ranging from 2 h (Dellacassa *et al.* 1990) to 30 h (Bignell *et al.* 1995). Times are usually between 3 and 24 h but in the overwhelming majority of cases little or no explanation is given in the published literature for the use of particular distillation times and one must assume that the cut-off values represent apparent cessation of oil accumulation. Furthermore, in apparatus such as the Clevenger, Likens-Nickerson and modified Dean-Stark, the condensed aqueous distillate passes continually through the column of collected essential oil, thereby disturbing it and often carrying oil globules back into the still; accurate measurement of the volume of oil collected is therefore difficult. An oil receiver developed by McKern and Smith-White (1948), later modified by Hughes (1970) in order to permit the simultaneous collection of heavier-than-water oils or oil constituents, eliminates most of the problems associated with the earlier types of apparatus. In particular, the end point of the distillation is accurately determined from a distillation curve constructed by plotting the volume of essential oil vs time. This method was used by Lassak (1990, 1992) who showed that hydrodistillation of most of the *Eucalyptus* species tested required quite long distillation times for complete oil extraction; three examples are shown in Figure 18.1.

Complete essential oil extraction is necessary if the real quantitative composition of the oil, as present in the foliage, is sought. It is well known that the composition of the oil varies with distillation time (e.g. Koedam *et al.* 1979) and any practising essential oil chemist will have observed that essential oils change in colour and viscosity as the distillation progresses. In eudesmol-rich eucalypts, for example, the oils tend to become progressively more viscous – and often solidify in the receiver – towards the end of the distillation. Table 18.1 illustrates, in the case of two *Eucalyptus* species, how different distillation times can affect the overall composition of the oil.

## Solvent extraction methods

Solvent extraction is a technique relatively rarely used for the recovery of eucalyptus oil. Weston (1984) showed that, in contrast to conventional water distillation, the dichloromethane-extracted

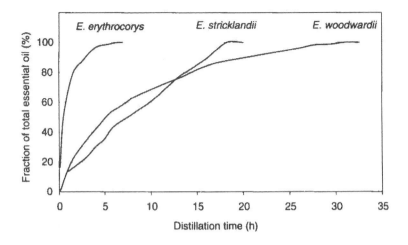

*Figure 18.1* Distillation curves for three species of *Eucalyptus*.

*Table 18.1* Influence of distillation time on the composition of eucalyptus oil (in terms of sesquiterpenoids)

| Species | Distillation time (h) | Sesquiterpenoids | Reference |
|---|---|---|---|
| E. globulus | 2 | Traces | Dellacassa et al. (1990) |
|  | 6 | Significant amounts | Li and Madden (1995) |
|  | 8 | Significant amounts | Boland et al. (1991) |
| E. nitens | 3 | None | Franich (1986) |
|  | 8 | Significant amounts | Boland et al. (1991) |

oil of *E. delegatensis* contained little more than traces of α-phellandrene. He suggested that the α-phellandrene normally produced was an artifact resulting from the dehydration of *trans*-piperitol. It would be interesting to check this by applying the same procedure to other hydrodistilled eucalyptus oils rich in α-phellandrene such as those obtained from the non-cineole forms of *E. dives* and *E. radiata* subsp. *radiata*.

Although ethanolic extraction of eucalyptus oils has no commercial significance, Ammon *et al.* (1985) have suggested its use for the rapid and accurate determination of terpenes, particularly 1,8-cineole, in the foliage of eucalyptus. The method has been applied to several *Eucalyptus* species by Brooker *et al.* (1988), Doran and Brophy (1990) and Grayling and Brooker (1996) and appears to give quite satisfactory results. However, the claim that it is a rapid method is perhaps somewhat exaggerated in view of the very long extraction times required for complete extraction of the oils (at least two weeks in the cases cited). Its suitability for analysing *Eucalyptus* species such as *E. cloeziana* (α-pinene chemotype), which have leaf oils composed almost entirely of monoterpenoid hydrocarbons (Boland *et al.* 1991), also needs to be tested to see whether extraction is quantitative – monoterpenoid hydrocarbons are not very soluble in ethanol containing even small amounts of water.

Supercritical fluid extraction (SFE) is a relatively recent extension of the 'normal' solvent extraction technique. Supercritical $CO_2$ behaves like a lipophilic solvent but by varying the temperature and pressure selectivity can be achieved. The low extraction temperatures normally used, as well as the absence of an acidic aqueous phase, reduce the likelihood of artifact formation from any chemically labile plant constituents which may be present, thus yielding a more 'nature-like' extract. However, as Della Porta *et al.* (1999) have pointed out, the optimum extraction conditions must be chosen in order to avoid the co-extraction of undesirable high molecular weight compounds. Whilst SFE methods are suitable for the production of some natural flavours and fragrances they have not been applied to any significant extent to the extraction of eucalyptus oils. One of the probable reasons for this is the relatively high cost of SFE, which can be justified in the case of high-priced flavour and fragrance materials but not with the cheaper eucalyptus oils. Apart from a brief mention in connection with the extraction of an unnamed species of *Eucalyptus* (Hawthorne *et al.* 1989) the only substantial investigations have been those of Milner *et al.* (1997) and Della Porta *et al.* (1999). Milner *et al.* (1997) found that the qualitative and quantitative compositions of the sesquiterpenoid fractions present in supercritical $CO_2$-extracted leaf oils of *E. varia* subsp. *varia* and *E. sparsa* were similar to those of leaf oils obtained by vacuum distillation, but quite different from hydrodistilled oils. The main differences with the latter were the high levels of bicyclogermacrene and low levels of certain sesquiterpenoid alcohols such as α- and β-eudesmol, globulol and viridiflorol in the SFE-extracted oils, the reverse of what is found in the hydrodistilled oils. Della Porta *et al.* (1999) extracted *E. globulus* leaves and obtained an oil richer in 1,8-cineole than one produced using conventional hydrodistillation.

*Vacuum distillation*

Another interesting development is the extraction of eucalyptus leaf oils by means of vacuum distillation. This rather novel procedure, developed by Inman *et al.* (1991), involves the extraction of the oil from plant material which has been frozen with liquid nitrogen and reduced to a fine powder; it is then subjected to a vacuum and the volatiles condensed on a gold-plated copper rod maintained at around $-75°C$. This method has been used by Bignell *et al.* (e.g. 1997 and other references of theirs cited therein) to analyse a large number of *Eucalyptus* species. The chemical compositions of the oils obtained in this manner showed some significant differences with respect to hydrodistilled oils (Bignell *et al.* 1995, 1996) and led the authors and Milner *et al.* (1997) to suggest that vacuum-distilled oils reflect more closely the composition of the oils present in the living leaf tissue. In particular, they found that bicyclogermacrene levels were much higher in vacuum-distilled oils, while hydrodistilled oils contained more sesquiterpenoid alcohols (as in SFE-extracted oils). They proposed that the compositional differences observed in hydrodistilled oils might be due to hydrogen ion initiated reactions of bicyclogermacrene, as suggested by Tressl *et al.* (1983) in the case of hops (*Humulus lupulus*).

Despite the claims that vacuum distillation produces oils more akin to the original '*in vivo*' oils, some of the results presented by Bignell *et al.* (1995, 1996) appear to contradict this. The chemical changes undergone by bicyclogermacrene during hydrodistillation appear to be beyond dispute but the ratio of $C_{15}:C_{10}$ compounds (i.e. total sesquiterpenoids vs total monoterpenoids) in the vacuum-distilled oils and in hydrodistilled oils of the same samples of foliage should be approximately the same. In fact, in the case of *E. sparsa* and *E. rudis*, they are very dissimilar (Table 18.2). One other issue needs to be addressed: the possible loss of certain volatile constituents such as isovaleraldehyde (3-methylbutanal). Despite the fact that this compound occurs in many *Eucalyptus* species, and is an undesirable constituent of several commercial eucalyptus oils such as *E. globulus* and *E. smithii* (see Appendix 4), it has not been reported in vacuum-distilled oils of the type produced by Bignell and her co-workers.

Table 18.2 Comparison of vacuum-distilled and hydrodistilled oils obtained by Bignell and co-workers in terms of monoterpenoid ($C_{10}$) and sesquiterpenoid ($C_{15}$) content

| Species | Vacuum-distilled oil | | | Hydrodistilled oil | | |
|---|---|---|---|---|---|---|
| | Total $C_{10}$ (%) | Total $C_{15}$ (%) | $C_{15}/C_{10}$ ratio | Total $C_{10}$ (%) | Total $C_{15}$ (%) | $C_{15}/C_{10}$ ratio |
| E. sparsa[a] | 23.1 | 67.2 | 2.9 | 33.1 | 59.2 | 1.8 |
| E. rudis[b] | 8.2 | 83.2 | 10.1 | 19.3 | 66.2 | 3.4 |

a Bignell et al. 1995.
b Bignell et al. 1996.

*Other methods*

Hellyer (1963) was able to remove the essential oil from the oil glands of individual *E. dives* (piperitone chemotype) leaves by inserting a glass capillary directly into the oil gland under a microscope using a micromanipulator. Capillary GC showed that the gas chromatograms of the capillary-isolated oil and of the hydrodistilled oil from the same batch of leaves were almost identical, both qualitatively and quantitatively. Of all the extraction techniques available, this is probably the one most likely to yield artifact-free oils. Using the same technique, Malingré et al. (1969) successfully isolated the essential oil of *Mentha aquatica*. Milner et al. (1997) mention unpublished work on the manually removed oils from *E. torquata*, *E. sparsa* and *E. woodwardii* and their similarity to vacuum-distilled oils (e.g. Bignell et al. 1995). Unfortunately, the method is very slow and the glands of many *Eucalyptus* species are very small and difficult to handle.

It is clear that the trend is towards milder methods of essential oil extraction since such oils should represent more accurately the oils originally present in the leaves. (Note, however, that the question as to whether the market would accept an oil that has a different composition – and therefore, inevitably, different aroma characteristics – to the traditional one, albeit that it is closer to the intrinsic oil, would remain to be answered). However, there are two aspects to be considered, a quantitative one – has all the oil been not only extracted but quantitatively recovered? – and a qualitative one – how does the chemical composition of the extracted oil compare with that of the oil contained in the oil glands? The method of extraction should also be relevant to the goals of the investigation. None of the methods referred to above are likely to satisfy all these requirements.

It can be argued that cohobative hydrodistillation is the simplest and most suitable method for the laboratory extraction of eucalyptus oils, particularly if directed towards the evaluation of their eventual commercial potential, and if it is standardised then oil yield values and oil compositional data obtained by different groups of workers should be directly comparable. Such standardisation could be achieved by adopting the Hughes-modified McKern/Smith-White type of oil receiver, together with the distillation curve technique for monitoring the progress of the oil extraction. Using typical sample sizes of 300–1000 g of plant material, the method provides quite accurate information on both oil yields and quality (chemical composition). The fact that some chemical changes can occur (due to a combination of elevated temperature and weakly acidic conditions inside the distillation vessel) does not detract from the usefulness of this method *if the quality and characteristics of the oil are what the industry demands*. The most commonly encountered changes relate to the decomposition of terpenic esters (Pickett et al. 1975), the transformation of sabinene and sabinene hydrates to terpinen-4-ol (Southwell and Stiff 1990 and references therein), the Cope rearrangement of hedycaryol to elemol (Jones and Sutherland

1968), of germacrene C to δ-elemene (Morikawa and Hirose 1969) and of bicyclogermacrene to bicycloelemene, as well as its isomerisation to derivatives of aromadendrene and alloaromadendrene such as globulol, ledol, etc. (Nishimura *et al.* 1969). Hydrodistillation offers further advantages: oils can be isolated, and their yields determined, without recourse to expensive instrumentation, which may not be readily accessible in small field laboratories in some developing countries; and it allows the early elimination of samples which do not show promise as far as oil yields are concerned.

As a final word of caution, the isolation of an artifact-free leaf oil does not guarantee that its analysis will provide an artifact-free composition. If gas chromatography is used for analysis, as is commonly the case, the high injector temperatures employed are likely to give rise, even if only partially, to certain types of reactions (such as Cope rearrangements and polymerisations) which will distort the genuine composition of the oil. The combination of a mild extraction method coupled with a mild analytical technique, perhaps HPLC, would be ideal. Unfortunately, HPLC is not as yet either sufficiently sensitive or sufficiently discriminatory to be able to separate complex mixtures of terpenoids.

## Analysis of essential oils

### *Identification of essential oil constituents*

With the development of instrumental techniques it has been possible to dispense with the need to identify oil constituents by isolation and measurement of their physical constants and by the properties of their crystalline derivatives. The use of capillary gas chromatography (GC) and spectral techniques such as mass spectrometry (MS), especially when combined (GC-MS), has been described elsewhere in this volume (Chapter 5) and has enabled the routine examination of essential oils to be achieved more simply, rapidly and reliably than ever before.

However, a few words of caution are necessary. Many of the published eucalyptus oil analyses have been carried out using just one type of GC column, often a polar one such as BP 20 or equivalent. Such columns do not separate α-pinene from α-thujene (or only very incompletely) and this probably explains why the latter compound has been only rarely reported in the literature. In the author's experience α-thujene is frequently found in eucalyptus oils, albeit in very small amounts. The use of a second, apolar, column usually solves this problem and may bring other examples of multiple peaks to light. It is good practice, always, to use two columns of different polarity in any essential oil work. This has long been recognised by the International Standardization Organization (ISO), which includes both types of chromatogram in all its standards.

Monoterpenoid esters can sometimes prove troublesome as their mass spectra do not always exhibit a molecular ion. The monoterpenoid ketone cryptone ($C_9H_{14}O$, mol. wt. 138) is a comparatively unusual eucalyptus oil constituent and has been wrongly reported as menthyl acetate in the leaf oils of *E. benthamii* var. *benthamii*, *E. dives* (piperitone form), *E. elata*, *E. polybractea*, *E. propinqua* and *E.* sp. aff. *propinqua* (Boland *et al.* 1991), partly because its molecular ion at $z/e$ 138 was thought to be derived from the loss of acetic acid from the ephemeral molecular ion of menthyl acetate, $C_{12}H_{22}O_2$, mol. wt. 198 (J. Brophy pers. comm.). In most cases, any uncertainties can be avoided by co-injection of the test sample with the authentic compound. It is the author's firm belief that this check with authentic compounds is essential as an adjunct to GC-MS generated data, especially when the constituent in question is not commonly found in oils of *Eucalyptus*.

From a chemotaxonomic, as well as an economic, point of view it is disappointing that so many eucalyptus oil constituents, particularly sesquiterpenoids, remain unidentified. It is to be hoped that this will be addressed by future workers since minor constituents can sometimes exhibit interesting, and potentially valuable, bioactive properties. Two such examples are the steam-volatile allelopathic substances, *cis*- and *trans*-($\pm$)-*p*-menthane-3,8-diol, present in *E. citriodora* leaves (Nishimura *et al*. 1982).

The ready availability of chiral stationary phases permits the determination of the enantiomeric composition of essential oils. This technique has been rapidly gaining in importance but, for some reason, has not been applied to any significant extent to eucalyptus oils, perhaps because the main constituent of most commercial oils is the optically inactive 1,8-cineole. Chiral analysis has important commercial, as well as scientific, applications. It can be used to establish the genuineness of particular essential oils as the enantiomeric ratio of a given unsymmetrical terpenoid appears to be species-specific. In such cases, chiral analysis can help in revealing adulteration with other essential oils or with isolates which are either synthetic or derived from another plant species. It can also be useful in toxicological studies of essential oils since enantiomers of some compounds exhibit different toxicities: *l*-pulegone, for example, is less toxic than *d*-pulegone (Opdyke 1978). Finally, chiral analysis would be of considerable use in the elucidation of essential oil biosynthesis.

## Chemical variation within a species

The concept of chemical variants (also called chemical forms or chemotypes) has been discussed elsewhere (e.g. Chapters 4 and 5) but a few points are worth reiterating. The concept has been based on the existence of discontinuities in the chemical composition of certain eucalyptus oils. However, the development of more sensitive analytical methods, such as capillary GC, has indicated that in some of those cases previously considered to be examples of qualitative variation, the variation was only quantitative, as in *E. dives* (Hellyer *et al*. 1969). Before proposing the existence of chemical variants one also needs to be satisfied that the sampling procedures used have been sound. The age of the leaves sampled and the sample size (in terms of number of trees) are just two important considerations. Immature leaf tips of the citronellal form of *E. citriodora*, for example, contain less citronellal than young mature leaves (Penfold *et al*. 1953 and references therein). Sampling a large number of individual trees – collected at random from different locations and spanning, as far as possible, the whole of the known natural distribution of the species – is extremely important since in some *Eucalyptus* species a particular oil constituent can vary *continuously* between very wide limits. Southwell (1973), for example, found that 1,8-cineole varied from 1 to 70 per cent in the leaf oil of *E. punctata*. Characterisation of a species by sampling an inadequate number of trees can introduce a *false discontinuity* and result in the postulation of chemical variants where none actually exist.

One more word of warning: hybridisation is not uncommon amongst eucalypts and the unintended inclusion of unrecognised hybrid trees in an essential oil investigation will distort the results and render them useless for chemotaxonomic purposes. In later years, it became common practice at the Museum of Applied Arts and Sciences in Sydney to grow progeny from the seed of wild growing trees and to investigate their leaf essential oils only if *all* trees thus raised were morphologically identical, indicative of an absence of hybridisation.

## Search for new commercial uses of *Eucalyptus* extracts

### Essential oils

Most published work has focused on the chemistry of leaf oils, and this is likely to continue to be the case. All *Eucalyptus* leaf oils are very complex mixtures of terpenoids (mainly mono- and sesquiterpenoids), certain shikimate-derived compounds such as β-triketones and some phenolics, and, to a much smaller extent, some relatively rare non-terpenoid aliphatic compounds.

*Flavour, fragrance and medicinal use*

In order to be of any commercial interest, yields of leaf oil should be at least 1.5–2 per cent based on the mass of the fresh plant material (leaves and terminal branchlets) and there should preferably be one main oil constituent – one which has an established market and is present in reasonably high concentrations (normally not less than 45–70 per cent of the oil, depending on the particular constituent). In the great majority of *Eucalyptus* species these two requirements are not satisfied simultaneously: either the oil yield is too low or the chemical composition is unsuitable, or both. In the case of cineole-rich oils, recent literature indicates that only a few species with possible commercial potential have been identified (Table 18.3), additional to those which already are, or have once been, used commercially for oil production (Lassak 1988, Coppen and Hone 1992).

A non-cineole eucalypt with some commercial potential is *E. nova-anglica*, which contains a leaf oil rich in nerolidol (77 per cent), a sesquiterpene rarely found in eucalypts (Boland *et al.* 1991, Brophy *et al.* 1992); oil yield is approximately 2.7 per cent (fresh weight basis). Another non-cineole/non-citronellal species whose commercial potential *has* recently been realised is *E. olida*, previously referred to as *Eucalyptus* sp. nov. aff. *campanulata* (Curtis *et al.* 1990); its leaf oil, produced in Australia, contains E-methyl cinnamate to the extent of about 95 per cent.

The greater part of the present world eucalyptus oil production of about 2500–3000 t/year is of the 1,8-cineole type. Most 1,8-cineole type oils are used medicinally whilst smaller quantities are consumed by the flavour and fragrance industries. Existing eucalyptus plantations established throughout many warmer regions of the world – not only for essential oil production but, more importantly, as a renewable source of firewood – are more than adequate in satisfying present

*Table 18.3* Cineole-rich *Eucalyptus* species with commercial potential

| Species | Oil yield[a] (%) | 1,8-Cineole content of oil (%) | Reference |
|---|---|---|---|
| E. badjensis | 2.8 | 70 | Boland *et al.* (1991) |
| E. bakeri | 1.8–3.0 | 85–96 | Brophy and Boland (1989) |
| E. brownii | 1.9–2.3 | 80–89 | Boland *et al.* (1991) |
| E. globulus subsp. maidenii | 2.2–2.8 | 46–70 | Boland *et al.* (1991) |
| E. globulus subsp. pseudoglobulus[b] | 4.0–5.6[c] | 47–69 | Boland *et al.* (1991) |
| E. kochii subsp. kochii | 2.3–5.5[c] | 83–94 | Gardner and Watson (1947/48), Brooker *et al.* (1988) |
| E. kochii subsp. plenissima | 2.2–8.6[c] | 83–95 | Gardner and Watson (1947/48), Brooker *et al.* (1988) |
| E. loxophleba | 2.4 | 67 | Boland *et al.* (1991) |
| E. nichollii | 1.7–2.3 | 84 | Boland *et al.* (1991) |
| E. pumila | 3.8–5.8[c] | 80–90 | Boland *et al.* (1991) |
| E. salubris | 1.4–2.3 | 78 | Brophy and Lassak (1991) |
| E. saxatilis | 3.5–5.1[c] | 64–79 | Boland *et al.* (1991) |
| E. sturgissiana | 1.1–2.5 | 80–90 | Boland *et al.* (1991) |
| E. subcrenulata | 2.5–4.6[c] | 61–66 | Boland *et al.* (1991) |

a Fresh weight basis except where indicated.
b Oil from this species has probably been produced and marketed under the name *E. globulus*.
c Dry weight basis.

world demand for cineole-rich oils. In addition to these stocks, however, huge numbers of eucalypts are being planted to satisfy an ever-growing demand for paper pulp, artificial board and timber for general construction purposes. They are also being planted to combat soil salination due to rising ground water levels in deforested arid or semi-arid areas (for example, in Western Australia). Where these eucalypts are suitable oil-bearing species the potential exists, therefore, for a manyfold increase in eucalyptus oil production, although the supply could not be matched by an increase in demand, at least in terms of existing markets.

## Fuel use

Concerns raised in the 1970s about decreasing world supplies of fossil fuels led to the examination of plants as a potential source of hydrocarbons which could be used to extend, or even substitute for, fuels normally derived from crude petroleum. Eucalyptus oils of the 1,8-cineole type ranked fairly high on the list of possible substitute fuels (Nishimura and Calvin 1979, Nishimura *et al.* 1980, Ammon *et al.* 1985, Beckmann 1988, Duffy 1993, etc.). It is quite surprising that this kind of thinking persisted into the 1990s in view of earlier reports which questioned the ability to produce liquid fuels from plants in quantities sufficient to meet projected demand (Gartside 1977, Coffey and Halloran 1979).

*E. polybractea* leaf oil yields of up to 10 t/km$^2$ annually have been achieved in Australia on selected plantations and under ideal conditions. In South Africa, where the higher oil-yielding *E. radiata* subsp. *radiata* ('*E. australiana*' of commerce) is being grown, annual oil yields of about 20 t/km$^2$ are believed to be achievable (Davis 1998). Small-scale experiments conducted in South Africa with the same species have indicated that still higher annual oil yields, up to 45 t/km$^2$ might be possible (Donald 1980). World crude oil consumption in 1990 was estimated to be close to 3,000 million tonnes. Even if one sought to replace just a fraction, say 10 per cent, of that figure with eucalyptus oil, perhaps as an additive to petrol/ethanol fuel mixtures, and using an annual yield of 20 t/km$^2$, the quantities of oil needed would require an arable and well watered growing area of millions of square kilometres. It is reasonable to assume that a significant proportion of the eucalyptus oil so produced would finish up being used to power the heavy machinery required for soil preparation, planting of seedlings, harvesting of the foliage, etc. The production cost of eucalyptus oil from even the most highly mechanised commercial plantations is still of the order of US$2–3/kg of oil produced. This is at least ten times the production cost of petrol (gasoline). Even if the recently introduced 'carbon credits' were taken into account it is still most unlikely that eucalyptus oil could ever become a viable substitute for petroleum-based fuels on a global scale.

## Use as a solvent/cleaning agent

A more promising future for eucalyptus oils appears to lie in their use as industrial solvents. Eucalyptus oils, particularly those containing piperitone or α-phellandrene as their major components, have been known for a long time to be excellent solvents for paints, resins, varnish, grease, gums, tar, etc. (Penfold and Morrison 1950). In the past, cineole or cineole-rich eucalyptus oils, with or without addition of phellandrene, have also been suggested for use as paint removers and clothes cleaners. At the time, their relatively high price prevented them from being used on a large scale but they have enjoyed something of a rebirth as a popular wool wash in Australia in recent years.

Recent concerns about the harmful effects of chlorinated solvents such as chloroform, trichlorethylene and 1,1,1-trichloroethane on human health and the environment have led to a search for chlorine-free solvents of comparable solvent power. 1,8-Cineole and cineole-rich eucalyptus oils appear to fulfil most of the requirements sought of such solvents: biodegradability, low toxicity, ability to dissolve grease, no adverse effect on the Earth's ozone layer and low chemical reactivity, as well as a pleasant odour. Industrial trials conducted in Western Australia have been successful and cineole is now routinely used as a degreasing agent (Barton 1989, Barton and Knight 1997). An added advantage is the ease with which it can be recovered from the degreasing solutions, either by direct distillation or by steam distillation. Its main disadvantage is its low flash point (about 48°C for 1,8-cineole and 44°C for eucalyptus oil containing 80–85 per cent cineole) and relatively high boiling point (176–177°C).

*Chemical and microbiological transformations as a route to commercially useful products*

A very large body of literature exists on the chemical reactions and transformations which terpenoids undergo. However, only very few such reactions have resulted in commercially useful products which can be obtained in sufficient yield and purity to warrant large-scale industrial production. Citronellal, a major constituent of the leaf oil of *E. citriodora*, as well as of several plant species belonging to other genera, is the starting material for the manufacture of hydroxy-dihydrocitronellal (commercially known as 'hydroxycitronellal'), an important perfumery compound. *l*-Piperitone from *E. dives* leaf oil was used until relatively recently as the starting material for the synthesis of *l*-menthol used in flavours. Derivatives of 1,8-cineole have not yet found commercial application.

Both piperitone and α-phellandrene, major constituents of *E. dives* (piperitone form), *E. radiata* subsp. *radiata* (the chemical form once called '*E. phellandra*') and several other species, contain reactive functional groups which lend themselves to chemical manipulation. Piperitone can be reduced to a mixture of *cis*- and *trans*-piperitols, which yield fragrant esters on esterification with low molecular weight aliphatic carboxylic acids. α-Phellandrene can be converted to Diels–Alder adducts which may have commercial potential as plasticisers. It might also be possible to react α-phellandrene with vinylic co-monomers to produce polymers with novel and desirable properties.

Transformations of chemical structures can also be achieved by the action of microorganisms. Early examples were the use of bottom yeast to convert citronellal to citronellol in 59 per cent yield (Mayer and Neuberg 1915) and the conversion of citronellal to a mixture of citronellol and citronellic acid using *Acetobacter xylinum* (Molinari 1929). More recently, racemic piperitone has been converted to a mixture of ($\pm$)-*trans*-6-hydroxy-*p*-menth-1-en-3-one (minor product) and ($\pm$)-7-hydroxy-*p*-menth-1-en-3-one (major product) by organisms such as *Fusarium*, *Proactinomyces roseus* and *Aspergillus niger* (Lassak *et al*. 1973). A more interesting development is the conversion of 1,8-cineole to a mixture of 2α- and 2β-hydroxy-1,8-cineole and 2-keto-1,8-cineole by *Pseudomonas flava* (MacRae *et al*. 1979, Carman *et al*. 1986).

A whole series of metabolites of 1,8-cineole is produced by the gut flora of the brushtail possum, *Trichosurus vulpecula*, a herbivorous Australian marsupial which feeds partly on eucalyptus leaves (see also Chapters 5 and 15). Compounds identified so far (Figure 18.2) include: 9-hydroxy-1,8-cineole and 1,8-cineol-9-oic acid (Flynn and Southwell 1979, Carman and Klika 1992); 7-hydroxy-1,8-cineole (Bull *et al*. 1993); 2α-hydroxy-1,8-cineole, 2α,9-dihydroxy-1,8-cineole and 2α,10-dihydroxy-1,8-cineole (Carman *et al*. 1994); 2α,4-dihydroxy-1,8-cineole (Carman and Rayner 1994, 1996); 3α-hydroxy-1,8-cineole (Carman *et al*. 1994, Carman and Rayner 1996); and 2α,7-dihydroxy-1,8-cineole and 7,9-dihydroxy-1,8-cineole (Carman and Garner 1996). To the author's knowledge, none of these piperitone and 1,8-cineole metabolites

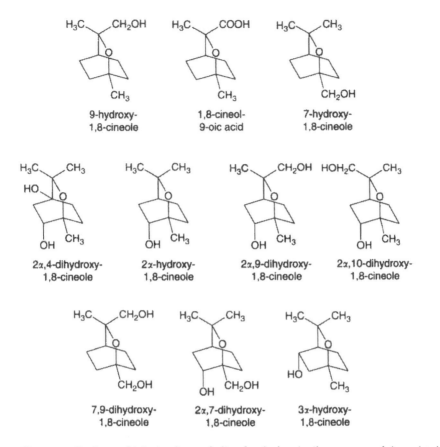

*Figure 18.2* Products of 1,8-cineole metabolism by the brushtail possum worth investigating for potential applications in medicine or perfumery.

have yet been investigated for potential applications in medicine or perfumery but it would be of some interest to do so.

To summarise, a large effort has been put into the investigation of the chemical composition of eucalyptus leaf oils in an attempt to find new uses or sources of them, beyond the traditional ones. Despite the scientific value of this work, very little of commercial benefit has resulted from it, the most notable exceptions being the discovery of the *E*-methyl cinnamate-rich leaf oil of *E. olida* and its subsequent commercialisation on a modest scale (as a flavour additive) and the use of cineole-rich eucalyptus oils as industrial degreasing agents. In the latter case, only time will tell whether this use can be sustained in view of the present high price of the oils (relative to alternative solvents).

In order to justify and recoup the substantial financial investment required for the establishment of eucalyptus plantations, both current and planned, new and substantial uses for the oils will need to be found. Whilst it is not easy, and perhaps unwise, to make predictions for the future, it is the author's view that the microbiological transformation of readily available eucalyptus oil constituents may hold the key to the achievement of these goals, particularly if the metabolites show pharmacological activity (antibacterial, fungistatic, etc.).

*Non-volatile constituents*

*Rutin*

Apart from small amounts of kino – resinous exudates of the bark and wood of many *Eucalyptus* species – the only other non-volatile constituent of eucalyptus foliage which has been produced commercially is rutin (3,5,7,3',4'-pentahydroxyflavone-3-rutinoside). Rutin is used medicinally and its aglycone, quercetin, is a powerful antioxidant. The traditional Australian source of this compound was *E. macrorhyncha* although, elsewhere, it was obtained from the flower buds of *Sophora japonica* and buckwheat. Australian rutin production fluctuated between 13.5 and 18 t per year and lasted from the mid-1950s to approximately the end of the 1960s. There are other eucalypts which contain substantial amounts of rutin in their foliage, notably *E. youmanii* and *E. delegatensis* (Humphreys 1964). Whilst there is virtually no rutin production from eucalypts today, increased European interest in this medicinal compound, particularly in France and Germany, might rekindle the industry in Australia.

In *E. macrorhyncha* the greatest concentration of rutin is found in the youngest leaves, up to 25 per cent on a dry weight basis. However, the overall yields of rutin obtained in commercial operations were about 10 per cent of dry foliage (Humphreys 1964). Experiments conducted at the Castle Hill plantation of the Sydney Museum of Applied Arts and Sciences suggested that *E. youmanii* might be a better species as overall rutin yields could be as high as 20 per cent (dry weight) during the summer months (Small 1979). *E. delegatensis* foliage contains less rutin, 6–7 per cent on average. However, this species is also an important timber tree and utilisation of waste by-product foliage in any new venture would have cost and conservation advantages over, say, *E. youmanii*.

*Other bioactive non-volatile compounds*

Although a great deal of work remains to be done, research carried out since the 1970s has indicated that the potential for eucalyptus to be used to combat some of the most serious and widespread illnesses and diseases may rest with some of its non-volatile constituents, rather than with the more familiar volatile oils. Ghisalberti (1996) and Singh *et al.* (1999) have drawn attention to this in their reviews of the non-volatile constituents of eucalypts. In tropical areas where water-borne trematode infections such as schistosomiasis and fascioliasis are found, it may not even be necessary to isolate the active constituents in order to take advantage of their properties. Hammond *et al.* (1994) have proposed that the natural fall of leaves from *Eucalyptus* species with molluscicidal properties planted in appropriate places could effect self-delivery.

Egawa *et al.* (1974) noticed the almost total absence of microorganisms (fungi, actinomycetes, bacteria) on the surface of, and inside, the leaves of *E. gunnii*. They ascribed this to the presence of three antifungal compounds: gallic acid and two incompletely characterised compounds, a monoformyl-trihydroxybenzene and a diformyl-trihydroxybenzene previously isolated from this species by Kobayashi *et al.* (1972). Several other *Eucalyptus* species were subsequently found to contain antifungal compounds, though no attempts were made at their identification (Egawa *et al.* 1977).

Since then, a substantial amount of work has been carried out on the isolation and structure elucidation of novel bioactive metabolites present in the foliage of *Eucalyptus* species. The stilbenes polydatin (piceid), polydatin 6''-O-(E)-p-coumarate and rhaponticin from *E. rubida* all exhibit attachment-inhibiting activity towards the blue mussel, *Mytilus edulis galloprovincialis* (Yamashita *et al.* 1989, Etoh *et al.* 1990), and indicate the potential of such compounds as antifouling agents for protecting ships hulls and other structures in a marine environment. However, the great majority of the other bioactive metabolites so far identified in eucalypts have proved to be acylphloroglucinols (Ghisalberti 1996, Singh et al. 1999).

Three phloroglucinol-derived peroxides, G-1, G-2 and G-3, isolated from *E. grandis*, possess root growth-inhibiting properties (Crow et al. 1971) whilst grandinol, a derivative of formylphloroglucinol occurring in *E. grandis* and *E. perriniana*, possesses both root growth-inhibiting and bactericidal properties (Crow et al. 1977, Nakayama et al. 1990). Two new acylphenones from *E. robusta*, 2,6-dihydroxy-3,5-dimethyl-4-methoxy-butyrophenone and its 2'-methylbutanoyl homologue, exhibit phosphodiesterase-inhibiting activity (Cheng and Snyder 1991). Grandinal, a phloroglucinol dimer from *E. grandis* (Singh et al. 1997), and two flavonoid glycosides from *E. resinifera*, resinosides A and B (Hyodo et al. 1992), exhibit blue mussel attachment-inhibition activity.

Perhaps the most exciting advances in this field relate to the discovery of four groups of novel, highly bioactive acylphloroglucinol derivatives peculiar to the genus *Eucalyptus*: the robustadials, sideroxylonals, macrocarpals and euglobals. These have been described earlier in this volume (Chapter 12) and embrace a number of activities, any or all of which could, in the longer term, be exploited commercially. Robustadials A and B, with a combined phloroglucinol–monoterpene structure, are antimalarial constituents of *E. robusta* leaves (Xu et al. 1984, Cheng and Snyder 1988). The sideroxylonals, dimers of a diformylphloroglucinol moiety isolated from *E. sideroxylon* and *E. grandis*, are antibacterial but also show strong attachment-inhibiting activity towards the blue mussel; sideroxylonal A is one of the most powerful antifouling agents known (Satoh et al. 1992, Singh et al. 1996).

The macrocarpals, present in the leaves of *E. macrocarpa*, *E. globulus* and *E. amplifolia*, have a combined isopentyldiformylphloroglucinol–sesquiterpene structure and exhibit a range of biological activities. They are all strongly antibacterial against Gram-positive bacteria, including ones which cause dental diseases, and some have recently been found to be active against Gram-negative bacteria (Murata et al. 1990, Yamakoshi et al. 1992, Osawa et al. 1995, 1996). Some (macrocarpals A, B, D and G) have been found to inhibit aldose reductase and may find application in the treatment of complications resulting from diabetes (Murata et al. 1992). Others (from *E. globulus*) are inhibitors of HIV-RTase and have potential in the fight against AIDS (Nishizawa et al. 1992). The antibacterial and antiviral properties of macrocarpals appear to be due more to the diformylphloroglucinol moiety than to the different sesquiterpenoid groups attached to it, although Osawa et al. (1996) have suggested that the antibacterial potency may be regulated by the structure of the sesquiterpene. Macrocarpals A, B, E, am-1 and H possess blue mussel attachment-inhibiting activity, although it is less than that of sideroxylonal A (Singh and Etoh 1995).

The euglobals are the largest, and structurally most variable, of these four groups of *Eucalyptus* metabolites. They are formyl phloroglucinol adducts with either a mono- or a sesquiterpenoid moiety attached. Euglobals have been isolated from *E. globulus* (e.g. Kozuka et al. 1982), *E. grandis* (e.g. Takasaki et al. 1994a, Umehara et al. 1998), *E. amplifolia*, *E. blakelyi* (Takasaki et al. 1994b), *E. incrassata* (e.g. Takasaki et al. 1997) and *E. tereticornis* (Kokumai et al. 1991) but have been detected in many more species (Chapter 12). Many euglobals show strong inhibition of Epstein-Barr virus activation induced by 12-O-tetradecanoylphorbol-13-acetate, a tumour-promoting substance, and this gives rise to hope that they might, in the future, play some role in cancer prevention (Takasaki et al. 1995).

All the species from which these bioactive metabolites have so far been isolated belong to the *Eucalyptus* subgenus *Symphyomyrtus* (Pryor and Johnson 1971). In view of the potential of these compounds for disease prevention or treatment, it would be profitable to widen the search to other species of the subgenus *Symphyomyrtus*, both within and outside those Sections/Series in

Table 18.4 Distribution of bioactive non-volatile compounds[a] in *Eucalyptus*

| Classification of Pryor & Johnson[b] | | Classification of Chippendale | Species | Compound type |
|---|---|---|---|---|
| Section | Series | Series (no.) | | |
| Transversaria | Salignae | Transversae (no. 42) | E. grandis | Grandinol, grandinal, sideroxylonals, euglobals, peroxides |
| | | | E. robusta | Phenones, robustadials |
| | | Annulares (no. 43) | E. resinifera | Resinosides A, B |
| Bisectaria | Macrocarpae | Curviptera (no. 62) | E. macrocarpa | Macrocarpals |
| Dumaria | Incrassatae | Tetrapterae (no. 68) | E. incrassata | Euglobals |
| Exsertaria | Tereticornes | Exsertae (no. 72) | E. amplifolia | Macrocarpals, euglobals |
| | | | E. blakelyi | Euglobals |
| | | | E. tereticornis | Euglobals |
| Maidenaria | Viminales | Viminales (no. 77) | E. globulus | Macrocarpals, euglobals |
| | | | E. gunnii | ? |
| | | | E. perriniana | Grandinol |
| | | | E. rubida | Stilbenes |
| | | | E. viminalis | Macrocarpals |
| Adnataria | Melliodorae | Melliodorae (no. 91) | E. sideroxylon | Sideroxylonals |

a Excluding rutin. See also Figure 12.5, for details of other Sections and Series within *Symphyomyrtus* in which euglobals have been detected.
b All within sub-genus *Symphyomyrtus*.

which these compounds have already been encountered. Table 18.4 shows the distribution of the different types of bioactive non-volatile compounds referred to above. It includes, for reference, the alternative classification of *Eucalyptus* employed by Chippendale (1988), which does not recognise *Symphyomyrtus*.

A major drawback in the commercial exploitation of these bioactive metabolites from natural sources is that the majority of them occur in the buds and foliage of eucalypts in extremely small amounts (at most 0.1 per cent of fresh plant material, and normally in amounts one or two orders of magnitude less than this). Huge amounts of eucalyptus foliage would be required to prepare even small quantities of the desired compound. Handling of such enormous tonnages of plant material, disposal of waste, and extraction and purification of the final product would be prohibitively costly and make the whole operation uneconomic. Synthesis, which has already achieved some modest success in the laboratory, is probably the only solution to any future large-scale industrial production.

## Concluding remarks

The botanical classification of *Eucalyptus* has often been fraught with difficulties. Attempts at providing chemical assistance have, on the whole, been unconvincing although there have been a few exceptions. Brooker and Lassak (1981) showed that *E. ovata* and *E. brookeriana*, superficially very similar, could be distinguished by the chemical characteristics of their leaf oils. The seedling leaf oils of the Tasmanian and mainland Australian populations of *E. delegatensis* differ significantly from each other in their 4-phenylbutan-2-one content (Boland *et al*. 1982) and this fact contributed to the separation of the species into two subspecies, *delegatensis* and *tasmaniensis*

(Boland 1985). The leaf oils of *E. salubris* var. *salubris* and *E. salubris* var. *glauca* were sufficiently different from each other to suggest that they might be different species (Brophy and Lassak 1991). Independent of this work the variety *glauca* was separated from *E. salubris* and designated *E. ravida* (Johnson and Hill 1991). However, Carman's (1992) attempt at producing a computer program which would show that a unique compound could be assigned to each *Eucalyptus* species was of little assistance as the great majority of *Eucalyptus* species contain the same compounds. Furthermore, it did not take into account the existence of chemical variants within the same species.

A novel and fundamental approach is being taken at the present time with the whole family Myrtaceae, including, of course, *Eucalyptus*, by far the largest of its genera: DNA sequencing studies are being undertaken (Ladiges *et al.* 1995). The results of these studies will no doubt point to new relationships within the genus and indicate new directions for future chemical research and the search for useful bioactive compounds.

## References

Abbott, P.S. (1989) Commercial eucalyptus oil production. In *Proc. Eucalyptus Oil Production Seminar*, Gnowangerup, Western Australia, February 1989, Misc. Publ. 9/89, Agdex 184/500, Dep. Agric., Western Australia, pp. 29–48, 52.

Ammon, D.G., Barton, A.F.M., Clarke, D.A. and Tjandra, J. (1985) Rapid and accurate determination of terpenes in the leaves of *Eucalyptus* species. *Analyst*, 110, 921–924.

Barton, A. (1989) Commercial possibilities for high-cineole Western Australian eucalyptus oil. In *Proc. Eucalyptus Oil Production Seminar*, Gnowangerup, Western Australia, February 1989, Misc. Publ. 9/89, Agdex 184/500, Dep. Agric., Western Australia, pp. 21–28.

Barton, A.F.M. and Knight, A.R. (1997) High-cineole eucalyptus oils in degreasing applications. *Chem. Aust.*, 64(1), 4–6.

Beckmann, R. (1988) Oil from eucalypts. *Ecos*, 56, 7–8, 10.

Bignell, C.M., Dunlop, P.J. and Brophy, J.J. (1997) Volatile leaf oils of some south-western and southern Australian species of the genus *Eucalyptus* (Series I). Part XVIII. A – Subgenus Monocalyptus. B – Subgenus Symphyomyrtus, (i) Section Guilfoyleanae, (ii) Section Bisectaria, Series Accedentes, Series Occidentales, Series Levispermae, Series Loxophlebae, Series Macrocarpae, Series Orbifoliae, Series Calycogonae, (iii) Section Dumaria, Series Incrassatae and Series Ovulares. *Flavour Fragr. J.*, 12, 423–432.

Bignell, C.M., Dunlop, P.J., Brophy, J.J. and Jackson, J.F. (1995) Volatile leaf oils of some south-western and southern Australian species of the genus *Eucalyptus*. Part VI – Subgenus Symphyomyrtus, Section Adnataria. *Flavour Fragr. J.*, 10, 359–364.

Bignell, C.M., Dunlop, P.J., Brophy, J.J. and Jackson, J.F. (1996) Volatile leaf oils of some south-western and southern Australian species of the genus *Eucalyptus*. Part VII – Subgenus Symphyomyrtus, Section Exsertaria. *Flavour Fragr. J.*, 11, 35–41.

Boland, D.J. (1985) Taxonomic revision of *Eucalyptus delegatensis* R.T. Baker (Myrtaceae). *Aust. For. Res.*, 15, 173–181.

Boland, D.J., Brophy, J.J., Flynn, T.M. and Lassak, E.V. (1982) Volatile leaf oils of *Eucalyptus delegatensis* seedlings. *Phytochemistry*, 21, 2467–2469.

Boland, D.J., Brophy, J.J. and House, A.P.N. (eds) (1991) *Eucalyptus Leaf Oils. Use. Chemistry, Distillation and Marketing*, ACIAR CSIRO, Inkata Press, Melbourne.

Brooker, M.I.H., Barton, A.F.M., Rockel, B.A. and Tjandra, J. (1988) The cineole content and taxonomy of *Eucalyptus kochii* Maiden & Blakely and *E. plenissima* (Gardner) Brooker, with an appendix establishing these two taxa as subspecies. *Aust. J. Bot.*, 36, 119–129.

Brooker, M.I.H. and Lassak, E.V. (1981) The volatile leaf oils of *Eucalyptus ovata* Labill. and *E. brookerana* A.M. Gray (Myrtaceae). *Aust. J. Bot.*, 29, 605–615.

Brophy, J.J. and Boland, D.J. (1989) Leaf essential oil of *Eucalyptus bakeri*. In D.J. Boland (ed.), *Trees for the Tropics*, ACIAR, Canberra, pp. 205–207.

Brophy, J.J. and Lassak, E.V. (1991) Steam volatile leaf oils of some Western Australian *Eucalyptus* species. *Flavour Fragr. J.*, 6, 265–269.

Brophy, J.J., Lassak, E.V. and Boland, D.J. (1992) The leaf essential oils of *Eucalyptus nova-anglica* Deane & Maiden. *J. Essent. Oil Res.*, 4, 29–32.

Bull, S.D., Carman, R.M., Carrick, F.N. and Klika, K.D. (1993) 7-Hydroxy-1,8-cineole and 7-cineolic acid – 2 new possum urinary metabolites. *Aust. J. Chem.*, 46, 441–447.

Carman, R.M. (1992) The expression of distinctive secondary metabolites by different *Eucalyptus* spp. *Aust. J. Chem.*, 45, 1919–1921.

Carman, R.M. and Garner, A.C. (1996) 7,9-Dihydroxy-1,8-cineole and $2\alpha,7$-dihydroxy-1,8-cineole: two new possum urinary metabolites. *Aust. J. Chem.*, 49, 741–749.

Carman, R.M., Garner, A.C. and Klika, K.D. (1994) 2,9-Dihydroxy and 2,10-dihydroxy-1,8-cineole. Two new possum urinary metabolites. *Aust. J. Chem.*, 47, 1509–1521.

Carman, R.M. and Klika, K.D. (1992) Partially racemic compounds as brushtail possum urinary metabolites. *Aust. J. Chem.*, 45, 651–657.

Carman, R.M., MacRae, I.C. and Perkins, M.V. (1986) The oxidation of 1,8-cineole by *Pseudomonas flava*. *Aust. J. Chem.*, 39, 1739–1746.

Carman, R.M. and Rayner, A.C. (1994) $2\alpha,4$-dihydroxy-1,8-cineole. A new possum urinary metabolite. *Aust. J. Chem.*, 47, 2087–2097.

Carman, R.M. and Rayner, A.C. (1996) Chiral $2\alpha,4$-dihydroxy-1,8-cineole as a possum urinary metabolite. *Aust. J. Chem.*, 49, 1–6.

Cheng, Q. and Snyder, J.K. (1988) Revised structures of robustadials A and B from *Eucalyptus robusta*. *J. Org. Chem.*, 53, 4562–4567.

Cheng, Q. and Snyder, J.K. (1991) Two new phloroglucinol derivatives with phosphodiesterase inhibitory activity from the leaves of *Eucalyptus robusta*. *Z. Naturforsch., B: Chem. Sci.*, 46, 1275–1277.

Chippendale, G.M. (1988) *Myrtaceae – Eucalyptus, Angophora*. In A.S. George (ed.), *Flora of Australia*, Vol. 19, Australian Government Publishing Service, Canberra.

Coffey, S.C. and Halloran, G.M. (1979) Higher plants as possible sources of petroleum substitutes. *Search*, 10, 423–428.

Coppen, J.J.W. and Hone, G.A. (1992) *Eucalyptus Oils: A Review of Production and Markets*, NRI Bulletin 56, Natural Resources Institute, Chatham, UK.

Crow, W.D., Nicholls, W. and Sterns, M. (1971) Root inhibitors in *Eucalyptus grandis*: naturally occurring derivatives of the 2,3-dioxabicyclo[4.4.0]decane system. *Tetrahedron Letts.*, 1353–1356.

Crow, W.D., Osawa, T., Paton, D.M. and Willing, R.R. (1977) Structure of grandinol: a novel root inhibitor from *Eucalyptus grandis*. *Tetrahedron Letts.*, 1073–1074.

Curtis, A., Southwell, I.A. and Stiff, I.A. (1990) *Eucalyptus*, a new source of E-methyl cinnamate. *J. Essent. Oil Res.*, 2, 105–110.

Davis, G.R. (1998) An "essential" oil. *The Australian Standard* (Sep.), 26–27.

Della Porta, G., Porcedda, S., Marongiu, B. and Reverchon, E. (1999) Isolation of eucalyptus oil by supercritical fluid extraction. *Flavour Fragr. J.*, 14, 214–218.

Dellacassa, E., Menendéz, P., Moyna, P. and Soler, E. (1990) Chemical composition of *Eucalyptus* essential oils grown in Uruguay. *Flavour Fragr. J.*, 5, 91–95.

Donald, D.G.M. (1980) The production of cineole from *Eucalyptus*: a preliminary report. *S. Afr. For. J.*, (114), 64–67.

Doran, J.C. and Brophy, J.J. (1990) Tropical red gums – a source of 1,8-cineole-rich eucalyptus oil. *New Forests*, 4, 157–178.

Duffy, J. (1993) Mallees offered as farm saviour. *The West Australian*, (24 July), 13.

Egawa, H., Furusawa, I., Akai, S., Kobata, K., Fumoto, J., Kobayashi, A. and Koshimizu, K. (1974) *Proc. Kansai Plant Prot. Soc., Tokyo*, 16, 42.

Egawa, H., Tsutsui, O., Tatsuyama, K. and Hatta, T. (1977) Antifungal substances found in leaves of *Eucalyptus* species. *Experientia*, 33, 889–890.

Etoh, H., Yamashita, N., Sakata, K., Ina, H. and Ina, K. (1990) Stilbene glucosides isolated from *Eucalyptus rubida* as repellents against the blue mussel *Mytilus edulis*. *Agric. Biol. Chem.*, 54, 2443–2444.

Flynn, T.M. and Southwell, I.A. (1979) 1,3-Dimethyl-2-oxabicyclo [2.2.2]-octane-3-methanol and 1,3-dimethyl-2-oxabicyclo [2.2.2]-octane-3-carboxylic acid, urinary metabolites of 1,8-cineole. *Aust. J. Chem.*, 32, 2093–2095.

Franich, R.A. (1986) Essential oil composition of juvenile leaves from coppiced *Eucalyptus nitens*. *Phytochemistry*, 25, 245–246.

Gardner, C.A. and Watson, E.M. (1947/48) The Western Australian varieties of *Eucalyptus oleosa* and their essential oils. *J. Proc. Roy. Soc. W. Aust.*, 34, 73–86.

Gartside, G. (1977) The energy costs of prospective fuels. *Search*, 8, 105–111.

Ghisalberti, E.L. (1996) Bioactive acylphloroglucinol derivatives from *Eucalyptus* species. *Phytochemistry*, 41, 7–22.

Giles, R. (1998) *The Dinkum Oil* [Oil Mallee Association of W.A. Inc. Newsletter], (March), 11.

Grayling, P.M. and Brooker, M.I.H. (1996) Evidence for the identity of the hybrid *Eucalyptus* 'brachyphylla' (Myrtaceae) from morphology and essential oil composition. *Aust. J. Bot.*, 44, 1–13.

Hammond, J.A., Fielding, D. and Nuru, H. (1994) *Eucalyptus*: a sustainable self-delivery molluscicide? *Vet. Res. Comm.*, 18, 359–365.

Hawthorne, S.B., Miller, D.J. and Krieger, M.S. (1989) Coupled SFE-GC: a rapid and simple technique for extracting, identifying and quantitating organic analytes from solids and sorbent resins. *J. Chromatogr. Sci.*, 27, 347–354.

Hellyer, R.O. (1963) Unpublished records, Museum of Applied Arts & Sciences, Sydney, Australia.

Hellyer, R.O., Lassak, E.V., McKern, H.H.G. and Willis, J.L. (1969) Chemical variation within *Eucalyptus dives*. *Phytochemistry*, 8, 1513–1514.

Hill, K.D. and Johnson, L.A.S. (1995) Systematic studies in the eucalypts. 7. A revision of the bloodwoods, genus *Corymbia* (Myrtaceae). *Telopea*, 6, 185–504.

Hughes, A. (1970) A modified receiver for heavier than water essential oils. *Chem. Ind.*, 1536.

Humphreys, F.R. (1964) The occurrence and industrial production of rutin in southeastern Australia. *Econ. Bot.*, 18, 195–253.

Hyodo, S., Etoh, H., Yamashita, N., Sakata, K. and Ina, K. (1992) Structure of resinosides from *Eucalyptus resinifera* as repellents against the blue mussel *Mytilus edulis*. *Biosci. Biotech. Biochem.*, 56, 138.

Inman, R.B., Dunlop, P. and Jackson, J.F. (1991) Oils and waxes of eucalypts. Vacuum distillation method for essential oils. In H.F. Linskens and J.F. Jackson (eds), *Modern Methods of Plant Analysis, New Series*, Vol. 12, Springer-Verlag, Berlin, pp. 195–203.

Johnson, L.A.S. and Hill, K.D. (1991) Systematic studies in the eucalypts – 2. A revision of the gimlets and related species: *Eucalyptus* extracodical series Salubres and Annulatae (Myrtaceae). *Telopea*, 4, 201–222.

Jones, R.V.H. and Sutherland, M.D. (1968) Hedycaryol, the precursor of elemol. *J. Chem. Soc., Chem. Comm.*, 1229–1230.

Kobayashi, A., Koshimizu, K., Mitsui, T., Egawa, H. and Fukami, H. (1972) Paper presented at Meeting of Japanese Agric. Soc., Tokyo, 1972.

Koedam, A., Scheffer, J.J.C. and Svendsen, A.B. (1979) Comparison of isolation procedures for essential oils. II. Ajowan, caraway, coriander and cumin. *Z. Lebensm. Untersuch. Forsch.*, 168, 106–111.

Kokumai, M., Konoshima, T., Kozuka, M., Haruna, M. and Ito, K. (1991) Euglobal-T1, a new euglobal from *Eucalyptus tereticornis*. *J. Nat. Prod.*, 54, 1082–1086.

Kozuka, M., Sawada, T., Mizuta, E., Kasahara, F., Amano, T., Komiya, T. and Goto, M. (1982) The granulation-inhibiting principles from *Eucalyptus globulus* Labill. III. The structures of euglobal-III, -IVb and -VII. *Chem. Pharm. Bull.*, 30, 1964–1973.

Ladiges, P.Y., Udovicic, F. and Drinnan, A.N. (1995) Eucalypt phylogeny – molecules and morphology. *Aust. Syst. Bot.*, 8, 483–497.

Lassak, E.V. (1988) The Australian eucalyptus oil industry, past and present. *Chem. Aust.*, 55, 396–398.

Lassak, E.V. (1990) Water distillation extraction curves and their significance. Paper presented at 2nd Malaysian Internat. Conf. on Essential Oils and Aroma Chemicals, Kuala Lumpur, December 1990.

Lassak, E.V. (1992) The usefulness of water distillation curves in essential oil extraction practice. Paper presented at 7th Asian Symp. on Medicinal Plants and Spices, Manila, February 1992.

Lassak, E.V., Pinhey, J.T., Ralph, B.J., Sheldon, T. and Simes, J.J.H. (1973) Extractives of fungi. V. Microbial transformation products of piperitone. *Aust. J. Chem.*, 26, 845–854.

Li, H. and Madden, J.L. (1995) Analysis of leaf oils from a *Eucalyptus* species trial. *Biochem. Syst. Ecol.*, 23, 167–177.

MacRae, I.C., Alberts,V., Carman, R.M. and Shaw, I.M. (1979) Products of 1,8-cineole oxidation by a pseudomonad. *Aust. J. Chem.*, 32, 917–922.

Malingré, T.M., Smith, D. and Batterman, S. (1969) Isolation and gas chromatographic analysis of the essential oils from individual labiatous glandular hairs. *Pharm. Weekblad*, 104, 429–435.

Mayer, P. and Neuberg, C. (1915) Photochemical reductions. XII. Transformation of citronellal to citronellol. *Biochem. Z.*, 71, 174.

McKern, H.H.G. and Smith-White, S. (1948) An improved distillation apparatus for the determination of the essential oil content of plant material. *Aust. Chem. Inst., J. Proc.*, 15, 276–278.

Milner, C.P., Trengove, R.D., Bignell, C.M. and Dunlop, P.J. (1997) Supercritical $CO_2$ extraction of the essential oils of *Eucalyptus*: a comparison with other methods. In H.F. Linskens and J.F. Jackson (eds), *Modern Methods of Plant Analysis, New Series*, Vol. 19, Springer-Verlag, Berlin, pp. 141–158.

Molinari, E. (1929) Biochemical dismutation and studies on acetic acid fermentation. *Biochem. Z.*, 216, 187–215.

Morikawa, K. and Hirose, Y. (1969) Germacrene-C, precursor of δ-elemene. *Tetrahedron Letts.*, 1799–1801.

Murata, M., Yamakoshi, Y., Homma, S., Aida, K., Hori, K. and Ohashi, Y. (1990) Macrocarpal A, a novel antibacterial compound from *Eucalyptus macrocarpa*. *Agric. Biol. Chem.*, 54, 3221–3226.

Murata, M., Yamakoshi, Y., Homma, S., Arai, K. and Nakamura, Y. (1992) Macrocarpals, antibacterial compounds from *Eucalyptus*, inhibit aldose reductase. *Biosci. Biotech. Biochem.*, 56, 2062–2063.

Nakayama, R., Murata, M., Homma, S. and Aida, K. (1990) Antimicrobial compounds from *Eucalyptus perriniana*. *Agric. Biol. Chem.*, 54, 231–232.

Nishimura, H. and Calvin, M. (1979) Essential oil of *Eucalyptus globulus* in California. *J. Agric. Food Chem.*, 27, 432–435.

Nishimura, H., Kaku, K., Nakamura, T., Fukuzawa, Y. and Mizutani, J. (1982) Allelopathic substances, (±)-*p*-menthane-3,8-diols isolated from *Eucalyptus citriodora* Hook. *Agric. Biol. Chem.*, 46, 319–320.

Nishimura, H., Paton, D.M. and Calvin, M. (1980) *Eucalyptus radiata* oil as a renewable biomass. *Agric. Biol. Chem.*, 44, 2495–2496.

Nishimura, K., Shinoda, N. and Hirose, Y. (1969) A new sesquiterpene, bicyclogermacrene. *Tetrahedron Letts.*, 3097–3100.

Nishizawa, M., Emura, M., Kan, Y., Yamada, H., Ogawa, K. and Hamanaka, N. (1992) Macrocarpals; HIV-RTase inhibitors of *Eucalyptus globulus*. *Tetrahedron Letts.*, 2983–2986.

Opdyke, D.L.J. (1978) Monographs on fragrance raw materials. *d*-Pulegone. *Food Cosmet. Toxicol.*, 16, 867–868.

Osawa, K., Yasuda, H., Morita, H., Takeya, K. and Itokawa, H. (1995) Eucalyptone from *Eucalyptus globulus*. *Phytochemistry*, 40, 183–184.

Osawa, K., Yasuda, H., Morita, H., Takeya, K. and Itokawa, H. (1996) Macrocarpals H, I and J from the leaves of *Eucalyptus globulus*. *J. Nat. Prod.*, 59, 823–827.

Penfold, A.R., McKern, H.H.G. and Willis, J.L. (1953) Studies in the physiological forms of the Myrtaceae. Part VI. An examination of the progeny obtained from *Eucalyptus citriodora* Hook. var. A. *Researches on Essential Oils of the Australian Flora*, Museum of Applied Arts and Sciences, Sydney, 3, 15–20.

Penfold, A.R. and Morrison, F.R. (1950) *Uses of Commercial Eucalyptus Oils*, Museum of Applied Arts and Sciences Bulletin No. 17 (3rd edn), Government Printer, Sydney.

Pickett, J.A., Coates, J. and Sharpe, F.R. (1975) Distortion of essential oil composition during isolation by steam distillation. *Chem. Ind.*, 571–572.

Provatoroff, N. (1972) Some details of the distillation of spice oils. In *Proc. Conf. on Spices*, London, April 1972, Tropical Products Institute, London (now Natural Resources Institute, Chatham), pp. 173–181.

Pryor, L.D. and Johnson, L.A.S. (1971) *A Classification of the Eucalypts*, Australian National University Press, Canberra.

Satoh, H., Etoh, H., Watanabe, N., Kawagishi, H., Arai, K. and Ina, K. (1992) Structures of sideroxylonals from *Eucalyptus sideroxylon*. *Chem. Letts.*, 1917–1920.

Singh, A.K., Khare, M. and Kumar, S. (1999) Non-volatile constituents of eucalypts: a review on chemistry and biological activities. *J. Med. Arom. Plant. Sci.*, 21, 375–407.

Singh, I.P. and Etoh, H. (1995) New macrocarpal-am-1 from *Eucalyptus amplifolia*. *Biosci. Biotech. Biochem.*, 59, 2330–2332.

Singh, I.P., Hayakawa, R., Etoh, H., Takasaki, M. and Konoshima, T. (1997) Grandinal, a new phloroglucinol dimer from *Eucalyptus grandis*. *Biosci. Biotech. Biochem.*, 61, 921–923.

Singh, I.P., Takahashi, K. and Etoh, H. (1996) Potent attachment-inhibiting and -promoting substances for the blue mussel, *Mytilus edulis galloprovincialis*, from two species of *Eucalyptus*. *Biosci. Biotech. Biochem.*, 60, 1522–1523.

Small, B.E.J. (1979) Unpublished records, Museum of Applied Arts & Sciences, Sydney, Australia.

Southwell, I.A. (1973) Variation in the leaf oil of *Eucalyptus punctata*. *Phytochemistry*, 12, 1341–1343.

Southwell, I.A. and Stiff, I.A. (1990) Differentiation between *Melaleuca alternifolia* and *M. linariifolia* by monoterpenoid comparison. *Phytochemistry*, 29, 3529–3533.

Takasaki, M., Konoshima, T., Kozuka, M., Haruna, M., Ito, K. and Shingu, T. (1994a) Structures of euglobals-G1, -G2, -G3, -G4 and -G5 from *Eucalyptus grandis*. *Chem. Pharm. Bull.*, 42, 2591–2597.

Takasaki, M., Konoshima, T., Kozuka, M., Haruna, M., Ito, K. and Yoshida, S. (1994b) Four euglobals from *Eucalyptus blakelyi*. *Chem. Pharm. Bull.*, 42, 2177–2179.

Takasaki, M., Konoshima, T., Kozuka, M. and Tokuda, H. (1995) Anti-tumour promoting activities of euglobals from *Eucalyptus* plants. *Biol. Pharm. Bull.*, 18, 435–438.

Takasaki, M., Konoshima, T., Kozuka, M., Haruna, M. and Ito, K. (1997) Euglobals, -In-2 and -In-3, new euglobals from *Eucalyptus incrassata*. *Natural Medicines*, 51, 486–490.

Tressl, R., Engel, K.H., Kossa, M. and Köppler, H. (1983) Characterization of tricyclic sesquiterpenes in hop (*Humulus lupulus*). *J. Agric. Food Chem.*, 31, 892–897.

Umehara, K., Singh, I.P., Etoh, H., Takasaki, M. and Konoshima, T. (1998) Five phloroglucinol-monoterpene adducts from *Eucalyptus grandis*. *Phytochemistry*, 49, 1699–1704.

Weston, R.J. (1984) Composition of essential oil from leaves of *Eucalyptus delegatensis*. *Phytochemistry*, 23, 1943–1945.

Xu, R.S., Snyder, J.K. and Nakanishi, K. (1984) Robustadials A and B from *Eucalyptus robusta*. *J. Am. Chem. Soc.*, 106, 734–736.

Yamakoshi, Y., Murata, M., Shimizu, A. and Homma, S. (1992) Isolation and characterization of macrocarpals B-G, antibacterial compounds from *Eucalyptus macrocarpa*. *Biosci. Biotech. Biochem.*, 56, 1570–1576.

Yamashita, N., Etoh, H., Sakata, K., Ina, H. and Ina, K. (1989) New acylated rhaponticin isolated from *Eucalyptus rubida* as a repellent against the blue mussel *Mytilus edulis*. *Agric. Biol. Chem.*, 53, 2827–2829.

# Appendices

# Appendix 1. Sources of eucalyptus seed

The Australian Tree Seed Centre in Canberra is a national and international tree seed bank with a focus on Australian trees and shrubs. The Centre supplies authenticated seed for research and tree improvement programmes, together with advice on species and provenance selection and other information. A leaflet giving details of Australian private seed suppliers is available from the Centre or the information may be obtained from their Internet website:

Australian Tree Seed Centre
CSIRO Forestry and Forest Products
PO Box E4008
Kingston
ACT 2604
Australia
Tel.: +61-2-6281 8211
Fax: +61-2-6281 8266
E-mail <atsc@ffp.csiro.au>
Website: www.ffp.csiro.au/tigr/atscmain/index.htm

The website also has a searchable seed list which gives details of species and provenances available.

Many of the seed suppliers listed specialise in eucalypts and some offer bulk or individual tree collections from specified provenances. Catalogues and price lists are available from suppliers on request. The Centre points out that there is no tree seed certification scheme in Australia and recommend that

> the purchaser ascertain details of the seed origins before entering into any purchase agreement. The minimum details which might be expected are the precise locality of collection including latitude and longitude coordinates, altitude of the collecting site, year of harvesting and number of trees sampled.

Details are also provided of three State Government seed sources (Queensland, Tasmania and Western Australia).

# Appendix 2. Estimates of eucalypt plantations worldwide

The estimates for the areas of eucalyptus plantations given below (Table A2.1) are taken from two sources, Davidson (1995) and FAO (unpubl.). Much of Davidson's data is drawn from FAO sources, which has also been published (Pandey 1995) and provides data up to 1990. The second source is data made available by FAO in early 1999 which gives the position in 1995; it is based on further, unpublished reports by Pandey for FAO.

The intention, here, is simply to give some quantitative indication of the extent to which eucalypts have been planted worldwide. In theory, the amount of 'waste' biomass that is available for purposes other than that for which the trees are planted, particularly 'waste' leaf, is vast and constitutes a massive reservoir of potential for utilisation. In practice, of course, many of the species planted are not oil-bearing ones or ones which might be sources of other commercially useful medicinal or aromatic compounds. Even where the main species planted has a proven 'track record' in terms of, say, oil production – such as the large plantings of *E. globulus* in Chile, Portugal and Spain – the size of the industry is not commensurate with the size of the resource. The economics of production are not favourable when compared with prevailing prices. Nevertheless, costs and prices do not remain static and conditions in the future may make production of by-products more attractive.

An indication in Table A2.1 of the main species grown in each country would have been useful but there is insufficient reliable, up-to-date information available to do this. For some countries the predominant species can be stated with confidence or they have been detailed elsewhere in this book, for example, *E. globulus* in Chile, Portugal and Spain, *E. camaldulensis* in Nepal, Thailand and Vietnam, '*Eucalyptus* hybrid' in India, *E. grandis* in South Africa, etc. But for many countries this is not the case and it could be misleading to 'second guess' or use out-of-date information.

The estimates given below should be treated with caution. Davidson (1995) emphasises that the 'majority of figures are approximate and reliability varies'. The later FAO data (unpubl.) contain 'reduction factors' to be applied to the primary data shown in the third column of Table A2.1; the more reliable the data are, the nearer the factor is to 1.0. For Central and South America the factor for the region is approximately 0.88, for Africa it is about 0.84 and for Asia it is approximately 0.64. Worldwide, the total area of eucalypts could be as much as 25 per cent less than the figures in column three of the table indicate, caused in large part by the uncertainty in the area for India. Estimates given in Chapters 8 and 11 for the areas of eucalypts in China and India, respectively, are significantly higher than those given here, illustrating, again, the difficulty of acquiring reliable data. For India, and some other countries, eucalypts may not be planted in blocks that enable the trees to be included in the definition of 'plantations' for counting purposes. A later FAO source (Brown 2000) gives figures for the eucalyptus plantation

Table A2.1 Estimates of eucalypt plantations worldwide

| Region/Country | Area of eucalypts ('000 ha)[a] | |
|---|---|---|
| | 1990 | 1995 |
| Total | 13,414 | 14,619 |
| Of which: | | |
| *Africa* | | |
| Algeria | 30 | 39 |
| Angola | 135 | 128 |
| Benin | — | 6 |
| Burkina Faso | 7 | 14 |
| Burundi | 40 | 42 |
| Cameroon | 13 | 12 |
| Cape Verde | — | 2 |
| Central African Rep. | 2 | — |
| Chad | 1 | 2 |
| Comores | 1 | — |
| Congo | 35 | 48 |
| Congo, Dem. Rep.[b] | 20 | 12 |
| Ethiopia | 95 | 145 |
| Gabon | 2 | 3 |
| Ghana | 14 | — |
| Kenya | 17 | 17 |
| Lesotho | — | 3 |
| Libya | 26 | — |
| Madagascar | 130 | 151 |
| Malawi | 30 | 24 |
| Mali | 5 | 14 |
| Mauritius | 3 | 4 |
| Morocco | 200 | 187 |
| Mozambique | 14 | 14 |
| Namibia | — | 1 |
| Niger | 2 | 3 |
| Nigeria | 11 | 13 |
| Rwanda | 60 | 124 |
| Senegal | 40 | 52 |
| Sierra Leone | — | 2 |
| South Africa | 538 | 557 |
| Sudan | 23 | 76 |
| Swaziland | — | 32 |
| Tanzania | 25 | 4 |
| Togo | 10 | 19 |
| Tunisia | 42 | 35 |
| Uganda | 10 | 3 |
| Zambia | 26 | 7 |
| Zimbabwe | 30 | 10 |
| Total Africa | 1637 | 1805 |
| *Asia* | | |
| Bangladesh | — | 31 |
| China, PR | 670 | 663 |
| India | 4800 | 5063 |
| Indonesia | 80 | 99 |
| Laos | — | 3 |
| Malaysia | 8 | 9 |
| Myanmar | 25 | 49 |
| Nepal | 5 | 11 |

Table A2.1 (Continued)

| Region/Country | Area of eucalypts ('000 ha)[a] | |
|---|---|---|
| | 1990 | 1995 |
| Pakistan | 29 | 210 |
| Philippines | 10 | 177 |
| Sri Lanka | 45 | 35 |
| Taiwan | 4 | — |
| Thailand | 62 | 130 |
| Vietnam | 245 | 792 |
| *Total Asia* | 5983 | 7272 |
| *Pacific* | | |
| Australia | 75 | 160 |
| New Zealand | 22 | — |
| Papua New Guinea | 10 | 15 |
| Solomon Islands | — | 3 |
| *Total Pacific* | 107 | 178 |
| *North America* | | |
| USA | 110 | — |
| *Total N. America* | 110 | — |
| *Caribbean* | | |
| Cuba | 35 | 47 |
| Haiti | 2 | — |
| *Total Caribbean* | 37 | 47 |
| *Central America* | | |
| Costa Rica | 10 | 9 |
| El Salvador | 2 | 1 |
| Guatemala | 6 | 10 |
| Honduras | 1 | — |
| Mexico | 38 | — |
| Nicaragua | 6 | 6 |
| *Total C. America* | 63 | 26 |
| *South America* | | |
| Argentina | 236 | 249 |
| Bolivia | — | 15 |
| Brazil | 3617 | 3123 |
| Chile | 180 | 245 |
| Colombia | 31 | 60 |
| Ecuador | 44 | 66 |
| Paraguay | 8 | 7 |
| Peru | 211 | 314 |
| Uruguay | 160 | 278 |
| Venezuela | 70 | 71 |
| *Total S. America* | 4557 | 4428 |
| *Mediterranean* | | |
| Israel | 10 | — |
| Italy | 40 | — |
| Portugal | 500 | 403 |
| Spain | 350 | 460 |
| Turkey | 20 | — |
| *Total Mediterranean* | 920 | 863 |

a Source figures rounded to nearest 1000 ha.
b Democratic Republic of Congo, formerly Zaire.
— Indicates no estimate given in source data.

resources of India and Brazil of 3.1 and 2.7 million ha, respectively, that is, lower than the figures in Table A2.1, although 1995 is still used as the basis for the estimates.

Despite these reservations and difficulties the data in Table A2.1 represent the best quantitative picture available of the world's eucalypt plantings.

## References

Brown, C. (2000) *The Global Outlook for Future Wood Supply from Forest Plantations*, FAO Global Forest Products Outlook Study Working Paper Series, GFPOS/WP/03, Food and Agriculture Organization of the United Nations, Rome.

Davidson, J. (1995) Ecological aspects of eucalypt plantations. In K. White, J. Ball and M. Kashio (eds), *Proc. Regional Expert Consult. on Eucalyptus*, Bangkok, October 1993, Vol. 1, Food and Agriculture Organization of the United Nations, Regional Office for Asia and the Pacific, Bangkok, pp. 35–72.

Pandey, D. (1995) *Forest Resources Assessment 1990: Tropical Forest Plantation Resources*, FAO Forestry Paper 128, Food and Agriculture Organization of the United Nations, Rome.

# Appendix 3. Advice to a prospective new producer of eucalyptus oil or other leaf extractive[1]

To anyone contemplating the production of eucalyptus oil – or any other leaf isolate or extractive – for the first time, there are a number of points which have been discussed or alluded to in the main body of the book which, in an industrial profile, bear repeating here. They apply whether the extractive is intended for the domestic market or for export.

Firstly, before undertaking a detailed technical and financial appraisal the intending producer needs to ascertain the market for the product or, in the case of eucalyptus oil, the type(s) of oil to be produced. This will require an up-to-date assessment of the domestic, regional and international markets. For the cineole-containing, medicinal type of oil – the biggest one in volume and value terms – the price of Chinese 80 per cent oil will be an important indicator of the sort of price levels with which the new producer will have to compete.

Someone contemplating production of eucalyptus oil or other leaf extractive in a developing country may be prompted to do so by the existence of a ready source of suitable 'waste' leaf, that is, leaf which is available from an operation in which eucalypts are being grown primarily for their wood. *Eucalyptus globulus*, grown in Portugal for pulp production, but whose leaves yield a medicinal type of oil, and *E. citriodora*, grown in Brazil for charcoal production, and whose leaves yield a perfumery oil, are examples of eucalypts which are utilised in this way. But it is vital to the success of this type of operation that the logistics of collection are adequately considered. The distillery or extraction facility should not be sited too distant from the source of leaf and the security and continuity of supply of leaf should be assured. Although the landowners where the trees are grown may be happy to be rid of the foliage and not impose any great price on its removal, it should not be regarded as a near zero-cost raw material for the distiller. Leaf collection and transportation costs are a major part of the overall costs of such an enterprise.

If intending producers do not wish to be dependent on others for 'waste' leaf then they may decide to grow their own, multipurpose trees from which 'waste' leaf can be obtained. Besides having control over the supply of leaf to the distillery or factory, such a situation means that they are not reliant on one product only for a source of income. *E. smithii* in southern Africa is an example of a dual-purpose eucalypt which has been grown simultaneously for timber and oil production.

Alternatively, the trees may be grown specifically for oil (or other leaf extractive) and the leaf repeatedly harvested under a short-rotation coppice system of management. Harvesting of the leaf on a 12–20 month cycle may provide a year-round supply of material for distillation or extraction. Intensive cultivation is necessary and a combination of high biomass (leaf) and high oil or extractive yield from the leaf is desirable. Correct species (and provenance) selection is, therefore, vitally important. Having regard to the species of eucalypt that may already be used

---

[1] Adapted from Coppen and Hone (1992) by permission of the Natural Resources Institute, University of Greenwich.

by others, information should be sought on their suitability for local conditions. Field trials should then be established to determine growth and product characteristics for a number of them, including different provenances, if possible. If selected or superior seed is available then this should be tested. It is only by such means that a reasonable idea of oil or extractive yield on a 'per hectare per year' basis can be obtained and, therefore, the likely returns on investment. *E. smithii* and *E. dives* in South Africa, *E. globulus*, *E. citriodora* and *E. staigeriana* in South America, and *E. polybractea* in Australia are examples of eucalypt species grown specifically for oil.

The advantages to be gained from giving adequate attention to selection of seed for planting cannot be over-emphasised. Over-hasty planting of a species or provenance which is unsuited to local conditions, or does not produce an oil or extractive of acceptable quality or yield, will result in failure.

The distillery or factory, too, is important and due regard should be given to its design, construction and operation if, and when, a decision is made by the intending producer to commence investment. In the case of oil production, hard-won gains in oil yield in the selection of planting stock may be lost by poorly designed equipment or poor distillation practice.

The possibility of generating extra revenue from the production process by making use of 'waste' by-products should not be ignored. Wood left over from harvesting leaf, as well as the spent leaf from the processing, are potential sources of such income.

Product quality, particularly composition in the case of an oil, should be determined by a public analyst or other competent organisation. For medicinal oils the crude oil is usually rectified before sale to the consumer and if this is not to be done by the new producer himself then it is to a rectifier (who may be another, larger producer) that he should turn first for an opinion on his oil. A crude medicinal oil should probably have a 1,8-cineole content of 65 per cent or higher for the rectifier to consider purchasing it. In the case of *E. citriodora* oil, it should contain at least 65 per cent citronellal.

Ennever (1967) has described the problems and pitfalls associated with marketing essential oils, either new ones, or existing ones being offered by new producers. The trend in commerce today is one of globalisation and mergers, with many of the largest flavour and fragrance companies seeking security of supply by establishing production bases in China and elsewhere. The plethora of middlemen which traditionally characterised the essential oils industry is less true than it once was and this may make some of Ennever's comments appear less relevant. However, some of the advice he gave then, which ends on an encouraging note, *is* still relevant today. He observed that

> The production of essential oils calls for the utmost patience and the closest possible supervision at all stages.... General merchant houses, banks, agents, brokers, dealers, public warehouses and public analysts have all, to a greater or lesser extent, become participants in the marketing and distribution of essential oils. In theory, all these may appear unnecessary, but, in well proven practice, all have a valuable function in the process of channelling the established oils from the point of production to the point of use.... Marketing, as well as producing, essential oils is likely to prove very trying for the new producer and much patience, understanding and confidence will be needed for both. Given these, the production of some essential oils can be profitable and there is abundant evidence to support this claim.

### References

Coppen, J.J.W. and Hone, G.A. (1992) *Eucalyptus Oils: A Review of Production and Markets*, NRI Bulletin 56, Natural Resources Institute, Chatham, UK.

Ennever, W.A. (1967) Marketing essential oils. *Trop. Sci.*, 9, 136–143.

# Appendix 4. Composition of some commercially distilled eucalyptus oils

The data below (Tables A4.1 and A4.2) are from Coppen (unpubl.). The samples were acquired during visits by the author to commercial eucalyptus distilleries in Portugal, Spain, Brazil, Australia, South Africa, Swaziland and Zimbabwe during 1990/91. All the samples are freshly produced crude oils from the primary distillation of leaf, that is, before any rectification has taken place.

Gas chromatographic analysis was performed using a 20 m × 0.25 mm fused silica capillary column coated with a Carbowax 20M-type stationary phase and temperature programming 75–225°C at 4°C/min.

*Table A4.1* Composition of some medicinal, cineole-rich eucalyptus oils (per cent relative abundance)

|  | E. globulus Portugal | E. globulus Spain | E. globulus Brazil | E. polybractea Australia | E. radiata Australia | E. radiata South Africa | E. smithii Swaziland | E. cinerea Zimbabwe |
|---|---|---|---|---|---|---|---|---|
| Isovaleraldehyde | 0.2 | 0.1 | 1.3 | 0.1 | — | — | 1.4 | 1.2 |
| α-Pinene | 14.5 | 20.1 | 11.2 | 1.9 | 4.2 | 3.4 | 8.9 | 9.8 |
| α-Fenchene | tr | tr | tr | — | tr | tr | tr | tr |
| Camphene | 0.1 | 0.1 | 0.1 | tr | tr | tr | 0.1 | 0.1 |
| β-Pinene | 0.4 | 0.5 | 0.5 | 0.4 | 0.8 | 0.8 | 0.4 | 0.2 |
| Sabinene | tr? | tr? | tr? | 0.3 | 0.9 | 1.0 | — | — |
| Myrcene | 0.2 | 0.3 | 0.4 | 0.2 | 1.5 | 1.5 | 0.5 | 0.3 |
| α-Phellandrene | 0.2 | 0.4 | 0.2 | 0.1 | 0.3 | 0.5 | 0.4 | 0.8 |
| α-Terpinene | tr | 0.1 | tr | tr | 0.3 | 0.4 | tr | 0.1 |
| Limonene | 1.4 | 1.0 | 1.7 | 0.5 | 2.0 | 6.5 | 7.2 | 6.9 |
| 1,8-Cineole | 66.8 | 59.4 | 73.6 | 89.7 | 70.8 | 65.2 | 72.2 | 62.5 |
| γ-Terpinene | 1.0 | 0.6 | 1.2 | 0.2 | 0.7 | 1.0 | 1.0 | 0.3 |
| *trans*-β-Ocimene | 0.1 | 0.1 | 0.1 | tr | — | — | tr | tr |
| *p*-Cymene | 2.3 | 1.2 | 1.7 | 1.6 | 0.4 | 0.7 | 1.9 | 0.8 |
| Terpinolene | 0.1 | 0.3 | 1.1 | 0.1 | 0.2 | 0.2 | 0.1 | 0.2 |
| *p*-Cymenene | 0.1 | 0.1 | 0.1 | — | — | — | tr? | tr? |
| Linalool | — | — | tr | tr? | 0.3? | 0.4? | 0.1 | 0.1 |
| Pinocarvone | 0.5 | 0.4 | 0.2 | tr | — | — | tr | tr |
| Terpinen-4-ol | 0.5 | 0.1 | 0.5 | 0.8 | 1.6 | 1.6 | 0.5 | 1.1 |
| Aromadendrene | 2.2 | 4.6 | 0.3 | 0.1? | tr | tr | 0.1 | 0.3 |
| Alloaromadendrene | 0.6 | 0.9 | 0.1 | tr? | 0.1 | tr | tr | 0.2 |
| *trans*-Pinocarveol | 1.3 | 0.9 | 1.0 | 0.2 | tr | 0.1 | 0.1 | 0.1 |
| δ-Terpineol | 0.1 | tr | 0.1 | 0.1 | 0.2 | 0.2 | 0.2 | 0.2 |
| α-Terpineol | 2.6 | 3.0 | 1.0 | 0.5 | 11.5 | 11.0 | 2.1 | 8.0 |
| α-Terpinyl acetate | — | — | — | — | — | — | — | 3.1 |
| Viridiflorol | 0.1 | 0.1 | tr | tr? | — | tr? | tr | 0.1? |
| Spathulenol | tr | tr | tr | tr? | tr? | tr? | tr | 0.1? |
| γ-Eudesmol | 0.1 | tr | — | tr? | — | tr? | 0.1 | tr? |

(Continued)

Table A4.1 (Continued)

|  | E. globulus Portugal | E. globulus Spain | E. globulus Brazil | E. polybractea Australia | E. radiata Australia | E. radiata South Africa | E. smithii Swaziland | E. cinerea Zimbabwe |
|---|---|---|---|---|---|---|---|---|
| Agarospirol | tr? | tr? | tr? | — | — | tr? | 0.1 | tr? |
| α-Eudesmol | 0.2 | tr | tr | tr? | — | tr? | 0.3 | tr? |
| β-Eudesmol | 0.3 | 0.1 | tr | tr? | — | tr? | 0.5 | tr? |
| Others | 4.1 | 5.6 | 3.6 | 3.2 | 4.2 | 5.5 | 1.8 | 3.5 |

tr = trace (<0.05 per cent).
? Indicates that peak's identity has not been confirmed by GC-MS.

Table A4.2 Composition of some eucalyptus oils used in perfumery or as sources of chemical isolates (per cent relative abundance)

|  | E. citriodora Brazil | E. staigeriana Brazil | E. dives South Africa |
|---|---|---|---|
| α-Pinene | 0.2 | 3.1 | 4.3 |
| β-Pinene | 0.6 | 1.2 | tr |
| Sabinene | 0.1 | 0.1 | 0.1 |
| Myrcene | 0.2 | 0.9 | 2.0 |
| α-Phellandrene | — | 2.3 | 30.6 |
| α-Terpinene | — | 0.2 | 1.8 |
| Limonene | 0.2 | 26.8 | 0.4 |
| β-Phellandrene | — | — | 3.1 |
| 1,8-Cineole | 0.4 | 3.3 | ? |
| cis-β-Ocimene | 0.1 | 0.2 | 0.1? |
| γ-Terpinene | 0.1 | 2.6 | 1.7 |
| trans-β-Ocimene | tr | 0.2 | — |
| p-Cymene | — | 0.8 | 2.9 |
| Terpinolene | 0.1 | 10.8 | 3.0 |
| 6-Methyl-5-hepten-2-one | — | 0.2 | — |
| 2,6-Dimethyl-5-heptenal | 0.1 | — | — |
| p-Cymenene | — | 0.3 | 0.1 |
| Citronellal | 81.5 | 0.2 | — |
| Linalool | 0.4 | 1.3 | 0.9 |
| Isopulegol | 1.9 | — | — |
| Pulegol | 3.7 | — | — |
| β-Caryophyllene | 0.2 | 0.2 | 0.4 |
| Terpinen-4-ol | — | 0.8 | 4.3 |
| Citronellyl acetate | 0.4 | 0.3 | — |
| Neral | tr? | 9.6 | — |
| Methyl geranate | tr? | 4.7 | — |
| α-Terpineol | — | — | 1.1 |
| Piperitone | — | tr | 40.9 |
| Germacrene B | — | 3.1 | — |
| Geranial | tr? | 12.5 | — |
| Geranyl acetate | tr? | 4.6 | — |
| Citronellol | 6.5 | 0.5 | — |
| Nerol | 0.1 | 1.4 | — |
| Geraniol | 0.3 | 4.7 | — |
| Others | 2.9 | 3.1 | 2.3 |

tr = trace (<0.05 per cent).
? Indicates that peak's identity has not been confirmed by GC-MS.

# Appendix 5. Quality criteria and specifications of eucalyptus oils

## Introduction

Given the variability that can exist for eucalyptus oils, even when they are produced from the same botanical source, there is clearly a need for some sort of standard or specification for oils of commerce so that the buyer, or prospective buyer, knows what he can expect when he makes a purchase. The larger commercial producers, particularly those which export their oil, usually offer their own specifications and, if requested, these can be passed on by dealers or importers to their own customers. Once the link between producer and buyer is firmly established, and the latter has seen that successive consignments meet the needs and expectations of his end-user customers, subsequent orders can be placed on the basis of mutual trust. The end user or formulator will usually monitor purchases by undertaking appropriate quality control tests. For medicinal oils, 1,8-cineole content (among other things) is important, while for perfumery oils the content of some specific constituent such as citronellal and/or the overall fragrance characteristics are important.

National and international standards exist which formalise certain of these quality criteria and, if they choose to, producers can assert that their oil meets such standards while buyers can demand that what they purchase does so. For the low-volume, specialised types of oil, published standards do not exist and compliance with quality requirements is a matter between the buyer and seller.

Any prospective new producer of eucalyptus oil must be aware of the standards which exist for the type of oil he is hoping to produce. Some indication of the parameters which are quantified for the different types of oil is given below but full details should be obtained from the relevant national or international standards organisation or pharmacopoeia. Specifications are also subject to change and the current position should be checked with the same institutions. Trade and other organisations should also be able to offer advice. The addresses of many of these bodies are given in Appendix 7.

## Medicinal oils

### *Specifications of standards organisations*

The International Standardization Organization (ISO) is a worldwide federation of national standards institutes and has issued three standards covering different types of cineole-rich eucalyptus oil (ISO 1974, 1980, 1983). The standards specify those characteristics of the oils which can be quantified with a view to facilitating the assessment of their quality; they cannot adequately define those properties which involve a buyer's subjective judgement, such as odour. In many cases the ISO standards are formally approved by member bodies of individual countries and go on to be adopted as national standards. Australia has produced its own standards for eucalyptus oil (referred to below) but other eucalyptus oil producers such as Brazil and South Africa have not.

For all three ISO standards, a minimum cineole content or range of values is specified, together with ranges of values for relative density, refractive index and optical rotation, and the solubility in ethanol. ISO 770:1980 and ISO 4732:1983 refer to steam-distilled oils from *E. globulus*. The first of these stipulates a minimum cineole content of 70 per cent while the second describes the three types of rectified oil which are commonly available from Portugal: 70/75, 75/80 and 80/85 per cent cineole. The third standard, ISO 3065:1974, specifies requirements for Australian eucalyptus oil of 80–85 per cent cineole content and defines the origin of the oil simply as 'appropriate species of *Eucalyptus* of Australian origin'.

Standards Australia has recently published revised specifications for Australian eucalyptus oil (SA 1998). AS 2113.1-1998 lays down requirements for oil containing 70–75 per cent cineole and supersedes AS 2113-1977, while AS 2113.2-1998 refers to oil containing 80–85 per cent cineole and supersedes AS 2115-1977. Both detail physico-chemical requirements, with cineole contents in the defined ranges, as before, but the revised versions contain some additional material, including a gas chromatographic profile of the oils, a table showing the constituents and an appendix with information on flash points.

An Indian standard, IS 328:1992, exists for 'Oil of Eucalyptus Globulus', although the description of the oil allows it to be distilled from 'other cineole-containing species of eucalyptus' as well as *E. globulus* (BIS 1992). It has different ranges of values for relative density, refractive index and optical rotation to those of the analogous ISO standard and, importantly, a lower requirement for cineole: 60 per cent instead of 70 per cent.

The designations of the above standards, the minimum cineole content (or range of values) demanded by them, and the designations of the ISO standards which describe the methods of analysis to be followed, are given in Table A5.1. The minimum cineole content specified in the pharmacopoeia and food specifications referred to below are also included.

*Pharmacopoeia specifications*

Although conveniently termed medicinal eucalyptus oils, the standards described above make no reference to the use to which the oils are put. As noted below, some cineole-rich oils are used in foods. For strictly medicinal purposes the oils must comply with national or international pharmacopoeias. Compliance with the British Pharmacopoeia (BP) specification is often cited by producers and this states that eucalyptus oil must contain not less than 70.0 per cent 1,8-cineole (BP 1998). The specification is identical to that of the European Pharmacopoeia (EP 1998). The oil is defined as being obtained from 'various species of eucalyptus rich in 1,8-cineole' but goes on to say that 'the species used are *Eucalyptus globulus* Labillardière, *Eucalyptus fruticetorum* F. von Mueller (*Eucalyptus polybractea* R.T. Baker) and *Eucalyptus smithii* R.T. Baker'. In addition to giving a minimum cineole content, the specification lays down ranges for optical rotation, relative density and refractive index and states the solubility of the oil in alcohol. It also describes chemical tests which are to be carried out. These are intended to ensure that levels of aldehydes and phellandrene which might be present are below certain limits and these additional requirements distinguish the BP specification from ISO 770:1980 with which it is otherwise very similar. Several of the crude medicinal oils contain small amounts of isovaleraldehyde (see Appendix 4) and rectification serves both to increase their cineole content and to remove this undesirable constituent, thereby enabling them to meet BP standards.

Table A5.1 List of standards and specifications for cineole-rich eucalyptus oils, together with requirements for cineole content and methods for the determination of parameters referred to in the ISO standards

| Type/Designation | Type of oil/Analytical method | Cineole content (%)[a] |
|---|---|---|
| **Standards** | | |
| ISO 770:1980(E) | E. globulus | min. 70 |
| ISO 4732:1983(E) | Rectified E. globulus, Portugal: 70–75%, 75–80%, 80–85% | 70.0–74.9, 75.0–79.9, 80.0–85.0 |
| ISO 3065:1974(E) | Australian, 80–85% | 80–85 |
| AS 2113.1,2-1998 | Australian, 70–75%, 80–85% | 70–75, 80–85 |
| IS 328:1992 | E. globulus | min. 60 |
| **Pharmacopoeias** | | |
| British | Eucalyptus oil | min. 70.0 |
| European | Eucalyptus oil | min. 70.0 |
| Indian | Eucalyptus oil | min. 60 |
| Chinese | Oleum eucalypti | min. 70 |
| **Food Chemicals** | | |
| US | Eucalyptus oil | min. 70.0 |
| **Methods** | | |
| ISO 212:1973 | Sampling | |
| ISO 356:1996 | Preparation of test samples | |
| ISO 1202:1981 | Determination of 1,8-cineole content[b] | |
| ISO 279:1981 | Determination of relative density | |
| ISO 280:1976 | Determination of refractive index | |
| ISO 592:1981 | Determination of optical rotation | |
| ISO 7359:1985 | Gas chromatography on packed columns | |
| ISO 7609:1985 | Gas chromatography on capillary columns | |
| ISO 1271:1983 | Determination of carbonyl value (free hydroxylamine method)[c] | |

a On m/m or w/w basis.
b Measures the crystallisation temperature of a mixture of the test oil and o-cresol; the temperature depends on the 1,8-cineole content of the oil.
c Required for E. citriodora oil.

Other national pharmacopoeias that lay down standards for eucalyptus oil include those of Austria, Belgium, Brazil, the People's Republic of China, France, Germany, Hungary, India, Italy, Japan, Netherlands, Portugal and Switzerland. Although contained in the 21st (1985) edition of the United States Pharmacopoeia (16th edition National Formulary), eucalyptus oil was deleted from the 22nd/17th edition (1990) and remains absent from the 23rd/18th edition (1995). The Indian Pharmacopoeia defines eucalyptus oil in a similar way to the BP and has similar or identical values for the physico-chemical data, but, like the Indian standard referred to above (BIS 1992), it has a minimum cineole requirement of 60 per cent rather than 70 per cent (IP 1996). Like the BP it has tests for aldehydes and phellandrene.

The Pharmacopoeia of the People's Republic of China (PPRC 1992) demands a minimum cineole content of 70 per cent for eucalyptus oil but allows it to be produced from a non-eucalypt species ('by steam distillation from the plants of *Eucalyptus globulus* Labill. (Fam. Myrtaceae), *Cinnamomum camphora* (L.) Sieb. (Fam. Lauraceae) or other plants belong[ing] to the same genus of these two families'). Acceptable ranges for relative density and refractive index are given but, unlike the BP, no range for optical rotation is specified. A test for the absence of phellandrene is described and a heavy metals limit of 10 ppm is laid down.

*Specifications for use in foods*

Cineole-rich eucalyptus oil is sometimes used as a flavouring agent in foods and in the United States the Food Chemicals Codex (NAS 1996) lays down specifications with which it must comply. The 1996 version (4th edition) is identical to that in the 1981 3rd edition. The oil is described as being that distilled from *Eucalyptus globulus* (FEMA No. 2466) and 'other species of *Eucalyptus*' and must contain not less than 70.0 per cent cineole. In addition to specified limits for specific gravity and refractive index, the oil must pass tests that demonstrate nil (or low) amounts of heavy metals and phellandrene.

The Food Chemicals Codex also lays down specifications for eucalyptol (FEMA No. 2465), the trivial name given to the main constituent of medicinal type eucalyptus oil, 1,8-cineole (synonyms for which include cajeputol and 1,8-epoxy-*p*-menthane). An infra-red spectrum is given for identification purposes. Since eucalyptol is a single chemical, the specifications reflect its physico-chemical properties:

| | |
|---|---|
| Molecular weight, formula | 154.25, $C_{10}H_{18}O$ |
| Boiling point | 176°C |
| Specific gravity (25°C) | 0.921–0.924 |
| Refractive index (20°C) | 1.455–1.460 |
| Optical rotation | –0.5° to +0.5° |
| Solidification point | minimum 0°C |

## Perfumery oils

Of the perfumery oils, published standards exist only for *E. citriodora* oil. Other perfumery oils, such as that from *E. staigeriana*, are traded on the basis of sample assessment by the buyer.

The international standard for *E. citriodora* oil, ISO 3044:1997, sets maximum and minimum limits for relative density, refractive index and optical rotation (ISO 1997). The ranges of values for all three parameters are slightly different to those given in the previous version of the standard (ISO 3044:1974). Citronellal content is important and the oil must contain a minimum of 70 per cent carbonyl compounds expressed as citronellal to comply with the standard. In keeping with the modern trend for standards for essential oils to provide chromatographic information, ISO 3044:1997 shows typical gas chromatograms obtained on a polar and apolar column in a separate annex. A second annex provides flash point information.

An Indian standard exists for *E. citriodora* oil, IS 9257:1993. Physico-chemical data are slightly different to those given in the ISO standard but the minimum citronellal content remains 70 per cent (BIS 1993).

The Fragrance Materials Association of the United States (FMA) has a standard for *E. citriodora* oil (EOA 130) and this states that the aldehyde content, calculated as citronellal, should be in the range 65–85 per cent. Ranges of values for specific gravity, refractive index and optical rotation are also prescribed. There is also an FMA monograph for eucalyptol (1991 revision, replacing EOA 288).

## Other oils

No published standards exist for oil from *E. dives* (piperitone variant) and quality criteria are a matter for agreement between buyer and seller. If the oil is being utilised as a source of piperitone then the content of the latter is usually around 40 per cent or more.

## References

BIS (1992) *Indian Standard. Oil of Eucalyptus Globulus – Specification*, IS 328:1992, Bureau of Indian Standards, New Delhi, India, 4 pp.

BIS (1993) *Indian Standard. Oil of Eucalyptus Citriodora – Specification*, IS 9257:1993, Bureau of Indian Standards, New Delhi, India, 3 pp.

BP (1998) Eucalyptus oil. In *British Pharmacopoeia*, Vol. I, British Pharmacopoeial Commission, The Stationery Office, London, pp. 570–571.

EP (1998) Eucalyptus oil. In *European Pharmacopoeia*, 3rd edn, 1999 Supplement, Council of Europe, Strasbourg, France, pp. 489–490.

IP (1996) Eucalyptus oil. In *Indian Pharmacopoeia*, Vol. I, Controller of Publication, New Delhi, India, p. 310.

ISO (1974) *International Standard. Oil of Australian Eucalyptus, 80 to 85% Cineole Content*, ISO 3065-1974(E), International Organization for Standardization, Geneva, Switzerland, 2 pp.

ISO (1980) *International Standard. Oil of Eucalyptus Globulus*, ISO 770-1980(E), International Organization for Standardization, Geneva, Switzerland, 2 pp.

ISO (1983) *International Standard. Rectified Oil of Eucalyptus Globulus Labillardière, Portugal*, ISO 4732-1983(E), International Organization for Standardization, Geneva, Switzerland, 2 pp.

ISO (1997) *International Standard. Oil of Eucalyptus Citriodora Hook.*, ISO 3044-1997(E), International Organization for Standardization, Geneva, Switzerland, 5 pp.

NAS (1996) Eucalyptus oil. In *Food Chemicals Codex*, 4th edn, National Academy of Sciences, National Academy Press, Washington D.C., USA, pp. 138–139.

PPRC (1992) Oleum Eucalypti. In *Pharmacopoeia of the People's Republic of China*, English Edition, Guangdong Science & Technology Press, Guangzhou, China, p. 129.

SA (1998) *Oil of Australian Eucalyptus. Part 1: 70–75 Percent Cineole*, AS 2113.1-1998, and *Part 2: 80–85 Percent Cineole*, AS 2113.2-1998, Standards Australia, Strathfield, Australia.

# Appendix 6. Packaging and labelling requirements for the handling and transportation of eucalyptus oils[1]

## Introduction

The production and export of eucalyptus oil (and other essential oils) used to be simply a matter of producing it to the required quality and transporting it from A to B, with a minimum of documentation. However, the increasing attention being given worldwide – but particularly in Europe – to safety and environmental issues has seen an ever-burgeoning responsibility placed on the shoulders of producers, processors and importers to see that 'dangerous' substances and goods are adequately packaged and labelled, and shipped with a much more comprehensive set of documents. Such measures are designed to ensure the safe handling and transportation of materials that are actually or potentially dangerous substances. With a few exceptions, the main risk associated with essential oils is their flammability. In the case of eucalyptus oils, this risk is different for the cineole-rich medicinal oils and oil from *Eucalyptus citriodora*.

## Legislation

Within Europe, the most important legislation relating to dangerous substances which is relevant to essential oils are Council Directives 67/548/EEC and 79/831/EEC. The regulations relate to the storage, handling and use of such substances and require, for example, that every package shows the name and origin of the substance, the appropriate danger symbol (e.g. a flame in a red diamond indicating a flammable liquid) and standard codes indicating special risks and safety advice. These codes are of the form Rx (for risks) and Sx (for safety), where x is a number specific for a particular phrase. Prefixes 'Xi', 'Xn' or 'T' are added for substances which present an irritant or other harmful risk or are toxic. Substances deemed to be marine pollutants require special labelling.

None of the eucalyptus oils of international trade are perceived to present health or environmental risks and the only labelling required (valid in 1999) is 'R10' for the cineole-rich oils such as *E. globulus*. This indicates their flammability and is relevant for all oils with a flash point in the range 21–55°C. *E. citriodora* oil has a flash point of approximately 65°C and so does not require a flammability warning.

Recent legislation pertaining to the aspiration hazard of hydrocarbons (94/69/EC and 96/54/EC) has given rise to the need to label certain of such materials with the risk code R65

---

[1] Adapted from Coppen and Hone (1992) by permission of the Natural Resources Institute, University of Greenwich. Updated information is drawn from Green (1997) and Protzen (1998), for which both authors are thanked. Additional information provided by M. Irvine and D. Moyler is acknowledged. Aspects of legislation relating to consumer products containing eucalyptus oil are discussed in Chapter 16.

('Harmful; may cause lung damage if swallowed') and the safety phrase S62. This requirement is now impinging on essential oils and the European Flavour and Fragrance Association (EFFA) have decided to apply the R65 class to all essential oils and preparations with a total hydrocarbon content of greater than 10 per cent. For eucalyptus oils, the medicinal, cineole-rich ones will attract an R65/S62 classification but *E. citriodora* oil will not. It was expected that this labelling requirement would be enforced during 1999.

In the United States, drums of either of the two types of eucalyptus oil have to be labelled for worker protection with the phrase 'May irritate skin and eyes'. For transportation purposes cineole-rich oils are classified as flammable while *E. citriodora* oil is not classified (Chapter 16).

Within the European Union there is a special regulation concerning the protection of water resources, the Water Resources Act. This classifies substances in one of three classes, where Class 1 represents a weak hazard to water and Class 3 represents a strong hazard. Although so far only practised and enforced in Germany, the act is likely to be adopted across Europe in due course. Tests have indicated that eucalyptus oil falls into Class 1.

When dangerous 'substances' are transported they become 'goods' and when conveyed from one country to another they are subject to international regulations according to the means of conveyance. In the case of shipment by sea, regulations of the International Maritime Dangerous Goods (IMDG) code have to be observed. As with dangerous substances, dangerous goods have to be marked with warning labels.

To warn and inform people who are handling dangerous substances about the potential risks, Material Safety Data Sheets have been developed for use within the European Union. These detail the physical, chemical and toxicological properties of the substance in question, together with measures to be taken in the event of accidental spillage or other exposure, and have to be provided by importers for all dangerous goods prior to release of the consignment by Customs from the port of landing, and whenever they are offered for sale, transported and supplied. Advice and information on these aspects for essential oils is regularly published and updated by the International Fragrance Association and the International Organisation of the Flavour Industry. Another example of the way in which such information is compiled and disseminated is the CHIP (Chemical Hazard Information and Packing Regulations)-List, a database offered by the British Essential Oil Association which gives information on labelling (hazard symbols and risk and safety codes required, etc.) for individual essential oils.

## Packaging

The risks posed by handling and transportation of dangerous substances are considerably reduced by proper packaging. In the case of the cineole-rich eucalyptus oils and other essential oils in which the fire risk is the greatest hazard, this means using UN-approved drums and containers which meet the specifications for Packing Group III (flammable liquids with a flash point in the range 23–61°C and boiling point above 35°C). Drums must be metal with a non-removable head and must have been tested to specified pressure limits. For identification purposes the appropriate UN code is stamped on the container by its manufacturer.

European Directive 94/62/EC requires a stepwise reduction in the levels of heavy metals in packaging materials. The final limit is 100 ppm to be achieved by July 2001. In the context of essential oils this is likely to involve a switch from solvent- to water-based paints used on drums and is of more concern to companies in importing countries – where drums are colour coded to assist identification during storage and handling – rather than those at the point of origin of the oils.

## Documentation

The traditional requirements for documentation which accompanies shipments have been the commercial invoice, weight and origin certificate, Bill of Lading, insurance documents and a certificate of the product's quality. Nowadays, with the introduction of systems of classification and labelling has come the need to provide evidence of compliance in shipping documents. The invoice must now provide information on product identification, the UN product code, its flash point and any other hazards. For cineole-rich eucalyptus oils UN No. 1169 ('extracts, aromatic, liquid') or No. 1197 ('extracts, flavouring, liquid') are applicable, although UN No. 1993 ('flammable liquids, non-toxic, not otherwise specified') can be used alternatively.

As indicated earlier, within the European Union importers are obliged to submit Material Safety Data Sheets for consignments prior to release by Customs. It is also regarded as good commercial practice (and is becoming increasingly common) for the exporters themselves to include such data along with the other documentation.

## Whose responsibility?

Within the European Union, where much of the legislation has been promulgated, it is the importers of essential oils who are considered to be the producers (since they supply the goods to the market) and it is they, therefore, who are ultimately responsible for proper classification and labelling of goods. However, although it may be tempting for producers at origin to feel that they are absolved from taking the measures indicated above when they are exporting to Europe and elsewhere, they should recognise, particularly in these days of the 'global economy', that it is in their own interests to meet the appropriate packaging and labelling requirements. Regulations are being increasingly enforced by policing at the point of import and if consignments are found to be lacking in terms of correct packaging, labelling or documentation any penalties meted out to the importer will have adverse repercussions, in one way or another, for the producer/exporter.

The regulations discussed above are being continually assessed and revised as new facts or circumstances come to light; even changes of a general nature may still impinge on eucalyptus oil. More detailed and up-to-date information than can be given here can be obtained from national and international transportation authorities, trade associations, international organisations such as IFEAT, IFRA and IOFI, or the importers themselves. Addresses of some of these organisations are given in Appendix 7.

## References

Coppen, J.J.W. and Hone, G.A. (1992) *Eucalyptus Oils: A Review of Production and Markets*, NRI Bulletin 56, Natural Resources Institute, Chatham, UK.

Green, C.L. (1997) Export packaging and shipment of essential oils: requirements and regulations in the major markets. Paper presented at UNIDO Essential Oils Conference, Harare, Zimbabwe, November 1997.

Protzen, K.D. (1998) New developments in legislation regulations for the transportation of essential oils. In *Global Markets: Present and Future. Proc. IFEAT Internat. Conf. Essential Oils and Aromas*, London, November 1998, pp. 260–269.

# Appendix 7. Useful addresses

## Trade associations

*British Essential Oil Association* (BEOA)
15 Exeter Mansions
Exeter Road
London NW2 3UG
UK
Tel.: +44-20-8450 3713
Fax: +44-20-8450 3197

*Cosmetic, Toiletry and Fragrance Association* (CTFA)
1101 17th Street NW, Suite 300
Washington
DC 20036
USA
Tel.: +1-202-331 1770
Fax: +1-202-331 1969

*European Flavour and Fragrance Association* (EFFA)
Square Marie-Louise, 49
1000 Brussels
Belgium
Tel.: +32-2-238 9905
Fax: +32-2-230 0265

*Flavour and Extract Manufacturers' Association* (FEMA)
1620 I Street NW, Suite 925
Washington
DC 20006
USA
Tel.: +1-202-293 5800
Fax: +1-202-463 8998

*Fragrance Materials Association* (formerly Essential Oil Association of the USA)
Same contact details as for FEMA

*International Federation of Essential Oils and Aroma Trades* (IFEAT)
Federation House
6 Catherine Street
London WC2B 5JJ
UK
Tel.: +44-20-7836 2460
Fax: +44-20-7836 0580

*International Fragrance Association* (IFRA)
Rue Charles-Humbert, 8
1205 Geneva
Switzerland
Tel.: +41-22-321 3548
Fax: +41-22-781 1860

*International Organisation of the Flavour Industry* (IOFI)
Same address as for IFRA

## Standards organisations

Addresses are given for standards organisations in the major eucalyptus oil producing countries, together with those in the major markets of France, Germany, Japan, the United Kingdom and United States. All those listed below are members of ISO and the ISO standards for eucalyptus oil can be obtained from them, as well as ISO headquarters in Geneva. Some national bodies have standards for eucalyptus oil which are identical to the ISO ones. A few, such as the Australian and Indian bodies, issue their own standards.

Other ISO members from whom the ISO standards can be obtained are those in Albania, Algeria, Argentina, Armenia, Austria, Bangladesh, Belarus, Belgium, Bosnia and Herzegovina, Bulgaria, Canada, Colombia, Costa Rica, Croatia, Cuba, Cyprus, Czech Republic, Denmark, Ecuador, Egypt, Ethiopia, Finland, Ghana, Greece, Hungary, Iceland, Indonesia, Iran, Ireland, Israel, Italy, Jamaica, Kazakhstan, Kenya, Korea (Democratic People's Republic of), Korea (Republic of), Libya, Macedonia (former Yugoslav Republic of), Malaysia, Mauritius, Mexico, Mongolia, Morocco, Netherlands, New Zealand, Nigeria, Norway, Pakistan, Panama, Philippines, Poland, Romania, Russian Federation, Saudi Arabia, Singapore, Slovakia, Slovenia, Sri Lanka, Sweden, Switzerland, Syria, Tanzania, Thailand, Trinidad and Tobago, Tunisia, Turkey, Ukraine, Uruguay, Uzbekistan, Venezuela, Vietnam, Yugoslavia and Zimbabwe.

*International Standardization Organization* (ISO)
1 Rue de Varembé
Case Postale 56
1211 Geneva 20
Switzerland
Tel.: +41-22-749 0111
Fax: +41-22-733 3430;
  Sales: +41-22-734 1079
E-mail <sales@iso.ch>

*Australia*
Standards Australia
PO Box 1055
Strathfield
NSW 2135
Tel.: +61-2-9746 4700
Fax: +61-2-9746 8450
E-mail <intsect@standards.com.au>

*Brazil*
Associação Brasileira de Normas
  Técnicas
Av. 13 de Maio, no. 13, 28° Andar
CP 1680
20003-900 Rio de Janeiro, RJ
Tel.: +55-21-210 3122
Fax: +55-21-532 2143
E-mail <abnt@abnt.org.br>

*Chile*
Instituto Nacional de Normalización
Matías Cousiño 64, 6° Piso
Casilla 995 - Correo Central
Santiago
Tel.: +56-2-696 8144
Fax: +56-2-696 0247
E-mail <inn@huelen.reuna.cl>

*China, People's Republic of*
China State Bureau of Technical Supervision
4 Zhichun Road
Haidian District
PO Box 8010
Beijing 100088
Tel.: +86-10-6203 2424
Fax: +86-10-6203 1010

*France*
Association Française de Normalisation
Tour Europe
92049 Paris
Tel.: +33-1-4291 5555
Fax: +33-1-4291 5656

*Germany*
Deutsches Institut für Normung
10772 Berlin
Tel.: +49-30-260 10
Fax: +49-30-260 11231
E-mail <postmaster@din.de>

*India*
Bureau of Indian Standards
Manak Bhavan
9 Bahadur Shah Zafar Marg
New Delhi 110002
Tel.: +91-11-323 7991
Fax: +91-11-323 4062
E-mail <bisind@del2.vsnl.net.in>

*Japan*
Japanese Industrial Standards Committee
For sales information:
Japan Standards Association
1-24 Akasaka 4
Minato-ku
Tokyo 107
Tel.: +81-3-358 38003
Fax: +81-3-358 62029

*Portugal*
Instituto Português da Qualidade
Rua C à Avenida dos Três Vales
2825 Monte da Caparica
Tel.: +351-1-294 8100
Fax: +351-1-294 8101
E-mail <ipqmail@ipqm.ipqgtw-ms.mailpac.pt>

*South Africa*
South African Bureau of Standards
Private Bag X191
Pretoria 0001
Tel.: +27-12-428 7911;
  Sales: +27-12-428 6481
Fax: +27-12-344 1568;
  Sales: +27-12-428 6928
E-mail <sales@sabs.co.za>

*Spain*
Asociación Española de Normalización y
  Certificación
Fernández de la Hoz 52
28010 Madrid
Tel.: +34-1-432 6000
Fax: +34-1-310 4976

*United Kingdom*
British Standards Institution
389 Chiswick High Road
London W4 4AL
Tel. (Sales): +44-20-8996 9001
Fax (Sales): +44-20-8996 7001
E-mail <info@bsi.org.uk>

*United States*
American National Standards Institute
11 West 42nd Street, 13th Floor
New York
NY 10036
Tel.: +1-212-642 4900
Fax: +1-212-398 0023
E-mail <info@ansi.org>

## Other organisations

*British Pharmacopoeia Commission*
Market Towers
1 Nine Elms Lane
London SW8 5NQ
UK
Tel.: +44-20-7273 0561
Fax: +44-20-7273 0566

*Research Institute for Fragrance Materials* (RIFM)
2 University Plaza, Suite 406
Hackensack
NJ 07601
USA
Tel.: +1-201-488 5527
Fax: +1-201-488 5594
E-mail <help@rifm.org>

*Trade & Investment Information Centre*
[Trade statistics and other information]
Singapore Trade Development Board
230 Victoria Street
#07-00 Bugis Junction Office Tower
Singapore 188024
Singapore
Tel.: +65-433 4435
Fax: +65-337 5256

*Trade Partners UK Information Centre*
[Trade statistics and other information]
Kingsgate House
66-74 Victoria Street
London SW1E 6SW
UK
Tel.: +44-20-7215 5444
Fax: +44-20-7215 4231

*US Food and Drug Administration*
Office of Generic Drugs
MetroPark North
7500 Standish Place
Rockville
MD 20855
USA

# Subject index

Note: (1) Eucalyptus species (by common name and botanical name) are indexed separately, following the Subject Index. All other species, including bacterial and fungal species, are indexed here.

(2) Page numbers in bold indicate photographs.

(3) Countries are indexed in the context of cultivation of eucalypts and production of oil or other eucalyptus product, etc. Countries that are discussed in terms of trade and markets (Chapter 17) are also indexed, but not those which are only included as origins or destinations within statistical tables. Australia is only indexed here if the reference is more substantial than a passing mention of it (in a historical context, as the home of *Eucalyptus*, etc).

(4) A few proprietary products are indexed when there is some reference to them in the text. However, a great many more are listed in Tables 16.1, 16.2 and 16.3 and are not indexed here.

Abdominal pain 214, 261
*Acacia* 95
   *mearnsii* 39, 46
*Acanthoecia laminati* 208
*Acanthoscelides obtectus* 309, 310, 313–15
*Acetobacter xylinum* 394
Acylphenones 397
Acylphloroglucinols 106, 112, 275, 276, 278–81, 283–5, 325–7, 337–41, 397
Addresses 424–6
*Adoretus sinicus* 208
Adulteration, *see* Oil
*Aedes*
   *aegypti* 305, 307
   *albopictus* 305, 306, 308
   *vexans* 305
Aegle oil 292
Aflatoxin 292, 293
Agarospirol 415
Ageing 270
Agglomerone 109, 110, 114, 117, 134, 137
*Agrobacterium tumefaciens* 209
Agroforestry (involving *Eucalyptus* as an oil crop) 204, 255
AIDS 272, 397
Air fresheners 347
Airwaves chewing gum 361
Aldose reductase inhibition 270, 272, 274, 397
Algeria 217, 409
Allelopathy 317–20
Alloaromadendrene 123, 124, 128, 133, 141, 144, 151, 390, 414

*Alternaria*
   *alternata* 294
   *citrii* 292
Amyl acetate 143
Analgesic 355
Analytical techniques 94, 104, 105, 390, 391
Angelica oil 295, 356
Angola 216, 217, 220, 230, 409
*Angophora* 9
*Anomala* 208
*Anopheles*
   *funestus* 307
   *gambiae* 307
Antibacterial 270, 272, 274, 295–300, 361, 362, 397
Anticholestatic 262
Antidiarrhoeal 355
Antifeedant 339, 340
Antifouling agents 397
Antifungal 270, 292–5, 298, 299, 396
Anti-herbivore effect 335, 337
Anti-inflammatory 261, 270, 282–4, 355
Antimalarial 270, 274, 275, 305–7, 397
   *see also* Mosi-guard™; Mosquito
Antimicrobial 291–300, 360, 361
Antineuralgic 355
Antioxidants 112, 270–2, 396
Antiphlogistic 355
Antirheumatic 355
Antiseptic 214, 260, 261, 269, 291, 346, 355
Antispasmodic 355
Anti-tumour 270, 284–7
Antiviral 287, 355, 397

## Subject index

Apodophyllone 115, 135, 142
Argentina 28, 242, 248, 249, 410
Aromadendrene 107, 108, 114–51, 390, 414
Aromatherapy 108, 236, 269, 345, 346, 355, 356
Arthritis 350
Articular pain 269
Artifacts, *see* Distillation
Asarone 313
Ashes 9
*Aspergillus* 292
   *flavus* 293, 294
   *fumigatus* 292–4
   *nidulans* 292
   *niger* 294, 295, 299, 394
   *ochraceus* 299
   *oryzae* 292
   *repens* 293
   *terreus* 292
Asthma 260, 349
Athlete's foot 356
Australia 39–41, 44, 53, 57–9, 88, 174, 177, 183–7, 188–90, 191, 192, 193–201, 350, 375, 376, 410, 414, 417
Autan® 307
Ayurvedic medicine 261
*Azadirachta indica* 295, 309
   *see also* Neem

Babul stem and root borer 256
*Bacillus*
   *anthracis* 296
   *fumilis* 297
   *megaterium* 295
   *mycoides* 297
   *pumilus* 297
   *subtilis* 274, 296
   *thuringiensis* 316
Backache 355
Bacteria 274, 295–300, 361, 362
   *see also* named species
Bangladesh 409
Bark beetle, see *Phoracantha semipunctata*
Bark types 13–17
Basil oil 362
Bath additives 198, 351, 356–8, 360
Bay oil 295
Beetles 335–7
   see also *Acanthoscelides obtectus*; *Gonipterus scutellatus*; *Phoracantha semipunctata*; *Stegobium paniceum*; *Tribolium* spp.
Benin 219, 234, 409
Benzaldehyde 109, 110, 150, 151, 299, 325, 326
Benzoin 356
BEOA, *see* British Essential Oil Association
Bicycloelemene 390
Bicyclogermacrene 114, 115, 117–22, 124, 125, 127–35, 137, 139–41, 143–51, 326, 339, 388, 390
Biocidal products 363
Biogenesis 79

of acylyphloroglucinols 327, 338, 340
of eucalyptus oil constituents 110, 111,
of euglobals 281, 282
Biomass (of leaf), 220, 222, 225, 236
   correlation with
      oil yield 88
      stem parameters 87, 92
   yields of 194–6, 210, 211, 242, 244, 259
Biosynthesis, *see* Biogenesis
Bites, *see* Insect; Snake
Bitter almond oil 295
Blackbutts 9
Bleach products 347
Blight 256
Blisters (of skin) 345, 355
Bloodwoods 6, 7, 9, 15, 17, 19, 20, 26, 384
Boiler,
   in distillation 163, 170, 171, 173, 174, 176
   fuelwood for 233
   use of eucalyptus tannins for treating 199
   *see also* Leaf, use of spent leaf as fuel
Bolivia 248, 249, 378, 410
Borneol 127, 128, 135
*Botrytis cinerea* 209, 230
Boxes 7, 9, 28
*Brachytrupes portentosus*, see *Tarbinskiellus portentosus*
Brazil 5, 26, 28, 31, 53, 54, 57, 65, 68, 90, 91, 239–44, 245–7, 248, 368, 377, 378, 381, 410, 414, 415
Breeding
   programme 66–8, 75–8, 80, 84, 85, 87, 90, 91, 95, 98, 192, 200, 203, 211, 235, 236, 340, 341
   strategy 75–8, 80, 82, 85, 88, 95–8, 208
British Essential Oil Association 422, 424
British Pharmacopoeia, *see* Pharmacopoeia
Bronchitis 260, 269, 349–52
Bruises 352
Brushtail possum, *see* Possums
Burkina Faso 219, 409
Burma, *see* Myanmar
Burns 269, 345, 353, 355
Burundi 217–19, 409
Butyl butyrate 124, 140
Butyraldehyde 120
*Byssochlamys nivea* 293

δ-*Cadinene* 122, 130, 132, 134, 141, 145
Cadinols 130, 134
Cajeput oil 354, 355
Calamenene 130, 146
*Callosobruchus*
   *chinensis* 310, 311, 313
   *maculatus* 309, 310, 313, 314
Cameroon 409
Camphene 293, 414
Campholenic aldehyde 143, 144
Camphor 133, 213, 214, 293
   oil 349, 354, 355, 359
Cancer 270, 283
*Candida albicans* 292, 294, 295, 297, 299

Canker 220, 231, 235
Cape Verde 409
Capillary fragility 199, 261
Car-4-ene 143
Carbon-nutrient balance 327
Carcinogenesis 283, 284, 286
Carcinogenicity 272
*Carea subtilis* 208
Carissone 133
Carvacrol 128, 291, 293, 295, 298, 300, 320
Carvone 128, 141, 147, 299
β-Caryophyllene 116–18, 121, 125, 127–9, 131, 135, 137, 141, 147, 149, 150, 415
Caryophyllene oxide 120, 121, 132, 148
Case moth, see *Hyalarcta huebneri*
Cassia oil 346
*Casuarina* 95
Cataract 272
Catarrh 349–53
Cedar oil 349, 356
*Celosterna scabrator* 256
Central African Republic 409
Chad 409
*Chalia laminati*, see *Acanthoecia laminati*
Chemical
  forms 78–80, 95, *or*
    Chemical variants 194, 391
    Chemotypes 10, 24, 27, 30, 32, 89, 105, 108, 221
    Chemovars 105
    *see also* Physiological forms
  structures
    of allelochemicals 320
    of 1,8-cineole metabolites 395
    of mosquito repellents 306
    of non-terpenes 326
    of non-volatile constituents 113, 271, 273–5, 280, 281, 326
    of oil constituents 107, 109
Chest rub 349
Chewing gum 361, 362
Chickenpox 345
Chilblains 352
Chile 28, 53, 54, 248, 249, 378, 379, 410
China 26–8, 52, 53, 55, 56, 59, 66, 78, 102, 199, 202–4, 205–11, 212–4, 370–3, 379–82, 409, 413, 418
Chinese medicines 213, 214, 269, 274
Chinese Pharmacopoeia, see Pharmacopoeia
Chiral
  analysis 391
  *p*-menthane–3,8-diols 318
  metabolites of cineole 334
Cholera 261
Choleretic effect 262
Cholestasis 262
Cicatrisant 355
Cineole, see 1,8-Cineole
1,4-Cineole 366
1,8-Cineole 10, 24, 25, 27–30, 32, 33, 55, 56, 79–82, 84–9, 92, 93, 95, 97, 98, 103–8, 110–12, 114–51, 164, 184, 185, 196, 197, 201, 210, 213, 218–22, 234–6, 241, 242, 248, 257, 258–60, 295–7, 299, 300, 309, 311, 313, 315, 316, 319, 328, 329, 331–7, 340, 345–7, 355, 362, 366, 367, 378, 387, 388, 391, 393, 394, 413, 414–19
  derivatives, *see* Metabolism; Microbiological transformations
Cinnamaldehyde 299
*Cinnamomum*
  *camphora* 297, 306
    as source of 'eucalyptus oil' 213, 370, 420
  *tamala* 294
  *zeylanicum* 297, 298
Cinnamon oil 291, 295, 354
Citral 108, 125, 131, 142, 146, 220, 223, 241, 242, 299, 368
Citronella oil 292, 304, 305, 307
Citronellal 10, 26, 89, 107, 108, 117, 120, 184, 210, 213, 219, 223, 240–2, 246, 249, 258, 261, 295, 297, 299, 300, 317, 318, 334, 347, 348, 368, 391, 394, 415, 419
Citronellic acid 394
Citronellol 108–10, 120, 137, 261, 294, 295, 297, 298, 307, 317, 318, 368, 394, 415
Citronellyl acetate 144, 415
*Citrus limon* 298
*Cladosporium cladosporioides* 294
Classification
  of *Eucalyptus* 4, 7–10, 30, 105, 106, 276, 277, 398
  of eucalyptus oil (customs) 369
Cleaning agent 198, 359, 362, 393, 394
Climate,
  conditions/requirements 6, 58–62, 195, 196, 223, 227, 252–4
  mapping programs 58, 59
Clonal
  forestry 64–7, 90, 97, 203, 208
  programmes 85
*Clostridium sporogenes* 299
Clove oil 295, 346, 354
$CO_2$, effect on oil concentration 327, 328
Cold sores 345
Colds 198, 269, 345, 349–53
Colic 214
Colombia 65, 248, 378, 410
Common brushtail/ringtail possum, *see* Possums
Comores 409
Composition, *see* Oil; named *Eucalyptus* spp; Non-volatile constituents
Composition-activity relationships 298–300
Condensers, types 176, 177, 178
Congestion 214, 345, 350–3, 355, 357, 358, 360, 361
Conglomerone 118, 121
Congo, 53, 65, 203, 224, 236, 409
  Democratic Republic of 31, 217, 219, 220, 377, 409
Container size (for seedlings) 206

Coppice management 67, 68, 207, 208, 218, 227, 228, 244, 245, 255, 412
Coppicing 55, 186, 187, 196, 258, 259
 ability 90–2, 186, 227, 252, 255
 *see also* Harvesting; Lignotuber
*Corcyra cephalonica* 310, 311
*Corticium salmonicolor* 256
*Corymbia* 9, 26, 102, 105, 106, 276, 277, 384
*Corynebacterium* 297
Cosmetic, Toiletry and Fragrance Association 356, 424
Cosmetics 362, 363
Costa Rica 410
Costs 56, 65, 66, 75, 89, 91, 94, 165, 171, 175, 187, 191, 223–5, 232–5, 243, 244, 255, 257, 385, 388, 393, 398, 412
Coughs 345, 350–3
*p*-Coumaric acid 320
*Cryphonectria cubensis* 231
*Cryptococcus neoformans* 292
Cryptone 117, 119–23, 129–31, 134, 135, 137, 141, 142, 144, 148, 150, 151, 390
CTFA, *see* Cosmetic, Toiletry and Fragrance Association
Cuba 410
Cultivation
 in Africa 216–36
 in Australia 183–201
 in China 202–14
 in India 251–63
 in South America 239–49
 overview 52–68
Cuminal 119, 123, 135, 138, 141, 142, 148
*Curvularia lunata* 294
Cuts 198, 355
Cutworms 229
Cyanogenic glycosides 325
*Cylindrocladium quinqueseptatum* 62, 256
*Cymbopogon martinii* 294
*p*-Cymene 114–38, 140–51, 242, 260, 293, 295, 299, 334, 414, 415
*p*-Cymenene 414, 415
Cystitis 269

Damping-off 230
Dandruff 356
*Dappula tertius* 208
Decongest(ion/ant) 349, 353, 355, 357, 361
DEET 305, 307, 308
Degreasing agent 394, 395
Dehydroangustione 143
Democratic Republic of Congo, *see* Congo, Democratic Republic of
Dental caries 362
Deodorant 355
Depurative 355
*Dermataphagoides farinae* 310, 316
Dermatitis 269
Detergent 261, 356
Diabetic complications 272, 397
Diarrhoea 261, 269

Diet choice
 in insects 334–7
 in vertebrate herbivores 328–34
β-Diketones 270, 271
4,6-Dimethoxy-2-hydroxyacetophenone 112
2,6-Dimethyl-5-heptenal 415
Diphtheria 261
Diseases of eucalypts 62, 64, 193, 208, 209, 223, 230, 231, 256
Disinfectant 183, 185, 213, 214, 260, 261, 347, 359, 367, 368
Distillation, 161–80
 ancillary equipment 178–80
 artifacts due to 104, 325, 386–8
 curves 166, 167, 386, 387, 389
 in Australia 183, 185, 186, 188–90, 196, 197
 in Brazil 246, 247–9
 in China 204, 212, 213
 in India 257, 258
 in Swaziland 232, 233, 234
 laboratory *vs* other oil extraction methods 94, 103, 104, 385–90
 theory and practice 161–80
 times 165–70, 189, 234, 246, 248, 385–7
 using mobile still 174, 175, 188–90, 191
 vacuum 104, 106, 388, 389
Distillery, arrangement/design of 176, 190, 413
Diuretic 355
DNA sequencing 399
Documentation 423
Dysentery 214, 261, 269

Ecuador 410
Eczema 352, 359
EFFA, *see* European Flavour and Fragrance Association
Egypt 219
El Salvador 410
δ-Elemene 390
Elemol 135, 139, 143, 389
Ellagic acid 270–2
Embrocations 260
*Endothia gyrosa* 231
*Enterobacter* 296
Environment 36–48
*Environmental*
 factors 60–3, 87, 89, 93
 variance, $V_E$ 85
 *see also* Climate
EOA, *see* Fragrance Materials Association
*Ephestia kuhniella* 310, 312
*Epidermophyton*
 *floccosum* 292
 *rubrum* 295
Epstein-Barr virus activation inhibition 270, 283–5, 397
Equipment (for distillation) 171–80
Erosion,
 control of 191, 193, 234, 243, 244
 and eucalypts 39, 45–7, 226
*Erwinia carotovora* 297

Subject index 431

Erysipelas 269
*Escherichia coli* 295–9
Essential Oil Association, *see* Fragrance Materials Association
Ethiopia 55, 216, 217, 409
Eucalyptamint 353
Eucalyptin 112, 113
Eucalyptol, 103, 234, 248, 269, 291, 346, 380
 monograph 348, 419
 production of 197, 371
 *see also* 1,8-Cineole
Eucalyptone 273, 274
'Eucalyptus hybrid' 39, 56, 251–4, 257, 259, 260, 262, 295, 296, 379
Eucalyptus oil, *see* Oil
Eucamalol 306–8
Eudesmol/α-, β-, γ-eudesmol 108–10, 114–51, 262, 320, 388, 414
Eugenol 128, 214, 291, 297, 298
Euglobals 270, 275–87, 325–7, 397, 398
European Flavour and Fragrance Association 422, 424
European Pharmacopoeia, *see* Pharmacopoeia
European Union 374, 375
Euthymol™ 361
Evaporation (from eucalypts) 38–45
Expectorant 269, 353, 355
Export data, *see* Oil, trade

Farnesol 119, 121, 123, 127, 129, 136
Fascioliasis 396
Febrifuge 355
FEMA, *see* Flavour and Extract Manufacturers' Association
α-Fenchene 414
Fenchone 316
Fennel oil 361
Fertiliser, use of 42, 62–4, 195, 207, 226, 227, 243–5, 248, 259
Fever 269, 350, 352
 haemorrhoidal 261
 rheumatic 261
Fibrositis 345
Flammability, *see* Oil
Flash point, *see* Oil
Flavesone 123, 129
*Flavobacterium suaveolens* 295
Flavonoids 112, 269, 272
Flavour and Extract Manufacturers' Association 424
Flavours 108, 109, 198, 213, 361, 368, 377, 392, 395
Flotation agent, *see* Mineral flotation
FMA, *see* Fragrance Materials Association
*Fomes* 209
Food Chemicals Codex 419
Formulations, *see* Oil
'Four-around' planting 205
Fragrance Materials Association 419, 424
France 350
Frostbite 199

Fuel, use of eucalyptus oil as additive 201, 367, 393
 *see also* Leaf, use of spent leaf; Boiler
Fungal diseases, *see* Diseases
Fungi,
 food spoilage 292–4
 human pathogens 292
 plant pathogens 294, 295
 soil 295
 *see also* named species
Fungicides, use of 230, 256
2-Furoic acid 320
*Fusarium* 209, 230, 394
 *culmorum* 299
 *moniliforme* 294, 295
 *oxysporum* 292, 294
 *solani* 292, 294

Gabon 409
Gallic acid 396
*Ganoderma* 209, 256
Gargles 260, 353
Gas chromatography 94, 103–5, 317, 337, 390, 414, 417, 419
Gastrointestinal tract 261
GC, *see* Gas chromatography
GC-IR 103, 104
GC-MS 94, 103, 104, 317, 390
Genetic
 control 84
 correlation, $r_g$ 87, 88
 engineering 340
 gain 85, 91, 95, 97
 improvement 75–98, 249
 resources 78
 selection 65, 211
 variance, $V_G$ 85
*Geotrichum candidum* 293
Geranial 107, 108, 146, 222, 241, 368, 417
Geraniol 118, 119, 121, 125, 126, 132, 133, 137, 138, 142, 146, 150, 241, 294, 298, 300, 348, 415
Geranium oil 292, 346, 356
Geranyl acetate 25, 107, 108, 114, 126, 133, 136, 146, 220, 222, 223, 241, 257, 299, 368, 415
Germacrene B 415
Germacrene C 390
Germany 350
Germicidal effects 260, 328, 333
Ghana 219, 409
Ghost gums 9, 18, 20
Giddiness 214
Glands, *see* Oil
Gliders 112, 328, 333, 334
*Gliocladium roseum* 293
Globulol 107, 108, 114–30, 132–51, 388, 390
Glucosyltransferase inhibition 270, 273
*Gonipterus scutellatus* 216, 229, 230
Gram-negative bacteria 296, 299, 361, 397
Gram-positive bacteria 295–8, 361, 397

Grandinal 274, 327, 398
Grandinol 270, 398
Greater glider, see Gliders
*Grevillea* 54, 95
Growth
  prediction 60–2
  traits 80, 81, 85–8, 95, 96
Guaiol 109, 110, 116, 117, 120, 134, 145
Guatemala 31, 269, 410

Habit 6, 10–13, 24, 29, 30, 205
Haemorrhoids 199
Hair
  conditioner 358, 360, 361
  shampoo, see Shampoo
Haiti 410
Hand cleaner 198
Hand cleanser/gel/wash 357, 358, 360
Handling of eucalyptus oils 421–3
Harvesting (for oil),
  frequency between oil crops 68, 194, 196, 227, 234, 244, 245
  mechanical 24, 25, 161, 174, 187–9, 191, 232
  mechanised 244
  of *E. cinerea* in Zimbabwe 222
  of *E. citriodora* in Brazil 244, 245
  of *E. citriodora* in India 259
  of *E. dives* in South Africa 228
  of *E. globulus* in Brazil 246
  of *E. polybractea* in Australia 187–9
  of *E. smithii* in Swaziland 228, 231, 232
Hay fever 351, 353
Headache 214, 261, 352, 353, 355, 356, 358
Head lice 359
Healing (of sores, cuts, abrasions) 355
Heartrot fungus 256
Hedycaryol 389
HeLa cell inhibition 270, 274
*Helminthosporium compactum* 292
Hepatotoxicity 262
Herbicides, use of 193, 225, 226, 243, 256
Herbivores/herbivory 324–41
Heritability 24, 85–7, 92, 95, 236
History
  of oil production 22–4, 183–7, 209, 210, 217, 218, 240, 241
  of eucalypt plantations 202, 203, 216, 217, 239, 240, 251
HIV-RTase inhibition 270, 272, 397
Homeopathic preparation 352
Homoisobaeckeol 135
Honduras 410
Honey production 204
HPLC 390
  of euglobals 276, 278
Humulene 118
*Hyalarcta huebneri* 193
Hybridisation 18, 90, 251, 391
Hybrids 56, 65, 66, 203, 208, 224, 236, 240, 262
  see also 'Eucalyptus hybrid'

Hydrology 37, 38, 42–4, 46, 48, 56, 68
Hydroxycitronellal 240, 368, 394
Hydroxycitronellol 261
Hydroxydihydrocitronellal 261, 394
4-Hydroxy-tritriacontane–16,18-dione 112, 113, 271
Hypoglycemia 355

IFEAT, see International Federation of Essential Oils and Aroma Trades
IFRA, see International Fragrance Association
Import data, see Oil, trade
India 26, 28, 36, 39, 42–4, 47, 53, 55–7, 68, 251–63, 379, 409
Indian Pharmacopoeia, see Pharmacopoeia
Indigestion 261
Indonesia 4, 5, 52, 59, 62, 409
Industrial oils 213, 222, 259, 366, 368
  see also *E. dives*; *E. olida*; Mineral flotation; Solvent
Influenza 198, 269, 349, 352, 353
Inhalants 260, 349, 353
Insect
  bites 198, 214, 307, 345, 350, 352, 355, 359
  damage, correlation with oil 334–7
  pests 53, 193, 208, 227, 229, 230, 243, 244, 255, 256, 304–16, 324, 341
  repellent 198, 222, 304–7, 312, 314, 316, 346, 348, 349, 363
Insecticides, 229, 256, 308, 309, 316
  *Eucalyptus*, use as 269, 309, 312–16
Interception, see Rainfall
Internal reflux 163, 169, 170
International Federation of Essential Oils and Aroma Trades 423, 424
International Fragrance Association 347, 348, 422–4
International Organisation of the Flavour Industry 422–4
International Standardization Organization 104, 390, 416–19, 425
IOFI, see International Organisation of the Flavour Industry
Ironbarks 8, 15, 25, 26, 30, 31, 33
Irrigation 44, 192, 195, 200, 227, 244, 254, 259
ISO, see International Standardization Organization; Standards
Isoamyl alcohol 128
Isoamyl isovalerate 116, 118, 128, 136
Isoamyl phenyl acetate 114
Isobaeckeol 135
Isobicyclogermacral 109, 110, 115, 123, 127, 141, 143
Isobicyclogermacrene 125, 140
Isoeugenol 298
Isoflavesone 129
Isolates, chemical, 201, 368
  potential new sources of 108–10, 392
*p*-Isopropyl phenol 116

Subject index 433

4-Isopropylbenzyl alcohol 306, 307
Isopulegol 150, 307, 415
Isotorquatone 115, 135
Isovaleraldehyde 116, 118, 122, 123, 128, 135, 138, 142, 146, 197, 235, 388, 414, 417
Israel 36, 39, 410
Italy 351, 410
Itching 355

Jacksonone 130
Jamaica 269
Japan 369, 370
Jensenone 109, 110, 115, 130, 325, 326, 331, 339
Joint disorders/pain 345, 350–2
Juniper oil 354, 356, 357
*Juniperus communis* 292

Kangaroos 328
Kenya 55, 219, 409
Ketones,
   long-chain 112
   *see also* β-Diketones; β-Triketones; named ketones
Kino 15, 199, 231, 261
Koalas 112, 324, 325, 328, 329, 333, 334, 340

Labelling requirements 362, 363, 421–3
Land preparation 192, 206, 224, 225, 242, 243
Laos 59, 409
Laterticone 132
Lavender oil 346, 355, 356
LC/MS (of euglobals) 275–8
Leaf
   age 93
   biomass, *see* Biomass
   ontogeny 18, 92, 93
   use of spent leaf as
      fuel for distillation 175, 176, 190, 213, 233, 245, 246, 257
      mulch/source of nutrients 189, 199, 234, 245, 247
   venation 19, 20, 22, 27, 28
   wax 112, 271
Leaf-spot 231
Leaves,
   cost and logistics of collection 412
   damage by insects and/or mammals 193, 324–41
   diet of insects and mammals, *see* Diet choice
   moisture content 94, 103, 194
   use as insecticide 269, 309, 312–16
   use in traditional medicine 214, 261, 269, 274
Ledol 390
Leeches 305
Legislation 362, 363, 421–3
Lemon oil 346, 356
Lemongrass oil 292
Leptospermone 123, 127, 129, 135–7

Lesotho 409
Lesser grain borer, see *Rhyzopertha*
*Leuconostoc cremoris* 295
Libya 409
Light, effect on
   growth 61, 62
   oil production/concentration 93, 327, 328
Lignotuber 11–13, 24–6, 67, 187, 196
Limonene 105, 107, 108, 112, 114–24, 126, 127, 129–32, 134–42, 145–8, 150, 151, 197, 241, 299, 309, 337, 368, 414, 415
Linalool 114, 118, 133, 136, 138, 144, 316, 414, 415
Linalyl acetate 299, 316
*Listeria monocytogenes* 296
Listerine™ 353
Lovage oil 295

Macrocarpals 112, 113, 270, 272–4, 285, 325–7, 338, 339, 362, 397, 398
*Macrophomina phaseolina* 209, 292
Madagascar 203, 216, 217, 219, 409
Mahoganies 7, 9, 19
Malaria, see *Aedes*; *Anopheles*; antimalarial; mosquito repellen(t)cy
Malawi 55, 217, 230, 409
Malaysia 409
*Malbranchea pulchella* 295
Mali 409
Mallees 3, 11, 12, 15, 18, 24, 25, 27, 28, 30, 33, 187, 200, 222, 367
Marjoram oil 295, 356
Market research 412
Marketing 413
Markets for eucalyptus oil, *see* Oil
Marsupials 112, 328–34, 337–40
   *see also* Gliders; Koalas; Possums
Mass spectrometry
   of cryptone 390
   of euglobals 276, 278, 279
   of β-triketones 104
   *see also* GC-MS
Mauritius 409
Measles 356
Mechanical harvesting, *see* Harvesting
Medicinal oils/uses 22, 108, 185, 197, 198, 213, 260, 345–63, 366, 367, 416–19, 422
   see also *E. exserta*; *E. globulus*; *E. leizhou No. 1*; *E. polybractea*; *E. radiata*; *E. smithii*; other cineole-rich spp; named pharmacological activities, e.g. antibacterial, anti-cancer; Pharmacopoeia; Traditional medicine
*Melaleuca* 95, 171, 188, 337
   *alternifolia* 86, 88
   *cajeputi* 355
   *viridiflora* 297
*p*-Menth–2-en–1-ol(s) 105, 115, 118, 121, 123–5, 136–9, 142, 143, 146, 148, 149, 320
*Mentha piperita* 294

*p*-Menthane-3,8-diol(s) 305–7, 317–19, 348, 391
Menthol 186, 214, 316, 349, 354, 357, 361, 362, 368, 394
  production from *E. dives*/piperitone 27, 108, 186, 198, 368, 377
Mentho-Lyptus™ 353
Menthone 299
Menthyl acetate 134, 142, 390
Metabolism (of eucalyptus oil constituents)
  by humans 111
  by insects 112, 337
  by mammals 112, 333, 334, 394, 395
  *see also* Microbiological transformations
Methyl cinnamate 33, 107, 108, 110, 123, 124, 137, 198, 201, 368, 375, 392, 395
Methyl eudesmate 109, 110, 114, 122
Methyl eugenol 128, 298
Methyl geranate 146, 222, 241, 415
6-Methyl-5-hepten-2-one 415
Methyl salicylate 214, 350, 354, 362
4-Methylpent-2-yl acetate 133, 201
Mexico 410
Microbiological transformations 394, 395
*Micrococcus*
  *glutamicus* 296, 297
  *luteus* 298
  sp. 297
Micropropagation 65–7, 90, 91, 224, 235, 262
  *see also* Tissue culture
Microsporum
  *canis* 292
  *gypseum* 292, 295
Microwave extraction 103, 104
Midges 305, 307
Mineral flotation 185, 186, 198, 221
Mites 308, 310, 315, 316
  see also *Dermataphagoides*; *Tyrophagus*
Moisture
  content of leaves, *see* Leaves
  of soil, *see* Soil
Molluscicidal properties 396
Monkeys 329
*Monocalyptus* 9, 105, 106, 276–8
Monograph, *see* Eucalyptol; Oil
Morocco 53, 65, 216, 217, 219, 224, 409
Mosi-guard™ 304, 307, 346, 349
Mosquito
  bites 214, 307
  repellen(t)cy 26, 108, 110, 304–7, 346, 348, 349
  *see* also *Aedes*; *Anopheles*
Moths, see *Corcyra*; *Ephestia*; *Hyalarcta*; *Phthorimaea*; *Sitotroga*
Mouthwash 198, 261, 353, 361
Mozambique 409
Mucolytic effect 345
*Mucor*
  *hiemalis* 293
  *racemosus* 293
Mugwort oil 294

Mulch, *see* Leaf
Muscle
  ache/pain 198, 214, 350–2, 355, 357, 360
  rub 352, 353, 358
Musculoskeletal pain 350–3
Mustard oil 295
Mutagenicity 272
Muurolols 130
Myanmar 409
*Mycobacterium avium* 295
*Mycosphaerella muelleriana* 231
Myrcene 129, 133, 138, 139, 142, 414, 415
Myrtenol 129
Myrtle oil 356
Mysore gum, *see* 'Eucalyptus hybrid'
*Mytilus edulis galloprovincialis* 396

Namibia 409
Nappy rash 353
Nasal congestion 350–3, 357
Neem 309, 313
  oil 295, 305, 311, 312
Nepal 56, 409
Nepharcopathy 272
Nephritis 214
Neral 107, 108, 222, 241, 368, 415
Nerol 298, 300, 415
Nerolidol 123, 137, 138, 151, 392
Nerve pain 351
Nervous exhaustion 356
Neryl acetate 299
Neuralgia 345
Neuropathy 272
New Zealand 410
Nicaragua 410
Niger 409
Nigeria 55, 219, 220, 409
NMR (of euglobals) 279, 282
Non-volatile constituents 112, 113, 269–87, 396–8
Nose infections 350
Nursery 66, 193, 205, 206, 223, 230, 244, 256
Nutmeg oil 295, 349
Nutrients, 42, 43, 55, 62–4, 67, 68, 226, 227, 235, 242, 244
  effect on oil concentration 234, 327, 328
  *see also* Fertiliser
Nutrition 60, 62–4

*cis/trans*-β-Ocimene 123, 136–8, 149, 414, 415
*Ocimum*
  *americanum* 294, 295
  *gratissimum* 297
Oil (eucalyptus)
  adulteration of 197, 391
  ex 'camphor' 213, 370, 371, 379, 380
  composition 104, 114–51, 413–15
    seasonal variation 79, 93, 94
    tree-to-tree variation 80–2, 105, 391
    variation due to ontogeny 92
    within-tree variation 92, 93

Subject index 435

see also named *Eucalyptus* spp
correlation with concentration of
  acylphloroglucinols 338–40
distillation, see Distillation
ecological factors which affect concentration in
  *Eucalyptus* 327, 328
flammability 362, 421–3
flash point 197, 362, 394, 421–3
formulations 345–63
glands 6, 20–2, 26–33, 161, 337, 385, 389
grades 367, 380
markets for 365–82
monograph 348
optical rotation 104, 260, 261, 370, 417–19
plant sources 365, 366
price(s), 379–81
  Chinese (cineole) oil 380, 381
  *E. citriodora* oil 381
  eucalyptol 380
production estimates,
  Australia 375
  Bolivia 378
  Brazil 377, 378
  Central Africa 217, 377
  Chile 378
  China 209, 371
  Colombia 378
  India 257, 258, 379
  Paraguay 378
  Portugal 375
  South Africa 377
  Southern Africa 217, 376, 377
  Spain 375
  Swaziland 231, 376
  Uruguay 378
  Vietnam 371
  Zimbabwe 376
rectification 196, 197, 210, 217, 234, 236, 260, 367, 368, 377, 413, 417
refractive index 104, 213, 260, 261, 417–19
relative density 104, 213, 260, 261, 417, 418
role in determining diet of insects and
  mammals 328, 329, 331, 334, 335
solubility
  in alcohol 104, 261, 387, 417
  in water 178, 189, 298, 300, 356
specifications,
  *E. citriodora* 241, 261
  *E. globulus* 241, 260
  *E. staigeriana* 241
  for use in foods 418, 419
  see also Pharmacopoeia; Standards
standards, see Standards
storage after distillation 196, 235, 260
trade (international) 209, 369–82
traits 76, 80–2, 84–90, 92, 93, 95–7
yields, 103, 114–51, 194, 210–12, 219–21, 234, 241, 242, 246, 248, 257–60
  determination of 94, 103, 171

factors affecting 92–4, 194–6, 227, 234, 258–60
  per ha 91, 196, 210, 211, 227, 234, 246, 248, 257, 259, 262
  seasonal variation 93, 194, 211, 234, 258
  tree-to-tree variation 80–3, 258
  variation due to ontogeny 92
  variation with leaf age 93
  within-tree variation 92, 93
Olbas™ oil 353, 354
Oleanolate, cinnamic acid ester 112, 113
Optical rotation, see Oil
Oral health care 361, 362
Orange oil 292
*Origanum minitiflorum* 295
Origanum oil 291, 294
Oropharynx disorders 351, 352
*Oryzaephilus surinamensis* 311

Packaging requirements 356, 362, 421–3
*Paecilomyces variotii* 292, 293
Pakistan 410
Palmerosa oil 292
Palustrol 121, 133, 145
Papua New Guinea 52, 410
Papuanone 139
Paraguay 248, 370, 378, 379, 410
*Parthenium*, use of *Eucalyptus* to combat 319
Patchouli oil 292
Peace of Mind™ 362
*Pelochrista* 208
*Penicillium*
  *citrinum* 292, 295
  *clavigerum* 293
  *cyclopium* 293
  *italicum* 294
  *notatum* 293, 295
  *purpurogenum* 293
Peppermint oil 349, 354–6, 358, 361, 362
Peppermints 9, 19, 22, 24–30
Perfumery (and/or oils) 26, 108, 198, 213, 222, 240, 241, 261, 346–8, 366, 368, 394, 419
  see also *E. citriodora*; *E. macarthurii*; *E. staigeriana*
*Perga dorsalis* 193
Personal care products 356–61
Peru 55, 248, 410
*Pestalotiopsis versicolor* 295
Pesticides 243, 256
Pests
  of eucalypts 193, 208, 223, 227, 229, 230, 255, 256
  of stored products 307–16
  see also Insect pests
Pharmaceutical products 345–63
Pharmacopoeia, 417
  British 104, 185, 197, 260, 367, 417, 418, 426
  Chinese 213, 418

436  *Subject index*

European 417, 418
Indian 260, 418
United States 418
Pharyngitis 214
Phellandral 119, 123, 135, 141, 142
Phellandrene 24, 27, 29, 30, 105, 108, 114, 119, 126, 145, 148, 184, 185, 197, 213, 221, 222, 393, 417–19
α-Phellandrene 79, 92, 107, 108, 115, 116, 118, 119, 121–30, 132, 134–51, 184, 210, 213, 219, 319, 328, 387, 393, 394, 414, 415
β-Phellandrene 105, 115–18, 121–5, 131, 134, 137, 138, 145, 146, 149, 213, 415
Phenotypic
  correlation, $r_p$ 87, 88
  variance $V_P$ 85
  variation 80
4-Phenylbutan-2-one 123, 124, 320, 398
β-Phenylethyl phenylacetate 109–11, 114, 122
β-Phenyllactic acid 320
3-Phenylpropanal 132
Philippines 4, 52, 57, 59, 410
Phloracetophenone 2,4-dimethyl ether 116
Phloroglucinol derivatives 272–87
*Phoracantha semipunctata* 230
*Phthorimaea operculella* 310, 316
Physiological forms 10, 78, 105, 184
  *see also* Chemical forms
Phytohormones, extraction of 203
*Phytophthora* 230
  *capsici* 295
Pimento oil 295
Pine oil 292, 356
α-Pinene 30, 107, 108, 114–51, 259, 260, 297, 299, 319, 334, 337, 387, 390, 414, 415
β-Pinene 114, 117–23, 126, 127, 129, 130, 132–4, 137–9, 142–51, 259, 260, 299, 300, 319, 334, 414, 415
Pink disease 256
*trans*-Pinocarveol 114–16, 118, 120–36, 138–51, 414
Pinocarvone 128, 131, 133, 144, 149
*Pinus*
  *caribaea* 47, 292
  *patula* 46
  *pinaster* 41
  *radiata* 39, 89
  *roxburghii* 39
*cis-/trans*-Piperitol 105, 115, 118, 123–6, 138, 139, 142, 146, 149, 387, 394
Piperitone 24, 25, 27, 30, 79, 105, 107, 108, 115, 122–7, 137–40, 142, 144–7, 149–51, 184, 186, 210, 213, 217, 219, 221–3, 242, 299, 337, 368, 377, 394, 415
  derivatives, *see* Microbiological transformations
Plantations,
  advantages of, for oil production 191, 192
  areas of eucalypts (estimates by country/region) 53, 54, 203, 204, 216–18, 239, 240, 253, 370, 377, 408–11
  *see also* Cultivation; Harvesting

Planting 193, 206, 207, 225, 243, 244, 258
*Plasmodium berghei* 274
Pollination 82–5, 91, 95–7, 208
Polydatin 396
Portugal 25, 28, 40, 53, 54, 66, 67, 174, 187, 375, 410, 414
Possums 112, 328–34, 338–40, 394
Price(s)
  of Chinese (cineole) oil, *see* Oil
  of *E. citriodora* oil, *see* Oil
  of eucalyptol, *see* Oil
  of rutin 262
*Proactinomyces roseus* 394
Production of eucalyptus oil (estimates by country), *see* Oil
Progeny trials 80–3, 85, 86, 88, 89, 92, 93, 211
Propagation 244
  *see also* Micropropagation; Tissue culture; Vegetative propagation
*Proteus vulgaris* 296, 297
*Protoeucalyptus* 5–7
Provenance
  selection/trials 80, 89, 208, 211, 218, 220, 221, 252, 256
  variation 62, 80–2, 195, 218, 249
Provenance-progeny trials 78, 83, 87, 89
Prunasin 325, 326
Pseudomonas
  *aeruginosa* 296, 298, 299
  *flava* 394
  *mangifera indica* 297
  *solanacearum* 209, 297
Psyllids 230
Pulegol 415
Pulegone 391
Pyrethrum 309
*Pythium* 230

Quwenling 304, 305, 307, 346, 348

Rainfall, interception/water use by eucalypts 38, 39, 41–3, 45, 47
  *see also* Climate
Rashes 359
Rectification, *see* Oil
Refractive index, *see* Oil
Relative density, *see* Oil
Research Institute for Fragrance Materials 347, 348, 426
Research
  needs 54, 64, 111, 249, 384–99
    in Africa 235
    in Australia 200, 201
    in India 262
  trends 384–99
Resinosides 397, 398
Respiratory tract disorders/infections 261, 345, 350–2
Retinitis 261
Retinopathy 272
Rhaponticin 396

## Subject index 437

Rheumatic pain 214, 345, 350–2
Rheumatism 260, 261
Rhinitis 350
Rhinopharyngitis 350
*Rhizoctonia solani* 295
*Rhodopseudomonas sphaeroides* 300
*Rhyzopertha dominica* 309–11, 313
Ribbon gums 15
RIFM, *see* Research Institute for Fragrance Materials
Ringtail possum, *see* Possums
Robustadials 270, 274, 275, 397, 398
Robustaol A 270, 274, 275, 327
Root diseases 193, 209, 220, 230, 256
Rooting 65, 66, 90, 91, 224
Rosemary oil 293, 346, 356
*Rosmarinus officinalis* 293, 298
Rotation for oil production, *see* Harvesting, frequency
Rubbing oil 198
Rubefacient 260, 355
Rutin 112, 113, 199, 261, 262, 365, 396
Rwanda 55, 217, 219, 220, 409

Sabinene 389, 414, 415
*Saccharomyces cerevisiae* 294
Safety 226, 233, 347, 348, 362, 421–3
Salinity, control of, in Western Australia 24, 36, 40, 57, 200, 367
*Salmonella*
  *paratyphi* 297
  *typhi* 297
*Salvia fruticosa* 295
Sample preparation 103, 104
  *see also* Distillation, laboratory
Sampling 92–4, 391
Sandalwood oil 356
*Sarcina lutea* 296, 297
*Satureja thymbra* 295
Sawfly, see *Perga dorsalis*
Scabies 269
Scalp massage 198
Schistosomiasis 396
*Sclerotinia sclerotiorum* 295
*Sclerotium rolfsii* 209, 292, 294
Seasonal variation, *see* Oil yields/composition
Seed
  morphology 17, 18, 26, 27, 31
  orchards 78, 84, 91, 97, 195, 203, 205, 208, 249, 262
  production 91, 244
  selection 413
  suppliers 407
Seedlings 12, 13, 18, 65, 66, 89, 92, 193, 200, 206, 207, 223, 224, 229, 230, 243, 244, 256, 259, 317, 327, 398
Selection
  criteria 91, 92
  of herbivore-resistant genotypes 340, 341
  for oil 24, 61, 62, 75–98, 195, 200, 211, 236, 249, 262

Selfing 83, 84
Senegal 409
Separators 178, 179
Seychelles 31
Shampoo
  for carpets 198, 362
  for hair 358
  for horses 349
Shaving gel/oil 359
*Shigella*
  *nigesta* 297
  sp. 297
Shingles 345
Sideroxylonals 270, 274, 326, 327, 339, 340, 397, 398
Sierra Leone 409
Sinusitis 349–51
Site
  index 60, 61
  requirements 223
  selection 91, 242
    using models 61
  variation 89, 234, 257, 260
*Sitophilus*
  *granarius* 310, 312
  *oryzae* 310, 314
  *zeamais* 309, 310, 312
*Sitotroga cerealella* 310, 313
Skin
  disease 261
  disorders 350, 352, 353, 359
  infections 345, 359
Smoking withdrawal 353
Snake bites 269
Snoring 361
Snout beetle, see *Gonipterus scutellatus*
Snow gums 19, 30
Soap 213, 261, 297, 346, 347, 356, 367, 368
Soft tissue disorders 351, 352
Soil
  moisture 45
  properties, use in calculating site index 60
  types 10, 24, 26, 28, 60, 61, 63, 195, 220, 221, 234, 242, 252
  *see also* Erosion; Nutrients
Solomon Islands 410
Solubility, *see* Oil
Solvent
  extract 347
  extraction 94, 103, 386–8
  use of oil as industrial 57, 200, 367, 393, 394
Sore throats 345, 350, 353
South Africa 27, 30, 37, 44–6, 53, 54, 65, 84, 90, 91, 186, 187, 211, 216–22, 224, 227, 228–30, 376, 377, 409, 414, 415
Spacing 42, 193, 206, 207, 225, 243, 244, 255, 259
Spain 25, 28, 53, 57, 67, 187, 351, 375, 410, 416
Spathulenol 79, 80, 89, 109, 110, 114, 117, 119–23, 125–7, 129, 130, 132–7, 139–41, 143–51, 320, 414

### 438  Subject index

Spearmint oil 356
Species
  selection 57–62, 77
  trials 58, 62, 208, 211, 218, 231, 252
Specifications, *see* Oil; *see also* Pharmacopoeia; Standards
Spike lavender oil 354
*Sporothrix schenckii* 292
Sports injuries 350–2
Sprains 350, 352
Sri Lanka 410
*Stachybotrys* 293
Standards 104, 197, 213, 260, 261, 367, 382
  addresses of organisations 424–6
  Australian 197, 417, 418
  *E. citriodora* oil 197, 261, 419
  *E. globulus* oil 260, 417–19
  eucalyptol 419
  FMA 419
  Food Chemicals Codex 418, 419
  Indian 260, 261, 417–19
  ISO 104, 260, 367, 382, 390, 416–19
    methods 417, 418
  medicinal oils 416–19
  perfumery oils 419
*Staphylococcus*
  *albus* 297
  *aureus* 274, 295–9
Statistical analysis (of oil compositional data) 94, 95, 105
Steam
  distillation, *see* Distillation
  properties of 163, 169
  requirements 166, 168
*Stegobium paniceum* 310, 311
Stem canker 220, 231
Stilbenes 396
Stills,
  packing of 174, 213, 233, 246
  types of 171–5, 186, 188–90, 204, 212, 233, 257, 258
Stocking 193, 207, 225, 227, 259
  *see also* Spacing
Stomach-ache 214
Storage, *see* Oil, storage of
Stored products, use of *Eucalyptus* against pests of 307–17
Strains 350, 352
*Streptococcus*
  *mutans* 362
  *pyogenes* 296, 297
Stringybarks 7–9, 15, 19
Structure-activity relationships 298–300
Sudan 251, 409
Supercritical fluid extraction 104, 387, 388
Swamp gums 7
Swaziland 217, 218, 220, 221, 224, 225, 227, 228, 229–31, 232, 233–5, 376, 409, 414
Switzerland 352
*Symphyomyrtus* 9, 18, 106, 276–8, 283, 397

Synergistic effects 292, 300, 311
*Syzygium aromaticum* 298

Taiwan 410
Tannins, extraction of 199, 203
Tanzania 30, 218, 220, 409
*Tarbinskiellus portentosus* 208
Tasmanian pademelon 328
Tasmanone 109, 110, 114, 117, 119, 121, 132, 137, 147
Tea tree oil 86, 170, 171, 188, 356
  see also *Melaleuca*
Termites 229, 243, 256
Terpinen-4-ol 79, 115, 118, 122, 138, 140, 142–4, 146, 149, 151, 299, 300, 389, 414, 415
α-Terpinene 118, 122, 124, 134, 136, 138, 142, 146, 260, 414, 415
γ-Terpinene 116, 119, 120, 122, 124, 127–32, 134, 136, 138, 140, 142, 143, 150, 151, 414, 415
α-Terpineol 107, 108, 114–30, 133–40, 142–51, 297, 299, 300, 414, 415
δ-Terpineol 119, 135, 146, 414
Terpinolene 122, 124, 134, 138, 146, 241, 300, 414, 415
α-Terpinyl acetate 114, 116, 119, 122, 130, 135, 138, 142, 143, 145, 149, 299, 414
Thailand 53, 59, 60, 78, 89, 97, 98, 352, 410
Thin layer chromatography 276, 277
Throat disorders/infections 350–3
α-Thujene 127, 390
*Thymbra spicata* 295
Thyme oil 291, 293, 295, 346, 356, 361
Thymol 198, 291, 295, 298, 300, 320, 349, 361, 362
*Thymus capitatus* 293
Ticks 307
Tigerbalm™ 353–5
Tissue culture 65–7, 91, 262
  *see also* Micropropagation
TLC, *see* Thin layer chromatography
Tobacco mosaic disease 256
Togo 409
Tonsillitis 214, 269
Toothache 214, 269, 350
Toothpaste 361
Torquatone 109, 110, 114, 115, 118, 119, 121, 125, 129, 132, 135, 139, 143, 146–9, 151, 325, 326
Trade
  associations 348, 424
  in eucalyptus oil, *see* Oil
  statistics, *see* Oil
Traditional medicine 269
  *see also* Ayurvedic; Chinese; Unani
*Trametes cubensis* 256
Transpiration 37–40, 42, 45
Transportation of eucalyptus oils 421–3
Travel sickness 214

*Tribolium*
  *castaneum* 309, 310
  *confusum* 310–2
*Trichoderma viride* 295
*Trichophyton mentagrophytes* 292
*Trichosporium vesiculosum* 251
β-Triketones 104, 105, 110, 112
*n*-Tritriacontane–16,18-dione 112, 271
Tunisia 216, 217, 409
Turkey 410
Turpentine 295, 349, 354
Typhoid 261
*Tyrophagus*
  *longior* 310, 315
  *putrescentiae* 310, 315, 316

Uganda 36, 309, 409
Unani medicine 261
United Kingdom 352, 353
United States 353, 371, 372, 410
    Pharmacopoeia, *see* Pharmacopoeia
Upper respiratory tract, *see* Respiratory tract
Uroterpenol 111
Ursolic acid 112, 113, 262
Uruguay 54, 242, 248, 378, 410

Vacuum distillation, *see* Distillation
Vaginal disorders 261, 353
Valeraldehyde 120, 143
Var./Variety A, B, C, etc., *see* Physiological forms
Varicose veins 199
Vegetative propagation 64–7, 90, 91, 95, 96, 195, 200, 224, 263
Venezuela 269, 410
Verbenone 137, 139, 148, 299
Vermicide 261
Vermifuge 355
*Verticillium albo-atrum* 209
Vetivert oil 346
Vicks VapoRub™ 349–53

Vietnam 53, 59, 62, 371, 410
Viridiflorene 129, 141, 148
Viridiflorol 114, 118, 119, 123, 129, 132, 133, 137, 138, 145, 388, 414
Virus, *see* Epstein-Barr virus activation inhibition
Vulnerary 355

Wallabies 328, 333
Waste leaf, use of for distillation 25, 37, 40, 67, 205, 239–41, 248, 253, 257, 259, 375, 377, 408, 412
Water
  supply 195
  use by eucalypts 37, 40–5, 48
Wax, *see* Leaf
Weather, effect on biomass yields 195, 196
  *see also* Climate
Weeding/weeds/control 63, 64, 193, 207, 225, 226, 234, 243, 256, 259
  use of *Eucalyptus* to combat 317, 319
Weevils, see *Callosobruchus*; *Sitophilus*
Western Australia, multipurpose eucalypt project 24, 57, 200, 367, 393
White grub 229
Whooping cough 261
Wintergreen 349, 354–6
Wool wash 198, 346, 362, 393
Wounds 261, 353, 355

*Xanthomonas campestris* 297

Yields,
  biomass, *see* Biomass yields
  oil, *see* Oil yields

Zaire, former, *see* Democratic Republic of Congo
Zambia 220, 409
Zimbabwe 33, 59, 87, 97, 217, 219–22, 229, 376, 409, 414
*Zygorrhyncus* 294

# Eucalyptus species index

Note that occasionally a species is referred to by more than one common name. Page numbers in bold indicate photographs.

Alpine Ash, see *E. delegatensis*
Alpine Cider Gum, see *E. archeri*
Apple Box, see *E. bridgesiana*
Apple-topped Box, see *E. angophoroides*
Araluen Gum, see *E. kartzoffiana*
Argyle Apple, see *E. cinerea*

Baarla Marble Gum, see *E. gongylocarpa*
Badgingarra Box, see *E. absita*
Badgingarra Mallee, see *E. pendens*
Baeuerlen's Gum, see *E. baeuerlenii*
Bailey's Stringybark, see *E. baileyana*
Baker's Mallee, see *E. bakeri*
Bald Island Marlock, see *E. conferruminata*
Balladonia Gum, see *E. fraseri*
Balladonia Mallee, see *E. balladoniensis*
Ball-fruited Mallee, see *E. buprestium*
Bancroft's Red Gum, see *E. bancroftii*
Barber's Gum, see *E. barberi*
Barlee Box, see *E. lucasii*
Barren Mountain Mallee, see *E. approximans* subsp. *approximans*
Beaufort Inlet Mallet, see *E. newbeyi*
Bell-fruited Mallee, see *E. preissiana*
Benson's Stringybark, see *E. bensonii*
Beyer's Ironbark, see *E. beyeri*
Big Badja Gum, see *E. badjensis*
Bimble Box, see *E. populnea*
Black Box, see *E. largiflorens*
Black Gum, see *E. aggregata*
Black Ironbox, see *E. raveretiana*
Black Morrel, see *E. melanoxylon*
Black Peppermint, see *E. amygdalina*
Black Sally, see *E. stellulata*
Blackbutt, see *E. patens*, *E. pilularis*, *E. todtiana*
Blackbutt Mallee, see *E. zopherophloia*
Blackdown Ironbark, see *E. melanoleuca*
Blackdown Stringybark, see *E. sphaerocarpa*
Blackdown Yellow Bloodwood, see *E. bunites*
Black-stemmed Mallee, see *E. arachnaea*
Blakely's Red Gum, see *E. blakelyi*
Blaxland's Stringybark, see *E. blaxlandii*

Bloodwood, see *E. umbonata*
Blue Box, see *E. baueriana*
Blue Gum, see *E. leucoxylon* subsp. *leucoxylon*, subsp. *petiolaris*
Blue Mallet, see *E. gardneri*
Blue Mountains Ash, see *E. oreades*
Blue Mountains Mahogany, see *E. notabilis*
Blue Mountains Mallee Ash, see *E. stricta*
Blue-leaved Jarrah, see *E. marginata* subsp. *thalassica*
Blue-leaved Mallee, see *E. cyanophylla*, *E. polybractea*
Blue-leaved Stringybark, see *E. agglomerata*
Bogong Gum, see *E. chapmaniana*
Book-leaf Mallee, see *E. kruseana*
Boranup Mallee, see *E. calcicola*
Bowen Ironbark, see *E. drepanophylla*
Boyagin Mallee, see *E. exilis*
Brittle Gum, see *E. mannifera* subsp. *maculosa*, subsp. *praecox*
Broad-leaved Bloodwood, see *E. curtipes*
Broad-leaved Peppermint, see *E. dives*
Broad-leaved Red Ironbark, see *E. fibrosa* subsp. *fibrosa*
Broad-leaved Sally, see *E. moorei* var. *latiuscula*
Broad-leaved Stringybark, see *E. caliginosa*
Broad-leaved White Mahogany, see *E. umbra*
Brooker's Gum, see *E. brookeriana*
Brown Barrel, see *E. fastigata*
Brown Bloodwood, see *E. trachyphloia*
Brown Mallet, see *E. astringens*
Brown Stringybark, see *E. baxteri*, *E. capitellata*
Brown's Box, see *E. brownii*
Budawang Ash, see *E. dendromorpha*
Bull Mallee, see *E. behriana*
Bullich, see *E. megacarpa*
Bundy, see *E. goniocalyx*
Burdett's Mallee, see *E. burdettiana*
Burracoppin Mallee, see *E. burracoppinensis*

Cabbage Gum, see *E. amplifolia*
Cadaghi, see *E. torelliana*
Caesia, see *E. caesia* subsp. *caesia*
Caley's Ironbark, see *E. caleyi*

# Eucalyptus species index

Camden White Gum, see *E. benthamii* var. *benthamii*
Camden Woollybutt, see *E. macarthurii*
Camfield's Stringybark, see *E. camfieldii*
Candlebark, see *E. rubida*
Cape Le Grand Mallee, see *E. aquilina*
Cape York Red Gum, see *E. brassiana*
Cap-fruited Mallet, see *E. dielsii*
Capped Mallee, see *E. pileata*
Carbeen, see *E. tessellaris*
Carne's Blackbutt, see *E. carnei*
Cider Gum, see *E. gunnii*
Cleland's Blackbutt, see *E. clelandii*
Cliff Mallee Ash, see *E. cunninghamii*
Coast Grey Box, see *E. bosistoana*
Coastal Dune Mallee, see *E. foecunda*
Coastal White Mallee, see *E. diversifolia*
Coolibah, see *E. coolabah, E. microtheca*
Coral Gum, see *E. torquata*
Creswick Applebox, see *E. aromaphloia*
Crowned Mallee, see *E. coronata*
Cullen's Ironbark, see *E. cullenii*
Cup Gum, see *E. cosmophylla*
Curly Mallee, see *E. gillii*

Darwin Box, see *E. tectifica*
Darwin Stringybark, see *E. tetradonta*
Darwin Woollybutt, see *E. miniata*
Desert Bloodwood, see *E. terminalis*
Desert Mallee, see *E. trivalvis*
Deua Gum, see *E. wilcoxii*
Devil's Peak Box, see *E. desquamata*
Diehard Stringybark, see *E. cameronii*
Dongara Mallee, see *E. obtusiflora*
Dorrigo White Gum, see *E. benthamii* var. *dorrigoensis*
Dowerin Rose, see *E. pyriformis*
Drummond's Mallee, see *E. drumondii*
Dundas Blackbutt, see *E. dundasii*
Dundas Mahogany, see *E. brockwayi*
Dunn's White Gum, see *E. dunnii*
Dwarf Blue Mallee, see *E. densa* subsp. *improcera*
Dwyer's Red Gum, see *E. dwyeri*

Ettrema Mallee, see *E. sturgissiana*

Faulconbridge Mallee Ash, see *E. burgessiana*
Fine-leaved Ironbark, see *E. exilipes*
Finke River Mallee, see *E. sessilis*
Flooded Gum, see *E. grandis, E. rudis*
Fluted Horn Mallee, see *E. stowardii*
Forest Red Gum, see *E. tereticornis*
Fremantle Mallee, see *E. foecunda*
Fuchsia Gum, see *E. dolichorhyncha*
Fuchsia Mallee, see *E. forrestiana*
Fuzzy Box, see *E. conica*

Gilja, see *E. brachycalyx*
Gimlet, see *E. salubris*
Gippsland Blue Gum, see *E. globulus* subsp. *pseudoglobulus*
Gippsland Mallee, see *E. kitsoniana*
Gippsland Peppermint, see *E. croajingolensis*

Goldfield's Blackbutt, see *E. lesouefi*
Grampians Gum, see *E. alpina*
Granite Rock Box, see *E. petraea*
Green Mallee, see *E. viridis*
Green Mallet, see *E. clivicola*
Green-leaf Box, see *E. chlorophylla*
Green-leaved Ironbark, see *E. virens*
Grey Box, see *E. microcarpa, E. moluccana*
Grey Gum, see *E. canaliculata, E. propinqua, E. punctata*
Grey Ironbark, see *E. paniculata*
Grey Mallee, see *E. flindersii, E. morrisii*
Gully Gum, see *E. smithii*
Gum Coolibah, see *E. intertexta*
Gum-barked Coolibah, see *E. intertexta*
Gum-topped Ironbark, see *E. decorticans*
Gympie Messmate, see *E. cloeziana*

Hamersley Bloodwood, see *E. hamersleyana*
Heart-leaved Silver Gum, see *E. cordata*
Hillgrove Gum, see *E. michaeliana*
Hills Salmon Gum, see *E. tintinnans*
Howitt's Box, see *E. howittiana*

Illyarrie, see *E. erythrocorys*
Ironbark, see *E. dura, E. fusiformis, E. granitica, E. indurata, E. siderophloia, E. suffulgens*

Jarrah, see *E. marginata* subsp. *marginata*
Jillaga Ash, see *E. stenostoma*
Jinjulu, see *E. glomerosa*
Johnson's Mallee, see *E. johnsoniana*
Jounama Snow Gum, see *E. pauciflora* subsp. *debeuzevillei*
Jutson's Mallee, see *E. jutsonii*

Kamarooka Mallee, see *E. froggattii*
Kamerere, see *E. deglupta*
Kangaroo Island Mallee, see *E. phenax*
Kangaroo Island Mallee Ash, see *E. remota*
Kangaroo Island Narrow-leaved Mallee, see *E. cneorifolia*
Karri, see *E. diversicolor*
Katherine Box, see *E. distans*
Kingscote Mallee, see *E. rugosa*
Kondinin Blackbutt, see *E. kondininensis*
Kopi Mallee, see *E. striaticalyx*
Kybean Mallee Ash, see *E. kybeanensis*

Lakefield Coolibah, see *E. acroleuca*
Large-fruited Blackbutt, see *E. pyrocarpa*
Large-fruited Red Mahogany, see *E. pellita*
Large-fruited Yellowjacket, see *E. watsoniana*
Large-leaved Cabbage Gum, see *E. grandifolia*
Large-leaved Spotted Gum, see *E. henryi*
Laterite Mallee, see *E. lateritica*
Leichhardt's Yellowjacket, see *E. leichhardtii*
Lemon-flowered Mallee, see *E. woodwardii*
Lemon-scented Gum, see *E. citriodora*
Lemon-scented Ironbark, see *E. staigeriana*
Long-fruited Bloodwood, see *E. polycarpa*

442  Eucalyptus species index

Long-leaved Box, see *E. goniocalyx*, *E. nortonii*
Lucky Bay Mallee, see *E. ligulata*

Maiden's Gum, see *E. globulus* subsp. *maidenii*
Mallee Ash, see *E. approximans* subsp. *codonocarpa*
Mallee Box, see *E. cuprea*, *E. porosa*
Mallee Red Gum, see *E. flindersii*
Mallee Wandoo, see *E. capillosa* subsp. *polyclada*, *E. livida*
Mann Range Mallee, see *E. mannensis*
Manna Gum, see *E. viminalis*
Marri, see *E. calophylla*
Marsh Gum, see *E. paludicola*
Marymia Mallee, see *E. semota*
McKie's Stringybark, see *E. mckieana*
Mealy Stringybark, see *E. cephalocarpa*
Melville Island Bloodwood, see *E. nesophila*
Merrit, see *E. flocktoniae*
Messmate Stringybark, see *E. obliqua*
Migum, see *E. leucophloia*
Molloy Red Box, see *E. leptophleba*
Mongamulla Mallee, see *E. deuaensis*
Moonbi Apple Box, see *E. malacoxylon*
Moort, see *E. platypus* var. *platypus*
Moreton Bay Ash, see *E. tessellaris*
Morrisby's Gum, see *E. morrisbyi*
Mottlecah, see *E. macrocarpa* subsp. *macrocarpa*
Mount Buffalo Gum, see *E. mitchelliana*
Mount Day Mallee, see *E. incerata*
Mount Imlay Mallee, see *E. imlayensis*
Mount Lesueur Mallee, see *E. suberea*
Mountain Ash, see *E. regnans*
Mountain Coolibah, see *E. orgadophila*
Mountain Grey Gum, see *E. cypellocarpa*
Mountain Gum, see *E. dalrympleana*
Mountain Marri, see *E. haematoxylon*
Mountain Spotted Gum, see *E. mannifera*
Mountain Swamp Gum, see *E. camphora*

Napunyah, see *E. thozetiana*
Narrow-leaved Black Peppermint, see *E. nicholii*
Narrow-leaved Bloodwood, see *E. lenziana*
Narrow-leaved Mallee, see *E. subtilis*
Narrow-leaved Mallee Ash, see *E. apiculata*
Narrow-leaved Peppermint, see *E. radiata*, *E. radiata* subsp. *radiata*, *E. robertsonii*
Narrow-leaved Red Gum, see *E. seeana*
Narrow-leaved Red Ironbark, see *E. crebra*
Narrow-leaved Red Mallee, see *E. leptophylla*
Narrow-leaved Sally, see *E. moorei* var. *moorei*
Narrow-leaved Stringybark, see *E. oblonga*
Narrow-leaved White Mahogany, see *E. tenuipes*
Needlebark Stringybark, see *E. planchoniana*
New England Blackbutt, see *E. andrewsii*
New England Peppermint, see *E. nova-anglica*
Normanton Box, see *E. normantonensis*
North Twin Peak Island Mallee, see *E. insularis*
Northern Sandplain Mallee, see *E. gittinsii*
Northern Silver Mallet, see *E. recta*
Nundroo Mallee, see *E. calcareana*

Oil Mallee, see *E. kochii* subsp. *kochii*
Omeo Gum, see *E. neglecta*

Paluma Yellowjacket, see *E. leptoloma*
Papua Ghost Gum, see *E. papuana*
Parramatta Red Gum, see *E. parramattensis*
Peeneri, see *E. peeneri*
Peppermint Box, see *E. odorata*
Pigeon House Ash, see *E. triflora*
Pimpin Mallee, see *E. pimpiniana*
Pink Bloodwood, see *E. intermedia*
Pink Gum, see *E. fasciculosa*
Plunkett Mallee, see *E. curtisii*
Pokolbin Mallee, see *E. pumila*
Poplar Box, see *E. populnea*
Poplar Gum, see *E. platyphylla*
Port Jackson Mallee, see *E. obtusiflora*
Port Lincoln Mallee, see *E. conglobata*
Powder-bark Wandoo, see *E. accedens*
Prickly Bark, see *E. todtiana*
Privet-leaved Stringybark, see *E. ligustrina*
Purple-leaved Mallee, see *E. pluricaulis* subsp. *porphyrea*

Queensland Peppermint, see *E. exserta*
Queensland Western White Gum, see *E. argophloia*
Queensland White Stringybark, see *E. nigra*

Range Bloodwood, see *E. abergiana*
Rate's Tingle, see *E. brevistylis*
Red Bloodwood, see *E. erythrophloia*, *E. gummifera*
Red Box, see *E. polyanthemos*
Red Ironbark, see *E. sideroxylon*, *E. tricarpa*
Red Mahogany, see *E. resinifera*, *E. scias*
Red Mallee, see *E. oleosa*, *E. socialis*
Red Morrel, see *E. longicornis*
Red Stringybark, see *E. macrorhyncha*
Red Tingle, see *E. jacksonii*
Red-capped Mallee, see *E. dissimulata*
Red-flowered Mallee, see *E. erythronema* var. *erythronema*, var. *marginata*
Red-flowered Mallee Box, see *E. lansdowneana* subsp. *lansdowneana*
Red-flowered Moort, see *E. nutans*
Red-flowering Gum, see *E. ficifolia*
Redheart Moit, see *E. decipiens*
Redwood, see *E. transcontinentalis*
Ribbon Bark Mallee, see *E. sheathiana*
Rib-fruited Mallee, see *E. corrugata*
Ridge-fruited Mallee, see *E. angulosa*, *E. incrassata*
Risdon Peppermint, see *E. risdonii*
River Peppermint, see *E. elata*
River Red Gum, see *E. camaldulensis*
River Yate, see *E. macrandra*
Rough-barked Gimlet, see *E. effusa*
Rough-leaved Bloodwood, see *E. setosa*
Round-leaved Bloodwood, see *E. latifolia*
Round-leaved Gum, see *E. deanei*
Rudder's Box, see *E. rudderi*

Salmon Gum, see *E. salmonophloia*
Salt River Gum, see *E. sargentii*
Saltlake Mallee, see *E. rigens*
Sand Mallee, see *E. eremophila* subsp. *eremophila*
Sandplain Mallee, see *E. ebbanoensis*
Scaly Bark, see *E. squamosa*
Scarlet Pear Gum, see *E. stoatei*
Scribbly Gum, see *E. haemastoma*, *E. rossii*, *E. sclerophylla*, *E. signata*
Shining Gum, see *E. nitens*
Shiny-barked Gum, see *E. pachycalyx*
Shirley's Silver-leaved Ironbark, see *E. shirleyi*
Silver Box, see *E. pruinosa*
Silver Gimlet, see *E. campaspe*
Silver Mallee, see *E. falcata*
Silver Mallet, see *E. argyphea*, *E. ornata*
Silver Peppermint, see *E. tenuiramis*
Silver Princess, see *E. caesia* subsp. *magna*
Silver-leaved Ironbark, see *E. melanophloia*
Silver-leaved Mountain Gum, see *E. pulverulenta*
Silvertop Ash, see *E. sieberi*
Silvertop Stringybark, see *E. laevopinea*
Silver-topped Gimlet, see *E. ravida*
Slaty Box, see *E. dawsonii*
Small-fruited Bloodwood, see *E. dichromophloia*
Small-leaved Gum, see *E. parvifolia*
Small-leaved Mottlecah, see *E. macrocarpa* subsp. *elachantha*
Smithton Peppermint, see *E. nitida*
Smooth-barked Coolibah, see *E. intertexta*
Smooth-stemmed Bloodwood, see *E. bleeseri*
Snap and Rattle, see *E. celastroides* subsp. *celastroides*
Snow Gum, see *E. pauciflora* subsp. *niphophila*, subsp. *pauciflora*
Soap Mallee, see *E. diversifolia*
South Australian Grey Mallee, see *E. flindersii*
South Australian Mallee Box, see *E. porosa*
South Gippsland Peppermint, see *E. willisii*
Southern Blue Gum, see *E. globulus* subsp. *bicostata*
Southern Mahogany, see *E. botryoides*
Spearwood, see *E. doratoxylon*
Spinning Gum, see *E. perriniana*
Spotted Gum, see *E. maculata*
Square-fruited Ironbark, see *E. quadricostata*
Square-fruited Mallee, see *E. calycogona*, *E. tetraptera*
Steedman's Mallet, see *E. steedmanii*
Steel Box, see *E. rummeryi*
Stirling Range Mallee, see *E. erectifolia*
Strickland's Gum, see *E. stricklandii*
Sugar Gum, see *E. cladocalyx*
Suggan Buggan Mallee, see *E. saxatilis*
Swamp Gum, see *E. ovata*
Swamp Mahogany, see *E. robusta*
Swamp Mallet, see *E. spathulata* subsp. *grandiflora*, subsp. *spathulata*
Swamp Peppermint, see *E. rodwayi*
Swamp Stringybark, see *E. conglomerata*

Swamp Yate, see *E. occidentalis* var. *occidentalis*
Sydney Blue Gum, see *E. saligna*
Sydney Peppermint, see *E. piperita* subsp. *piperita*, subsp. *urceolaris*

Tallerack, see *E. tetragona*
Tallowwood, see *E. microcorys*
Tasmanian Alpine Yellow Gum, see *E. subcrenulata*
Tasmanian Blue Gum, see *E. globulus*, *E. globulus* subsp. *globulus*
Tasmanian Snow Gum, see *E. coccifera*
Tasmanian Yellow Gum, see *E. johnstonii*
Tenterfield Woollybutt, see *E. banksii*
Thin-leaved Stringybark, see *E. eugenioides*
Timor White Gum, see *E. alba*
Timor White Mountain Gum, see *E. orophila*
Tindale's Stringybark, see *E. tindaliae*
Tingiringi Gum, see *E. glaucescens*
Tooth-leaved Shining Gum, see *E. denticulata*
Tropical Red Box, see *E. brachyandra*
Tuart, see *E. gomphocephala*
Tumble-down Red Gum, see *E. dealbata*
Two-winged Gimlet, see *E. diptera*

Umbrawarra Gum, see *E. umbrawarrensis*
Urn Gum, see *E. urnigera*

Varnished Gum, see *E. vernicosa*
Victorian Eurabbie, see *E. globulus* subsp. *pseudoglobulus*
Victorian Silver Gum, see *E. crenulata*

Wabling Hill Mallee, see *E. argutifolia*
Wadbilliga Ash, see *E. paliformis*
Wallangarra White Gum, see *E. scoparia*
Wandi Ironbark, see *E. jensenii*
Wandoo, see *E. wandoo*
Warilu Blue-leaved Mallee, see *E. gamophylla*
Warted Yate, see *E. megacornuta*
Water Gum, see *E. leucoxylon* subsp. *petiolaris*
Wattle-leaved Peppermint, see *E. acaciiformis*
Weeping Box, see *E. patellaris*
Weeping Mallee, see *E. sepulcralis*
Western Mallee Red Gum, see *E. vicina*
Wetar White Gum, see *E. wetarensis*
Wheatbelt Wandoo, see *E. capillosa* subsp. *capillosa*
Whipstick Mallee Ash, see *E. multicaulis*
White Ash, see *E. fraxinoides*
White Box, see *E. albens*
White Mahogany, see *E. acmenoides*, *E. apothalassica*
White Mallee, see *E. cylindriflora*, *E. dumosa*, *E. gracilis*
White Peppermint, see *E. pulchella*
White Stringybark, see *E. globoidea*
Whitebark, see *E. apodophylla*
White's Ironbark, see *E. whitei*
White-topped Box, see *E. quadrangulata*
William's Stringybark, see *E. williamsiana*
Woila Gum, see *E. olsenii*

444  Eucalyptus species index

Wolgan Snow Gum, see *E. gregsoniana*
Woolbernup Mallee, see *E. acies*
Woollybutt, see *E. longifolia*
Wyola Mallee, see *E. wyolensis*

Yalata Mallee, see *E. yalatensis*
Yapunyah, see *E. ochrophloia*
Yarldarlba, see *E. youngiana*
Yarra Gum, see *E. yarraensis*
Yate, see *E. cornuta*
Yellow Bloodwood, see *E. aureola*, *E. dimorpha*, *E. eximia*
Yellow Box, see *E. melliodora*
Yellow Gum, see *E. leucoxylon* subsp. *leucoxylon*
Yellow Stringybark, see *E. muelleriana*
Yellow Tingle, see *E. guilfoylei*
Yellow-flowered Mallee, see *E. flavida*
Yellowjacket, see *E. peltata*, *E. petalophylla*, *E. scabrida*
Yellow-top Mallee Ash, see *E. luehmanniana*
Yertchuk, see *E. consideniana*
York, see *E. loxophleba* subsp. *loxophleba*
York Gum, see *E. loxophleba*
Yorrell, see *E. gracilis*, *E. yilgarnensis*
Youman's Stringybark, see *E. youmanii*
Yumbarra Mallee, see *E. yumbarrana*

'*Eucalyptus* hybrid' 39, 56, 251–4, 257, 259, 260, 262, 295, 296, 379
'*E. 12ABL*' 203, 204, 208
*E. abergiana* 114
*E. absita* 114
*E. acaciiformis* 114
*E. accedens* 114
*E. acervula*, see *E. eugenioides*
*E. acies* 114
*E. acmenoides* 7, 9, 114
*E. acroleuca* 114
*E. aequioperta* 114
*E. affinis* 296
*E. agglomerata* 114
*E. aggregata* 109, 114, 276, 277, 283
*E. alba* 5
*E. alba* × *E. camaldulensis* 296
*E. albens* 114
*E. albida* 114
*E. alpina* 114
*E. alpina* × *E. baxteri* 90
*E. amplifolia* 114, 202, 270, 272, 277–9, 296, 397, 398
*E. amygdalina* 29, 114, 115, 291, 294
 var. *nitida*, see *E. nitida*
*E. andreana*, see *E. elata*
*E. andrewsii* 33, 115
 subsp. *andrewsii* 115
 subsp. *campanulata* 115
*E. angophoroides* 7, 115
*E. angulosa* 115
*E. angustissima* 57, 115, 200
*E. annulata* 115

*E. annuliformis* 115
*E. apiculata* 115
*E. apodophylla* 115
*E. apothalassica* 7
*E. approximans* 11, 20
 subsp. *approximans* 115
 subsp. *codonocarpa* 115
*E. aquilina* 115
*E. arachnaea* 115
*E. archeri* 116
*E. argophloia* 116
*E. argutifolia* 116
*E. argyphea* 116
*E. aromaphloia* 116
*E. aspratilis* 116
*E. astringens* 13, 116, 294
*E. aureola* 116
*E. australiana*, see *E. radiata* subsp. *radiata*

*E. badjensis* 30, 116, 221, 230, 392
*E. baeuerlenii* 116
*E. baileyana* 9, 116
*E. bakeri* 106, 116, 211, 212, 392
*E. balanopelex* 116
*E. balladoniensis*
 subsp. *sedens* 116
*E. bancroftii* 116
*E. banksii* 116
*E. barberi* 116
*E. baueriana* 116, 117
*E. baxteri* 117
*E. behriana* 117
*E. bensonii* 117
*E. benthamii*
 var. *benthamii* 117, 390
 var. *dorrigoensis* 117
*E. beyeri* 7, 117
*E. bicolor*, see *E. largiflorens*
*E. bicostata*, see *E. globulus* subsp. *bicostata*
*E. blakelyi* 117, 270, 277–9, 337, 397, 398
*E. blaxlandii* 117
*E. bloeseri* 117
*E. bloxsomei* 117
*E. bosistoana* 117
*E. botryoides* 117, 202, 205, 277, 296
*E. brachyandra* 9
*E. brachycalyx* 117
*E. brachycorys* 117
*E. brachyphylla* 117
*E. brassiana* 27, 117
*E. brevipes* 118
*E. brevistylis* 118
*E. bridgesiana* 7, 118, 277, 335
*E. brockwayi* 118
*E. brookeriana* 7, 10, 118, 398
*E. brownii* 118, 392
*E. bunites* 118
*E. buprestium* 118
*E. burdettiana* 118

*E. burgessiana* 118
*E. burracoppinensis* 118

*E. cadens* 118
*E. caesia* 17, 118
  subsp. *caesia* 118
  subsp. *magna* 118
*E. calcareana* 118
*E. calcicola* 118
*E. caleyi* 118
*E. caliginosa* 118
*E. calophylla* 5, 6, 15, 16, 19, 20, 22, 40, 41, 119, 294
*E. calycogona* 119
  var. *calycogona* 119
*E. camaldulensis* 9, 10, 13, 15, 24, 27, 31, 32, 39, 42, 43, 47, 53, 55, 56, 58–62, 64, 65, 67, 68, 78–80, 83–90, 93, 95–7, 98, 106, 110, 112, 119, 199, 202–5, 208, 216, 219, 221, 223, 224, 230, 235, 236, 242, 248, 249, 251–4, 262, 271, 272, 277, 283, 292, 294–7, 306, 307, 310, 312, 317, 319, 320, 335, 336, 366
  var. *camaldulensis* 31, 119
  var. *obtusa* 31, 119
*E. camaldulensis* × *E. grandis* 65
*E. camaldulensis* × *E. tereticornis* 295, 297
*E. cambagei*, see *E. goniocalyx*
*E. cameronii* 119
*E. camfieldii* 119
*E. campanulata*, see *E. andrewsii* subsp. *campanulata*
*E.* sp. nov. aff. *campanulata*, see *E. olida*
*E. campaspe* 119
*E. camphora* 119
  subsp. *aquatica* 119
  subsp. *camphora* 119
  subsp. *humeana* 119
  subsp. *relicta* 119
*E. canaliculata* 119
*E. capillosa*
  subsp. *capillosa* 120
  subsp. *polyclada* 120
*E. capitellata* 8, 120
*E. captiosa* 120
*E. carnabyi* 120
*E. carnea*, see *E. umbra* subsp. *carnea*
*E. carnei* 120
*E. catenaria* 120
*E. celastroides*
  subsp. *celastroides* 120
  subsp. *virella* 120
*E. cephalocarpa* 120
*E. ceratocorys* 120
*E. chapmaniana* 120
*E. chickaballapur* 251
*E. chlorophylla* 120
*E. cinerea* 18, 32, 33, 120, 217, 219, 221, 222, 223, 242, 252, 270, 271, 277, 283, 366, 376, 414
*E. citriodora* 9, 10, 13, 14, 26, 55, 56, 58, 59, 79, 89, 90, 102, 105, 108, 110, 120, 167, 197, 198, 202–4, 205–211, 213, 217, 219, 222, 223, 234, 239–42, 244, 245, 246–9, 252–6, 258, 259, 261, 269–71, 277, 283, 292, 294–7, 299, 305–7, 310, 311, 313, 314, 316–19, 327, 346–9, 362, 363, 365, 366, 368, 370, 371, 377, 379, 381, 391, 394, 412, 413, 415, 419, 421, 422
*E. cladocalyx* 19, 41, 120, 294, 296, 325
*E. clarksoniana* 120
*E. clelandii* 121
*E. clivicola* 121
*E. cloeziana* 9, 121, 387
*E. cneorifolia* 25, 121, 366
*E. coccifera* 121
*E. coerulea*, see *E. caleyi*
*E. concinna* 121
*E. conferruminata* 121
*E. conglobata* 121
*E. conglomerata* 121
*E. conica* 121
*E. consideniana* 121
*E. coolabah*
  subsp. *coolabah* 121
  subsp. *microtheca* 122
*E. cooperiana* 122
*E. cordata* 122, 277–9, 283
*E. coriacea*, see *E. pauciflora*
*E. cornuta* 122
*E. coronata* 122
*E. corrugata* 122
*E. corymbosa*, see *E. gummifera*
*E. corynocalyx*, see *E. cladocalyx*
*E. cosmophylla* 122, 270, 271, 277, 283
*E. crebra* 122, 277, 296
*E. crenulata* 18, 109, 122
*E. cretata* 122
*E. croajingolensis* 122
*E. crucis*
  subsp. *crusis* 122
  subsp. *lanceolata* 122
*E. cullenii* 15, 17, 122
*E. cunninghamii* 122
*E. cuprea* 122
*E. curtipes* 123
*E. curtisii* 9, 123
*E. cyanophylla* 123
*E. cyclostoma* 123
*E. cylindriflora* 123
*E. cylindrocarpa* 123
*E. cypellocarpa* 123, 277

*E. dalrympleana* 123, 277, 296
  subsp. *heptantha* 123
*E. dawsonii* 123
*E. dealbata* 123, 294
*E. deanei* 123
*E. decipiens* 123
*E. decorticans* 123
*E. decurva* 123
*E. deglupta* 4, 9, 53, 65, 67, 123, 277, 295, 296

446  *Eucalyptus species index*

*E. delegatensis* 13, 123, 248, 277, 320, 327, 387, 396
  subsp. *delegatensis* 123, 124, 398
  subsp. *tasmaniensis* 123, 124, 398
*E. dendromorpha* 124
*E. densa* subsp. *improcera* 124
*E. denticulata* 124
*E. desmondensis* 124
*E. desquamata* 124
*E. deuaensis* 124
*E. dextropinea*, see *E. muelleriana*
*E. dichromophloia* 202
*E. dielsii* 124
*E. dimorpha* 124
*E. diptera* 124
*E. dissimulata* 124
*E. distans* 124
*E. diversicolor* 10, 13, 124, 294, 296
*E. diversifolia* 124
*E. dives* 9, 10, 24, 26, 27, 78, 79, 103, 105, 108, 124, 125, 164, 167, 178, 184–7, 191, 194, 198, 210–12, 217, 219, 221–3, 227, 228, 231, 232, 234–6, 252, 271, 277, 335, 365, 366, 368, 375, 377, 387, 389–91, 394, 413, 415, 419
*E. dolichorhyncha* 125
*E. doratoxylon* 125
*E. drepanophylla* 125
*E. drummondii* 125
*E. dumosa* 125
*E. dundasii* 125
*E. dunnii* 10, 125, 203, 240
*E. dura* 125
*E. duyeri* 12, 125

*E. ebbanoensis* 125
*E. effusa*
  subsp. *effusa* 125
  subsp. *exsul* 125
*E. elaeophloia* 125
*E. elaeophora*, see *E. goniocalyx*
*E. elata* 25, 30, 125, 126, 219, 222, 223, 277, 337, 390
*E. erectifolia* 126
*E. eremicola* 126
*E. eremophila* 126
  subsp. *eremophila* 126
*E. erythrandra* 126
*E. erythrocorys* 126, 387
*E. erythronema*
  var. *erythronema* 126
  var. *marginata* 126
*E. erythrophloia* 126
*E. eudesmioides*
  subsp. *eudesmioides* 126
*E. eugenioides* 126
*E. ewartiana* 126
*E. exilipes* 126
*E. exilis* 126
*E. eximia* 7, 126, 294
*E. exserta* 27, 55, 126, 127, 203–5, 209–11, 252, 269, 365, 366, 370

*E. falcata* 127
*E. famelica* 127
*E. fasciculosa* 20, 21, 127
*E. fastigata* 127, 277–9
*E. fibrosa* 127
  subsp. *fibrosa* 127
*E. ficifolia* 20, 127, 277, 283
*E. flavida* 127
*E. fletcheri*, see *E. baueriana*
*E. flindersii* 127
*E. flocktoniae* 127
*E. foecunda* 127
*E. formanii* 127
*E. forrestiana* 127
*E. fraseri* 127
*E. fraxinoides* 15, 127
*E. froggattii* 127
*E. fruticetorum*, see *E. polybractea*
*E. fusiformis* 127

*E. gamophylla* 127
*E. gardneri*
  subsp. *gardneri* 127
  subsp. *ravensthorpensis* 127
*E. georgei* 128
*E. gigantea*, see *E. delegatensis*
*E. gillenii* 128
*E. gillii* 128
*E. gittinsii* 128
*E. glaucescens* 128
*E. globoidea* 9, 128
*E. globulus* 10, 15, 27, 28, 39–41, 53, 55, 60–3, 65–8, 78, 80–3, 86, 87, 90, 91, 93, 103, 106, 108, 112, 167, 174, 184, 185, 187, 194, 197, 202–7, 209–12, 213, 216, 217, 221, 223, 230, 235, 239–42, 244, 246, 248, 251–4, 257, 258, 260, 263, 269–73, 275, 277–9, 283, 284, 291–9, 306, 310, 312–17, 319, 345–8, 355, 356, 359–62, 365–7, 370, 375, 377–9, 381, 387, 388, 397, 398, 412, 413, 414, 417–19, 421
  subsp. *bicostata* 28, 128, 202, 209, 221, 230, 231, 234, 258, 277
  subsp. *globulus* 27, 58, 59, 128, 219, 242, 366, 370
  subsp. *maidenii* 28, 128, 203–7, 210, 211, 217, 219, 221, 223, 230, 252, 254, 277, 294, 366, 370, 392
  subsp. *pseudoglobulus* 28, 128, 392
*E. glomerosa* 128
*E. gomphocephala* 128, 216, 294
*E. gongylocarpa* 128, 129
*E. goniantha* 129
  subsp. *goniantha* 129
*E. goniocalyx* 129, 296
*E. gracilis* 129
*E. grandifolia* 129
*E. grandis* 10, 13, 19, 42, 43, 45, 46, 53, 56, 62, 63, 65, 68, 84, 90, 95, 110, 129, 203, 209, 216, 224, 227, 236, 240, 252, 254, 263, 270, 274, 277–9, 295, 296, 397, 398

*E. grandis* × *E. urophylla* 65, 66
*E. granitica* 129
*E. gregsoniana* 11, 129
*E. griffithsii* 129
*E. grossa* 129
*E. guilfoylei* 9, 129
*E. gullickii*, see *E. mannifera* subsp. *gullickii*
*E. gummifera* 129
*E. gunnii* 90, 129, 205, 209, 271, 277, 283, 396, 398

*E. haemastoma* 129, 277, 331
*E. haematoxylon* 129
*E. halophila* 129
*E. hamersleyana* 130
*E. hebetifolia* 130
*E. hemilampra*, see *E. resinifera*
*E. hemiphloia*, see *E. moluccana*
*E. henryi* 26, 130
*E. histophylla* 130
*E. horistes* 24, 57, 130, 200
*E. howittiana* 9
'*E. hybrida*' 313

*E. imlayensis* 130
*E. incerata* 130
*E. incrassata* 130, 270, 277–9, 397, 398
*E. indurata* 130
*E. insularis* 130
*E. intermedia* 130
*E. intertexta* 130
'*E. irbyi*' 130

*E. jacksonii* 130
*E. jensenii* 130, 325, 339
*E. johnsoniana* 130
*E. johnstonii* 130
*E. jutsonii* 130

*E. kartzoffiana* 130
*E. kessellii* 131
*E. kingsmillii* 131
  subsp. *alatissima* 131
'*E. kirtoniana*' 131, 296
*E. kitsoniana* 131
*E. kochii* 21, 22, 24, 33, 86, 89, 92, 106, 108, 200, 236, 366, 367
  subsp. *kochii* 33, 57, 67, 131, 200, 392
  subsp. *plenissima* 33, 57, 108, 131, 200, 392
*E. kondininensis* 131
*E. kruseana* 18, 131
*E. kumarlensis* 10, 11, 131
*E. kybeanensis* 131

*E. lactea*, see *E. mannifera* subsp. *praecox*
*E. laevopinea* 131, 296
*E. lane-poolei* 131
*E. lansdowneana*
  subsp. *albopurpurea* 131
  subsp. *lansdowneana* 131

*E. largiflorens* 131
'*E. laseroni*' 132
*E. latens* 132
*E. lateritica* 132
*E. latifolia* 132
*E. lehmannii* 18, 132, 296
*E. leichhardtii* 132
*E. leizhou No. 1* 56, 132, 203, 204, 206, 207, 209–11, 366
*E. lenziana* 132
*E. leptocalyx* 132
*E. leptoloma* 7, 132
*E. leptophleba* 132, 209
*E. leptophylla* 132
*E. leptopoda* 132
  subsp. *elevata* 132
*E. lesouefii* 132
*E. leucophloia* 132
*E. leucoxylon* 25, 41, 132, 277, 296
  subsp. *leucoxylon* 132
  subsp. *petiolaris* 132
*E. ligulata* 132, 133
*E. ligustrina* 133
*E. lindleyana*, see *E. elata*
*E. linearis*, see *E. pulchella*
*E. livida* 11, 12, 133
*E. longicornis* 133
*E. longifolia* 133, 294, 296
*E. loxophleba* 18, 133, 201, 392
  subsp. *gratiae* 133
  subsp. *lissophloia* 24, 133, 200
  subsp. *loxophleba* 133
*E. lucasii* 133
*E. lucens* 133
*E. luehmanniana* 133

*E. macarthurii* 108, 133, 198, 211, 212, 217, 220, 222, 223, 252, 257, 277, 294, 366, 368
*E. macrandra* 133
  subsp. *olivacea* 133
*E. macrocarpa* 270, 272, 397, 398
  subsp. *elachantha* 133
  subsp. *macrocarpa* 133
*E. macrorhyncha* 7, 13–15, 112, 199, 255, 261, 262, 277, 294, 396
  subsp. *cannonii* 133
  subsp. *macrorhyncha* 134
*E. maculata* 10, 11, 26, 41, 134, 202, 295, 296
'*E. maculata citriodon*' 305
*E. maculosa*, see *E. mannifera* subsp. *maculosa*
*E. maidenii*, see *E. globulus* subsp. *maidenii*
*E. malacoxylon* 134
*E. mannensis* 134
*E. mannifera* 134, 271, 276, 277, 283
  subsp. *gullickii* 134
  subsp. *maculosa* 134
  subsp. *praecox* 134
*E. marginata* 10, 40, 41, 134
  subsp. *marginata* 134
  subsp. *thalassica* 134

*'E. marsdenii'* 134
*E. mckieana* 134
*E. megacarpa* 134
*E. megacornuta* 134
*E. melanoleuca* 134
*E. melanophitra* 134
*E. melanophloia* 33, 134, 135, 296
*E. melanoxylon* 135
*E. melliodora* 135, 277, 294, 296, 335, 336
*E. merrickiae* 135
*E. michaeliana* 135
*E. micranthera* 135
*E. microcarpa* 40, 135
*E. microcorys* 9, 22, 23, 135, 202, 277, 296
*E. microtheca* 7, 135, 255
*E. mimica* 22, 23
*E. miniata* 9, 135
*E. mitchelliana* 135
*E. moluccana* 135, 277
*E. moorei* 135
    var. *latiuscula* 135
    var. *moorei* 135
*E. morrisbyi* 135
*E. morrisii* 27, 135
*E. muelleri*, see *E. johnstonii*
*E. muelleriana* 7, 19, 136
*E. multicaulis* 136
*E. myriadena* 136

*E. naudiniana*, see *E. deglupta*
*E. neglecta* 136
*'E. nepeanensis'* 136
*E. nesophila* 136
*E. newbeyi* 136
*E. nicholii* 25, 136, 392
*E. nigra* 136
*E. niphophila*, see *E. pauciflora* subsp. *niphophila*
*E. nitens* 10, 13, 54, 66, 84, 90, 91, 93, 136, 203, 248, 327, 387
*E. nitida* 26, 136
*E. normantonensis* 136
*E. nortonii* 136
*E. notabilis* 136
*E. nova-anglica* 136, 137, 392
*E. numerosa*, see *E. elata*
*E. nutans* 137

*E. obliqua* 7, 9, 18, 33, 39, 137
*E. oblonga* 105, 108, 137
*E. obtusiflora* 137
*E. occidentalis* 137, 294
    var. *occidentalis* 137
*E. ochrophloia* 137
*E. odorata* 137
*E. oldfieldii* 137
*E. oleosa* 18, 137, 294
    var. *borealis*, see *E. horistes*
    var. *glauca*, see *E. transcontinentalis*
    var. *kochii*, see *E. kochii* subsp. *kochii*
    var. *longicornis*, see *E. longicornis*
    var. *obtusa*, see *E. oleosa*

*E. olida* 33, 108, 137, 184, 198, 201, 366, 368, 375, 392, 395
*E. olsenii* 138
*E. orbifolia* 138
*E. oreades* 138
*E. orgadophila* 138, 325
*E. ornata* 138
*E. orophila* 138
*E. ovalifolia*, see *E. polyanthemos*
    var. *lanceolata*, see *E. polyanthemos*
*E. ovata* 7, 10, 18, 138, 277, 325, 329, 331, 338, 339, 398
*E. ovata* × *E. crenulata* 90
*E. ovularis* 138
*E. oxymitra* 138

*E. pachycalyx* 138
*E. pachyloma* 138
*E. pachyphylla* 138
*E. paliformis* 138
*E. paludicola* 138, 139
*E. paludosa*, see *E. ovata*
*E. paniculata* 7, 8, 202, 294, 296
*E. papuana* 139
*E. paracolpica* 139
*E. parramattensis* 139
*E. parvifolia* 139, 271, 277, 283, 296
*E. patellaris* 139
*E. patens* 139
*E. pauciflora* 19, 30, 40, 139, 209, 271, 277, 283, 320, 334, 335
    subsp. *debeuzevillei* 139
    subsp. *niphophila* 139, 283
    subsp. *pauciflora* 139
*E. peeneri* 139
*E. pellita* 18, 19, 53, 63, 83, 139, 292, 296
*E. peltata* 26
*E. pendens* 139
*E. perangusta* 140
*E. percostata* 140
*E. perriniana* 140, 270, 271, 277, 283, 397, 398
*E. petalophylla* 140
*E. petraea* 140
*E. phaenophylla* 140
    subsp. *interjacens* 140
*E. phellandra*, see *E. radiata*
*E. phenax* 140
*E. phlebophylla*, see *E. pauciflora*
*E. pileata* 140
*E. pilularis* 8, 9, 13, 140, 277
*E. pimpiniana* 140
*E. piperita* 22, 102, 105, 140, 183, 242, 294
    subsp. *piperita* 140
    subsp. *urceolaris* 140
*E. planchoniana* 140
*E. platycorys* 141
*E. platyphylla* 141
*E. platypus* 141
    var. *heterophylla* 141
    var. *platypus* 141
*E. plenissima*, see *E. kochii* subsp. *plenissima*

*E. pluricaulis*
  subsp. *pluricaulis* 141
  subsp. *porphyrea* 141
*E. polyanthemos* 141, 325, 339
*E. polybractea* 9, 24, 25, 28, 29, 53, 57, 63, 80, 82, 86, 88, 92, 106, 108, 141, 166, 167, 170, 185, 188, 189, 192–7, 199, 200, 222, 234, 271, 276, 277, 365, 366, 375, 390, 393, 413, 414, 417
*E. polycarpa* 7, 19
*E. populifolia*, see *E. populnea*
*E. populnea* 141
*E. porosa* 141
*E. praetermissa* 141
*E. preissiana* 141
  subsp. *lobata* 141
*E. propinqua* 18, 19, 141, 390
*E.* sp. aff. *propinqua* 141, 390
*E. pruiniramis* 141
*E. pruinosa* 33, 141
*E. pseudoglobulus*, see *E. globulus* subsp. *pseudoglobulus*
*E. pterocarpa* 141
*E. pulchella* 30, 142
*E. pulverulenta* 142, 271, 277, 306
*E. pumila* 142, 392
*E. punctata* 105, 142, 277, 296, 391
*E. pyriformis* 142
*E. pyrocarpa* 6, 7

*E. quadrangulata* 142
*E. quadrans* 142
*E. quadricostata* 142

*E. radiata* 9, 24, 26, 27, 29, 30, 67, 79–83, 91, 105, 108, 142, 185, 191, 194, 197, 198, 211, 212, 217, 220, 221, 223, 224, 227, 231, 234–6, 269, 277, 283, 299, 306, 345, 366, 375–7, 414
  subsp. *radiata* 30, 143, 387, 393, 394
  subsp. *robertsonii*, see *E. robertsonii*
  subsp. *sejuncta* 29, 143
  var. *australiana*, see *E. radiata* subsp. *radiata*
'*E. rariflora*' 143
*E. raveretiana* 9, 143
*E. ravida* 143, 399
*E. recta* 143
*E. redacta* 143
*E. redunca* 143
*E. regnans* 13, 39, 67, 84, 91, 143, 248, 255, 271, 277, 283, 296, 335
*E. remota* 143
*E. resinifera* 8, 41, 143, 277, 397, 398
*E. rhodantha* 143
*E. rhombica* 143
*E. rigens* 143
*E. rigidula* 143
*E. risdonii* 32, 143, 144
*E. robertsonii* 29, 144, 185, 277, 296
*E. robusta* 9, 144, 202, 214, 216, 269–71, 274, 275, 277, 283, 296, 397, 398

*E. rodwayi* 144
*E. rossii* 144
*E. rostrata*, see *E. camaldulensis*
*E. rubida* 144, 271, 277, 296, 396, 398
*E. rubiginosa* 9, 144
*E. rudderi* 144
*E. rudis* 144, 202, 205, 277, 296, 388, 389
*E. rugosa* 144
*E. rummeryi* 7
*E. rupicola*, see *E. cunninghamii*
*E. rydalensis*, see *E. aggregata*

*E. salicifolia*, see *E. amygdalina*
*E. saligna* 8, 10, 41, 144, 203, 214, 216, 240, 277
*E. salmonophloia* 25, 144
*E. salubris* 144, 392, 399
  var. *glauca*, see *E. ravida*
*E. santalifolia*, see *E. diversifolia*
*E. sargentii* subsp. *sargentii* 145
*E. saxatilis* 145, 392
*E. scabrida* 145
*E. scias* 145
*E. sclerophylla* 145
*E. scoparia* 145
*E. scyphocalyx* 145
*E. seeana* 145
*E. semiglobosa* 145
*E. semota* 145
*E. sepulcralis* 145
*E. serpentinicola* 145
*E. sessilis* 145
*E. setosa* 145
*E. sheathiana* 145
*E. shirleyi* 145
*E. siderophloia* 145
*E. sideroxylon* 25, 31, 145, 185, 220, 221, 223, 252, 270, 274, 277, 294, 296, 336, 397, 398
  subsp. *tricarpa*, see *E. tricarpa*
*E. sieberi* 145
*E. sieberiana*, see *E. sieberi*
*E. signata* 146, 277
*E. smithii* 10, 30, 91, 92, 106, 108, 146, 167, 187, 197, 203, 210–12, 217, 218, 220, 223, 224, 227, 228, 230–2, 233–6, 252, 254, 255, 269, 345, 365, 366, 370, 376, 377, 388, 412, 413, 414, 417
*E. socialis* 146, 277
*E. sparsa* 106, 108, 146, 388, 389
*E. sparsifolia*, see *E. oblonga*
*E. spathulata* 146
  subsp. *grandiflora* 146
  subsp. *spathulata* 146
*E. sphaerocarpa* 146
*E. squamosa* 146
*E. staeri* 146
*E. staigeriana* 9, 26, 30, 31, 108, 146, 184, 198, 220, 222, 223, 241, 242, 244, 248, 347, 366, 368, 377, 413, 415, 419
*E. stannicola* 146
*E. steedmanii* 146
*E. stellulata* 146, 277, 334

*E. stenostoma* 146
*E. stjohnii*, see *E. globulus* subsp. *pseudoglobulus*
*E. stoatei* 146
*E. stowardii* 147
*E. striaticalyx* 147
   subsp. *beadellii* 147
   subsp. *canescens* 147
   subsp. *gypsophila* 147
   subsp. *striaticalyx* 147
*E. stricklandii* 147, 387
*E. stricta* 147
*E. stuartiana*, see *E. bridgesiana*
   var. *cordata*, see *E. cinerea*
*E. sturgissiana* 106, 108, 147, 392
*E. subangusta*
   subsp. *cerina* 147
   subsp. *pusilla* 147
   subsp. *subangusta* 147
*E. subcrenulata* 147, 392
*E. suberea* 147
*E. sublucida* 147
*E. subtilior* 147
*E. subtilis* 147
*E. suffulgens* 147
*E. suggrandis* 147
*E. synandra* 147

'*E. taeniola*'– 147
*E. talyuberlup* 147
*E. tectifica* 148
*E. tenera* 148
*E. tenuipes* 9, 148
*E. tenuiramis* 148
*E. tenuis* 148
*E. terebra* 148
*E. tereticornis* 10, 18, 27, 53, 56, 90, 148, 202, 203, 205, 240, 251, 259, 260, 262, 263, 270, 277–9, 295–7, 310, 313, 314, 327, 328, 379, 397, 398
*E. tereticornis* × *E. camaldulensis* 262, 295, 296
*E. tereticornis* × *E. grandis* 224
*E. terminalis* 7, 20, 255
*E. tessellaris* 9, 20, 148
*E. tetragona* 148
*E. tetraptera* 148
*E. tetrodonta* 148
*E. thozetiana* 148
*E. tindaliae* 148
'*E. tingbaensis*' 148
*E. tintinnans* 148
*E. todtiana* 148, 149
*E. torelliana* 90, 149
*E. torquata* 149, 325, 389
*E. trachyphloia* 149
*E. transcontinentalis* 149
*E. tricarpa* 25

*E. triflora* 149
*E. trivalvis* 149
*E. tumida* 149

*E. umbellata*, see *E. tereticornis*
*E. umbonata* 149
*E. umbra* 149
   subsp. *carnea* 149
*E. umbrawarrensis* 149
*E. uncinata* 149
'*E. unialata*' 149
*E. urnigera* 149
*E. urophylla* 5, 53, 56, 62, 63, 90, 149, 150, 203, 204, 209, 236, 240
*E. urophylla* × *E. grandis* 203, 204, 209, 236

*E. varia*
   subsp. *salsuginosa* 150
   subsp. *varia* 150, 388
*E. vegrandis* 150
*E. vernicosa* 150
*E. vicina* 150
*E. viminalis* 15, 29, 150, 240, 242, 252, 255, 270, 271, 277, 283, 294, 296, 306, 320, 325, 329, 330, 335, 398
*E. virens* 150
'*E. virgata*' 150
*E. viridis* 25, 28, 150, 185, 194, 252, 366
'*E. vitrea*'– 150

*E. wandoo* 41, 150, 199
*E. watsoniana*
   subsp. *capillata* 150
   subsp. *watsoniana* 151
*E. websteriana* 151
*E. wetarensis* 151
*E. whitei* 151
*E. wilcoxii* 151
*E. wilkinsoniana*, see *E. eugenioides*
*E. williamsiana* 151
*E. willisii* 19, 20, 30, 151
*E. woodwardii* 151, 387, 389
*E. woollsiana*, see *E. microcarpa*
*E. wyolensis* 151

*E. xanthonema* 151

*E. yalatensis* 151
*E. yangoura*, see *E. globoidea*
*E. yarraensis* 151, 325
*E. yilgarnensis* 151
*E. youmanii* 112, 151, 199, 255, 261, 262, 396
*E. youngiana* 151
*E. yumbarrana* 151

*E. zopherophloia* 151